Nitrogen Cycling in Bacteria

Molecular Analysis

Edited by

James W. B. Moir

Department of Biology
University of York
Heslington, York
UK

Caister Academic Press

Caister Academic Press
Norfolk, UK

www.caister.com

British Library Cataloguing-in-Publication Data
A catalogue record for this book is available from the British Library

ISBN: 978-1-904455-86-8

Cover design adapted from Figure 1.3

Printed and bound in Great Britain

Contents

Contributors

Elizabeth M. Baggs
Institute of Biological and Environmental Sciences
University of Aberdeen
Aberdeen, UK

e.baggs@abdn.ac.uk

Shilpa Bali
Department of Biochemistry
University of Oxford
Oxford, UK

shilpa.bali@bioch.ox.ac.uk

Eulogio J. Bedmar
Departamento de Microbiología del Suelo y Sistemas Simbióticos
Estación Experimental del Zaidín (EEZ-CSIC)
Granada, Spain

ejbedmar@eez-csic.es

Eric S. Boyd
Department of Chemistry and Biochemistry & Astrobiology Biogeocatalysis Research Center
Montana State University
Montana, USA

eboyd@montana.edu

Purificación Cabello
Departamento de Botánica
Ecología y Fisiología Vegetal
Área de Fisiología Vegetal
Edificio Celestino Mutis
Universidad de Córdoba
Spain

bu1cahap@uco.es

Francisco Castillo
Departamento de Bioquímica y Biología Molecular
Edificio Severo Ochoa
Universidad de Córdoba
Spain

bb1carof@uco.es

María J. Delgado
Departamento de Microbiología del Suelo y Sistemas Simbióticos
Estación Experimental del Zaidín (EEZ-CSIC)
Granada, Spain

mdelgado@eez.csic.es

Anne-Lise Ducluzeau
Laboratoire de Bioénergétique et Ingénierie des Protéines (UPR 9036), CNRS
Marseille, France

ducluzeau@ifr88.cnrs-mrs.fr

Stuart J. Ferguson
Department of Biochemistry
University of Oxford
Oxford, UK

stuart.ferguson@bioch.ox.ac.uk

Trinity Hamilton
Department of Chemistry and Biochemistry & Astrobiology Biogeocatalysis Research Center
Montana State University
Montana, USA

thamilton@chemistry.montana.edu

Robert van Lis
Laboratoire de Bioénergétique et Ingénierie des Protéines (UPR 9036)
Marseille, France

rvanlis@yahoo.fr

Víctor M. Luque-Almagro
Departamento de Bioquímica y Biología Molecular
Edificio Severo Ochoa
Universidad de Córdoba
Spain

bb2lualv@uco.es

James W. B. Moir
Department of Biology
University of York
Heslington, York
UK

jm46@york.ac.uk

Conrado Moreno-Vivián
Departamento de Bioquímica y Biología Molecular
Edificio Severo Ochoa
Universidad de Córdoba
Spain

bb1movic@uco.es

Wolfgang Nitschke
Laboratoire de Bioénergétique et Ingénierie des Protéines (UPR 9036)
Marseille, France

wolfgang.nitschke@wanadoo.fr

John W. Peters
Department of Chemistry and Biochemistry & Astrobiology Biogeocatalysis Research Center
Montana State University
Montana, USA

john.peters@chemistry.montana.edu

Laurent Philippot
INRA
University of Burgundy
Soil and Environmental Microbiology
Dijon, France

philippo@dijon.inra.fr

David Richardson
School of Biological Sciences
Norwich Research Park
University of East Anglia
Norwich, UK

d.richardson@uea.ac.uk

M. Dolores Roldán
Departamento de Bioquímica y Biología Molecular
Edificio Severo Ochoa
Universidad de Córdoba
Spain

bb2rorum@uco.es

Luis M. Rubio
Centro de Biotecnológica y Genómica de Plantas UPM-INIA
Universidad Politécnica de Madrid
Madrid, Spain

lm.rubio@upm.es

Cristina Sánchez
Departamento de Microbiología del Suelo y Sistemas Simbióticos
Estación Experimental del Zaidín (EEZ-CSIC)
Granada, Spain

christina.sanchez@eez-csic.es

Barbara Schoepp-Cothenet
Laboratoire de Bioénergétique et Ingénierie des Protéines (UPR 9036)
Marseille, France

schoepp@ifr88.cnrs.mrs.fr

Jörg Simon
Institute of Microbiology and Genetics
Department of Biology
Technische Universität Darmstadt
Darmstadt, Germany

simon@bio.tu-darmstadt.de

Stephen Spiro
Department of Molecular and Cell Biology
University of Texas at Dallas
Texas, USA

stephen.spiro@utdallas.edu

Rob J.M. van Spanning
Department of Molecular Cell Physiology
VU University of Amsterdam
Amsterdam, Netherlands

rob.van.spanning@falw.vu.nl

Marc Strous
Max Planck Institute for Marine Microbiology
Bremen, Germany
Center for Biotechnology
University of Bielefeld
Bielefeld, Germany

mstrous@mpi-bremen.de

Other Books of Interest

Microbial Biofilms	2012
Bacterial Glycomics	2012
Non-coding RNAs and Epigenetic Regulation of Gene Expression	2012
Brucella	2012
Molecular Virology and Control of Flaviviruses	2012
Bacterial Pathogenesis	2012
Bunyaviridae	2011
Emerging Trends in Antibacterial Discovery	2011
Epigenetics	2011
Metagenomics	2011
The Biology of Paramyxoviruses	2011
Helicobacter pylori	2011
Microbial Bioremediation of Non-metals	2011
Lactic Acid Bacteria and Bifidobacteria	2011
Essentials of Veterinary Parasitology	2011
Hepatitis C: Antiviral Drug Discovery and Development	2011
Streptomyces: Molecular Biology and Biotechnology	2011
Alphaherpesviruses: Molecular Virology	2011
Recent Advances in Plant Virology	2011
Vaccine Design: Innovative Approaches and Novel Strategies	2011
Salmonella: From Genome to Function	2011
Insect Virology	2010
Environmental Microbiology: Current Technology and Water Applications	2010
Sensory Mechanisms in Bacteria: Molecular Aspects of Signal Recognition	2010
Bifidobacteria: Genomics and Molecular Aspects	2010
Molecular Phylogeny of Microorganisms	2010
Nanotechnology in Water Treatment Applications	2010
Iron Uptake and Homeostasis in Microorganisms	2010
Caliciviruses: Molecular and Cellular Virology	2010
Epstein-Barr Virus: Latency and Transformation	2010
Lentiviruses and Macrophages: Molecular and Cellular Interactions	2010
RNA Interference and Viruses: Current Innovations and Future Trends	2010
Retroviruses: Molecular Biology, Genomics and Pathogenesis	2010
Frontiers in Dengue Virus Research	2010
Metagenomics: Theory, Methods and Applications	2010
Microbial Population Genetics	2010
Borrelia: Molecular Biology, Host Interaction and Pathogenesis	2010
Aspergillus: Molecular Biology and Genomics	2010
Environmental Molecular Microbiology	2010
Neisseria: Molecular Mechanisms of Pathogenesis	2010
Lab-on-a-Chip Technology: Biomolecular Separation and Analysis	2009
Lab-on-a-Chip Technology: Fabrication and Microfluidics	2009

Preface

Nitrogen is the fourth most abundant element in biomass and is present in large amounts around the earth's surface, notably making up around 80% of the earth's atmosphere. However, the availability of nitrogen in a suitable form is a major limitation for life on earth, because most nitrogen is found in the form of dinitrogen gas, which is inaccessible to eukaryotes and many bacteria. Dinitrogen is both thermodynamically and kinetically stable, due to the very strong nitrogen–nitrogen triple bond. Furthermore, microbial redox processes lead to the continual replenishment of the dinitrogen pool, through the processes of denitrification and anaerobic ammonium oxidation (anammox), thus depleting the stock of bioavailable N.

Nitrogen is a highly unusual element in that it can exist stably at a wide range of redox states, ranging from –3 (in ammonia, NH_3) to +5 (in nitrate, NO_3^-), and at all intermediate redox states: –2 (in hydrazine, N_2H_4), –1 (in hydroxylamine, NH_2OH), zero (in dinitrogen), +1 (in nitrous oxide, N_2O), +2 (in nitric oxide, NO), +3 (in nitrite, NO_2^-), +4 (in nitrogen dioxide, NO_2). Enzymes have evolved to catalyse transformations between these different nitrogen redox states, and these processes have been harnessed by microorganisms (mainly prokaryotes) to drive energy generation through respiration. The major well established biological N-cycle processes are summarized in Fig. 1.

The central thermodynamic messages of the nitrogen cycle are (i) nitrogen redox interconversions can be used to drive biochemical energy production and (ii) N_2 is the sink. This is encapsulated in the anammox process, in which the intuitively unlikely reaction of nitrite and ammonia to generate dinitrogen is thermodynamically favourable and enables the highly specialized anammox bacteria to harvest biochemical energy using this process.

The aim of this book is to provide both a snapshot of current nitrogen cycle research and a broader perspective on the state of our understanding of this key biogeochemical cycle. Recent developments, including, of course, the explosion of genome sequence information are having a massive impact on our appreciation of the diversity of approaches that distinct organisms use in carrying out nitrogen cycling processes. Biochemistry, bioinformatics, cell biology, microbial physiology and environmental analysis are all approaches which are leading to enhanced understanding, and the ways in which these approaches are informing us about the evolution, organization, regulation, mechanism, and biological roles of nitrogen cycle processes are covered here. Work with well-established model organisms continues to give insight into nitrogen cycle processes, whilst much progress has also been made in the analysis of newly identified organisms and processes.

Chapter 1 focuses on the evolution of nitrogen cycle processes, and illustrates the ancient nature of nitrogen redox biochemistry, and how the evolution of denitrification likely preceded even oxygen respiration. Chapters 2 and 3 consider the organization of respiratory processes involved in nitrogen cycle transformations from the perspectives of multiprotein complex formation

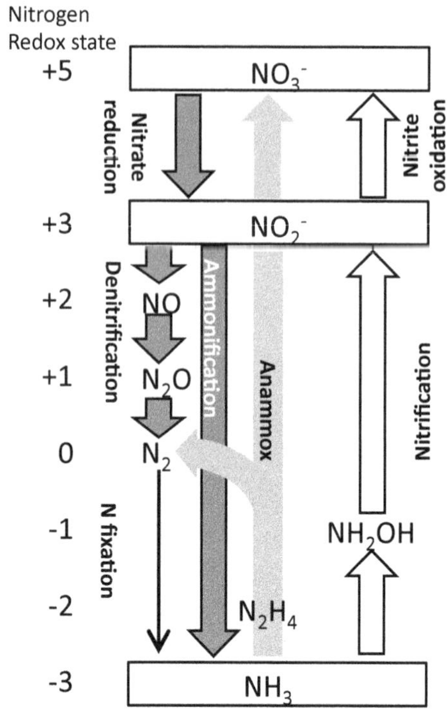

Figure 1 The biological nitrogen cycle. The major established nitrogen cycling processes are shown, linked to the nitrogen cycle redox state changes with which they are associated. Reductive processes are shown with dark arrows, oxidative processes with white arrows and combined oxidative and reductive process of anammox is shown as a grey arrow. The line arrow indicates nitrogen fixation, which is the only energy-consuming reaction; all the other processes can be associated with energy production.

(Chapter 2) and electron transfer (Chapter 3). Chapter 4 outlines the state of our understanding of the biochemical mechanism of nitrogenases. The function of nitrogen fixation is to provide nitrogen in a suitable form for assimilation. The processes of nitrogen transport and assimilation are considered in Chapter 5. The most striking finding of the last 20 years in nitrogen cycle research was the discovery of the anammox process. Current understanding of anammox and discussion of a new denitrification-independent biological mechanism for dinitrogen generation are covered in Chapter 6. In Chapter 7, the regulation and biochemical mechanism of denitrification is considered. Chapter 8 discusses the assembly of nitrogen cycle enzymes, which are characterized by the possession of metal-containing cofactors. Nitric oxide (NO) is both an intermediate of the nitrogen cycle in bacteria and an important biological messenger in mammalian systems. How bacteria deal with NO is dealt with in Chapter 9. Chapter 10 discusses recent findings related to denitrification in bacteria in the rhizosphere. Chapter 11 focuses on the production of the greenhouse gas nitrous oxide from terrestrial environments, and finally Chapter 12 investigates the relevance of nitrogen cycling processes to bacteria in the human body, including in pathogens.

What is evident is that nitrogen cycle research is continually producing new surprises in the form of new mechanisms, novel metabolic processes and unexpected physiological roles for nitrogen cycle processes. Much progress has been made and yet there remain important questions about the fundamental molecular basis of nitrogen cycling, which have important implications from the arena of human health and disease through to issues related to the global environment and climate change.

James Moir

The Nitrogen Cycle in the Archaean: an Intricate Interplay of Enzymatic and Abiotic Reactions

1

Robert van Lis, Anne-Lise Ducluzeau, Wolfgang Nitschke and Barbara Schoepp-Cothenet

Abstract

On modern planet earth, a multitude of nitrogen cycle enzymes equilibrate the atmospheric reservoir of dinitrogen with the more oxidized and more reduced nitrogen compounds essential for life. The enzymes are elaborate entities and the reactions performed are complicated and in cases energetically challenging. Nitrogen, however, must have been a crucial element already at life's very beginnings which raises the question how the primordial nitrogen cycle of emerging life in the Archaean – necessarily using simpler and likely fewer enzymes – may have evolved into the very complex network of present planet earth.

To address this question, we have analysed molecular phylogenies of the presently known enzymes involved in the present day nitrogen cycle. The results collected and presented in this chapter indicate that in the Archaean, the enzymatic part of this cycle was restricted to a partial segment of the modern energy conserving denitrification pathway and that abiotic redox conversions of nitrogen specific to the geoenvironment of the Archaean were the evolutionary precursors of many reactions now requiring enzyme catalysis. As found in recent years for other core metabolic processes, the biological nitrogen cycle appears to be evolutionarily rooted in inorganic chemistry.

Introduction

The statement that nitrogen (N) is an indispensable ingredient of life is commonplace. Nitrogen indeed plays a myriad of differing roles from being an essential constituent of life's building blocks to providing a substrate for energy conversion, i.e. a bioenergetic fuel. Correspondingly, the nitrogen atoms present in the various nitrogen-containing compounds used by life come in a wide range of formal reduction states. In amino acids, this formal reduction state is –3, i.e. similar to that in the inorganic compound ammonia, whereas nitrogen enters the bioenergetic denitrification chain as nitrate (NO_3^-) with a formal reduction state of +5 (see Fig. 1.1 for biologically relevant nitrogen compounds and N's charges therein). Concerning its range of accessible redox states covering 8 electron equivalents, nitrogen even beats sulfur in redox versatility. By far the largest reservoir of nitrogen on our planet is up in the air in the form of dinitrogen gas (N_2). Since N_2 has a formal charge of zero, life needed to find ways to either tap into geochemical redox transformations of N_2 or to develop enzymatic mechanisms for biologically producing compounds with the required reduction states of N. On our present planet, the steady state equilibrium between atmospheric N_2 and the other (both more reduced and more oxidized) versions of this element appears to be fully dependent on the activity of biological processes. The nitrogen cycle (see Fig. 1.2), just as the majority of earth's geochemical cycles, has thus become inseparable from life as recognized already more than 20 years ago in James Lovelock's Gaia hypothesis (Lovelock, 1988). However, things certainly have not always been this way. Earth did not start out as Gaia but as an inorganic disequilibrium system. This system was certainly 'poised' to give birth to life (Russell and Hall, 1997; Nitschke and Russell, 2010) but

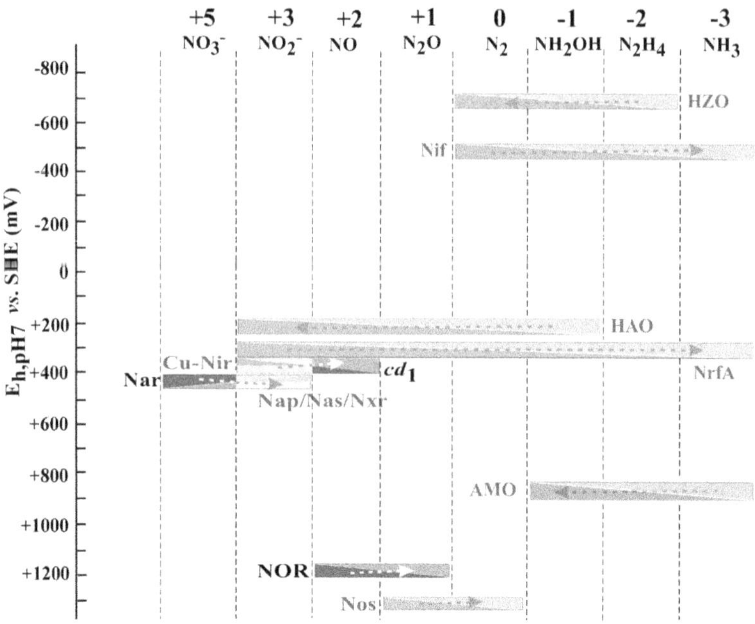

Figure 1.1 Schematic representation of the various redox transitions playing a role in the modern nitrogen cycle and being catalysed by enzymes as indicated in the figure. The horizontal dimension indicates the changes in nominal oxidation state of the nitrogen atom and the vertical axis represents the redox midpoint potential (at pH 7 and with respect to the standard hydrogen electrode, SHE) at which the respective transformations occur. Bold characters and darker shadings denote enzymes which are suggested by the phylogenetic evidence discussed in the text to have originated prior to the Archaea/Bacteria divergence.

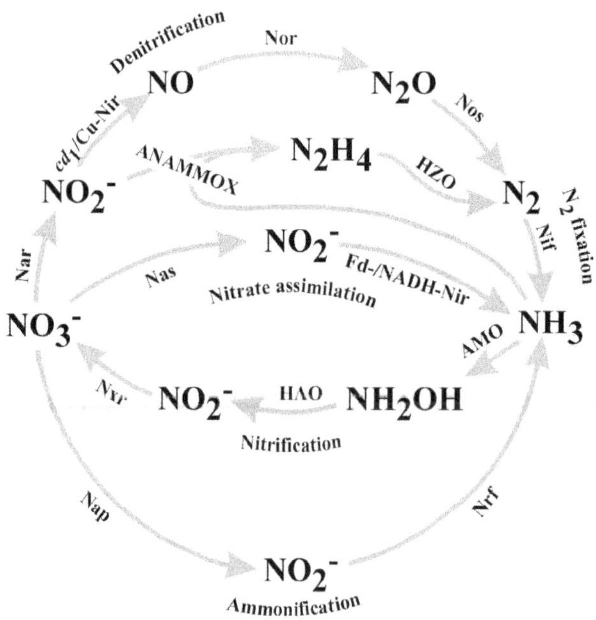

Figure 1.2 Partial reactions of the nitrogen cycle on present-day earth as they are known so far.

the takeover from a geochemical to a biochemical nitrogen cycle was likely a stepwise phenomenon.

Now, do we have means to go back in time and deduce how this gradual transition from geochemistry to biology was played out? Clues to the ancient past of the nitrogen cycle can in fact be gathered from the two disciplines involved, that is, biology and geochemistry. In the realm of geochemistry, we have to take into account experimentally observed inorganic synthesis pathways for the relevant nitrogen compounds viable under ambient conditions of the early earth. In biology, the most pertinent information will probably come from the phylogeny of enzymes involved in nitrogen compounds' transformations. Such protein (or gene) phylogenies can provide hints to whether an enzyme, and hence likely the pathway it belongs to, has been present and functioning already in the forebears of all living cells, now frequently called the Last Universal Common Ancestor (LUCA), whether it has been recruited from another pathway prior to or after LUCA or finally whether it represents an evolutionary innovation of more recent times. Since the major geochemical and biological transitions of our planet are certainly linked, one can then use the (relatively accurate) palaeogeochemical datings to calibrate the (quite unreliable, Graur and Martin, 2004) phylogenetic clocks, at least for the cases of the most incisive transitions in earth's geochemical past. One such major overturning of earth's environment, both geochemically and biologically, occurred when molecular oxygen (O_2) started to accumulate in the atmosphere (Nitschke *et al.*, 1998; Xiong *et al.*, 2000; Baymann *et al.*, 2001; Tomitani *et al.*, 2006; Allen *et al.*, 2007) between 2 and 2.5 billion years ago and it is very likely that the nitrogen cycle in general as well as its biological declinations will have undergone substantial modifications during this 'revolutionary' interval in earth's history. The appearance of atmospheric oxygen also is used to define the end of the Achaean and the beginning of the Proterozoic era of earth's history.

In this contribution we will assemble currently available data on the origin and evolution of key enzymes operating in the modern nitrogen cycle and try and distinguish parts of the cycle introduced or modified when oxygen levels started to rise from those which have been in place in the Archaean era or even already in LUCA, i.e. prior to the divergence of domains more than 3.4 billion years ago.

Enzymes rather than pathways

The modern nitrogen cycle is conveniently represented by schemes topologically equivalent to that shown in Fig. 1.2. Specific arcs on and in this cycle correspond to well-identified pathways classified according to their role in specific microorganisms and labelled nitrate respiration, ammonification, anaerobic ammonia oxidation (ANAMMOX) and so on (Madigan *et al.*, 2008). Correspondingly, treatments of the nitrogen cycle in scholarly articles or textbooks generally proceed along this classification scheme or deal separately with one or the other of the pathways. Frequently, the authors of such articles cannot avoid cross-referencing other parts of the cycle since specific structural units play a role in its various arcs. However, the cross-relationships of the respective enzymes are rarely exploited. The authors of this contribution admit that they do not belong to any of the genuine nitrogen cycle communities. Instead, our research interests focus on the relatedness and phylogenetic relationship of the protein modules spanning up the manifold of bioenergetic systems and it has in the recent past become increasingly obvious that only a very limited number of modules are needed to build up the multitude of energy-conserving systems, a fact for which we have coined the term 'redox enzyme construction kit' (Baymann *et al.*, 2003). For example, the [2Fe–2S] protein called 'Rieske protein' is a component of enzymes as different as mitochondrial complex III and prokaryotic arsenite oxidase (Duval *et al.*, 2010) and the 'molybdo(Mo)-di-pterin' protein module catalyses a multitude of different enzymatic reactions (Duval *et al.*, 2008), among which many occur in the nitrogen cycle as will be detailed below. While we might therefore not always be up to the standards of the nitrogen cycle community when it comes to details and terminology, we hope to compensate this shortcoming by our somewhat deviant approach concentrating on enzyme superfamilies, phylogenies thereof and energetic constraints governing specific enzymes' reactions. With respect to that latter point, the electrochemical properties of the various nitrogen compounds' redox couples (Fig. 1.1) play a crucial role and will be taken into consideration. Rather than going through the various pathways one by one, we have thus chosen to proceed by enzyme superfamilies.

The geochemical argument against the presence of certain enzymes in the Archaean

Inorganic chemists have in the past frequently argued that bioavailability must vary enormously between elements as a result of their strongly differing solubilities (Williams, 1996; Anbar, 2008) and that furthermore these solubilities change substantially in response to the oxidation state of the element and nature of available counterions (Anbar, 2008). In this context, it has been proposed that a number of elements were unlikely to have played a biological role in the Archaean, prior to the accumulation of oxygen in the atmosphere and its influence on ocean chemistry. This line of reasoning is now frequently put forward by biologists to discredit whole enzyme families as possibly participating in metabolism in the Archaean. Whereas we are reticent to subscribe to this approach in its generalizing version, we think that the argument probably holds when it comes to copper. Cu is in fact in pole position among the elements unlikely to have been soluble in the Archaean. Its solubility is estimated to plunge by more than 10 (!) orders of magnitude under the conditions of the Achaean ocean (Anbar, 2008) whereas for most other incriminated elements this decrease hardly exceeds a factor of 100. However, in the context of this contribution, we prefer to rely on the messages sent by the phylogenetic trees of the different enzymes rather than resorting to bioavailability arguments. It is nevertheless reassuring that for the case of copper enzymes involved in the nitrogen cycle, the phylogenetic evidence never is in conflict with the solubility reasoning as will be shown below.

The phylogenetic approach towards deducing ancestral enzymes

In a number of articles dealing with the evolution of the nitrogen cycle, the presence of well-separated clusters of Archaeal and Bacterial sequences in phylogenetic trees, and sometimes even the mere existence of both Archaeal and Bacterial representatives, is taken as evidence for the occurrence of the respective enzyme in LUCA. However, early but post-divergence (of Archaea and Bacteria), horizontal gene transfer from one domain to the other will yield equivalent phylogenies. Whenever possible, we will therefore try and root the respective trees by paralogous sequences and a position of the root in between Archaea and Bacteria is used as a criterion for likely pre-divergence origins. Rooting in fact allows deciding from which point (node on the tree) branches diverge in time. Furthermore, the general congruency of the analysed tree with current species trees will also be taken into account. There is, of course, no perfectly 'vertical-inheritance' tree but in the case of rampant horizontal gene transfer in a given tree, we will remain cautious towards interpreting this tree as indicating pre-divergence origins, even if the root-criterion mentioned above is fulfilled. The trees obtained on the Rieske/cyt*b* complexes (Lebrun *et al.*, 2006) and on arsenate reductase (Duval *et al.*, 2008) illustrate extreme examples for high likeliness of a pre-LUCA emergence or later appearance followed by extensive horizontal gene transfer, respectively. It is, however, imperative to keep in mind that phylogenetic analysis deals with probabilities and must be taken as such.

Nitrate reductases

Nitrate reductases, Nar, Nap, Nas and Nxr, are all members of the vast superfamily of the Mo-dipterin enzymes, in the past also referred to as the DMSO-reductase superfamily but now frequently called the Complex iron–sulfur molybdoenzyme (CISM) family (Rothery *et al.*, 2008; Refojo *et al.*, 2010). Members of this superfamily in general catalyse one and two electron redox conversions of an enormous variety of substrates ranging from the reduction of CO_2 to formate (Jormakka *et al.*, 2002) to the oxidation of arsenite to arsenate

(Ellis *et al.,* 2001). However, not only a plethora of substrates are dealt with by this enzyme superfamily, the redox versatility of the reactions it catalyses is at least as impressive. The redox midpoint potential of the CO_2/formate couple at pH7 is at −430 mV, whereas the chlorate/chlorite couple, the substrate of another member of the superfamily, chlorate reductase (Thorell *et al.,* 2003), titrates at +1 V! The enzyme superfamily thus covers a range of almost 1.5 V in tolerated redox reactions, which, to our knowledge, makes it the unchallenged champion in redox versatility observed so far. The nitrate to nitrite conversion catalysed by nitrate reductase occurs at a midpoint potential of +440 mV (see Fig. 1.1).

Several molecular phylogenies of parts of the superfamily have been published previously (Rothery *et al.,* 2008; Stolz *et al.,* 2006). We cannot help noticing that the majority of these trees suffer from a methodological flaw. In the algorithm used for tree reconstruction, the possibility of multiple substitutions accounting for differences between pairs of sequences was not taken into account, a parameter indispensable when comparing sequences from representatives of all prokaryotic domains and ever more so when comparing paralogous systems. This error led to very deep branches, badly resolved clades and in the worst cases wrong clustering. Calculating phylogenies with the multiple substitution option activated, by contrast, yields well-resolved clades each corresponding to the know groups of enzymes. Corresponding trees can be found in Duval *et al.* (2008) or Refero *et al.* (2010). Unfortunately, the choice of sequences considered by Refero *et al.* strongly undersamples archaeal representatives and the obtained tree therefore doesn't permit to assess the presence or absence of a given enzyme in LUCA. In the tree shown in Duval *et al.* (2008), several subfamilies profoundly resemble the tree of species, in particular with respect to the deep, basal bifurcation into archaeal and bacterial sequences, indicating that the genes coding for the enzyme are predominantly inherited in a vertical manner and that the enzyme was already present in LUCA.

What do phylogenies of nitrate reductases tell us about the evolutionary history of this enzyme? Microbiological, biochemical and enzymatic evidence has previously established that nitrate reductases come in four different versions, *i.e.* the membrane-attached, proton-translocating Nar, the periplasmic Nap, the cytoplasmic Nas and the membrane-attached nitrite oxidoreductase (Nxr, a nitrate reductase working in reverse). Nar and Nap are involved in denitrification and dissimilatory nitrate respiration whereas Nas participates in assimilatory reactions furnishing reduced N-compounds for nitrogen anabolism (for a review see Gonzales *et al.* (2006) and Chapter 5). The least studied of these enzymes, Nxr, has only been observed in nitrite oxidizing α- and γ-proteobacteria (Vanparys *et al.,* 2007) and terminates the nitrification pathway (Fig. 1.2). In addition to the denitrification chain, the Nar enzyme has also been found to function in nitrate respiration of *E. coli*, a bioenergetic pathway oxidizing formate and reducing nitrate and involving only the two enzymes formate dehydrogenase and nitrate reductase linked by quinones (Rothery *et al.,* 2001). Although the majority of our present structural and functional understanding of Nar has certainly been gained *via* the study of this specific system, we would argue that the *E. coli*-type nitrate respiration is a rare case likely restricted to Enterobacteria. In this reaction scheme, only a very small fraction of the electrochemical energy contained in the formate/nitrate redox couple (with a ΔE_m exceeding 800 mV!) is actually converted into chemiosmotic energy and an organism probably has to dwell in a heterotrophic substrate bounty to endure such low energy efficiency of its bioenergetic chain.

In addition to the Mo-di-pterin catalytic subunit common to all nitrate reductases, these enzymes are complemented by a variety of accessory redox proteins. Nar- and Nxr-type enzymes contain tetracubane iron–sulfur protein subunits as well as *b*-type cytochromes (albeit of variable structures) whereas Nap and Nas may feature dihaem cytochromes or ferredoxins as additional subunits (Rothery *et al.,* 2008).

A number of phylogenetic analyses have been carried out in the past. Whereas a comprehensive set of Nar, Nap and Nas sequences are treated by Philippot (2002) and are interpreted as indicating the presence of Nar already in LUCA, the mutual phylogenetic relationship of these subgroups, *i.e.* their composite tree, is not explored. From their analysis of a composite phylogenetic tree of nitrate reductases, Stolz and Basu (2002) concluded that, although Nar is observed in Archaea, prokaryotic nitrate reductases were acquired predominantly by horizontal gene transfer. We therefore decided to revisit the combined phylogeny of nitrate reductases taking advantage of the explosion in number of available sequences since the last studies were performed. A phylogenetic tree analysing exclusively Nxr sequences has been reported by Vanparys *et al.* (2007). Owing to the very restricted species distribution we consider the Nxr group as a post-LUCA enzyme and omit it from the tree of the Mo-di-pterin subunits discussed below.

Fig. 1.3 shows the composite phylogeny reconstructed for the catalytic subunit of Nar, Nas and Nap (i.e. NarG, NasA, and NapA) rooted by the corresponding subunit of formate dehydrogenase which itself shows all characteristics of a pre-LUCA enzyme (Nitschke and Russell, 2010). The obtained tree features a very distinct clustering of Nar on one and a mixture of Nas and Nap on the other side. The Nar cluster falls into two major clades corresponding to Archaea (dark shading) and Bacteria (lighter shading) and the Nap/Nas/Fdh outgroup roots Nar right in between these two clades. The Nar subtree moreover corresponds reasonably to 16S r-RNA species trees. According to the criteria outlined above, NarG thus qualifies as a likely pre-LUCA enzyme. This is not the case for the Nas and Nap enzymes. First of all, the tree suggests a very close relationship between Nas and Nap [as already noticed previously by Stolz and Basu (2002) as well as Jepson *et al.* (2006)], so close that Nap appears as a subgroup of Nas. Nevertheless, with respect to function, Nap distinguishes itself from Nas by its periplasmic localization evidenced by the TAT motif in all sequences sampled. The whole Nas/Nap subtree is a mess when it comes to species phylogenies. The branching topology of clades and the structure within clades substantially

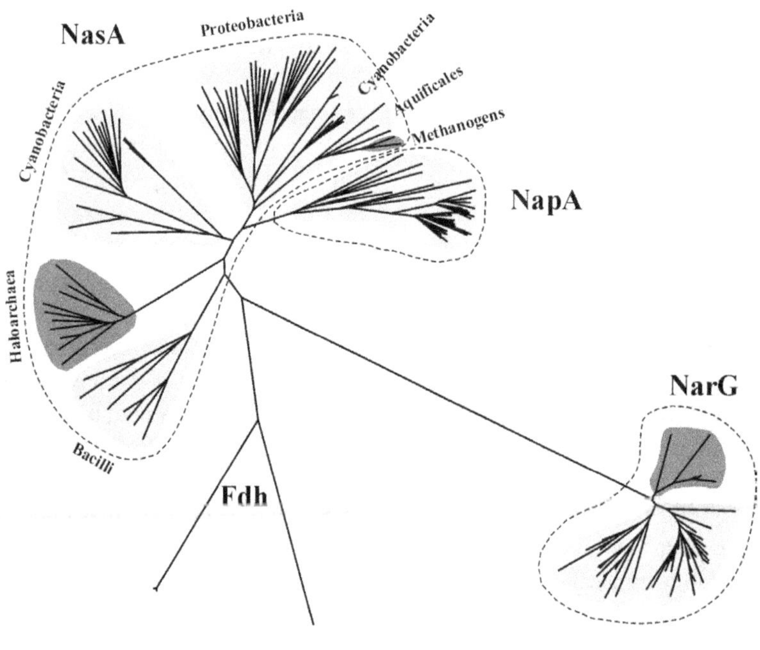

Figure 1.3 Schematic phylogenetic tree as reconstructed using the neighbour-joining algorithm from amino acid sequences of the catalytic Mo-di-pterin subunit in Nar-, Nap- and Nas-type nitrite reductases. The tree is rooted by the sequence of formate dehydrogenase, likely to be the most ancestral subgroup of the entire superfamily (Nitschke and Russell, 2010). The possibility of multiple substitutions is accounted for and tree building parameters are as detailed in Duval *et al.* (2008). Archaeal and bacterial representatives are distinguished by darker and lighter shading, respectively. Trees detailing all species and bootstrap values underlying this and the following figures can be obtained from the authors upon request.

deviates from 16S r-RNA species trees and the Archaea are only represented by the haloarchaeal phylum and a few methanogens. Whereas in species trees, the haloarchaeal clade roots within the methanogens, in Fig. 1.3 these two phyla are disconnected and separated by many bacterial branches. Most importantly, the root of the Nas/Nap subtree does not separate Archaea from Bacteria. The phylogenetic structure of the Nas/Nap clade therefore indicates an origin from a gene paralogous to Nar in LUCA which was not inherited into the Archaea but evolved within the bacterial domain to eventually become horizontally transferred into Haloarchaea and methanogens in two independent events.

This molecular phylogeny thus strongly indicates that the Nar system, but not Nap and Nas, already operated in LUCA. From a purely phylogenetic point of view, it would therefore be tempting to infer that the Nar enzyme functions in all prokaryotes just as it does in the well-characterized model organisms *E. coli* or *P. denitrificans* (see, for example, Gonzalez *et al.*, 2006). This clearly is not the case! As shown in Fig. 1.4 (left scheme), the catalytic Mo-di-pterin subunit of Nar in Bacteria is oriented towards the cytoplasm – as it makes sense one might argue, since the nitrate to nitrite conversion consumes $2H^+$ and thus contributes productively to the build-up of the chemiosmotic potential across the cytoplasmic membrane. However, the enzyme performing the reduction of nitrite (NO_2^-) to nitric oxide (NO) is periplasmic (see Fig. 1.4 and the discussion below). The product of Nar in a denitrification chain such as that of *P. denitrificans* therefore has to be transported across the membrane to serve as substrate for nitrite reductases. Transmembrane transport for denitrification and assimilation is carried out

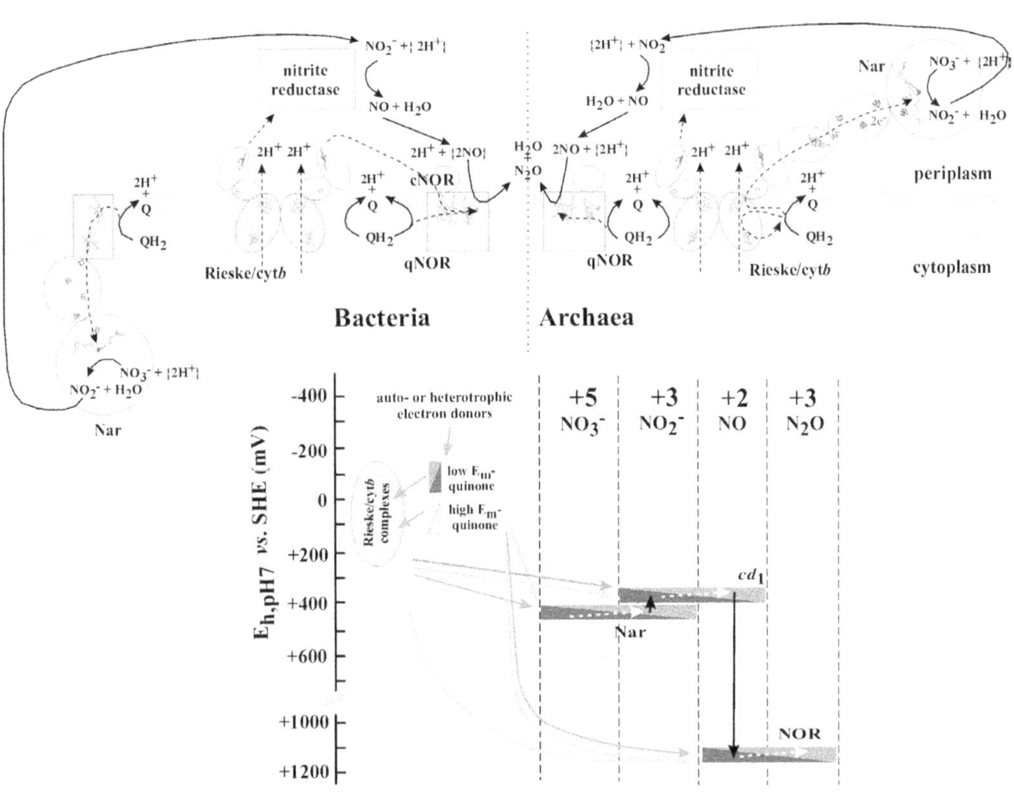

Figure 1.4 The differing organisation of the nitrate to nitrous oxide segment in the denitrification chain of Bacteria (top, left scheme) and Archaea (top, right scheme). Chemiosmotically 'counterproductive' H^+ ions are denoted $\{H^+\}$. The electrochemical pathways of the corresponding reactions are indicated in the bottom part of the figure. The redox ranges of pool quinones and the Rieske/cytb complex are additionally indicated to position the redox conversions of nitrogen compounds with respect to possible donors and acceptors.

by members of the Major Facilitator Superfamily and ABC transporter-type complexes (Moir and Wood, 2001; Wood *et al.*, 2002). As shown in Fig. 1.4 (left scheme), the electrons required to reduce nitrate to nitrite are delivered via the quinone pool and are transferred from the membrane-integral cytochrome *b*-subunit (NarI) through a chain of iron–sulfur clusters (NarH) to the catalytic centre on NarG. At least this is how the reaction occurs in the characterized species (see Gonzalez *et al.*, 2006).

As discussed above, the archaeal *narG* genes clearly form a sister-cluster to their bacterial counterparts (see Fig. 1.3). However, the operon organization of the Nar enzyme in the majority of Archaea is substantially different from that in Bacteria in that the membrane-integral NarI subunit is replaced by a soluble haem protein (Fig. 1.4, right scheme). Even more astonishingly, archaeal NarG sequences feature a TAT translocation signal in their N-terminal sequence stretches, a characteristic not found in bacterial Nars (Gonzalez *et al.*, 2006; Martínez-Espinosa *et al.*, 2007). It therefore is inevitable that the membrane-polarity of the NarG subunit is reversed in Archaea as compared to Bacteria. So do Archaea make their Nar enzyme work in a counterproductive fashion? Not sure. First of all, as mentioned above, this spatial organization of enzymes avoids having to translocate nitrate from one side of the membrane to the other. As pointed out previously (Martínez-Espinosa *et al.*, 2007; Yoshimatsu *et al.*, 2007; Bonete *et al.*, 2008), the integration of the Nar operon in Haloarchaea in a gene cluster which also comprises the Rieske/cyt*b* complex suggests that, in contrast to the bacterial cases, electrons delivered by the quinone pool in Archaea first pass through the Rieske/cyt*b* complex, which pumps $4H^+$ per $2e^-$ transferred, before entering the chain of redox centres leading up to the catalytic molybdenum site in NarG (Fig. 1.4, right). For simplicity and in the absence of structural information, the archaeal and bacterial Rieske/cyt*b* complexes in Fig. 1.4 have been represented by the proteobacterial enzyme. However, the nature of the membrane-extrinsic haem subunit varies significantly between phyla as discussed in detail in Kramer *et al.* (2008). The redox potential scheme shown in the middle panel of Fig. 1.4 emphasizes that both pathways are energetically viable and that the net difference in chemiosmotic build-up between both systems is marginal.

Surprisingly, the decisive hint to this significantly different organization of the archaeal Nar enzyme comes from the Haloarchaea. Indeed, the Haloarchaea feature a gene cluster encompassing both the Rieske/cyt*b* and the Nar operons in a consecutive fashion. This genetic association of operons makes a functional association likely which in turn rationalizes the inverted sidedness of the archaeal NarG protein as detailed above. However, the phylogenies of both cytochrome *b* and the Rieske subunits of Rieske/cyt*b* complexes (Nitschke, unpublished) show that this enzyme has been horizontally transferred into the Haloarchaea from a source situated within the domain Bacteria. In genome comparison studies it has indeed been observed that Haloarchaea have exchanged genes quite promiscuously with halophilic members of the Bacteria (Mongodin *et al.*, 2005). As will be discussed below, further components of the denitrification chain such as Cu-containing nitrite reductase, NO reductase and N_2O reductase also result from horizontal gene transfer from the domain Bacteria into the Haloarchaea. The Rieske/cyt*b* – Nar gene cluster observed in Haloarchaea therefore cannot be an entity vertically inherited from more ancestral Archaea. Does this mean that the organization of the chain as proposed by Martínez-Espinosa *et al.* (2007) and shown in Fig. 1.4 should not be generalized to all Archaea? The simplicity of this scheme makes us tend to assume that the reaction chain depicted in Fig. 1.4, right panel is indeed viable for the entire archaeal domain. To be confident, however, whether this scheme represents THE archaeal mode of Nar-function, the viability of this model must be demonstrated for at least one member of the Archaea in which the Rieske/cyt*b* complex has been vertically inherited. The nature of the Nar-cyt*b* subunit moreover appears to vary between archaeal species. In addition to the soluble haem protein observed in haloarchaeal Nar and many other representatives of the Archaea (Fig. 1.4, right scheme), archaeal Nars containing a membrane-integral cytochrome *b*-subunit homologous to NarI do seem to exist (Cabello *et al.*, 2004).

Another question arising from the dichotomy of NarG's membrane-sidedness concerns the makeup of the ancestral chain. Was it of either the archaeal or the bacterial type or were both systems operating in LUCA (possibly to fulfil different metabolic needs)? Further biochemical studies of diverse bacterial and archaeal species may help to shine light on this question.

Cyt cd_1-type nitrite reductase

Cyt cd_1, the product of the *nirS* gene, is a soluble periplasmic enzyme catalysing the one-electron reduction of nitrite to NO. Cyt cd_1 (and alternatively Cu-containing nitrite reductases, see the following section) thus connect up the two transmembrane enzymes of the bioenergetic denitrification chain, i.e. Nar-type nitrate and NO reductase (Fig. 1.2). Cyt cd_1 is a haem protein containing a structurally unique porphyrin, haem d_1, located in the centre of the catalytic, β-propeller-shaped large domain of the enzyme together with a *c*-type haem bound by a peripheral domain with a type I-cytochrome fold (Brown *et al.*, 2001).

In addition to the *nirS* gene, two further homologues have been described (Palmedo *et al.*, 1995; Kawasaki *et al.*, 1997), denoted *nirF* and *nirN*, the products of which are considered to be involved in maturation of cyt cd_1 (deBoer *et al.*, 1994; Zajicek *et al.*, 2009). In composite phylogenies of *nirS*/*F*/*N*-homologous genes retrieved from sequenced genomes, three distinct subtrees corresponding to the three functionally distinguishable groups are observed with *nirF* and *nirN* genes being restricted to Bacteria (Ducluzeau *et al.*, unpublished).

Two histidine residues located within the large, d_1 binding domain (His327 and His369 in the *Pseudomonas aeruginosa* numbering) are considered to be intimately involved in the catalytic activity of the enzyme (Brown *et al.*, 2001). In the case of NirN, the first of these two histidine residues is not conserved and in NirF, both appear to be absent. We have therefore used the presence of both these histidine residues as a criterion distinguishing 'true' cd_1 enzymes from their maturation factors. Fig. 1.5 shows the corresponding phylogenetic tree of cyt cd_1 nitrite reductases.

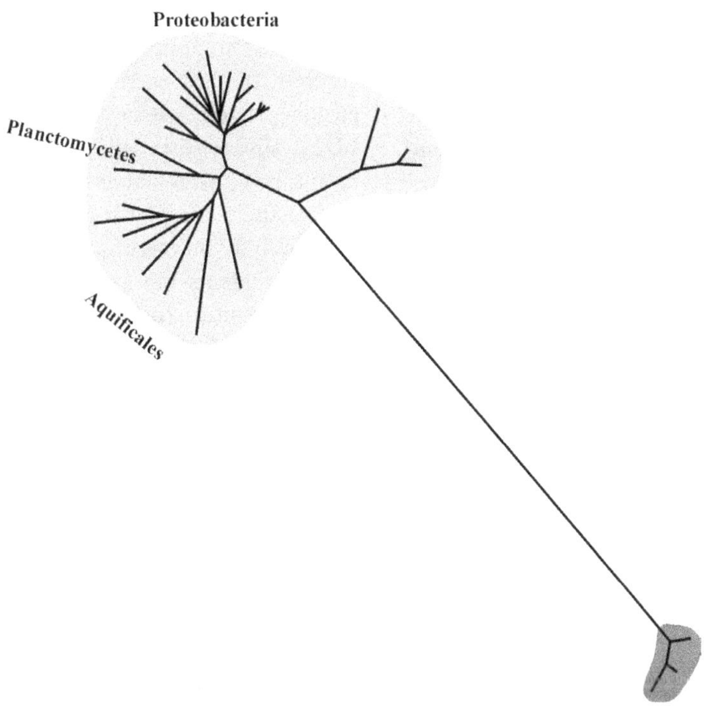

Figure 1.5 Schematic phylogenetic tree of NirS sequences as defined in the text. Phylogenetic reconstruction was performed as indicated in the legend to Figure 3 and the same shading pattern was used to distinguish archaeal from bacterial branches.

This tree is topologically similar to those reported previously (Jones *et al.*, 2008) but contains more archaeal and early branching bacterial representatives while at the same time being streamlined by the omission of numerous proteobacterial sequences which tend to distort the schematic representation due to the strong oversampling of proteobacterial genomes in the databases. The obtained phylogenies strongly resemble 16S r-RNA trees and in particular feature a deep Archaea/Bacteria cleavage. Unfortunately, the NirF and NirN sequences are only restricted to the bacterial domain and can't therefore serve to root the cd_1 tree. So far, the final decisive criterion for a *bona fide* pre-divergence enzyme, i.e. a rooting in between the archaeal and bacterial clusters, is missing for cd_1 nitrite reductase. However, the tree does not contain features supporting a post-LUCA appearance. We therefore conclude that cyt cd_1-type nitrite reductase probably is an enzyme that was present in LUCA, although the evidence is less compelling than for example for the Nar or the NOR enzymes.

Copper-containing nitrite reductases

Copper-containing nitrite reductase (NirK or Cu-Nir; see Figs. 1.1 and 1.2) is a multicopper enzyme made up from a compilation of cupredoxin units (see Nojiri *et al.*, 2007, and references to numerous previous structures therein). Archaeal and bacterial representatives have been identified and phylogenetic trees have been reconstructed for this family (Treusch *et al.*, 2005; Jones *et al.*, 2008). All published trees show the following common features.

1. Only a very small number of archaeal representatives have been identified so far.
2. These archaeal sequences form two clades distant from each other (Treusch *et al.*, 2005), with one of the two containing exclusively Haloarchaea and clustering with proteobacteria.
3. The bacterial parts of the tree show evidence for pervasive horizontal gene transfer (Treusch *et al.*, 2005; Jones *et al.*, 2008). Since our own trees only reproduce the phylogenies cited above, we refer the reader to the original articles for details.

The phylogenetic data thus support a post-LUCA emergence of NirK-type nitrite reductases and the acquisition of the gene by Haloarchaea through horizontal gene transfer from the bacterial domain.

Two substantially distinct enzymes, cd_1-Nir and Cu-Nir, thus catalyse the same step in the denitrification pathway, *i.e.* the reduction of nitrite to nitric oxide. One of these two, cd_1-Nir, appears to have been around already since LUCA but has become a relatively rare enzyme in extant species. The other one, Cu-Nir, by contrast, shows a wide species distribution but, according to its molecular phylogeny, likely has emerged much later than cd_1-Nir. These findings suggest Cu-Nir to be a newcomer endowed with some kind of properties conferring evolutionary fitness strongly superior to that of cd_1-Nir. So why did dentrification not start out with Cu-Nir? As we have mentioned early on in this contribution, Cu is extremely unlikely to have been bioavailable prior to the rise of environmental oxidation state entailed by oxygenic photosynthesis. The copper cofactor thus wasn't an option for enzymes in LUCA. By contrast, the phylogenies of many bioenergetic enzymes suggest haems as redox cofactors which LUCA already disposed of. The biosynthesis of haems, however, is a multistep process and therefore costly in energy. Cu-Nir doesn't require the complex biosynthesis of a haem cofactor and thus likely represents the 'cheap' way of converting nitrite into nitric oxide. The fitness superiority of Cu-Nir might therefore have bioenergetic roots.

As emphasized by Figs 1.1 and 1.4, the redox midpoint potential of the NO_2^-/NO couple is almost equipotential and even slightly lower than that of the NO_3^-/NO_2^- couple. In order to avoid product inhibition of nitrate reductase, nitrite reductase must be kept strongly poised towards the forward reaction. NO certainly is a gas and therefore constantly removed from the reaction medium. However, NO is also about 10 times more soluble in water than is O_2 and may thus accumulate to relatively high values. More importantly, at higher concentrations NO reacts deleteriously with proteins and thus becomes toxic to life. For these reasons, it is important for organisms to avoid accumulation of this nitrogen compound, a task which is carried out in denitrifying prokaryotes by the enzyme NO reductase.

Nitric oxide reductase (NOR)

NO reductase (NOR) is a membrane-integral enzyme belonging to the superfamily generally referred to as the haem–copper oxidases (HCO). From the moment NOR was recognized as part of this superfamily, the name HCO had become obsolete, yet never abandoned. NOR does not contain copper but instead harbours an iron atom in the so-called dinuclear site, which additionally contains a haem in all members of the superfamily. Moreover, the name 'oxidase' historically comes from the 'cytochrome *c*-oxidase' function of the archetypal member of the superfamily, *i.e.* mitochondrial complex IV. Many prokaryotic members of the family have since been observed to oxidize quinols rather than cytochrome *c* and a more appropriate and more and more frequently employed term therefore is 'O_2 reductases'. Now, NOR is no O_2 reductase either and the whole superfamily is desperately looking for a new term to be proposed.

Just like O_2 reductases, NORs exist in two different forms, using either soluble cytochromes or reduced quinones (quinols) as electron donors for the conversion of NO to nitrous oxide (N_2O) which are consequently called cNOR and qNOR, respectively. So far, cNORs seem to be restricted to proteobacteria and a very small number of other members of the domain Bacteria (some of which clearly have acquired the gene from proteobacteria via horizontal gene transfer) whereas qNORs are found in diverse members of the Bacteria and the Archaea.

The first phylogenetic trees of the entire superfamily have been reported in 2000 (Hendriks *et al.*, 2000) and the somewhat enigmatic topology of these trees has since that time given rise to a multitude of differing rationalizations (Hendriks *et al.*, 2000; Pereira *et al.*, 2001; Brochier-Armanet *et al.*, 2009; Gribaldo *et al.*, 2009). Taking advantage of a recent major breakthrough in our understanding of structure and function of O_2 reductases (Hemp *et al.*, 2005; Rauhamäki *et al.*, 2006), we have put forward an evolutionary scenario in which the mentioned enigmatic aspects are resolved (Ducluzeau *et al.*, 2009). The part of this scenario relevant to the question of the nitrogen cycle and the evolution thereof goes as follows: qNOR was the ancestral enzyme of the superfamily present in LUCA whereas cNOR and O_2 reductase activities were later inventions. This finding further corroborates the conclusion already suggested by the phylogenies of Nar and cd_1-nitrite reductase detailed above, that at least a segment of the modern bioenergetic respiration of oxidized nitrogen compounds already operated in the Archaean. The phylogeny of the whole HCO superfamily furthermore ties up NO and O_2 respiration and indicates that aerobic respiration arose post-LUCA from its ancestor, the respiration of nitrogen oxides which itself probably is almost as old as life itself (Ducluzeau *et al.*, 2009).

A phylogeny of qNORs is shown in Fig. 1.6. The tree shows a clearcut cleavage into archaeal and bacterial subtrees with only one group of Haloarchaea violating this dichotomy. Just as in the case of the Rieske/cyt*b* complex, qNOR of Haloarchaea appears to be the result of horizontal gene transfer. The electron and substrate transfers involving NOR in Archaea and Bacteria are depicted in Fig. 1.4.

N_2O reductase

N_2O reductase, the product of the *nosZ* gene, is the periplasmic enzyme closing the cycle towards the atmospheric reservoir *via* the two-electron reduction of N_2O to N_2 (Figs 1.1 and 1.2). The electron transfer and catalytic centres of N_2O reductase are exclusively made up of copper, with a dinuclear 'Cu_A' centre in a cupredoxin-type domain and a tetranuclear 'Cu_Z' cluster in the centre of the large catalytic domain of the enzyme which has a β-propeller-type fold (Haltia *et al.*, 2003).

The phylogenetic relationships of N_2O reductases have been investigated many years ago based on a relatively small set of sequences and deep insights into the evolutionary links of its domains to structural units of other bioenergetic enzymes (Zumft, 1992; Saraste *et al.*, 1996; Zumft and Kroneck, 2007) have been elaborated. Fig. 1.7 shows a schematic version of the phylogenetic

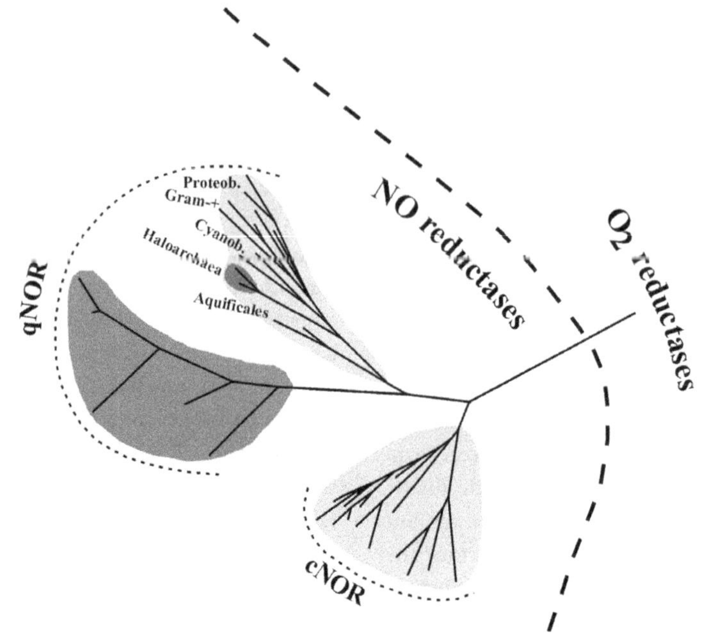

Figure 1.6 Schematic phylogenetic tree of qNOR and cNOR sequences. Sequences and multiple alignments correspond to those used in Ducluzeau *et al.* (2009), only enriched by 2 further haloarchaeal representatives. The group of haloarchaeal qNORs now contains four members. The phylogenetic analysis corresponds to that described in the legend to Fig. 1.3.

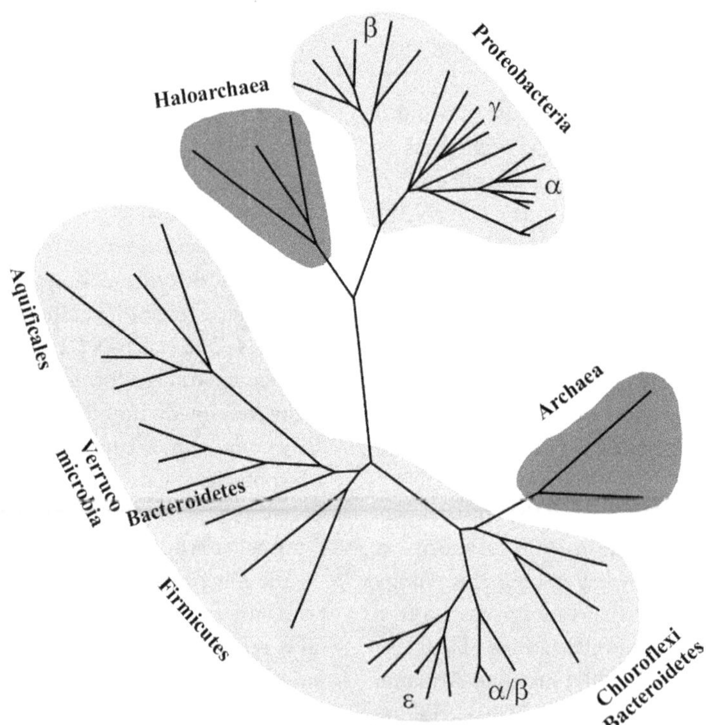

Figure 1.7 Schematic phylogenetic tree of NosZ sequences. Phylogenetic reconstruction was performed as indicated in the legend to Fig. 1.3 and the same shading pattern was used to distinguish archaeal from bacterial branches.

tree reconstructed from the presently available ensemble of NosZ sequences. Only five archaeal representatives were found falling into two distinct clusters and emerging at distant nodes in a tree dominated by bacterial sequences. The proteobacterial clades are split into two distinct islands, one of which contains sequences from the α-, β- and ε-subgroups whereas the other one is made up from the α-, β- and γ-subgroups with a topology strikingly different from species trees in that the α- and γ- rather than the β- and γ-clusters appear as sister clades. The evolutionarily deeply branching Chloroflexi and Aquificales are separated by Bacilli and proteobacteria, two phylogenetically quite distant clades. The whole tree shown in Fig. 1.7 thus shows strictly no resemblance to species phylogenies and we therefore conclude that NosZ is a post-LUCA enzyme that probably has appeared in the Bacteria and that has been imported into the Archaea and in particular the Haloarchaea via horizontal gene transfer. In 2007, Zumft and Kroneck published a phylogenetic tree rather similar to that of Fig. 1.7, although with a lower species coverage. These authors interpreted their phylogeny as indicating an early origin of *nosZ* and, since the gene is found both in Archaea and Bacteria, even its presence in LUCA (Zumft and Kroneck, 2007). As mentioned at the beginning of this contribution, the mere existence of a gene in both domains is not proof of a pre-divergence origin. The topology of the bacterial NosZ phylogeny is as different from 16S r-RNA trees in Zumft and Kroneck (2007) as it is in Fig. 1.7 and therefore also argues in favour of a post-divergence origin for the NosZ enzyme.

The NrfA/HAO/HZO superfamily of multihaem cytochromes

The enzyme dissimilatory nitrite reductase, the product of the *nrfA* gene, catalyses the six-electron reduction of nitrite to ammonia (Fig. 1.1) in the ammonification pathway (Fig. 1.2) as well as numerous partial reduction steps of this reaction (Einsle *et al.*, 2002). Structural similarities between NrfA and further enzymes of the nitrogen cycle, that is, hydroxylamine (NH_2OH) oxidoreductase (HAO, involved in the nitrification chain) and hydrazine (N_2H_4) oxidoreductase (HZO, an anammox enzyme), have been proposed in the past and the existence of a novel superfamily comprising these three enzymes has been proposed (Bergmann *et al.*, 2005; Klotz and Stein, 2007; Klotz *et al.*, 2008). A composite phylogeny of the different subgroups of the family has been reconstructed (Klotz *et al.*, 2008). This composite phylogeny consists almost entirely of bacterial species and the less than a handful of archaeal representatives do not cluster together. Our own analysis only corroborates the published trees and we therefore refer the reader to the original articles cited above. Furthermore, the pattern of species distribution doesn't match 16S r-RNA trees and the available data thus argue against a role of the superfamily in LUCA.

The AMO/pMMO superfamily of multicopper proteins

Ammonia monooxygenase (Amo) is a multicopper enzyme closely related to particulate methane monooxygenase (pMMO), a crucial enzyme in the metabolism of methanotrophic proteobacteria. A 3D structure of pMMO has been solved (Lieberman and Rosenzweig, 2005) demonstrating a structural link of its membrane-external subunit AmoB to subunit II of SoxB- and SoxM-type O_2 reductases and to the N-terminal domain of N_2O reductase with all these three protein units belonging to the cupredoxin 'megafamily' (Ducluzeau, 2009).

In the course of the last 10 years it has become obvious that aerobic ammonia oxidation mediated by Amo-like proteins is not, as previously thought, restricted to proteobacteria but also occurs in numerous Archaea (Nicol and Schleper, 2006). Astonishingly and in apparent contradiction to the argument that Cu-containing enzymes necessarily have arisen after the Archaea/Bacteria divergence and even after the rise of O_2 tensions in the atmosphere (see above as well as Klotz and Stein, 2007), published phylogenetic trees based on AmoA sequences indicate a very deep cleavage between Archaea and Bacteria (Treusch *et al.*, 2005), a feature generally taken to indicate pre-LUCA origins for the enzyme. However, other characteristics of this tree and of the corresponding multiple alignments as well as further circumstantial evidence strongly argue against this scenario.

1 Only α-/β- and γ- proteobacterial sequences have been detected so far.
2 Only a very restricted subclass of the Crenarchaeota seems to contain what the authors consider as AmoAs and the mutual sequence identities are correspondingly very high.
3 By contrast, the sequence similarity between archaeal and bacterial representatives is extremely marginal.

Intriguingly, many residues ligating metals in *bona fide* AmoA from proteobacteria are not conserved in the crenarchaeal sequences pretended to represent AmoA. We would therefore argue that, if ever the archaeal proteins turned out to be related to bacterial AmoA, they rather look like paralogues than orthologues. From the survey of the available literature, it appears to us that the functional role of the crenarchaeal 'AmoA' proteins presently is rather obscure (for a recent review, see Junier *et al.*, 2010). In this context, it is worthwhile mentioning that the second enzyme required to complete the AMO reaction, i.e. hydroxylamine oxidoreductase (HAO, see above) appears to be absent from the respective crenarchaeal species. Conservation of the Cu_A-binding residues in AmoB has been taken to argue in favour of an orthologue nature of the bacterial and archaeal genes (Treusch *et al.*, 2005). This argument, however, is not valid since, as mentioned above, AmoB is a dimer of cupredoxins and all cupredoxins with dinuclear Cu_A-sites would need to conserve the sequence positions serving as ligands for the metal centres.

Despite the fact that the case for Amo may seem more controversial than for other reactions of the cycle, our conclusion from the available data is that this enzyme family has emerged after the divergence of Archaea and Bacteria. According to the solubility argument for copper outlined in the beginning of this contribution and, even more importantly, as a result of the fact that Amo uses molecular oxygen as a substrate, the enzyme arguably appears as a good candidate for a post-O_2 innovation.

Sirohaem nitrite reductases

Nitrate assimilation has been studied extensively in higher plants, fungi and algae, but is only scarcely characterized in Archaea and Bacteria (to the exception of cyanobacteria, see Flores *et al.*, 2005). In some species, the genes and gene clusters coding for the assimilatory nitrate reduction systems have been sequenced, whereas no functional data are available for most of these cases. Assimilatory nitrite reductases come in two different types. The ferredoxin (Fd)-dependent enzyme is typical of photosynthetic organisms but is also found in non-photosynthetic Bacteria and Archaea (Alcantara-Hernandez *et al.*, 2009). The NAD(P)H-dependent enzyme is found in many heterotrophic organisms (for a recent review see Richardson *et al.*, 2001). Both types show structural similarities to the sirohaem-containing sulfite reductases (Crane *et al.*, 1995). Assimilatory nitrite reductases from prokaryotic and eukaryotic organisms strongly resemble each other but substantial differences exist between the ferredoxin- and the NAD(P)H-dependent enzymes, no matter whether they are of prokaryotic or eukaryotic origin. Ferredoxin-dependent nitrite reductases harbour siroheme and a [4Fe–4S] centre as prosthetic groups, whereas NAD(P)H-dependent nitrite reductases additionally contain FAD and [2Fe–2S] centres.

NADH-dependent nitrite reductase can be heterodimeric (with the *nirD* gene product as a separate subunit, e.g. in *Bacillus subtilis*) or monomeric (with NirD fused to the C-terminus of NasB as for example in *Klebsiella oxytoca*).

Phylogenetic analysis of the Fd-dependent nitrite reductase shows that the sequences of Archaea do not cluster together and that furthermore the entire tree does not correspond to species phylogenies (Alcantara-Hernandez *et al.*, 2009). Fd-dependent nitrite reductase therefore probably isn't a pre-divergence enzyme. No phylogenetic analysis has as yet been carried out on NADH-dependent nitrite reductases. We therefore BLAST searched sequence databases using the *Klebsiella oxytoca* NasB sequence as query and found only bacterial sequences suggesting the conclusion that the NADH-dependent nitrite reductase is also an innovation of the bacterial domain.

Nitrogenase

Life on this planet is nitrogen-limited as humans have experienced since they invented agriculture. The enzyme permitting life to tap into the enormous reservoir of nitrogen contained in atmospheric nitrogen gas is nitrogenase (Nif; see Fig. 1.2), a protein complex containing in the majority of cases composite iron/molybdenum clusters as catalytic centres (Seefeld *et al.*, 2009). The chemical transformation performed by nitrogenase, i.e. the reduction of N_2 to ammonia, is frequently cited as one of the most energy-demanding reactions in biology, consuming 16 ATP on its way from substrate to product. Until only a few years ago, nitrogenase was considered as necessarily having been present in the common ancestor of Archaea and Bacteria (Fani *et al.*, 2000; Leigh, 2000; Berman-Frank *et al.*, 2003). Two main reasons have been put forward to argue in favour of an ancestral nitrogenase. The first one essentially relies on the above mentioned dependence of life on supply of biologically fixed nitrogen, tacitly extrapolating from the present day nitrogen cycle back to the Archaean. The second line of reasoning is based on the notorious sensitivity of nitrogenase towards molecular oxygen and the argument invokes that the enzyme therefore had to have originated in an environment devoid of oxygen, i.e. in the archaean era (Fani *et al.*, 2000). We consider this latter argument invalid due to the following reason: a multitude of enzymes are known which are O_2 sensitive in anaerobes but adapted to cope with oxygen in aerobic species. Nitrogenase had 2 billion years of leisure to develop mechanisms rendering it less O_2 sensitive but did not manage and its parent organisms instead had to resort to compartmenting the sites of nitrogen fixation away from O_2. To us this means that there must be a fundamental incompatibility between O_2 and nitrogen fixation and we would argue that this incompatibility plainly results from electrochemistry. As shown in Fig. 1.2, the reduction of N_2 to ammonia occurs at an ambient redox potential (at physiological pH values) below −500 mV, i.e. even lower than the reduction of protons to molecular hydrogen! Not only is the nitrogenase reaction therefore among the most energy-intensive catalyses in biology, it also certainly is one of the most reducing redox reactions and the latter probably explains the former. Incidentally, we would like to mention that this fact is brought home by the observation that nitrogenase produces H_2 during catalysis, probably as an inevitable by-product of the extremely reducing levels on which the partial reduction reactions, in the course of which eight electrons are transferred, need to proceed. Given this electrochemical particularity of the reaction from N_2 to ammonia, it is inevitable that the presence of the strong and aggressive oxidant O_2 is deleterious to this catalysis.

So what do phylogenies of the protein subunits of nitrogenase suggest concerning its evolutionary ancestry? The most comprehensive analysis to date of genes involved in nitrogen fixation and the phylogenetic relationship thereof has been reported in 2004 (Raymond *et al.*, 2004). The results of this study challenge the previous paradigm of a pre-LUCA nitrogenase. The catalytic subunit of nitrogenase is found to be a subclade of a large superfamily of paralogous genes comprising the cluster of the BChl-reductases (as already recognized by Burke *et al.*, 1993) and a clade of so far unknown function which, however, certainly is not nitrogen fixation (Staples *et al.*, 2007). The nitrogenase subfamily is characterized by evidence for extensive horizontal gene transfer and is devoid of any clear signature of a deep Archaea/Bacteria cleavage. All these elements are in favour of a more recent emergence of nitrogenase from an enzyme pre-dating nitrogenase but performing a function different from N_2-fixation.

Nitrogen cycle enzymes in LUCA; who's in and who's out?

The phylogenies of the ensemble of enzymes discussed above allow assessing the likely inventory of N-cycle enzymatic reactions in LUCA. The summary of the corresponding results has been overlain on the schematic representation of the modern cycle and coded by differently shaped arrows in Fig. 1.8. Continuous arrows denote enzymes which their phylogenies indicate to have been present in LUCA, whereas dashed and dotted arrows denote likely later inventions. Of course and as always when dealing with deep phylogenies, the presence of now-extinct or not

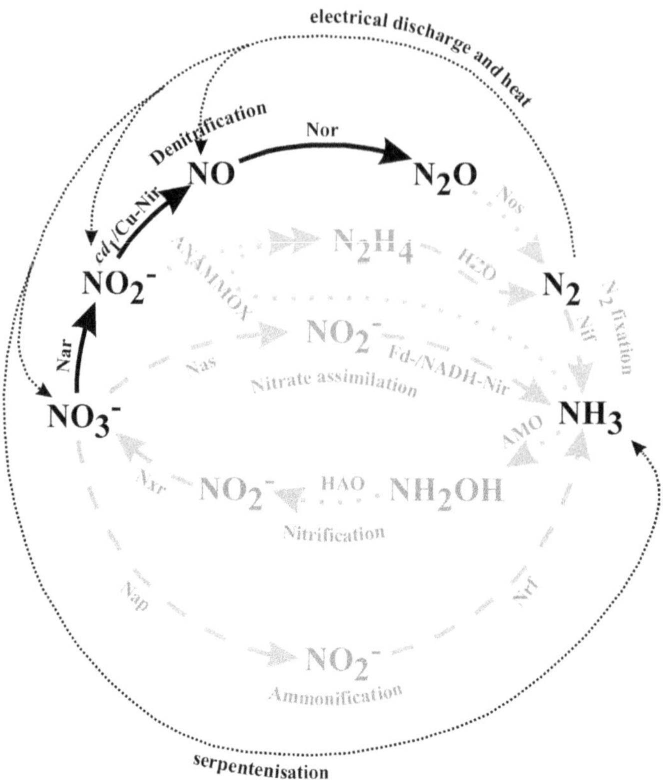

Figure 1.8 Redrawing of the scheme shown in Fig. 1.2 but distinguishing likely pre-divergence enzymes and reaction schemes (continuous lines, bold letters) from those for which the phylogenetic evidence argues for a later emergence (light letters). These latter reactions are further subdivided into potentially pre-O_2 (light dashed lines) and likely post-O_2 (light dotted lines) transformations. Abiotic generation pathways for oxidised and reduced nitrogen compounds from atmospheric N_2 are additionally indicated in the figure by bold dotted arrows.

yet discovered enzymes cannot be ruled out. The conclusions put forward in this contribution therefore reflect the presently most parsimonious scenario and specific details may be overturned by the simple discovery of a hitherto unknown enzyme. That this is not merely a hypothetical possibility has been emphasized by the unexpected coming on stage of the anammox reaction only 15 years ago (Mulder *et al.*, 1995) and more recent newly identified nitrogen transformations (see Chapter 6).

The astonishing take-home message from Fig. 1.8 is that only a small fraction of the present day nitrogen cycle appears to have been functional in LUCA. The principal raison d'être of enzymatic transformations of nitrogen compounds in LUCA appears to have been energy conversion, *i.e.* anaerobic respiration of nitrogen oxides up to N_2O gas which will ultimately have escaped to the atmosphere. This raises the question of where these mildly to strongly oxidized nitrogen oxides have come from in the first place.

In Fig. 1.8 we have chosen to distinguish enzymes which may have existed prior to the appearance of O_2 but after the Archaea/Bacteria divergence, that is, during the late Archaean (dashed lines) from those which rely entirely on copper as active centres or use O_2 as a substrate and which are thus good candidates for innovations having occurred during the Archaean/Proterozoic transition (dotted lines). It is noteworthy that this distinction is not based on phylogeny but exclusively on metal content as proposed in the past (Anbar, 2008; Klotz and Stein, 2007) and on mechanistic considerations.

The other startling outcome of the phylogenetic data is that no obvious biological source to produce the bioessential most-reduced nitrogen compound, ammonia, appears to have been available in the early Archaean since neither nitrogenase nor any other enzyme of the extant nitrogen cycle performing reduction of more oxidized nitrogen compounds to ammonia are predicted to have been present in LUCA.

We are therefore left with the dual problem that life seems to have used nitrogen in both its most oxidized and its most reduced forms without having developed means to produce either of them. As we will show in the following section, this 'problem' can be resolved when turning to present knowledge of palaeogeochemistry.

Abiotic production of nitrogen oxides and ammonia

Under conditions of the primordial atmosphere, likely composed mainly of carbon dioxide and dinitrogen, electrical discharge and high temperatures induced by volcanism, bolide impact and lightning will produce high quantities of NO according to the reaction scheme detailed in Ducluzeau *et al.* (2009) (and see Martin *et al.*, 2007). This NO will further react to HNO and rain out into the hadean ocean, where it yields nitrate and nitrite (Mancinelli and McKay, 1988; Summers and Khare, 2007; Ducluzeau *et al.*, 2009). We have previously estimated that as high as micromolar concentrations of nitrogen oxyanions are likely to accumulate in the hadean ocean within only a million years (Ducluzeau *et al.*, 2009) and thus to grow into a viable substrate for bioenergetic respiration of NO_3^-, NO_2^- and NO. Part of these nitrogen oxides-containing waters as well as dissolved N_2 will percolate into the crust where they will become reduced to ammonia via serpentenization (Shock, 1992; Russell and Hall, 1997). These reactions have been shown to proceed with high efficiencies in settings that reproduce the conditions of hadean hydrothermal vents (Brandes *et al.*, 1998; Schoonen and Xu, 2001; Smirnov *et al.*, 2008). A fraction of this NO_3^- and NO_2^- may also have been reduced to N_2O gas (Samarkin *et al.*, 2010) escaping back to the atmosphere.

Whereas the oxidized nitrogen compounds would therefore have been supplied to the hadean and early archaean ocean 'from above', reduced ammonia would leak from earth's interior into these oceans through hydrothermal vents. This scenario recalls the image of an earth as a giant protocell, negative in the interior, positive on the outside and with a strong electrochemical disequilibrium across the boundary (Russell and Hall, 1997; Nitschke and Russell, 2010). Both extreme nitrogen species of the cycle are thus likely to have been provided by this inorganic protocell even before the emergence of life and the nascent metabolism could immediately take advantage of these species for anabolism and for energy conversion.

Palaeogeochemists have argued in the past that cessation of continuous bolide bombardment, dwindling CO_2 contents of the atmosphere, reduced volcanism and incident of lightning in the Archaean may have induced a 'nitrogen crisis' (Navarro-Gonzáles *et al.*, 2001; Thomazo *et al.*, 2010) for early life driving the evolutionary appearance of the enzyme nitrogenase. Increasing oxygen concentrations and the resulting instability of ammonia may subsequently have induced the need for a further takeover of abiotic, inorganic reaction schemes by dedicated enzymes. At the same time, the appearance of a novel redox player on the stage of enzymatic reactions, *i.e.* copper, may have facilitated emergence of the required new enzymes as well as having allowed the replacement of 'older' and maybe biosynthetically more costly systems (e.g. those relying on porphyrins) by Cu-based catalysis.

It is obvious from what we have outlined in this contribution that the phylogenies of the considered nitrogen cycle enzymes are fully in line with the likely palaeogeochemical history of the early earth.

Acknowledgements

The authors would like to thank Mike Russell (JPL, Pasadena, USA) for many stimulating discussions and suggestions as well as for critical reading of the manuscript. We are furthermore grateful to Frauke Baymann (BIP, Marseilles, France) for help with the phylogenetic trees. Further stimulating input by Axel Magalon (LCB, Marseilles/France), Pascal Philippot (IPG, Paris, France) and Ariane Atteia (BIP, Marseilles, France) is gratefully acknowledged. Our work on the enzymes involved in the nitrogen cycle was supported by the French Agence National pour la Recherche (ANR) *via* the projects ANR-06-BLAN-0384 (NOtoOO) and ERMMoE.

References

Alcántara-Hernández, R.J., Valenzuela-Encinas, C., Zavala-Díaz de la Serna, F.J., Rodriguez-Revilla, J., Dendooven, L., and Marsch, R. (2009). Haloarchaeal assimilatory nitrate-reducing communities from a saline alkaline soil. FEMS Microbiol. Lett. *298*, 56–66.

Allen, J.F., and Martin, W. (2007). Evolutionary biology: out of thin air. Nature *445*, 610–612.

Anbar, A.D. (2008). Oceans. Elements and evolution. Science *322*, 1481–1483.

Baymann, F., Brugna, M., Mühlenhoff, U., and Nitschke, W. (2001). Daddy, where did (PS)I come from? Biochim. Biophys. Acta *1507*, 291–310.

Baymann, F., Lebrun, E., Brugna, M., Schoepp-Cothenet, B., Giudici-Orticoni, M.-T., and Nitschke, W. (2003). 'The Redox Protein Construction Kit; Pre-LUCA Evolution of Energy Conserving Enzymes. Phil. Trans. R. Soc. Lond. B *358*, 267–274.

Bergmann, D.J., Hooper, A.B., and Klotz, M.G. (2005). Structure and sequence conservation of hao cluster genes of autotrophic ammonia-oxidizing bacteria: evidence for their evolutionary history. Appl. Environ. Microbiol. *71*, 5371–5382.

Bonete, M.J., Martínez-Espinosa, R.M., Pire, C., Zafrilla, B., and Richardson, D.J. (2008). Nitrogen metabolism in haloarchaea. Saline Systems *4*, 9.

Brandes, J.A., Boctor, N.Z., Cody, G.D., Cooper, B.A., Hazen, R.M., and Yoder, H.S.jr (1998). Abiotic nitrogen reduction on the early Earth. Nature *395*, 365–367.

Brochier-Armanet, C., Talla, E., and Gribaldo, S. (2009). The multiple evolutionary histories of dioxygen reductases: Implications for the origin and evolution of aerobic respiration. Mol. Biol. Evol. *26*, 285–297.

Brown, K., Roig-Zamboni, V., Cutruzzola', F., Arese, M., Sun, W., Brunori, M., Cambillau, C., and Tegoni, M. (2001). Domain swing upon His to Ala mutation in nitrite reductase of *Pseudomonas aeruginosa*. J. Mol. Biol. *312*, 541–554.

Burke, D.H., Hearst, J.E., and Sidow, A. (1993). Early evolution of photosynthesis: clues from nitrogenase and chlorophyll iron proteins. Proc. Natl. Acad. Sci. U.S.A. *90*, 7134–7138.

Cabello, P., Roldán, M.D., and Moreno-Vivián, C. (2004). Nitrate reduction and the nitrogen cycle in archaea. Microbiology *150*, 3527–3546.

Chèneby, D., Hallet, S., Mondon, M., Martin-Laurent, F., Germon, J.C., and Philippot, L. (2003). Genetic characterization of the nitrate reducing community based on *narG* nucleotide sequence analysis. Microb. Ecol. *46*, 113–121.

Crane, B.R., Siegel, L.M., and Getzoff, E.D. (1995). Sulfite reductase structure at 1.6 Å: evolution and catalysis for reduction of inorganic anions. Science *270*, 59–67.

de Boer, A.P., Reijnders, W.N., Kuenen, J.G., Stouthamer, A.H., and van Spanning, R.J. (1994). Isolation, sequencing and mutational analysis of a gene cluster involved in nitrite reduction in *Paracoccus denitrificans*. Antonie Van Leeuwenhoek *66*, 111–127.

De Vries, S., and Schröder, I. (2002). Comparison between the nitric oxide reductase family and its aerobic relatives, the cytochrome oxidases. Biochem. Soc. Trans. *30*, 662–667.

Dalsgaard, T., Thamdrup, B., and Canfield, D.E. (2005). Anaerobic ammonium oxidation (anammox) in the marine environment. Res. Microbiol. *156*, 457–464.

Duval, S., Ducluzeau, A.-L., Nitschke, W., and Schoepp-Cothenet, B. (2008). Enzyme phylogenies as markers for the oxidation state of the environment; the case of respiratory arsenate reductase and related enzymes. BMC Evol. Biol. 8:206.

Duval, S., Santini, J.M., Nitschke, W., Hille, R., and Schoepp-Cothenet, B. (2010). The small subunit AroB of arsenite oxidase: lessons on the [2Fe–2S]-Rieske protein superfamily. J. Biol. Chem., in press, doi: 10.1074/jbc.M110.113811.

Ducluzeau, A.-L. (2009). Origin of aerobic respiratory enzymes. PhD Thesis. Université de Provence-Victor Hugo, Marseille, France.

Ducluzeau, A.-L., van Lis, R., Duval, S., Schoepp-Cothenet, B., Russell, M.J., and Nitschke, W. (2009). Was NO the first deep electron sink? Trends in Biochem. Sci. *34*, 9–15.

Einsle, O., Messerschmidt, A., Huber, R., Kroneck, P.M., and Neese, F. (2002). Mechanism of the six-electron reduction of nitrite to ammonia by cytochrome c nitrite reductase. J. Am. Chem. Soc. *124*, 11737–11745.

Ellis, P.J., Conrads, T., Hille, R., and Kuhn, P. (2001). Crystal structure of the 100 kDa arsenite oxidase from *Alcaligenes faecalis* in two crystal forms at 1.64 Å and 2.03 Å. Structure *9*, 125–132.

Fani, R., Gallo, R., and Liò, P. (2000). Molecular evolution of nitrogen fixation: The evolutionary history of the *nifD*, *nifK*, *nifE*, and *nifN* genes. J. Mol. Evol. *5*, 1–11.

Flores, E., Frias, J.E., Rubio, L.M., and Herrero, A. (2005). Photosynthetic nitrate assimilation in cyanobacteria. Photosyn. Res. *82*, 117–133.

Garvin, J., Buick, R., Anbar, A.D., Arnold, G.L., and Kaufman, A.J. (2009). Isotopic evidence for an aerobic nitrogen cycle in the latest archaean. Science *323*, 1045–1048.

Gonzales, P.J., Correia, C., Moura, I., Brondino, C.D., and Moura, J.J.G. (2006). Bacterial nitrate reductases: Molecular and biological aspects of nitrate reduction. J. Inorg. Biochem. *100*, 1015–1023.

Graur, D., and Martin, W. (2004). Reading the entrails of chicken: molecular timescales of evolution and the illusion of precision. Trends Genet. *20*, 80–86.

Gregory, L.G., Bond, P.L., Richardson, D.J., and Spiro, S. (2003). Characterization of a nitrate-respiring bacterial community using the nitrate reductase gene (*narG*) as a functional marker. Microbiology *149*, 229–237.

Gribaldo, S., Talla, E., and Brochier-Armanet, C. (2009). Evolution of the haem copper oxidases superfamily: a rooting tale. Trends Biochem Sci. *34*, 375–378.

Haltia, T., Brown, K., Tegoni, M., Cambillau, C., Saraste, M., Mattilas, K., and Djinovic-Carugo, K. (2003). Crystal structure of nitrous oxide reductase from *Paracoccus denitrificans* at 1.6 Å resolution. Biochem. J. *369*, 77–88.

Hemp, J. *et al.* (2005). Helix switching of a key active-site residue in the cytochrome *cbb*3 oxidases. Biochemistry *44*, 10766–10775.

Hendriks, J.,Oubrie, A., Castresana, J., Urbani, A., Gemeinhardt, S., and Saraste, M. (2000). Nitric oxide reductases in Bacteria. Biochim. Biophys. Acta *1459*, 266–273.

Jepson, B.J.N., Marietou, A., Mohan, S., Cole, J.A., Butler, C.S., and Richardson, D.J. (2006). Evolution of the soluble nitrate reductase: defining the monomeric periplasmic nitrate reductase subgroup. Biochem. Soc. Trans. *34*, 122–126.

Jones, C.M., Stres, B., Rosenquist, M., and Hallin, S. (2008). Phylogenetic analysis of nitrite, nitric oxide and nitrous oxide respiratory enzymes reveal a complex evolutionary history for denitrification. Mol. Biol. Evol. *25*, 1955–1966.

Jormakka, M., Törnroth, S., Byrne, B., and Iwata, S. (2002). Molecular basis of proton motive force generation: structure of formate dehydrogenase-N. Science *295*, 1863–1868.

Junier, P., Molina, V., Dorador, C., Hadas, O., Kim, O.-S., Junier, T., Witzel, K.-P., and Imhoff, J. (2010). Phylogenetic and functional marker genes to study ammonia-oxidising microorganisms (AOM) in the environment. Appl. Microbiol. Biotechnol. *85*, 425–440.

Kawasaki, S., Arai, H., Kodama, T., and Igarashi, Y. (1997). Gene cluster for dissimilatory nitrite reductase (nir) from Pseudomonas aeruginosa: sequencing and identification of a locus for heme d1 biosynthesis. J. Bacteriol. *179*, 235–242.

Klotz, M.G., and Stein, L.Y. (2007). Nitrifier genomics and evolution of the nitrogen cycle. FEMS Microbiol. Lett. *278*, 146–156.

Klotz, M.G., Schmid, M.C., Strous, M., op den Camp H.J.M., Jetten, M.S.M., and Hooper, A.B. (2008). Evolution of an octahaem cytochrome c protein family that is key to aerobic and anaerobic ammonia oxidation by bacteria. Environ. Microbiol. *10*, 3150–3163.

Kramer, D.M., Nitschke, W., and Cooley, J.W. (2008). The cytochrome bc_1 and related *bc* complexes: The Rieske/Cytochrome *b* complex as the functional core of a central electron/proton transfer complex. in: The Purple Phototrophic Bacteria (Hunter, C.N., Daldal, F., Thurnauer, M.C., and Beatty, J.T., eds.). Springer, pp. 451–473.

Lebrun, E., Santini, J.M., Brugna, M., Ducluzeau, A.-L., Ouchane, S., Schoepp-Cothenet, B., Baymann, F., and Nitschke, W. (2006). The Rieske protein; a case study on the pitfalls of multiple sequence alignments and phylogenetic reconstruction. Mol. Biol. Evol. *23*, 1180–1191.

Berman-Frank, I., Lundgren, P., and Falkowski, P. (2003). Nitrogen fixation and photosynthetic oxygen evolution in cyanobacteria. Res. Microbiol. *154*, 157–164.

Leigh, J.A. (2000). Nitrogen fixation in methanogens: The archaeal perspective. Curr. Issues. Mol. Biol. *2*, 125–131.

Lieberman, R.L., and Rosenzweig, A.C. (2005). Crystal structure of a membrane-bound metalloenzyme that catalyses the biological oxidation of methane. Nature *434*, 177–82.

Lovelock, J.E. (1988). The Ages of Gaia. WW Norton & Co, UK

Madigan, M.T., Martinko, J.M., Dunlap, P.V., and Clark, D.P. (2008). Brock Biology of microorganisms. (12th Edition), B. Cummings, USA.

Mancinelli, R.L., and McKay, C.P. (1988). The evolution of nitrogen cycling. Orig. Life Evol. Biosph. *18*, 311–325.

Martin, R.S., Mather, T.A., and Pyle, D.M. (2007). Volcanic emissions and the early Earth atmosphere. Geochim. Cosmochim. Acta *71*, 3673–3685.

Martínez-Espinosa, R.M., Dridge, E.J., Bonete, M.J., Butt, J.N., Butler, C.S., Sargent, F., and Richardson, D.J. (2007). Look on the positive side! The orientation, identification and bioenergetics of 'Archaeal' membrane-bound nitrate reductases. FEMS Microbiol. Lett. *276*, 129–139.

Moir, J.W., and Wood, N.J. (2001). Nitrate and nitrite transport in bacteria. Cell Mol. Life Sci. *58*, 215–224.

Mongodin, E.F., Nelson, K.E., Daugherty, S., DeBoy, R.T., Wister, J., Khouri, H., Weidman, J., Walsh, D.A., Papke, R.T., Sanchez Perez, G., Sharma, A.K., Nesbø, C.L., MacLeod, D., Bapteste, E., Doolittle, W.F., Charlebois, R.L., Legault, B., and Rodriguez-Valera, F. (2005). The genome of *Salinibacter ruber*: convergence and gene exchange among hyperhalophilic bacteria and archaeal. Proc. Natl. Acad. Sci. U.S.A. *102*, 18147–18152.

Mulder, A., van de Graaf, A.A., Robertson, L.A., and Kuenen, J.G. (1995). Anaerobic ammonium oxidation discovered in a denitrifying fluidized bed reactor. FEMS Microbiol. Ecol. *16*, 177–184.

Navarro-González, R., McKay, C.P., and Mvondo, D.N. (2001). A possible nitrogen crisis for Archaean life due to reduced nitrogen fixation by lightning. Nature *412*, 61–64.

Nicol, G.W., and Schleper, C. (2006). Ammonia-oxidising crenarchaeota: important players in the nitrogen cycle? Trends Microbiol. *14*, 207–212.

Nitschke, W., Mühlenhoff, U., and Liebl, U. (1998). Evolution. In: Photosynthesis: a Comprehensive Treatise, Raghavendra, A., ed. (Cambridge University Press, UK), pp. 285–304.

Nitschke, W., and Russell, M.J. (2010). Just like the universe the emergence of life had high enthalpy and low entropy beginnings. J. Cosmology (in press).

Nojiri, M., Xie, Y., Inoue, T., Yamamoto, T., Matsumura, H., Kataoka, K., Deligeer, Yamaguchi, K., Kai, Y., and Suzuki, S. (2007). Structure and function of a hexameric copper-containing nitrite reductase. Proc. Natl. Acad. Sci. U.S.A. *104*, 4315–4320.

Palmedo, G., Seither, P., Körner, H., Matthews, J.C., Burkhalter, R.S., Timkovich, R., and Zumft, W.G. (1995). Resolution of the nirD locus for heme d1 synthesis of cytochrome cd1 (respiratory nitrite reductase) from *Pseudomonas stutzeri.* Eur. J. Biochem. *232,* 737–746.

Pereira, M.M., Santana, C., and Teixeira, M. (2001). A novel scenario for the evolution of haem–copper oxygen reductases. Biochim. Biophys. Acta *1505,* 185–208.

Philippot, L. (2002). Denitrifying genes in bacterial and archaeal genomes. Biochim. Biophys. Acta *1577,* 355–376.

Pino, C., Olmo-Mira, F., Cabello, P., Martínez-Luque, M., Castillo, F., Roldán, M.D., and Moreno-Vivián, C. (2006). The assimilatory nitrate reduction system of the phototrophic bacterium *Rhodobacter capsulatus* E1F1. Biochem.Soc. Trans. *34,* 127–129.

Rauhamäki, V., Baumann, M., Soliymani, R., Puustinen, A., and Wikström, M. (2006). Identification of a histidine–tyrosine cross-link in the active site of the cbb_3-type cytochrome *c* oxidase from *Rhodobacter sphaeroides.* Proc. Natl. Acad. Sci. U.S.A. *103,* 16135–16140

Raymond, J., Siefert, J.L., Staples, C.R., and Blankenship, R.E. (2004). The natural history of nitrogen fixation. Mol. Biol. Evol. *21,* 541–554.

Refojo, P.N., Sousa, F.L., Teixeira, M., and Pereira, M.M. (2010). The alternative complex III: A different architecture using known building modules. Biochim. Biophys. Acta, in press.

Richardson, D.J., Berks, B.C., Russell, D.A., Spiro, S., and Taylor, C.J. (2001). Functional, biochemical and genetic diversity of prokaryotic nitrate reductases. Cell. Mol. Life Sci. *58,* 165–178.

Rosca, V., Duca, M., de Groot, M.T., and Koper, M.T.M. (2009). Nitrogen cycle electrocatalysis. Chem. Rev. *109,* 2209–2244.

Rothery, R.A., Blasco, F., Magalon, A., and Weiner, J.H. (2001). The diheme cytochrome *b* subunit (Narl) of *Escherichia coli* nitrate reductase A (NarGHI): structure, function, and interaction with quinols. J. Mol. Microbiol. Biotechnol. *3,* 273–283.

Rothery, R.A., Workun, G.J., and Weiner, J.H. (2008). The prokaryotic complex iron–sulfur molybdoenzyme family. Biochim Biophys Acta. *1778,* 1897–1929.

Russell, M.J., and Hall, A.J. (1997). The emergence of life from iron monosulphide bubbles at a submarine hydrothermal redox and pH front. J. Geol. Soc. Lond. *154,* 377–402.

Samarkin, V.A., Madigan, M.T., Bowles, M.W., Casciotti, K.L., Priscu, J.C., McKay, C.P., and Joye, S.B. (2010). Abiotic nitrous oxide emission from the hypersaline Don Juan Pond in Antarctica. Nature Geoscience, in press, doi:10.1038/ngeo847.

Saraste, M., Castresana, J., Higgins, D., Lübben, M., and Wilmanns, M. (1996). Evolution of cytochrome oxidase. in: Origin and Evolution of Biological Energy Conversion (Baltscheffsky, H., ed.), pp. 255–290.

Schoonen, M.A.A., and Xu, Y. (2001). Nitrogen reduction under hydrothermal vent conditions: implications for the prebiotic synthesis of C-H-O-N compounds. Astrobiology *1,* 133–142.

Seefeld, L.C., Hoffman, B.M., and Dean, D.R. (2009). Mechanism of Mo-dependent nitrogenase. Annu. Rev. Biochem. *78,* 701–722.

Shock, E.L. (1992). Chemical environments of submarine hydrothermal systems; marine hydrothermal systems and the origin of life. Origins of Life and Evolution of the Biosphere *22,* 67–107.

Smirnov, A., Hausner, D., Laffers, R., Stongin, D.R., and Schoonen, M.A.A. (2008). Abiotic ammonium formation in the presence of Ni-Fe metals and alloys and its implications for the hadean nitrogen cycle. Geochem. Trans. *9,* 5.

Staples, C.R., Lahiri, S., Raymond, J., von Herbulis, L., Mukhophadhyay, B., and Blankenship, R.E. (2007). Expression and association of group IV nitrogenase NifD and NifH homologs in the non-nitrogen-fixing archaeon *Methanocaldococcus jannaschii.* J. Bacteriol. *189,* 7392–7398.

Stolz, J.F., Basu, P., Santini, J.M., and Oremland, R.S. (2006). Arsenic and selenium in microbial metabolism. Annu. Rev. Microbiol. *60,*107–130.

Stolz, J.F., and Basu, P. (2002). Evolution of nitrate reductase: molecular and structural variations on a common function. Chembiochemistry 3, 198–206.

Thomazo, C., Ader, M., and Philippot, P. (2010). Isotopic evidence for a nitrogen crisis triggering biological nitrogen fixation 2.7 Gyr ago. Geobiology, submitted.

Thorell, H.D., Stenklo, K., Karlsson, J., and Nilsson, T. (2003). A gene cluster for chlorate metabolism in *Ideonella dechloratans.* Appl. Environ. Microbiol. *69,* 5585–5592.

Tomitani, A., Knoll, A.H., Cavanaugh, C.M., and Ohno, T. (2006). The evolutionary diversification of cyanobacteria: molecular-phylogenetic and paleontological perspectives. Proc. Natl. Acad. Sci. U.S.A. *103,* 5442–5447.

Treusch, A.H., Leininger, S., Kletzin, A., Schuster, S.C., Klenk, H.-P., and Schleper, C. (2005). Novel genes for nitrite reductase and Amo-related proteins indicate a role of uncultivated mesophilic crenarchaeota in nitrogen cycling. Environ. Microbiol. *7,* 1985–1995.

Vanparys, B., Spieck, E., Heylen, K., Wittebolle, L., Geets, J., Boon, N., and De Vos, P. (2007). The phylogeny of the genus *Nitrobacter* based on comparative rep-PCR 16S rRNA and nitrite oxidoreductase gene sequence analysis. Syst. Appl. Microbiol. *30,* 297–308.

Williams, R.J.P. (1996). From dihydrogen to dioxygen: Evolution of biological oxidation-reduction. In: Origin and Evolution of Biological Energy Conversion, Baltscheffsky, H., ed., pp. 109–141.

Wood, N.J., Alizadeh, T., Richardson, D.J., Ferguson, S.J., and Moir, J.W. (2002). Two domains of a dual-function NarK protein are required for nitrate uptake, the first step of denitrification in *Paracoccus pantotrophus.* Mol. Microbiol. *44,* 157–170.

Xiong, J., Fischer, W.M., Inoue, K., Nakahara, M., and Bauer, C.E. (2000). Molecular evidence for the early evolution of photosynthesis. Science *289,* 1724–1730.

Yoshimatsu, K., Araya, O., and Fujiwara, T. (2007). *Haloarcula marismortui* cytochrome *b*-561 is encoded by the *narC* gene in the dissimilatory nitrate reductase operon. Extremophiles *11*, 41–47.

Zajicek, R.S., Bali, S., Arnold, S., Brindley, A.A., Warren, M.J., and Ferguson, S.J. (2009). d(1) haem biogenesis – assessing the roles of three *nir* gene products. FEBS J. *276*, 6399–6411.

Zumft. W.G., Dreusch, A., Löchelt, S., Cuypers, H., Friedrich, B., and Schneider, B. (1992). Derived amino acid sequences of the *nosZ* gene (respiratory N_2O reductase) from *Alcaligenes eutrophus*, *Pseudomonas aeruginosa* and *Pseudomonas stutzeri* reveal potential copper-binding residues. Implications for the Cu_A site of N_2O reductase and cytochrome *c*-oxidase. Eur. J. Biochem. *208*, 34–40.

Zumft, W.G., and Kroneck, P.M. (2007). Respiratory transformation of nitrous oxide (N_2O) to dinitrogen by Bacteria and Archaea. Adv. Microb. Physiol. *52*, 107–227.

Redox Complexes of the Nitrogen Cycle

2

David Richardson

Abstract

The redox reactions of the nitrogen cycle comprise a large number of oxidative and reductive reactions that are catalysed by wide variety of enzymes with different catalytic centres. These enzymes are frequently organized as multiprotein complexes and in some recently emerging cases enzymes catalysing different reactions of the N-cycle appear to form super-complexes. This review will survey some of these N-cycle protein complexes.

Introduction

The inorganic nitrogen cycle is driven by a diverse range of redox reactions that are catalysed by enzyme complexes that often comprise an integral electron transfer chain in addition to an active site for the N-substrate reduction or oxidation. As studies of the biochemistry of the N-cycle have embraced increasingly diverse microbial phylogenies widespread evidence of large multi-redox centred N-cycle enzyme complexes is emerging. Recent examples include the structurally resolved forty eight haem homo-hexameric nitrite reductase complex of halophile *Thioalkalivibrio nitratireducens* (Polyakov *et al.*, 2009) or the octa-heteromeric NADH-quinone-nitrate reductase complex suggested by two-hybrid studies of the thermophile *Thermus thermophilus* (Cava *et al.*, 2008a). For recent reviews on N-cycle enzymology the reader is referred to Bothe *et al.* (2007) (Bothe *et al.*, 2007), but in this chapter I will focus on exploring the extent to the way in which N-cycle enzymes form functional multi-redox centre/multipolypeptide complexes in a range of Bacterial and Archaeal species.

The oxido-reductase enzyme complexes of denitrification and ammonification

Bacterial and Archaeal denitrification is commonly defined as the respiratory reduction of the water-soluble N-oxyanions nitrate (NO_3^-) and nitrite (NO_2^-) to the gaseous N-oxides nitric oxide (NO) and nitrous oxide (N_2O), which may be further reduced to dinitrogen (N_2) (for a review see van Spanning *et al.*, 2007). One of the paradigm organisms for studying bacterial denitrification has been *Paracoccus denitrificans,* for which the basic organization of the anaerobic denitrification system is shown in Fig. 2.1. Nitrate is imported into the cell by a transporter and reduced to nitrite at the cytoplasmic face of the plasma membrane by the membrane-bound nitrate reductase (Nar). The nitrite is then exported to the periplasmic compartment and reduced to nitric oxide by a periplasmic nitrite reductase (Nir). The nitric oxide is further reduced to nitrous oxide by a membrane-bound NO reductase (Nor) and finally this N_2O is reduced to dinitrogen gas by the nitrous oxide reductase (Nos).

In anaerobic respiratory nitrate ammonification bacteria reduce nitrate via nitrite to ammonium rather than to dinitrogen gas (for a review see Mohan and Cole, 2007). *Escherichia coli* is a prototype nitrate-ammonifying bacterium. It has two periplasmic enzymes that can together reduce nitrate to ammonium, a periplasmic nitrate

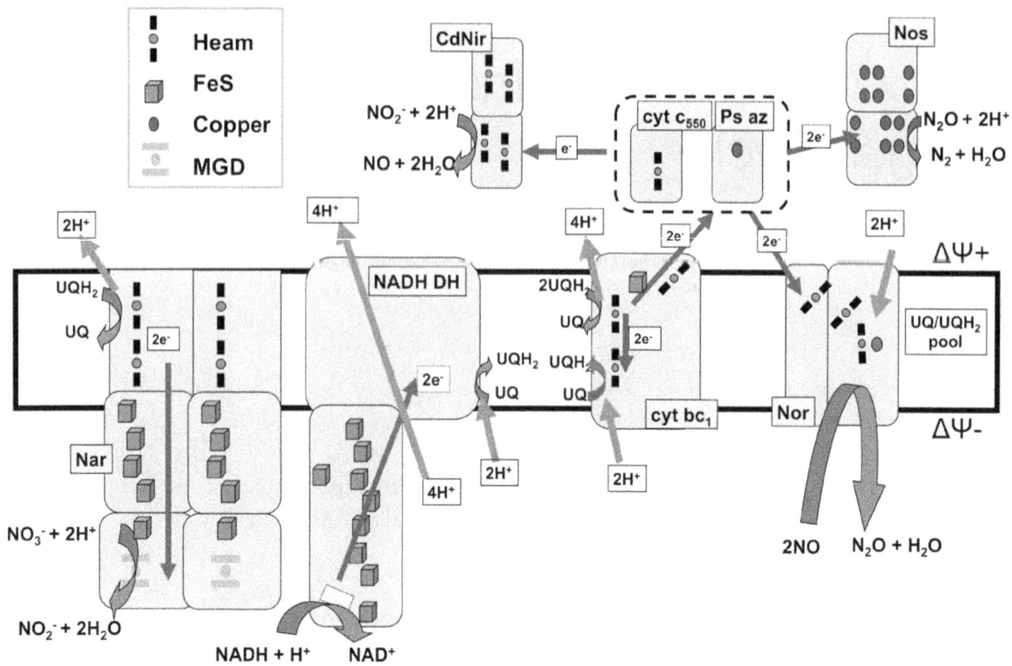

Figure 2.1 The denitrification pathway of *Paracoccus denitrificans*. The Nar, cdNir, Nor and Nos complexes all have two parallel electron transport pathways and two active sites for substrate reduction, but for clarity only the reaction for one of the two active sites is illustrated.

reductase (Nap) that reduces nitrate to nitrite and a periplasmic nitrite reductase (Nrf) that reduces nitrite to ammonium. The structures of the two enzymes involved in this process have been determined and this system is most active under carbon-rich nitrate-limited conditions. *E. coli* also contains nitrate and nitrite reductases with active sites in the cytoplasm that can reduce nitrate through to ammonium, the structurally defined Nar-type nitrate reductase, homologous to that in *P. denitrificans*, and the cytoplasmic nitrite reductase (NirB). This system is most active under nitrate-rich carbon-limited conditions. Some ammonifiers only have one of these systems, for example *Campylobacter jejuni*, *Haemophilus influenzae* and *Wollinella succinogens* only have the periplasmc system, whereas *Klebsiella pneumoniae* only has the cytoplasmic system. In the denitrification and ammonification systems of *P. denitrificans* and *E. coli* the key enzymes appear to be operationally independent, but many of them are complex multi-subunit systems with integrated electron transport chains (Figs 2.1 and 2.2). However evidence is emerging that the relatively simple picture of these independently operating systems may be more complex in some prokaryotic N-cycle systems.

The membrane-bound nitrate reductase (Nar) complexes

The so-called membrane-bound nitrate reductase, or Nar, is common to both denitrification and ammonification (Figs 2.1 and 2.2) and has been crystallographically resolved from *E. coli* (Bertero *et al.*, 2003; Jormakka *et al.*, 2004). In the crystal structure Nar forms a hetero-hexameric complex that binds 16 redox centres. These are organized into two electron transfer chains, each mediating quinol oxidation and nitrate reduction in a $[NarG]_2[NarH]_2[NarI]_2$ complex. Each chain comprises two haems, five iron–sulfur clusters and the active-site molybdenum cofactor (Fig. 2.3). The NarG subunit binds a molybdenum *bis* molybdopterin guanine dinucleotide (Mo-*bis*-MGD) cofactor at its catalytic site and an [4Fe–4S] cluster, the NarH subunit binds one [3Fe–4S] and three [4Fe–4S] clusters, and the NarI integral membrane subunit (with five

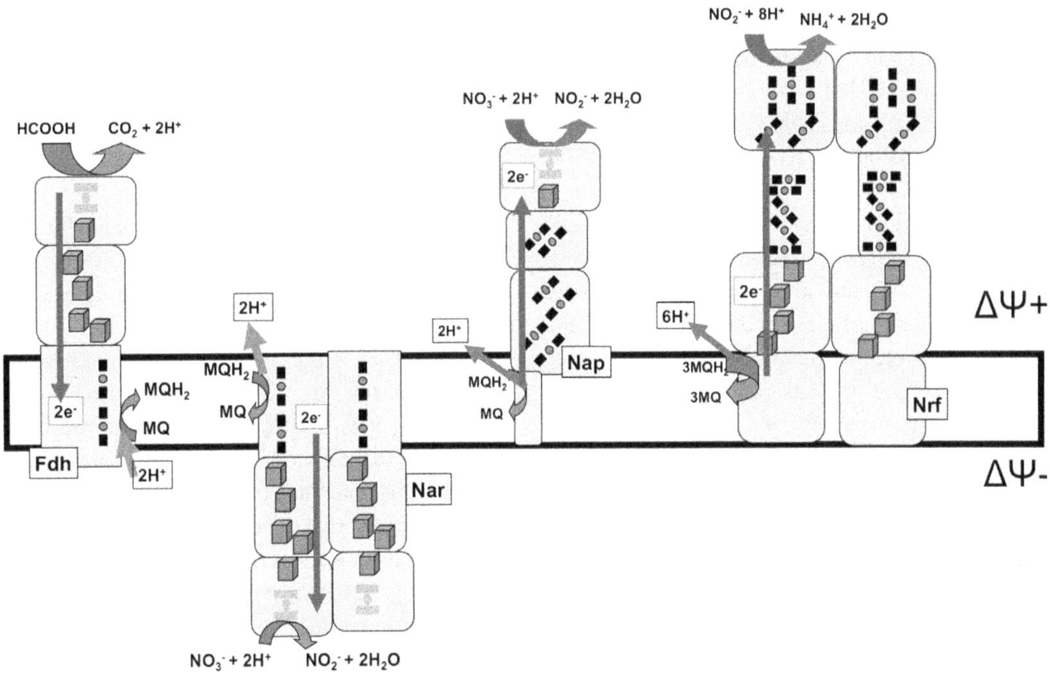

Figure 2.2 The nitrate respiratory ammonification pathways of *Escherichia coli*. The Nar, and Nrf complexes each have two parallel electron transport pathways and two active sites for substrate reduction, but for clarity only the reaction for one of the two active sites is illustrated. The full Nap system is shown in Fig. 2.5. Note that the formate dehyrogenase (Fdh) is actually a trimer of heterotrimers, but only one heterotrimer is shown.

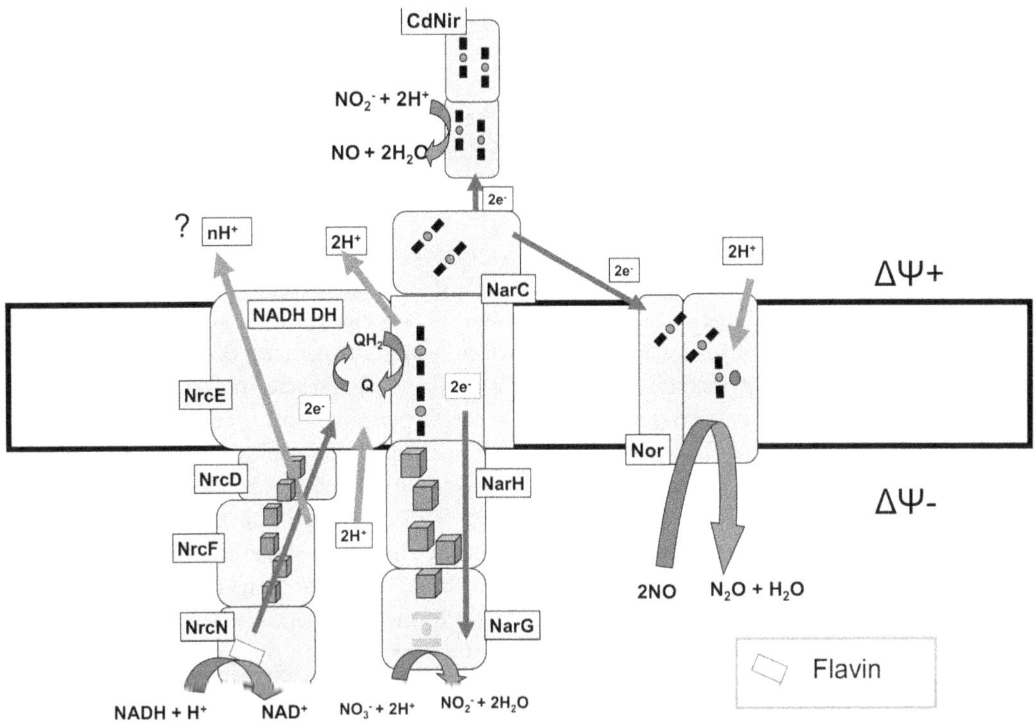

Figure 2.3 The NADH nitrate and possible denitrification supercomplex of *Thermus thermophilus*. The stoichiometry of Nrc and Nar subunits is unknown, but is likely that the Nar subunits form a $[NarGHI]_2$ complex as illustrated in Figs. 2.1 and 2.22 with NarC therefore also forming a dimer.

transmembrane helices) binds two haems: one low potential located at the periplasmic side, and one high potential located at the cytoplasmic side. NarG and NarH associate with NarI at the membrane potential-*negative* cytoplasmic face (Ψ−) of the cytoplasmic membrane. Hence this system an be referred to as an nNar system (Martinez-Espinosa *et al.*, 2007). The electron transport chains span ~10 nm and conserves the redox energy in the quinol-nitrate couple as a transmembrane proton-electrochemical gradient via a redox loop mechanism [see Chapter 3 and Simon *et al.* (2008) for a recent review]. In brief, the oxidation of quinol takes place at the periplasmic side of NarI where the protons are released and the two electrons are moved from the periplasmic-face haem to cytoplasmic-face haem. This charge separation makes the enzyme electrogenic in that it contributes to two charge separations across the membrane ($2q^+$), during transfer of two electrons from quinol to nitrate ($2q^+/2e^-$).

Whilst it is certainly a substantial complex, it may be that the *E. coli* nNar system will prove to be a relatively simple example of a Nar system. Evidence to suggest this comes first from the hyperthermophile *Thermus thermophilus* in which it has been shown that DNA encoding an NADH-dependent respiratory Nar system can be readily moved amongst *T. thermophilus* strains and a gene cluster encoding a nNar system similar to that described above for *E. coli* has been identified on this DNA (Cava *et al.*, 2008b). However, this gene cluster codes for additional redox proteins that include the subunits of an NADH dehydrogenase and a dihaem cytochrome c (NarC) (Fig. 2.3). Work using a two-hybrid approach to assess cellular protein–protein interactions strongly suggests that the NADH dehydrogenase and the nitrate reductase physically interact to form a large NADH-UQ-nitrate reductase complex (Cava *et al.*, 2008a) (Fig. 2.3). The mobility of the DNA encoding this complex in principle makes a proton-motive NADH-nitrate reductase system readily available to the microbial community in an environment populated by *T. thermophilus.* However, the complexity of this Nar system may extend still further. The extra dihaem cytochrome associated with the Nar system that is absent from *E. coli* and *P. denitificans* Nar appears to also have a role in mediating electron transfer to the nitrite reductase and nitric oxide reductase (Cava *et al.*, 2008a). It is not yet clear if this is a direct electron transfer, but it seems probable that is and so the possibility of a large NADH-dependent super-denitrification-complex emerges (Fig. 2.3).

There is also some evidence that denitrification supercomplexes may operate in some Archaea. Classically, the 'Bacterial' Nar system is oriented such that nitrate reduction takes place on the inside of this membrane. However, the active site subunit of the 'Archaeal' Nar systems has a twin arginine ('RR') motif, which is a suggestion of translocation to the outside of the cytoplasmic membrane via the twin arginine translocase (TAT) system that exports folded proteins. These types of nitrate reductase are then active at the membrane potential *positive* face of the cytoplasmic membrane and so have been referred to as pNar (Martinez-Espinosa *et al.*, 2007). In many denitrifying species of bacteria that have nNarG, the nitrite produced from nitrate is further reduced to nitric oxide, nitrous oxide and dinitrogen in a series of enzyme-catalysed reactions of denitrification. In each case, these reactions take place in the periplasm or towards the periplasmic face of the membrane and can be coupled to energy conservation through the protonmotive activity of the cytochrome bc_1 complex (QH_2-cytochrome *c* oxidoreductase) (Fig. 2.1). Like NarI, this complex binds two haems: one at either side of the membrane. However, rather than oxidizing one quinol per turnover (like NarI), it oxidizes two quinols at the periplasmic face (ΔΨ+) and moves two electrons to the cytoplasmic (ΔΨ−) face where it reduces one quinone. Overall, this so-called Q-cycle translocates two positive charges per quinol oxidized ($2q^+/2e^-$) and so it is bioenergetically equivalent to nNarI. This then makes the reduction of nitrate, nitrite, nitric oxide and nitrous oxide bioenergetically equivalent, despite the active site of the nNar system being on the ΔΨ− side of the membrane and the active sites of the nitrite, nitric oxide and nitrous oxide reductases being on the ΔΨ+ side (Fig. 2.1; for more detail on protonmotive electron transport systems see Simon *et al.*, 2008).

At present, it is not known how electrons move from the Q-pool to pNarG. However, the genetic context of the *Haloferax mediterranei* pNar and *Haloarcula marismortui* pNar provides an interesting possibility. Recent analysis of the genes of the *pnar* cluster revealed that one of the genes encodes a protein (NarC) that has sequence similarity with the di-*b*-haem subunits of the cytochrome bc_1 complex (it should be noted that this 'NarC' is entirely unrelated to the 'NarC' from *T. Thermophilus* which is a di-*c*-haem cytochrome). Adjacent to *narC* is a gene (*narB*) that is predicted to encode a Rieske iron–sulfur protein also of the type found in the cytochrome bc_1 complex. This then raises the possibility that the *H. mediterranei* and *H. marismortui* pNar systems maintain bioenergetic parity with the nNar systems by being coupled to a protonmotive Q-cycle mechanism via a NarBC QH_2-cytochrome *c* oxidoreductase system that may make a physical interaction with the nitrate reductase components (Fig. 2.4). Evidence that this might be the case came for the observation that nitrate reduction in *H. mediterranei* is sensitive to the cytochrome bc_1 complex inhibitor antimycin A, which is not the case for the cytochrome bc_1 complex-independent nNar systems (Martinez-Espinosa *et al.*, 2007). Since *H. mediterranei* and *H. Marismortui* also have enzymes for both nitrite and nitric oxide reduction then by analogy to the prototype *P. denitrificans* dentrification network it is highly likely that these also draw electrons from the NarCB system so that the possibility of an composite NarCB-dependent nitrate/nitrite/NO reductase system emerges (Fig. 2.4).

The periplasmic nitrate reductase (Nap) complexes

The catalytic subunit of the periplasmic NO_3^- Reductase (Nap) binds *bis*-MGD and a [4Fe–4S] cluster. It is sometimes found in a tight complex with a dihaem cytochrome c, a complex for which a crystal structure has been solved from *Rhodobacter sphaeroides* (Arnoux *et al.*, 2003). In some bacteria, such as *P. denitrificans*, the induction of the *nap* gene cluster does not follow that of anaerobic denitrification. Instead the cluster is maximally expressed during aerobic respiration, especially when grown on reduced carbon substrates such as butyrate, and it plays a role in redox balancing and energy dissipation (Richardson, 2000). In many species though Nap is synthesized anaerobically and can partake in

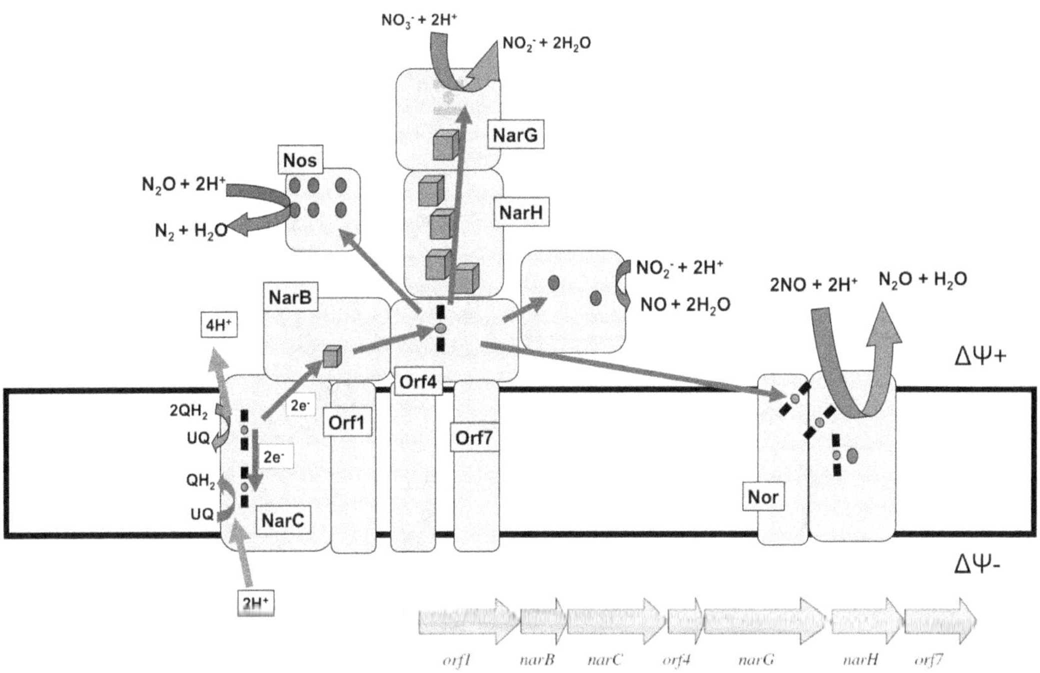

Figure 2.4 The Q-cycling Nar complex of *Halomediterranei marismortui*. The CuNir complex is most likely a homotrimer and Nor and Nos are most likely dimeric as illustrated in Fig. 2.1 for *P. denitrificans*.

denitrification (e.g. *Bradyrhizobium japonicum*) or ammonification (e.g. *E. coli*) (Potter *et al.*, 2001). In some of the simplest Nap systems electron transfer from quinol to NapA requires NapC, which is a membrane anchored tetra-haem quinol dehydrogenase. However, the situation can apparently become much more complex. In *E. coli* NapA is coded for in a large operon that includes genes for integral membrane and periplasmic iron–sulfur proteins (NapG and NapH) and cytoplasmic proteins NapD and NapF. A series of two-hybrid interaction studies have suggested that NapC dimerizes in the membrane and also interacts with NapH, which does not dimerize (Brondijk *et al.*, 2004; Nilavongse *et al.*, 2006). This leads to the suggestion of a NapCH complex (Fig. 2.5). The complexity is though further increased by suggestions that NapH and NapG interact making an electron transfer supercomplex that can channel electrons from both menaquinol and ubiquinol to NapA (Brondijk *et al.*, 2002; Brondijk *et al.*, 2004). Intriguingly NapH is predicted to bind iron–sulfur clusters at the cytoplasmic side of the membrane. It is not clear what the function of these is, but the clue may lie in NapF. Although NapA is a periplasmic protein two-hybrid studies have shown that NapF and NapA can interact in the cytoplasm. This is likely to be prior to the export of holo-NapA via the TAT system. The evidence suggests that NapF binds a labile iron–sulfur cluster and is a maturation protein required for the [4Fe–4S] assembly on the NapA scaffold (Nilavongse *et al.*, 2006; Olmo-Mira *et al.*, 2004). It may be that this assembly process needs to be charged by electrons coming for the quinol pool via NapH. NapA also interacts with another redox maturation protein NapD, for which the structure has recently been solved (Maillard *et al.*, 2007). Thus all of the protein–protein interactions illustrated in Fig. 2.5 have been demonstrated or are extremely likely and thus reflect a highly complex composite Nap system for electron transfer and redox protein assembly.

The NO-genic nitrite reductase complexes

The reduction of nitrite to nitric oxide is catalysed by either the cytochrome cd_1 (cdNir) or the Cu nitrite reductase (CuNir) ($NO_2^-+2H^++e^- \rightarrow NO+H_2O$) both of which are structurally defined (Godden *et al.*, 1991; Hough *et al.*, 2005; Williams *et al.*, 1997). They draw electrons from the cytochrome bc_1 complex via small cytochromes or cupredoxins (e.g. Fig. 2.1 for *P. denitrificans* and cdNir). The CuNirs are homotrimers which bind

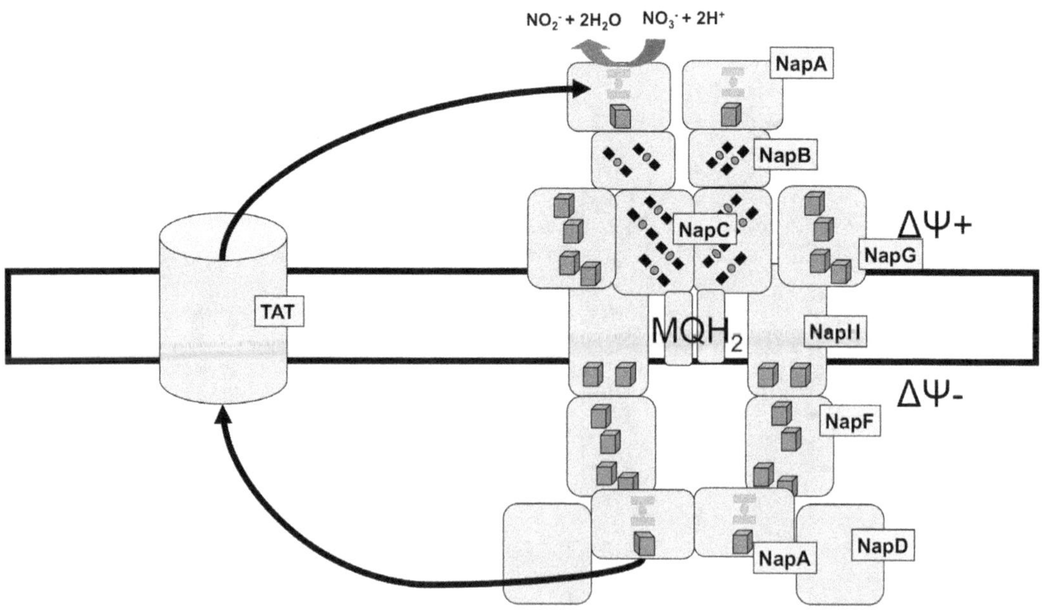

Figure 2.5 The possible protein–protein interactions operational in the *E. coli* Nap system. NapA may interact with NapF and NapD during assembly in the cytoplasm prior to export via the TAT system.

three type I and three type II copper centres. The type I copper is co-ordinated by one cysteine, one methionine and two histidine residues. The type II coppers lie at the subunit–subunit interfaces and each is ligated by three histidine residues, two being provided by one subunit and the third from the adjacent subunit. Thus the CuNir protomer can only be functional as a part of a trimeric complex. The type II copper centres are the sites of substrate binding and the type I copper plays a role in electron transfer from the soluble electron donors to the active site. Recently a crystal structure of a CuNir from *Achromobacter xylosoxidans* complexed with a mono-haem electron donor redox partner has emerged (Nojiri *et al.*, 2009). It has also been noted that some CuNirs have a C-terminal extension that is predicted to bind a *c*-type haem. These cytochrome *c* CuNirs would thus be predicted to be heterotrimeric complexes that bind three haems in addition to three Type 2 coppers and three Type 1 coppers. Structural confirmation of this is though awaited. Another subgroup of CuNirs can be found in *Neiserria* species where the CuNirs have a lipidated N-terminal extension which is suggested to anchor the enzyme to the outer-membrane. The key Cu-coordinating ligands are all conserved and so it is likely to be a trimeric complex that is anchored to the outer-membrane. Electron transfer to this outer-membrane CuNir is via an inner-membrane cytochrome bc_1 type quinol dehydrogenase periplasmic *c*-type cytochromes so that an inter-membrane quinol-nitrite reductase system must be operating (Deeudom *et al.*, 2008). In the case of *H. mediterranei* and *H. marismortui* the nitrite reductase predicted to interact with the Nar system is of the CuNir type raising the possibility of CuNir being part of much larger super-complexes.

The cdNir is a functional dimer containing two c-type electron transfer haems and two d_1-type catalytic haems. Crystal structures from *Paracoccus pantotrophus* and *Pseudomonas aeruginosa* have emerged (Nurizzo *et al.*, 1997; Williams *et al.*, 1997). Interestingly though, these two homodimeric complexes display quite significant structural differences. In both cases, the d_1 haem is enclosed in an eight-bladed β-propeller structure, but the co-ordination of the haem d_1 Fe(III) differs. In *P. pantotrophus* cdNir the Fe(III) has a proximal histidine and distal tyrosine ligands both provided by the same polypeptide that forms the β-propeller. However, in *Ps. aeruginosa* cdNir the distal ligand is a hydroxide that is hydrogen-bonded to a tyrosine from the partner polypeptide chain of the homodimer. Differences are also apparent for the cytochrome *c* domain. Although the fold of this is similar in both enzymes, in the *P. pantotrophus* enzyme the *c*-type haem is co-ordinated by two His ligands, but in the *Ps. aeruginosa* enzyme it exhibits His/Met co-ordination. There is however substantial structural re-arrangement in the *P. pantotrophus* cdNir on reduction that renders it more similar to that of *Ps. Aeruginosa.* The tyrosine ligand (residue 25) is lost from the d_1 haem, thus allowing substrate binding. His^{17}, a *c*-type haem ligand on the same polypeptide loop as Tyr^{25} also dissociates and is replaced by the sulfur of a methionine, leading to a large conformational change (Nurizzo *et al.*, 1997). Experiments under turnover conditions have shown that the *c*-type haem maintains histidine plus methionine, rather than *bis*-histidinyl, co-ordination. In the context of the denitrification electron transfer system the cytochrome c domain of the cdNirs will form a complex with the electron donating periplasmic c-type cytrochromes or cupredoxins. Thus far such complexes have not been crystallized but models for them have emerged (Williams *et al.*, 1995).

The NH_4^+-genic nitrite reductase complexes

The enzyme that converts nitrite to ammonium in the respiratory ammonification pathways of the N-cycle is a penta-haem cytochrome *c* nitrite reductases (NrfA) (Mohan and Cole, 2007). This appears to have a tendency for forming very large multi-haem complexes with the record thus far being a 48 haem complex (Polyakov *et al.*, 2009). The basic protomer is a pentahaem protein that catalyses the six-electron reduction of nitrite to ammonium, but the putative reaction intermediates [nitric oxide ($NO^•$) and hydroxylamine (NH_2OH)] are alternative substrates, as is sulfite which yields sulfide as the product (Clarke *et al.*, 2008; Einsle *et al.*, 2002). In *E. coli* NrfA crystallizes as a homodimer, with each monomer containing five haems. All haems have *bis*-His ligands, except the active site haem, which is unusual as it is a high-spin species co-ordinated with a lysine residue on the proximal side and either water or nitrite on the distal side (Bamford *et al.*, 2002; Einsle *et al.*, 1999). One of the *bis*-His ligated haems is only 5.7 Å away from its counterpart in the other monomer, potentially allowing transfer of electrons between monomers of the homodimeric NrfA complex.

The key to coupling NrfA with energy conservation lies in the communication with the MQH_2 (menaquinol) pool and thereby electrogenic quinone reductases, such as formate dehydrogenase. Genetic studies suggest that menaquinol oxidation is catalysed by a NrfCD complex, that has significant sequence similarity to the PsrBC quinol oxidation system of the structurally defined polysulfide reductase (Jormakka *et al.*, 2008). The polysulfide reductase (PsrABC) is an integral membrane protein complex responsible for quinol-coupled reduction of polysulfide, a process important in extreme environments such as deep-sea vents and hot springs. The structure of polysulfide reductase from *T. thermophilus* revealed a chain of four iron sulfur clusters in the PsrB (NrfC homologue) subunit and a quinone binding site in the PsrC (NrfD homologue) subunit. It also revealed the presence of a water-filled cavity connecting the quinone-binding site on the periplasmic side to the cytoplasm, suggesting that the PsrBC, and so by implication NrfCD, might be protonmotive. There is experimental evidence that PsrABC might translocate 1 $q^+/2e^-$ and the structure suggests this might be via a conformational proton pump. The electron transfer mediator between the NrfCD complex and NrfA is the pentahaem NrfB protein, the structure of which has also been resolved (Clarke *et al.*, 2007). NrfA and NrfB may form a 20-haem hetero-tetrameric super-complex *in vivo*. Since the PsrBC complex is also dimeric then the possibility that the *E. coli* Nrf system operates as a protonmotive 8 subunit system that binds eight iron–sulfur clusters and 20 haems emerges.

In some species of bacteria, such as *Wolinella succinogenes* and the sulfate-respiring bacteria *Desulfovibrio desulfuricans*, the electron donor to NrfA is NrfH, a membrane-associated tetra-haem cytochrome of the NapC family (described in 'The Periplasmic nitrate reductase (Nap) complexes', above) that directly mediates electron transfer from the quinol pool to NrfA. NrfH forms a tight complex with NrfA and a recently solved structure of a *Desulfovibrio vulgaris* NrfAH complex showed an $\alpha_4\beta_2$ arrangement with two NrfA dimers bound to one NrfH dimer in a 28-haem complex (Rodrigues *et al.*, 2006). The complex is stabilized by each NrfA dimer providing a lysine ligand to a haem on the corresponding NrfH protein in the $(NrfA_2\text{–}NrfH)_2$ complex. However, the prize for the most haems in any NrfA complex thus far described goes to an octa-haem cytochrome *c* nitrite reductase that has been isolated from the haloalkaliphilic bacterium *Thioalkalivibrio nitratireducens* (Polyakov *et al.*, 2009). The octa-haem protomer is part of a class of Nrf in which there is a three-haem extension at the N-terminal end of the protein to the basic penta-haem NrfA described above. Significantly though, the *T. nitratireducens* NrfA exists as a stable 48-haem hexamer, in both the crystalline and solution states.

The respiratory nitric oxide reductases

The respiratory NO Reductase of *P. denitrificans* has been isolated as a four subunit protein $[NorB]_2[NorC]_2$ heterotetramer (Hendriks *et al.*, 1998). The NorB subunit contains 12 membrane-spanning α-helices and binds two *b*-haems and a non-haem iron, whilst the smaller NorC subunit, is a periplasmic *c* type cytochrome that is anchored to the membrane via a single α-helix. In *P. denitrificans* the function of NorC is to serve as the immediate electron acceptor for two periplasmic electron donors; pseudoazurin and cytochrome c_{550}. The catalytic subunit, NorB, is structurally homologous to the catalytic subunit (subunit I) of the respiratory haem–copper oxidases (HCuOs), e.g. cytochrome aa_3 oxidase (de Vries and Schroder, 2002). The HCuOs catalyse the four-electron reduction of dioxygen to water and terminate the aerobic respiratory chains of both bacteria and mitochondria. Much of the energy liberated in this exergonic reaction is conserved as a transmembrane electrochemical proton gradient. By contrast the NorCB complex is not a proton pump or electrogenic.

The principal difference between NorB and the haem–copper oxidases is in the elemental composition of the dinuclear centre which in NorB contains iron (Fe_B) rather than copper ion (Cu_B). Movement of protons to the dinuclear centre of HCuOs, which lies buried in the hydrophobic interior of the membrane, is via pathways formed by ionisable amino acid side chains – the D-channel and the K-channel. The amino acid residues that contribute to the D- and K-channels are not conserved in NorB. This is not surprising as NOR has no requirement to take protons from the cytoplasm for catalysis. However, homology modelling of *Pseudomonas stutzeri* NorB using the X-ray structure of *P. denitrificans* cytochrome aa3 suggested that the haem::Fe_B dinuclear centre, at which NO is reduced to N_2O, is also buried in the lipid bilayer (Reimann *et al.*, 2007). Consequently NorB is predicted to contain a channel that allow protons consumed in catalysis to move from the periplasmic face of the enzyme to the buried active site and experimental evidence is emerging in support of a glutamate-rich E-channel fulfilling this role (Flock *et al.*, 2008; Reimann *et al.*, 2007; Thorndycroft *et al.*, 2007).

There is another group of respiratory nitric oxide reductases that derives electrons from the quinol pool rather than the cytochrome c/pseudoazurin pool. These so-called Q-Nors are single subunit proteins in which the basic NorB polypeptide has an N-terminal extension that might contribute to forming a quinol binding site (Cramm *et al.*, 1997; de Vries and Schroder, 2002; de Vries *et al.*, 2003). A further sub-group has a di-copper Cu_A site fused to the Q-Nor polypeptide to form a Nor that is able to accepted electrons from both quinol and the cytochrome c/psuedoazurin pool (deVries *et al.*, 2007). Thus, there are different levels of complexity within the respiratory Nor group of enzymes. It is notable that the operons encoding the NorCB-type of Nor commonly have genes predicted to encode additional hydrophobic integral membrane proteins (NorD and NorE). At present it is not clear what these proteins are for, but it is notable that they are required for successful heterologous assembly of functional *P. denitrificans* NorCB in *E. coli* (Butland *et al.*, 2001). The structure of *P. denitrificans* cytochrome aa_3 oxidase revealed additional two subunits to the two-subunit core in a heterotetrameric complex (Iwata *et al.*, 1995). This raises the possibility that NorD and NorE are subunits of a heterotetrameric NorCBDE complex *in vivo* that have been lost during purification of the NorCB complex. This will require further study.

Nitrous oxide reductase

N_2O reductase is a copper containing enzyme, with two redox-active copper centres (Brown *et al.*, 2000a,b; Haltia *et al.*, 2003). Cu_A is the di-copper electron input site in which the two coppers are ligated by two equatorial cysteine sulfurs, two terminal histidine nitrogens and two axial ligands. Cu_z is a tetranuclear copper cluster residing in the centre of a seven-bladed β-propeller domain. There is an inorganic sulfur that bridges the four copper ions along with seven histidine residues. Significantly, the N_2O reductase is a functional homodimeric enzyme complex. In the monomer the distance from Cu_A to Cu_z is 40 Å, which is far too large for electron transfer. But in the dimer the Cu_A from one subunit is only 10Å away from the Cu_z of the partner subunit, which is close enough for electron transfer across the dimeric interface. It is therefore essential for catalytic function that N_2O reductase exists as a dimeric structure and so the functional complex is a 12-copper homodimer.

In *Paracoccus* and many other α-proteobacteria electron transfer to N_2O reductase occurs in the periplasm with electrons being drawn for the respiratory network at the level of the cytochrome bc_1 complex and the cytochrome c_{550}/pseudoazurin pool. However, such an electron transport chain to nitrous oxide reductase is not ubiquitous. In one subgroup of Nos the situation appears more complex. The Nos from *Wolinella succinogenes* has a C-terminal extension that is predicted to bind a *c*-type haem, suggesting that the functional enzyme will be a dihaem 12-copper complex. Significantly, though the genome of *W. succinogenes* does not have a cytochrome bc_1 complex. Instead this complex may draw electrons from the quinol pool via a homologue of the NapGH protein complex (NosGH), discussed earlier in the context of the periplasmic nitrate reductase, and two additional mono-haem c-type cytochromes the genes for which cluster with the *nosZ* gene. Thus a possible NosGH-cytochrome c-nitrous oxide super-complex, similar to the *E. coli* NapGH–NapC–nitrate reductase complex (Fig. 2.5) can be envisaged for the *W. succinogenes* Nos system (Kern and Simon, 2009; Simon *et al.*, 2004).

Many *nos* gene clusters, but notably not that of *W. succinogenes*, have a gene encoding for an integral membrane protein NosR that has the cytoplasmic iron–sulfur cluster domain present in NapH/NosH, but additionally has a periplasmic flavin-binding domain. In *Pseudomonas stutzeri* NosR is essential for NosZ activity and may play a role in electron transfer from quinol to the periplasmic NosZ which might then form a complex with NosR (i.e. this would be similar to the proposed NosGH–NosZ complex in *W. succinogenes*) (Wunsch and Zumft, 2005; Zumft, 2005). It is notable that NosR is also present in many Nos gene clusters in species of bacteria that do have a cytochrome bc_1 complex and so do not need NosR for electron transfer from quinol to NosZ (e.g. *P. denitrificans*). It is possible though that NosR has an additional role in reductively activating the Cu_Z site of NosZ or repairing it if is oxidatively damaged (for example by exposure to oxygen) (Wunsch and Zumft, 2005; Zumft, 2005). This would though also involve the formation of a NosR–NosZ complex in order for electron transfer to proceed.

Enzymes involved in aerobic nitrification

Nitrification represents an oxidative respiratory branch of the nitrogen cycle in which ammonia is oxidized to nitrate via hydroxylamine and nitrite (Arp *et al.*, 2007; Ferguson *et al.*, 2007). The process is usually associated with autotrophic growth and so requires that electrons extracted from the electron donor are divided between protonmotive force generating electron transfer to oxygen and protonmotive force consuming reverse electron transfer to NAD. The oxidation of ammonia to nitrite and the oxidation of nitrite to nitrate have never been found in the same organism.

Ammonium monoxygenase

In prototype ammonia-oxidizing bacteria, such as *Nitrosomonas europaea*, the oxidation of ammonia is catalysed by an ammonia monoxygenase that most likely draws electrons from the quinol pool (Fig. 2.6). There is no consensus on the nature and number of co-factors that this enzyme binds, but it is widely accepted that it is a copper-dependent enzyme. What is clear is that it is an enzyme complex. It probably consists of three subunits, AmoA, AmoB and AmoC. These subunits have homology to subunits of the related particulate methane monooxygenase (pMMO) of methanotrophs, which is an integral membrane redox protein that catalyses the oxidation of methane to methanol. The structures of pMMO from the methanotroph *Methylococcus capsulatus* and *Methylosinus trichosporium* OB3b have been determined (Balasubramanian *et al.*, 2010; Hakemian *et al.*, 2008; Rosenzweig, 2008). The enzyme is a nine-subunit heterotrimeric complex with an $\alpha_3\beta_3\gamma_3$ polypeptide arrangement. The pMmoB subunit (AmoB) resembles cytochrome *c* oxidase subunit II and binds a dinuclear copper, located in soluble region of this subunit, that has recently been shown to be the site of methane oxidation. A third metal centre is located within the membrane, coordinated between pMmoA and pMmoC (Fig. 2.6). This is occupied by zinc in the *M. capsulatus* pMmo structure and by copper in the *M. trichosporium* pMmo structure. It is not clear what metal(s) should be occupying this site *in vivo*, though it is possible to model a dinuclear iron site into it and this could be a site of electron input from the quinol pool. The key residues binding the metal centres of pMmo are conserved in the Amo polypeptides and so a similar structural organization is likely for both enzymes. Functionally, the Amo requires a quinol binding site in the membrane and an oxygenase site to oxygenate NH_3 to NH_2OH ($NH_3 + QH_2 + O_2 \rightarrow NH_2OH + H_2O + Q$) (Fig. 2.6), which must then be liberated into the periplasm so that it can be further oxidized by the hydroxylamine oxidoreductase (see 'Hydroxylamine oxidoreductase', below).

Hydroxylamine oxidoreductase

The product of ammonia oxidation is hydroxylamine, which is oxidized to nitrite by hydroxylamine oxidoreductase (HAO). The structure of *N. europaea* HAO has been resolved and

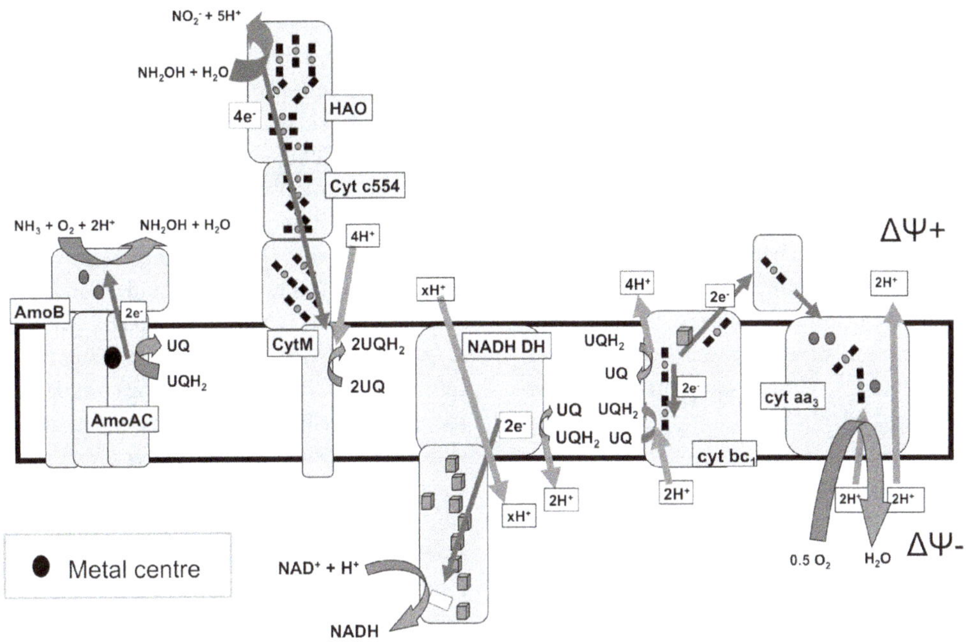

Figure 2.6 A schematic view of ammonia oxidation by *Nitrosomonas europaea.* Note the AMO is a $[AmoABC]_3$ complex, the HAO is a homotrimer and CytM may form a homodimer (see text for details of these complexes). For simplicity the scheme shows only the protomers. Hydroxylamine oxidation (NH_2OH) to nitrite yields four electrons. These are fed into the ubiquinol pool via the NapC homologue CytM. This generates 2 molecules of UQH_2 one of these is used to drive the AMO. The other is used to drive the cytochrome bc_1 complex. The electrons are then fed to the cytochrome aa_3 oxidase where oxygen is reduced and a proton electrochemical gradient is generated. This drives reverse electron transfer through the NADH dehydrogenase to generate NADH from NAD^+ a small amount of ubiquinol is tapped from the ubiquinol pool as an electron source for this reaction, but the majority (~95%) of the reductant generated from hydroxylamine oxidation is used to drive a AMO and cytochrome oxidase.

reveals a trimeric enzyme complex that has twenty four covalently attached *c*-type haems (Igarashi *et al.*, 1997). The three of these haems, the P460 centres, are penta-coordinate and sit at the bottom of a pocket, and are the active sites for hydroxylamine binding. The oxidation of hydroxylamine probably occurs via two successive abstractions of electron pairs, with the concomitant release of five protons. The structure shows that haem P460 is located close to an aspartate residue, a histidine residue and covalently attached to a tyrosine residue, all of which are potential candidates as proton acceptors. The structure reveals that adjacent to haem P460 are two other haems that may be required to rapidly accept a pair of electrons removed from the substrate and so help to prevent slower single electron transfer that could result in toxic reactive nitrogen intermediate (e.g. nitric oxide) being released. The additional *c*-haems within HAO provide a number of possible routes for onward electron transfer, ultimately resulting in the reduction of the tetra-haem cytochrome c_{554}, for which the structure has been solved and a model for a 36-haem ($[HAO]_3$ [Cyt C554]) electron transfer complex proposed (Iverson *et al.*, 1998).

There is a clear evolutionary relationship between HAO and the NrfA nitrite reductase described in 'The NH_4^+-genic nitrite reductase complexes', above. Despite significant amino acid sequence similarity, haems 1–5 of NrfA can be superimposed on to haems 4–8 of the octahaem HAO (hydroxylamine oxidoreductase) subunit and the remaining haems of HAO can be superimposed on to the three of the haems of NrfB in the NrfAB complex (Einsle *et al.*, 1999). All eight haems of octa-haem Nrf can be superimposed onto HAO (Polyakov *et al.*, 2009). In the case of the 36-haem HAO cytochrome c_{554} complex this has to mediate electron flow from hydroxylamine to the quinone pool (Fig. 2.6). The likely quinone reductase is a membrane-anchored tetra-haem cytochrome (CytM) that is a member of the NapC/NrfH family (see 'The oxido-reductase enzyme complexes of denitrification and ammonification', above) (Kim *et al.*, 2008). This draws another parallel between the HAO and Nrf systems. Since the NapC/NrfH proteins appear to operate as octa-haem dimers, the possibility of the operation of a 46 haem hydroxylamine-$[HAO]_3[Cytc_{554}]_3[CytM]_2$-quinone reductase system arises.

Nitrite oxido-reductase

Nitrite oxido-reductase (Nxr) oxidizes nitrite to nitrate in *Nitrobacter* spp. The enzyme has yet to be studied in detail at a biochemical or structural level, but sequence analysis indicates that it is a membrane-bound enzyme homologous to the membrane-bound nitrate reductase, in that it comprises a Mo catalytic α subunit and a FeS containing a β-subunit (Poly *et al.*, 2008; Starkenburg *et al.*, 2008b). Therefore the recent structures of membrane-bound nitrate reductase may also provide some mechanistic information as to how nitrite can be oxidized to nitrate. Recent genome sequences of nitrite-oxidizing bacteria provide an opportunity to speculate on the nature of the complexes that the Nxr enzyme might form (Poly *et al.*, 2008; Starkenburg *et al.*, 2006; Starkenburg *et al.*, 2008b). In *N. hamburgensis* the Nxr genes cluster with genes coding for cytochrome oxidase and quinol oxidase systems (Fig. 2.7). Two particular compelling observations emerge. The first is that the Nxr looks a very much like the *T. Thermophilus* Nar described in 'The membrane bound nitrate reductase (Nar) complexes' (above) in that it has both a NarI homologue and a homologue of the dihaem NarC (NxrC). In *T. thermophilus* there is good evidence the NarC mediates electron transfer from NarI to other redox complexes. The second compelling observation is that that the Nxr genes cluster with genes encoding a homologue of the cytochrome aa_3 oxidase. This raises the possibility that in *Nitrobacter* species NxrC mediates electron transfer from nitrite to oxygen at the interface of a protonmotive Nar-cytochrome aa_3 supercomplex. It is notable that early efforts to purify Nxr reported the presence of a-type haem (Yamanaka and Fukumori, 1988). This could have been contamination from co-purified cytochrome aa_3 oxidase. In order to grow autotrophically some electrons need to flow via reverse electron transfer to reduce NAD to NADH. This requires that electrons can enter the quinol pool as well as the cytochrome c pool. This would be possible via the NarI protein which would serve as a quinone reductase (i.e. the reverse of its role as a quinol dehydrogenase in nitrate reductase).

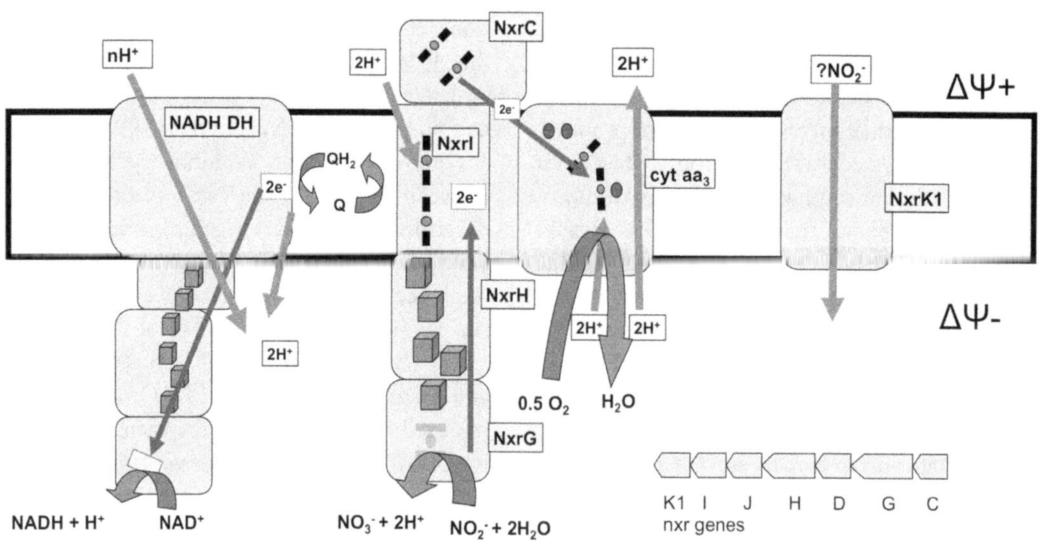

Figure 2.7 The possible arrangement of a nitrite oxidase complex in *Nitrobacter hamburgensis*.

It is pertinent to note though that some *Nitrobacter* species have been shown to grow anaerobically using nitrate as a terminal electron acceptor when coupled to the oxidation of simple organic compounds and thus it is likely that NxrGHI can operate as both quinol-nitrate oxidoreductase and nitrite-quinone oxidoreductase. The terminal product of denitrification in *Nitrobacter* is nitrous oxide and both *N. winogradskyi* and *N. hamburgensis* have a CuNir nitrite reductase (Starkenburg *et al.*, 2008a). The NO generated by this complex must be detoxified by further reduction to N_2O. A nitric oxide reductase gene cluster has been identified in *N. hamburgensis* that may serve to reduce NO. This is not present in *N. winogradskyi* but genes for other potential N_2O-genic cytoplasmic NO reductases such as flavo-haemoglobin are present (Poly *et al.*, 2008; Starkenburg *et al.*, 2006; Starkenburg *et al.*, 2008b).

Conclusions

Over the last years the structures of many of the key redox enzymes of the bioinorganic nitrogen cycle have emerged and for these a good understanding of the mechanism of catalysis at the active sites is developing. However, these enzymes can only function if integrated into cellular metabolism by an associated electron transport system. This requires that the polypeptide carrying the active site for the N–substrate interfaces with other proteins carrying electron transport chains. In some cases this integration is well resolved, but in other cases there is much work to be done in understanding the complexity of these systems and a function-structure level. We probably have still only exposed the tip of the iceberg as biochemists begin to address the bioinorganic N-cycle across a wide range of phyla and embrace newer reactions of the N-cycle.

Acknowledgements

I would like to thank the Royal Society and Wolfson Foundation for a merit award, current and past lab members for all their hard work and the BBSRC for project support in the N-cycle area.

References

Arnoux, P., Sabaty, M., Alric, J., Frangioni, B., Guigliarelli, B., Adriano, J.M., and Pignol, D. (2003). Structural and redox plasticity in the heterodimeric periplasmic nitrate reductase. Nat. Struct. Biol. *10*, 928–934.

Arp, D.J., Chain, P.S., and Klotz, M.G. (2007). The impact of genome analyses on our understanding of ammonia-oxidizing bacteria. Annu. Rev. Microbiol. *61*, 503–528.

Balasubramanian, R., Smith, S.M., Rawat, S., Yatsunyk, L.A., Stemmler, T.L., and Rosenzweig, A.C. (2010). Oxidation of methane by a biological dicopper centre. Nature *465*, 115–119.

Bamford, V.A., Angove, H.C., Seward, H.E., Thomson, A.J., Cole, J.A., Butt, J.N., Hemmings, A.M., and Richardson, D.J. (2002). Structure and spectroscopy of the periplasmic cytochrome c nitrite reductase from *Escherichia coli*. Biochemistry *41*, 2921–2931.

Bertero, M.G., Rothery, R.A., Palak, M., Hou, C., Lim, D., Blasco, F., Weiner, J.H., and Strynadka, N.C. (2003). Insights into the respiratory electron transfer pathway from the structure of nitrate reductase A. Nat. Struct. Biol. *10*, 681–687.

Bothe, H., Ferguson, S.J., and Newton, W.E., eds. (2007). The Biology of the Nitrogen Cycle (Amsterdam, Elsevier).

Brondijk, T.H., Fiegen, D., Richardson, D.J., and Cole, J.A. (2002). Roles of NapF, NapG and NapH, subunits of the Escherichia coli periplasmic nitrate reductase, in ubiquinol oxidation. Mol. Microbiol. *44*, 245–255.

Brondijk, T.H., Nilavongse, A., Filenko, N., Richardson, D.J., and Cole, J.A. (2004). NapGH components of the periplasmic nitrate reductase of Escherichia coli K-12: location, topology and physiological roles in quinol oxidation and redox balancing. Biochem. J. *379*, 47–55.

Brown, K., Djinovic-Carugo, K., Haltia, T., Cabrito, I., Saraste, M., Moura, J.J., Moura, I., Tegoni, M., and Cambillau, C. (2000a). Revisiting the catalytic CuZ cluster of nitrous oxide (N2O) reductase. Evidence of a bridging inorganic sulfur. J. Biol. Chem. *275*, 41133–41136.

Brown, K., Tegoni, M., Prudencio, M., Pereira, A.S., Besson, S., Moura, J.J., Moura, I., and Cambillau, C. (2000b). A novel type of catalytic copper cluster in nitrous oxide reductase. Nat. Struct. Biol. *7*, 191–195.

Butland, G., Spiro, S., Watmough, N.J., and Richardson, D.J. (2001). Two conserved glutamates in the bacterial nitric oxide reductase are essential for activity but not assembly of the enzyme. J. Bacteriol. *183*, 189–199.

Cava, F., Zafra, O., and Berenguer, J. (2008a). A cytochrome c containing nitrate reductase plays a role in electron transport for denitrification in *Thermus thermophilus* without involvement of the bc respiratory complex. Mol. Microbiol. *70*, 507–518.

Cava, F., Zafra, O., da Costa, M.S., and Berenguer, J. (2008b). The role of the nitrate respiration element of *Thermus thermophilus* in the control and activity of the denitrification apparatus. Environ. Microbiol. *10*, 522–533.

Clarke, T.A., Cole, J.A., Richardson, D.J., and Hemmings, A.M. (2007). The crystal structure of the pentahaem c-type cytochrome NrfB and characterization of its solution–state interaction with the pentahaem nitrite reductase NrfA. Biochem. J. *406*, 19–30.

Clarke, T.A., Holley, T., Hartshorne, R.S., Fredrickson, J.K., Zachara, J.M., Shi, L., and Richardson, D.J. (2008). The role of multihaem cytochromes in the respiration of nitrite in Escherichia coli and Fe(III) in Shewanella oneidensis. Biochem. Soc. Trans. *36*, 1005–1010.

Cramm, R., Siddiqui, R.A., and Friedrich, B. (1997). Two isofunctional nitric oxide reductases in *Alcaligenes eutrophus* H16. J. Bacteriol. *179*, 6769–6777.

de Vries, S., and Schroder, I. (2002). Comparison between the nitric oxide reductase family and its aerobic relatives, the cytochrome oxidases. Biochem. Soc. Trans. *30*, 662–667.

de Vries, S., Strampraad, M.J., Lu, S., Moenne-Loccoz, P., and Schroder, I. (2003). Purification and characterization of the MQH2:NO oxidoreductase from the hyperthermophilic archaeon Pyrobaculum aerophilum. J. Biol. Chem. *278*, 35861–35868.

Deeudom, M., Koomey, M., and Moir, J.W. (2008). Roles of c-type cytochromes in respiration in *Neisseria meningitidis*. Microbiology *154*, 2857–2864.

deVries, S., Suharti, and Pouvreau, L. (2007). Nitric oxide reductase: structural variations and catalytic mechanism. In The Biology of the Nitrogen Cycle, Bothe, H., Ferguson, S.J., and Newton, W.E., eds. (Amsterdam, Elsevier).

Einsle, O., Messerschmidt, A., Huber, R., Kroneck, P.M., and Neese, F. (2002). Mechanism of the six-electron reduction of nitrite to ammonia by cytochrome c nitrite reductase. J. Am. Chem. Soc. *124*, 11737–11745.

Einsle, O., Messerschmidt, A., Stach, P., Bourenkov, G.P., Bartunik, H.D., Huber, R., and Kroneck, P.M. (1999). Structure of cytochrome c nitrite reductase. Nature *400*, 476–480.

Ferguson, S.J., Richardson, D.J., and vanSpanning, R. (2007). The molecular biology and biochemistry of nitrification. In The Biology of the Nitrogen Cycle, Bothe, H., Ferguson, S.J., and Newton, W.E., eds. (Amsterdam, Elsevier).

Flock, U., Thorndycroft, F.H., Matorin, A.D., Richardson, D.J., Watmough, N.J., and Adelroth, P. (2008). Defining the proton entry point in the bacterial respiratory nitric-oxide reductase. J. Biol. Chem. *283*, 3839–3845.

Godden, J.W., Turley, S., Teller, D.C., Adman, E.T., Liu, M.Y., Payne, W.J., and LeGall, J. (1991). The 2.3 angstrom X-ray structure of nitrite reductase from *Achromobacter cycloclastes*. Science 253, 438–442.

Hakemian, A.S., Kondapalli, K.C., Telser, J., Hoffman, B.M., Stemmler, T.L., and Rosenzweig, A.C. (2008). The metal centers of particulate methane monooxygenase from Methylosinus trichosporium OB3b. Biochemistry 47, 6793–6801.

Haltia, T., Brown, K., Tegoni, M., Cambillau, C., Saraste, M., Mattila, K., and Djinovic-Carugo, K. (2003). Crystal structure of nitrous oxide reductase from *Paracoccus denitrificans* at 1.6 A resolution. Biochem. J. *369*, 77–88.

Hendriks, J., Warne, A., Gohlke, U., Haltia, T., Ludovici, C., Lubben, M., and Saraste, M. (1998). The active site of the bacterial nitric oxide reductase is a dinuclear iron center. Biochemistry 37, 13102–13109.

Hough, M.A., Ellis, M.J., Antonyuk, S., Strange, R.W., Sawers, G., Eady, R.R., and Samar Hasnain, S. (2005). High resolution structural studies of mutants provide insights into catalysis and electron transfer processes in copper nitrite reductase. J. Mol. Biol. *350*, 300–309.

Igarashi, N., Moriyama, H., Fujiwara, T., Fukumori, Y., and Tanaka, N. (1997). The 2.8 A structure of hydroxylamine oxidoreductase from a nitrifying chemoautotrophic bacterium, *Nitrosomonas europaea*. Nat. Struct. Biol. *4*, 276–284.

Iverson, T.M., Arciero, D.M., Hsu, B.T., Logan, M.S., Hooper, A.B., and Rees, D.C. (1998). Heme packing motifs revealed by the crystal structure of the tetra-heme cytochrome c554 from *Nitrosomonas europaea*. Nat. Struct. Biol. 5, 1005–1012.

Iwata, S., Ostermeier, C., Ludwig, B., and Michel, H. (1995). Structure at 2.8 A resolution of cytochrome c oxidase from *Paracoccus denitrificans*. Nature *376*, 660–669.

Jormakka, M., Richardson, D., Byrne, B., and Iwata, S. (2004). Architecture of NarGH reveals a structural classification of Mo-bisMGD enzymes. Structure *12*, 95–104.

Jormakka, M., Yokoyama, K., Yano, T., Tamakoshi, M., Akimoto, S., Shimamura, T., Curmi, P., and Iwata, S. (2008). Molecular mechanism of energy conservation in polysulfide respiration. Nat. Struct. Mol. Biol. *15*, 730–737.

Kern, M., and Simon, J. (2009). Electron transport chains and bioenergetics of respiratory nitrogen metabolism in *Wolinella succinogenes* and other Epsilonproteobacteria. Biochim. Biophys. Acta *1787*, 646–656.

Kim, H.J., Zatsman, A., Upadhyay, A.K., Whittaker, M., Bergmann, D., Hendrich, M.P., and Hooper, A.B. (2008). Membrane tetraheme cytochrome c(m552) of the ammonia-oxidizing nitrosomonas europaea: a ubiquinone reductase. Biochemistry 47, 6539–6551.

Maillard, J., Spronk, C.A., Buchanan, G., Lyall, V., Richardson, D.J., Palmer, T., Vuister, G.W., and Sargent, F. (2007). Structural diversity in twin-arginine signal peptide-binding proteins. Proc. Natl. Acad. Sci. U.S.A. *104*, 15641–15646.

Martinez-Espinosa, R.M., Dridge, E.J., Bonete, M.J., Butt, J.N., Butler, C.S., Sargent, F., and Richardson, D.J. (2007). Look on the positive side! The orientation, identification and bioenergetics of 'Archaeal' membrane-bound nitrate reductases. FEMS Microbiol. Lett. *276*, 129–139.

Mohan, S., and Cole, J.A. (2007). The dissimilatory reduction of nitrate to ammonia by anaerobic bacteria. In The Biology of the Nitrogen Cycle, Bothe, H., Ferguson, S.J., and Newton, W.E., eds. (Amsterdam, Elsevier).

Nilavongse, A., Brondijk, T.H., Overton, T.W., Richardson, D.J., Leach, E.R., and Cole, J.A. (2006). The NapF protein of the Escherichia coli periplasmic nitrate reductase system: demonstration of a cytoplasmic location and interaction with the catalytic subunit, NapA. Microbiology *152*, 3227–3237.

Nojiri, M., Koteishi, H., Nakagami, T., Kobayashi, K., Inoue, T., Yamaguchi, K., and Suzuki, S. (2009). Structural basis of inter-protein electron transfer for nitrite reduction in denitrification. Nature *462*, 117–120.

Nurizzo, D., Silvestrini, M.C., Mathieu, M., Cutruzzola, F., Bourgeois, D., Fulop, V., Hajdu, J., Brunori, M., Tegoni, M., and Cambillau, C. (1997). N-terminal arm exchange is observed in the 2.15 A crystal structure of oxidized nitrite reductase from *Pseudomonas aeruginosa*. Structure 5, 1157–1171.

Olmo-Mira, M.F., Gavira, M., Richardson, D.J., Castillo, F., Moreno-Vivian, C., and Roldan, M.D. (2004). NapF is a cytoplasmic iron–sulfur protein required for Fe–S cluster assembly in the periplasmic nitrate reductase. J. Biol. Chem. *279*, 49727–49735.

Poly, F., Wertz, S., Brothier, E., and Degrange, V. (2008). First exploration of Nitrobacter diversity in soils by a PCR cloning-sequencing approach targeting functional gene nxrA. FEMS Microbiol Ecol *63*, 132–140.

Polyakov, K.M., Boyko, K.M., Tikhonova, T.V., Slutsky, A., Antipov, A.N., Zvyagilskaya, R.A., Popov, A.N., Bourenkov, G.P., Lamzin, V.S., and Popov, V.O. (2009). High-resolution structural analysis of a novel octaheme cytochrome c nitrite reductase from the haloalkaliphilic bacterium *Thioalkalivibrio nitratireducens*. J. Mol. Biol. *389*, 846–862.

Potter, L., Angove, H., Richardson, D., and Cole, J. (2001). Nitrate reduction in the periplasm of gram-negative bacteria. Adv. Microb. Physiol. *45*, 51–112.

Reimann, J., Flock, U., Lepp, H., Honigmann, A., and Adelroth, P. (2007). A pathway for protons in nitric oxide reductase from *Paracoccus denitrificans*. Biochim. Biophys. Acta *1767*, 362–373.

Richardson, D.J. (2000). Bacterial respiration: a flexible process for a changing environment. Microbiology *146*, 551–571.

Rodrigues, M.L., Oliveira, T.F., Pereira, I.A., and Archer, M. (2006). X-ray structure of the membrane-bound cytochrome c quinol dehydrogenase NrfH reveals novel haem coordination. EMBO J. *25*, 5951–5960.

Rosenzweig, A.C. (2008). The metal centres of particulate methane mono-oxygenase. Biochem. Soc. Trans. *36*, 1134–1137.

Simon, J., Einsle, O., Kroneck, P.M., and Zumft, W.G. (2004). The unprecedented nos gene cluster of *Wolinella succinogenes* encodes a novel respiratory electron transfer pathway to cytochrome c nitrous oxide reductase. FEBS Lett. *569*, 7–12.

Simon, J., van Spanning, R.J., and Richardson, D.J. (2008). The organisation of proton motive and non-proton motive redox loops in prokaryotic respiratory systems. Biochim. Biophys. Acta *1777*, 1480–1490.

Starkenburg, S.R., Arp, D.J., and Bottomley, P.J. (2008a). Expression of a putative nitrite reductase and the reversible inhibition of nitrite-dependent respiration by nitric oxide in Nitrobacter winogradskyi Nb-255. Environ. Microbiol. *10*, 3036–3042.

Starkenburg, S.R., Chain, P.S., Sayavedra-Soto, L.A., Hauser, L., Land, M.L., Larimer, F.W., Malfatti, S.A., Klotz, M.G., Bottomley, P.J., Arp, D.J., *et al.* (2006). Genome sequence of the chemolithoautotrophic nitrite-oxidizing bacterium *Nitrobacter winogradskyi* Nb-255. Appl. Environ. Microbiol. *72*, 2050–2063.

Starkenburg, S.R., Larimer, F.W., Stein, L.Y., Klotz, M.G., Chain, P.S., Sayavedra-Soto, L.A., Poret-Peterson, A.T., Gentry, M.E., Arp, D.J., Ward, B., *et al.* (2008b). Complete genome sequence of Nitrobacter hamburgensis X14 and comparative genomic analysis of species within the genus *Nitrobacter*. Appl. Environ. Microbiol. *74*, 2852–2863.

Thorndycroft, F.H., Butland, G., Richardson, D.J., and Watmough, N.J. (2007). A new assay for nitric oxide reductase reveals two conserved glutamate residues form the entrance to a proton-conducting channel in the bacterial enzyme. Biochem. J. *401*, 111–119.

van Spanning, R., Richardson, D.J., and Ferguson, S.J. (2007). Introduction to the biochemistry and molecular biology of denitrification. In The Biology of the Nitrogen Cycle, Bothe, H., Ferguson, S.J., and Newton, W.E., eds. (Amsterdam, Elsevier).

Williams, P.A., Fulop, V., Garman, E.F., Saunders, N.F., Ferguson, S.J., and Hajdu, J. (1997). Haem-ligand switching during catalysis in crystals of a nitrogen-cycle enzyme. Nature *389*, 406–412.

Williams, P.A., Fulop, V., Leung, Y.C., Chan, C., Moir, J.W., Howlett, G., Ferguson, S.J., Radford, S.E., and Hajdu, J. (1995). Pseudospecific docking surfaces on electron transfer proteins as illustrated by pseudoazurin, cytochrome c550 and cytochrome cd1 nitrite reductase. Nat. Struct. Biol. 2, 975–982.

Wunsch, P., and Zumft, W.G. (2005). Functional domains of NosR, a novel transmembrane iron–sulfur flavoprotein necessary for nitrous oxide respiration. J. Bacteriol. *187*, 1992–2001.

Yamanaka, T., and Fukumori, Y. (1988). The nitrite oxidizing system of Nitrobacter winogradskyi. FEMS Microbiol. Rev. *4*, 259–270.

Zumft, W.G. (2005). Biogenesis of the bacterial respiratory CuA, Cu-S enzyme nitrous oxide reductase. J. Mol. Microbiol. Biotechnol. *10*, 154–166.

Organization of Respiratory Electron Transport Chains in Nitrate-reducing and Nitrifying Bacteria

3

Jörg Simon

Abstract

Nitrogen compounds serve as electron donor and electron acceptor substrates in several modes of microbial respiration such as nitrification, nitrate reduction, denitrification and nitrite ammonification. There are several well-established model bacteria for each of these processes and in many, though not all, cases the various dehydrogenases and reductases involved in the conversion of nitrogen compounds have been thoroughly characterized including the determination of high-resolution structure models. On the other hand, the architecture of complete respiratory electron transport chains is often less well known, especially with respect to donor:quinone dehydrogenase and quinol:acceptor reductase systems that connect the membranous quinone/quinol pool to the oxidative or reductive part of an electron transport chain. Notably, a rather limited number of redox-active protein modules has evolved that is employed by different bacteria in a versatile manner. Occasionally, bacterial species even display different electron transport chain set-ups despite using the same type of substrate-converting enzyme. This article highlights commonalities and differences in the organization of bacterial respiratory electron transport chains that are involved in environmentally important N-cycle processes and discusses the relevance of this knowledge in the context of microbial bioenergetics and (meta)genomics.

Introduction

The reactions of the biogeochemical nitrogen cycle are essential in sustaining life on earth (see Fig. 3.1 for an overview) (Richardson, 2000; Rudolf and Kroneck, 2005; Bothe *et al.*, 2007; Klotz and Stein, 2008). Especially, the participation of microorganisms is crucial and the most important bacterial groups involved are those that catalyse (I) aerobic chemolithotrophic nitrification, (II) anaerobic reduction of nitrate and/or nitrite to produce either nitric oxide (NO), nitrous oxide (N_2O), dinitrogen (N_2) or ammonium, (III) fixation of dinitrogen and (IV) anaerobic ammonium oxidation (anammox). Within this framework, three distinct types of anaerobic respiration have been recognized in the nitrogen cycle that depend on either nitrate and/or nitrite as electron acceptor substrate(s). These are denitrification, respiratory ammonification of nitrate or nitrite and anammox (Zumft, 1997; Simon, 2002; Kuenen, 2008). Depending on the organism, denitrification yields different gaseous products (NO, N_2O, N_2) in various amounts whereas, in the ammonification pathway, nitrite is reduced to ammonia without the notable release of intermediate products. In the anammox process, ammonium is used as electron donor for denitrifying, i.e. N_2-producing, nitrite reduction. Currently, no organism is known that catalyses more than one of these three respiratory modes.

The conversion of energy substrates in anaerobic respiration and in nitrification involves membrane-dependent respiratory electron transport chains that connect appropriate electron transport enzymes. In Gram-negative bacteria, these enzymes are either membrane-bound or reside in the periplasmic space. A key redox mediator within these respiratory chains is the membranous quinone/quinol pool that usually

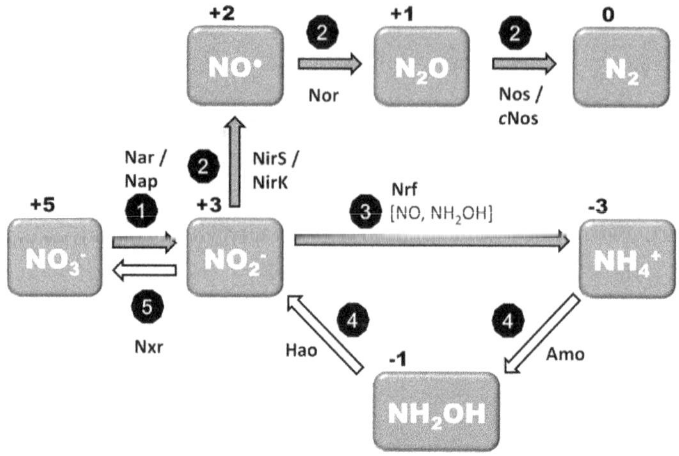

Figure 3.1 Conversion of nitrogen compounds that serve as microbial substrates in N-cycle processes relevant to the present chapter. Aerobic nitrification and anaerobic respiratory pathways are indicated by white and grey arrows, respectively. The oxidation state of nitrogen atoms is indicated above for the respective compound. Numbers in black decagons refer to the following processes: 1, respiratory nitrate reduction to nitrite; 2, denitrification of nitrite to N_2; 3, nitrite ammonification; 4, ammonium oxidation to nitrite; 5, nitrite oxidation to nitrate. These metabolic pathways are catalysed by distinct respiratory enzyme systems that are designated according to their terminal enzyme: Nar, membrane-bound nitrate reductase; Nap, periplasmic nitrate reductase; NirS, cytochrome cd_1 nitrite reductase; NirK, copper nitrite reductase; Nor, nitric oxide reductase; Nos, nitrous oxide reductase; cNos, cytochrome *c* nitrous oxide reductase; Nrf, cytochrome *c* nitrite reductase; Amo, ammonium monooxygenase; Hao, hydroxylamine oxidoreductase; Nxr, nitrite oxidoreductase. The compounds in brackets indicate that the Nrf system also reduces nitric oxide and hydroxylamine to ammonium. The anammox process, i.e. the comproportionation of nitrite and ammonium to form dinitrogen, as well as dinitrogen fixation are not shown.

consists of one or more of various forms of ubiquinones and menaquinones, depending on the organism. Quinones are small, freely diffusible, lipophilic and membrane-entrapped organic molecules that carry two electrons and two protons when fully reduced, i.e. in the quinol state. Despite the fact that the architecture of electron transport chains is highly diverse, from a bioenergetic point of view they all fulfil a common function in generating a proton motive force (*pmf*) across the coupling membrane (either the cytoplasmic or the anammoxosome membrane) through converting the free energy (ΔG) of a redox reaction catalysed by the electron transport chain. The *pmf* is then primarily used to phosphorylate ADP by a membrane-bound ATP synthase complex.

Over the past about 15 years, many high-resolution structures of terminal reductases involved in respiratory nitrogen metabolism of various nitrate-reducing, denitrifying, ammonifying or nitrifying bacteria have been obtained. Such enzymes include the membrane-bound nitrate reductase complex NarGHI, periplasmic nitrate reductase NapA and its cognate redox partner NapB, cytochrome cd_1 nitrite reductase NirS, Cu-containing nitrite reductase NirK, N_2O reductase NosZ and the ammonifying cytochrome *c* nitrite reductase NrfA (reviewed by Simon, 2002; Einsle and Kroneck, 2004; Fritz *et al.*, 2005; Tavares *et al.*, 2006). In addition, a structural model of the hydroxylamine oxidoreductase (Hao) from the nitrifying bacterium *Nitrosomonas europaea* is available (Igarashi *et al.*, 1997). All these enzymes are either membrane-bound or periplasmic metalloproteins (or protein complexes) that exchange electrons with the quinone/quinol pool either directly or by using suitable redox partner proteins like cytochromes or iron–sulfur proteins. In contrast to the detailed knowledge on the structure/function relationships of these individual respiratory enzymes, the enzymic composition of entire electron transport chains and the necessary protein-protein and protein–quinone/quinol interactions are less well understood. Such information, however, is decisive to elucidate the mechanism by which the electron transport from an electron donor to an electron acceptor substrate is coupled to *pmf* generation. In particular, there is currently only a limited understanding of the detailed function of donor:quinone dehydrogenases and quinol:acceptor reductases involved in electron transport from hydroxylamine to the quinone pool or from the quinol pool to the various terminal reductases of anaerobic respiration.

Based on some fundamental bioenergetic considerations, this chapter focuses on the organization of respiratory electron transport chains that catalyse (I) respiratory nitrate reduction to nitrite, (II) denitrification, (III) respiratory nitrite ammonification and (IV) nitrification, thereby pointing out that substrate-converting enzymes are connected to the quinone/quinol pool predominantly via some evolutionary conserved protein modules that use an astonishingly small number of redox cofactor types. Interestingly, however, there is an intriguing variety in the design of respiratory electron transport chains in different model bacteria that catalyse the same physiological reaction. In such cases, the architecture of respiratory chains is often modelled on the basis of genome sequences since the involved enzyme systems are regularly encoded by gene clusters that encode the above mentioned oxidoreductases (e.g. genomic *nar, nap, nir, nos, nrf* and *hao* loci). Taxonomically, most of the best-examined bacteria relevant to this article belong to the phylum Proteobacteria which is divided into alpha-, beta-, gamma-, delta- and epsilonproteobacterial classes. Only a small number of model organisms can be described here but emphasis is laid on the danger to generalize conclusions drawn from these few examples, especially when regarding the vast majority of uncultured bacteria whose (meta)genomes nowadays appear massively in the nucleotide data banks. Therefore, if available, data obtained from bacteria grouped outside the Proteobacteria are taken into consideration.

Note that a detailed structural and functional description of enzymes involved in the turnover of nitrogen compounds is beyond the scope of this article and the reader is referred to Chapter 2. Electron transport processes that are part of the assimilatory pathways of nitrate reduction and nitrite ammonification as well as of anammox and dinitrogen fixation are dealt with in other chapters of this book. N-cycle processes in Archaea are dealt with in Chapter 1.

Bioenergetic considerations

Respiratory processes serve to generate ATP by electron transport phosphorylation and the organization of electron transport chains usually reflects the bioenergetic basis of a given respiratory redox reaction. The free energy change at pH 7.0 ($\Delta G_0'$) of a respiratory process depends on the standard redox potentials at pH 7.0 (E_0') of the electron donor and acceptor substrates (and their respective reaction products) according to the equation $\Delta G_0' = -nF\Delta E_0'$, where n is the number of electrons transferred and F is the Faraday constant. Owing to their rather positive E_0' values, the diverse range of nitrogen compounds are well suited as terminal electron acceptors within a respiratory electron transport chain, especially when substrates like formate, hydrogen gas, NADH or succinate are used as electron donors in chemotrophic or lithotrophic energy-conserving pathways (Table 3.1). On the other hand, ammonium and nitrite are poor electron donors, thus explaining the fact that chemolithotropic nitrifying organisms usually use oxygen as sole terminal electron acceptor. Operating a nitrifying electron transport chain by a lithoautotroph also necessitates reversed electron transport in order to produce NADH (by NAD^+ reduction) for assimilatory CO_2 reduction.

As stated above, part of the free energy released during the catalysis of a respiratory redox reaction is conserved in the generation of a *pmf* across the coupling membrane. Such a *pmf* has both a chemical (ΔpH, dimensionless) and an electrical ($\Delta\psi$, dimension mV) component and the formula $pmf\ (\text{mV}) = \Delta\psi - 59\ \Delta\text{pH}$ describes this correlation in its simplest form. $\Delta\psi$ is defined as the electrical potential difference between the positively and negatively charged side of the membrane (usually a positive value) whereas the ΔpH is defined as the pH difference between both membrane sides (usually negative). Since the *pmf* has a value in the range of about 150–200 mV, a redox potential change

Table 3.1 Standard redox potentials at pH 7 (E_0') of selected redox pairs (values with respect to the standard hydrogen electrode). Nitrogen compounds are shown in bold

Redox pair (oxidized/reduced)	E_0' (V)
CO_2/formate	−0.43
$2H^+/H_2$	−0.42
NAD^+/NADH	−0.32
Menaquinone/menaquinol	−0.08
Fumarate/succinate	0.03
Nitrite/hydroxylamine	**0.1**
NO/hydroxylamine	**0.1**
Ubiquinone/ubiquinol	0.11
Nitrite/ammonium	**0.34**
Nitrite/NO	**0.36**
Nitrate/nitrite	**0.43**
Hydroxylamine/ammonium	**0.74**
O_2/H_2O	0.82
2 NO/N_2O	**1.18**
N_2O/N_2	**1.35**

(ΔE) of at least this magnitude is required during electron transfer from a donor to an acceptor substrate if this step is to be coupled to *pmf* generation. The ratio of protons (transported across the membrane) and electrons (transported from donor to acceptor), i.e. the H^+/e^- value, corresponds to the quotient of ΔE and the *pmf* (at a hypothetical 100% thermodynamic efficiency). There are different possibilities to generate the *pmf*: (1) proton pumping as exemplified by NADH:ubiquinone oxidoreductase (complex I-type NADH dehydrogenase) or haem copper oxidases, (2) movement of protons across the membrane via the so-called Q-cycle as catalysed by ubiquinol:cytochrome *c* oxidoreductase (cytochrome bc_1 complex) and (3) the electrogenic redox loop mechanism in which a separation of positive and negative charges across the membrane is achieved by different locations of the active sites for substrates and quinones in respiratory enzymes (or in respiratory electron transport chains). Classically, a redox loop consists of two quinone/quinol-reactive (multi-subunit) enzymes and a quinone species that serves as redox mediator between them. Bacterial redox loops are commonly found in anaerobic respiratory systems that are driven by ΔG values smaller than in heterotrophic aerobic respiration and, consequently, short electron transfer chains and few proton-pumping enzymes are employed in such cases (Simon *et al.*, 2008). Keeping these fundamental bioenergetic principles in mind, the design of some crucial quinone/quinol-reactive redox modules is discussed in the following section as these proteins play essential roles in the different N-cycle processes discussed afterwards.

Quinone/quinol-dependent electron transport modules relevant to N-cycle processes

An often overlooked but bioenergetically essential part of electron transport chains is the connection of substrate-converting enzymes to the quinone/quinol pool. Notably, nature apparently has evolved only a limited number of donor:quinone dehydrogenases and quinol:acceptor reductases that serve this purpose in electron transport chains relevant to this chapter. These taxonomically widespread quinone/quinol-reactive protein modules comprise dihaem cytochrome *b* membrane anchors, cytochromes *c* of the NapC/NrfH/Cyt. c_{m552} family, iron–sulfur proteins of the NapH/NosH family and members of the

NrfD/PsrC family that apparently lack any metal cofactors. The key properties of these protein classes are briefly presented in the following paragraphs (for reviews see Simon and Kern, 2008; Simon *et al.*, 2008).

Quinone/quinol-reactive dihaem *b*-type cytochromes

These membrane-bound cytochromes *b* are key components of multi-subunit complexes serving in aerobic and anaerobic respiratory electron transport chains of Eucarya, Bacteria and Archaea. Prominent examples are the cytochrome bc_1 complex, cytochrome bo_3 and cytochrome *bd* quinol oxidases as well as quinone-reactive membrane anchor subunits of Ni/Fe-hydrogenase, formate dehydrogenase, Nar-type nitrate reductase, succinate dehydrogenase and fumarate reductase. The haem *b* groups of these enzymes are located within the membrane plane, usually oriented to different sides of the membrane and axially ligated by histidine side chains that are provided by at least two different transmembrane helical regions (Fig. 3.2A). Furthermore, haem *b* is often involved in the composition of catalytic centres operative in quinone/quinol turnover. Redox partner proteins of such dihaem cytochromes *b* are mainly iron–sulfur proteins or *c*-type cytochromes. In many cases, electron transport catalysed by dihaem *b*-type cytochromes is coupled to the generation of a *pmf* across the membrane, either by proton pumping, the Q-cycle or the redox loop mechanism (Simon *et al.*, 2008).

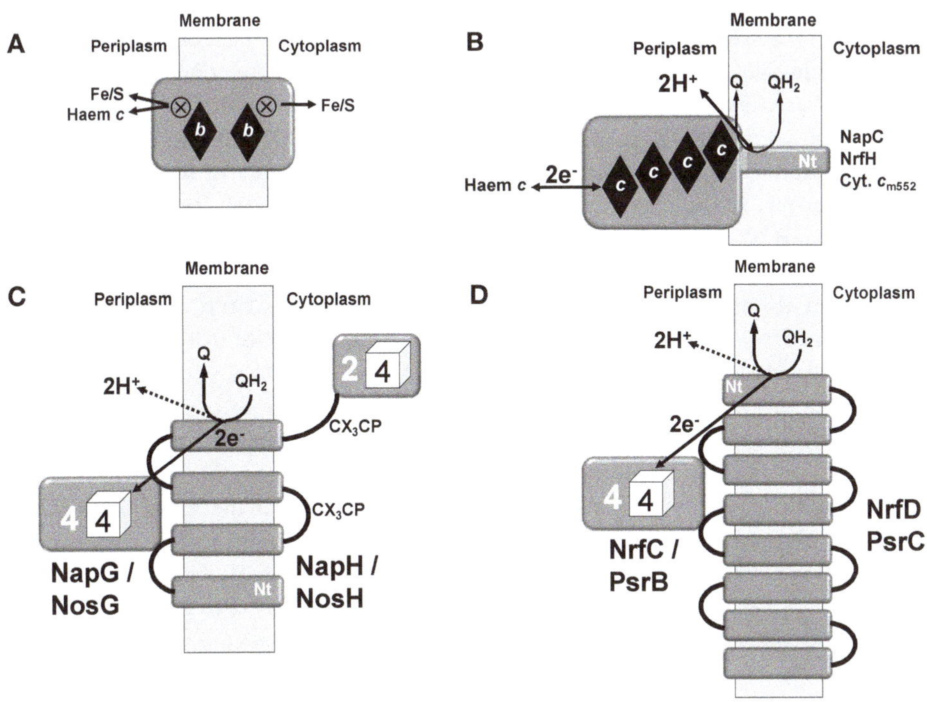

Figure 3.2 Selected bacterial quinone/quinol-reactive electron transport modules and their redox partner cofactors or proteins. The schematic architecture and cofactor content of dihaem cytochrome *b* membrane anchors (A), members of the NapC/NrfH/Cyt. c_{m552} family (B), the NapH/NosH family (C) and the NrfD/PsrC family (D) is shown. The ⊗ symbol denotes possible locations of quinone/quinol binding sites in dihaem cytochromes *b*. These proteins usually contain at least four transmembrane domains. Q/QH_2 designate any type of quinone/quinol. Diamonds and cubes depict haem *b* or *c* groups and [4Fe–4S]-type iron–sulfur centres, respectively. Broken arrows indicate that the side of the membrane to which protons are released is speculative. Nt indicates the N-terminus of a protein. See text for details on the architecture of specific proteins.

The NapC/NrfH/Cyt. c_{m552} family of multihaem *c*-type cytochromes

The proteins of this family are membrane-bound tetra- or pentahaem *c*-type cytochromes that comprise an N-terminal membrane-spanning helix and a globular cytochrome *c* domain situated at the outside of the bacterial membrane (Fig. 3.2B). Most members are tetrahaem cytochromes *c* consisting of about 175 amino acid residues. The NapC/NrfH/Cyt. c_{m552} family was previously referred to as the NapC/NirT family since NirT from *Pseudomonas stutzeri* was the first such protein to which a function had been assigned (in this case electron transport to periplasmic cytochrome cd_1 nitrite reductase) (Jüngst *et al.*, 1991). NapC and NrfH donate electrons, either directly or indirectly, to periplasmic nitrate reductase (NapA) and periplasmic cytochrome *c* nitrite reductase (NrfA), respectively, whereas Cyt. c_{m552} was proposed to feed electrons derived from hydroxylamine oxidation into the ubiquinone pool of nitrifying organisms (Cartron *et al.*, 2002; Kern *et al.*, 2008; Kim *et al.*, 2008, and references therein). In each case, multihaem *c*-type cytochromes act as redox partner proteins. The crystal structure of NrfH from *Desulfovibrio vulgaris* was solved as part of a quinol oxidizing $(NrfHA_2)_2$ complex (Rodrigues *et al.*, 2006, 2008). A putative menaquinol binding site was proposed at the periplasmic side of the membrane, situated in the vicinity of haem 1, that serves as an entrance to a densely packed electron-transferring haem *c* wire. It is therefore very likely that protons are exchanged with the periplasmic rather than the cytoplasmic space during quinone/quinol turnover. This would rule out the participation of NapC/NrfH/Cyt. c_{m552}-type cytochromes in *pmf* generation.

Quinol dehydrogenases of the NapH/NosH family

NapH/NosH-type proteins contain four hydrophobic segments that are likely to form transmembrane domains, two conserved CX_3CP signatures and two four-cysteine clusters typically involved in [4Fe–4S] centre formation (Fig. 3.2C). The NapH/NosH proteins are thought to form stable complexes with specific periplasmic redox partner proteins called NapG or NosG that are predicted to be poly-ferredoxins with four [4Fe–4S] iron–sulfur centres. Recently, the formation of a NapGH complex has been shown in the case of *Wolinella succinogenes*, a nitrate-ammonifying Epsilonproteobacterium (Kern and Simon, 2008). NapGH/NosGH complexes are assumed to catalyse ubi- or menaquinol oxidation and to transfer electrons to cognate periplasmic *c*-type cytochromes in electron transport chains leading to either periplasmic nitrate reductase (NapA) or cytochrome *c* nitrous oxide reductase (*c*NosZ). The orientation of the four transmembrane segments of NapH is depicted in Fig. 3.2C according to experimental evidence obtained for *E. coli* NapH using a series of alkaline phosphatase and β-galactosidase fusion proteins and implying that all four poly-cysteine motifs are located at the cytoplasmic side of the membrane (Brondijk *et al.*, 2004). The architecture and localization of the quinol-reactive site in NapH/NosH is not known and it cannot be excluded that more than one such site is present. Furthermore, the function of the cytoplasmic, potentially metal-binding CX_3CP motifs and the putative iron–sulfur centres has not been elucidated. The potential participation of NapH/NosH-type proteins in *pmf* generation has not been examined.

The NrfD/PsrC family of membrane-integral quinol dehydrogenases

NrfD is a polytopic membrane protein encoded in the *nrf* gene cluster of enteric bacteria. It was shown to be essential for formate-dependent nitrite reduction of *E. coli* and was predicted to traverse the membrane eight times with both the N- and C-termini located in the periplasm (Hussain *et al.*, 1994). According to the model presented in Fig. 3.2D, NrfD forms a complex with NrfC, which is predicted to bind four [4Fe–4S] clusters. The NrfCD complex possibly catalyses electron transfer via the pentahaem cytochrome *c* NrfB to cytochrome c nitrite reductase NrfA. Due to its proposed structure, NrfD is likely to be a quinol dehydrogenase but, similar to NapH/NosH, no functional or structural details are yet available. NrfD-type proteins are widespread in bacterial anaerobic electron transport chains with homologous proteins present, for example, in polysulfide and tetrathionate reductase systems. The NrfD homologue PsrC was shown experimentally to form a complex with the NrfC-type protein PsrB and the catalytic subunit PsrA in the membrane-bound polysulfide reductase complex of *W. succinogenes* (Klimmek *et al.*, 2004). PsrC is the membrane anchor of the complex and the two hydrophilic subunits PsrB and PsrA are oriented towards the periplasm. The crystal structure of PsrC from *Thermus thermophilus* confirmed the predicted presence of eight transmembrane helices and was found to contain a single quinol-binding site at the periplasmic membrane surface (Jormakka *et al.*, 2008). Genes encoding members of the NrfD/PsrC family are found in a variety of different bacterial genetic contexts, usually in combination with a gene encoding a polyferredoxin similar to NrfC/PsrB.

Respiratory nitrate reduction to nitrite

Nitrate is a bioenergetically favourable electron acceptor for respiratory processes in bacteria (Table 3.1) and nitrite is always the initial reduction product. In Bacteria, two different respiratory systems, Nar and Nap, have evolved that catalyse nitrate reduction in different cell compartments and differ fundamentally in the organization of the corresponding electron transport chains catalysing the oxidation of ubi- or menaquinol by nitrate (Richardson *et al.*, 2001). Depending on the primary reductant used in energy metabolism, there are numerous (*pmf*-generating or electroneutral) ways to feed electrons into the quinone pool which, apart from formate dehydrogenase, will not be described here in detail (see Simon *et al.*, 2008 for a recent review).

Electron transport to membrane-bound nitrate reductase (Nar system)

The membrane-bound nitrate reductase complex, NarGHI, is highly conserved in many Bacteria and was investigated in detail using enteric Proteobacteria like *E. coli* as model organisms. Especially, the respiratory electron transport system from the fermentation product formate to nitrate received a lot of attention since it forms the prototype of an electrogenic redox loop mechanism (Richardson and Sawers, 2002; Simon *et al.*, 2008 and references therein). In this redox loop, two electrogenic protein complexes, i.e. formate dehydrogenase (FdnGHI; called FdhABC in other Proteobacteria) and NarGHI, are electronically connected via the menaquinone/menaquinol pool in the membrane. The individual complexes catalyse reactions 3.1 and 3.2, respectively. Overall, both complexes produce a *pmf* by an electron transport chain operating at a H^+/e^- ratio of 2 (Fig. 3.3).

Formate dehydrogenase:

$$HCO_2^- \rightarrow CO_2 + 2\,e^- + H^+ \quad (3.1)$$

Nitrate reductase:

$$NO_3^- + 2\,e^- + 2\,H^+ \rightarrow NO_2^- + H_2O \quad (3.2)$$

Such a mechanism is enabled by the fact that formate is oxidized at the periplasmic side of the membrane while nitrate reduction takes place at the cytoplasmic side (thus requiring nitrate uptake and often nitrite export by specific transporters). The dihaem cytochrome *b* membrane anchors of both enzymes belong to the quinone/quinol-reactive protein family presented in Fig. 3.2A. Crystal structures of the enzymes have ultimately confirmed that both complexes contain a menaquinone/menaquinol-reactive site that is located at the membrane side opposite to the active site for formate or nitrate (Fig. 3.3). Thus, the cytochrome *b* membrane anchor proteins (FdnI and NarI) catalyse transmembrane electron transport involving both haem *b* groups that consequently are oriented to different sides of the membrane. The redox partner proteins of FdnI and NarI are the iron–sulfur proteins FdnH and NarH, respectively. In many Proteobacteria, the formate dehydrogenase complex may be substituted by a heterotrimeric Ni/Fe-hydrogenase which couples hydrogen oxidation to menaquinone reduction, thereby also generating a *pmf* (Simon *et al.*, 2008). This type of membrane-bound uptake hydrogenase contains a dihaem cytochrome *b* membrane anchor (HydC) which is similar to FdhC, an iron–sulfur protein (HydA) and a catalytic subunit (HydB) that contains a typical Ni/Fe centre (Gross *et al.*, 2004).

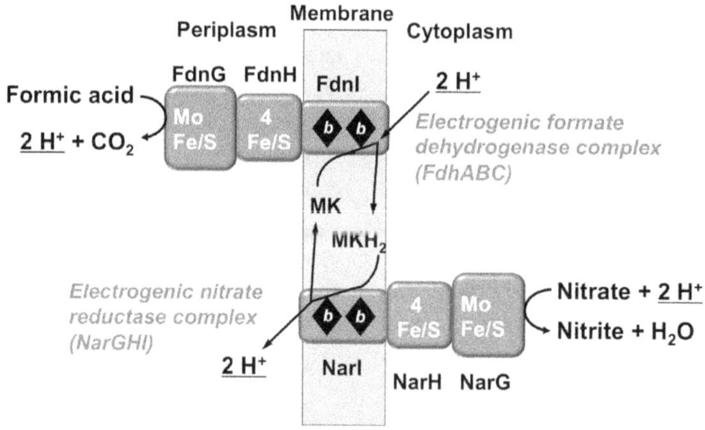

Figure 3.3 Architecture of the electrogenic electron transport chain from formate to nitrate in *E. coli*. Two individual reactions contribute to *pmf* formation according to the redox loop mechanism: (I) periplasmic formate oxidation by menaquinone and (II) menaquinol oxidation coupled to cytoplasmic nitrate reduction (the corresponding electrogenic protons are underlined). Note that the catalytic centres for formate and menaquinone in the FdnGHI complex as well as for menaquinol and nitrate in the NarGHI complex are situated on different sides of the membrane. The subunits FdnI and NarI are quinone/quinol-reactive dihaem cytochrome *b* membrane anchor modules that catalyse transmembrane electron transfer. Mo, molybdenum-*bis*-molybdopterin guanine dinucleotide cofactor; Fe/S, iron–sulfur centre.

Electron transport to periplasmic nitrate reductase (Nap system)

The NapA protein is the nitrate-reducing protein of the periplasmic nitrate reduction system of many different Proteobacteria that, in many cases (e.g. in *E. coli*), also contain the NarGHI complex. The Nap system also sustains reaction 3.2 but uses a different type of nitrate reductase that is exported to the periplasmic space by the twin-arginine translocation (Tat) complex (Potter *et al.*, 2001; Sargent, 2007). Bacterial *nap* gene clusters vary immensely and it seems that the *napA* and *napD* genes are the only open reading frames that are present throughout (Marietou *et al.*, 2005; Kern *et al.*, 2007; Kern and Simon, 2009a). This is not surprising as NapD was shown to be a chaperone that is required for directing cytoplasmic NapA to the Tat complex (Maillard *et al.*, 2007). In the following, the organization of the prototypical Nap systems of *W. succinogenes* (Fig. 3.4) and *E. coli* are discussed.

W. succinogenes grows by nitrate respiration using formate as electron donor and menaquinone as sole quinone-type redox mediator according to reactions 3.1 and 3.2. NapA, the only nitrate reductase in this organism, is encoded by the first gene of the *napAGHBFLD* gene cluster (Baar *et al.*, 2003; Simon *et al.*, 2003). A NapGH complex as depicted in Fig. 3.2C was shown to be essential for nitrate respiration and it is very likely that NapG transfers electrons to the dihaem cytochrome *c* NapB which was shown to form a complex with NapA in many other bacteria (Richardson *et al.*, 2001; Kern and Simon, 2008, 2009a). According to Fig. 3.4, menaquinol oxidation by nitrate would be electroneutral if the protons derived from menaquinol oxidation were released by NapH on the periplasmic side. Therefore, in contrast to the Nar system, reaction 3.1 catalysed by the electrogenic formate dehydrogenase complex (FdhABC) in combination with the Nap system in *W. succinogenes* would only produce a *pmf* with a H^+/e^- ratio of 1 (Kern and Simon, 2009a). However, electrogenic nitrate reduction by menaquinol cannot be ruled out at present as there is enough free energy between the menaquinone/menaquinol and nitrate/nitrite redox couples (standard redox potential difference of about 500 mV; Table 3.1) to allow for this. In such a case, a second quinol-binding site at the cytoplasmic side of the membrane might exist and some rather hydrophobic stretches of NapH

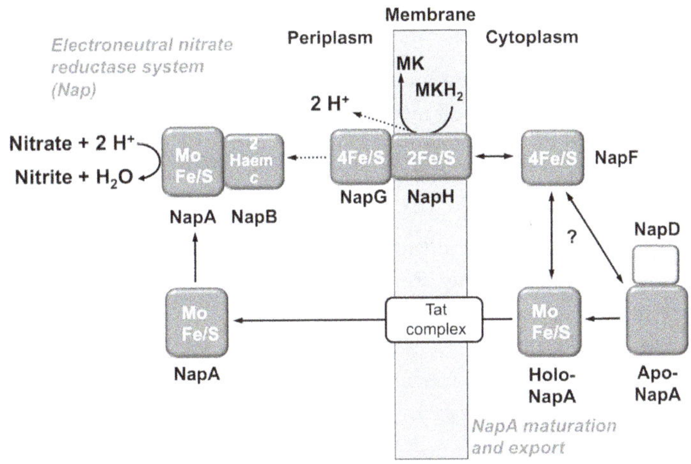

Figure 3.4 The respiratory Nap system of *W. succinogenes*. Nitrate is reduced by a periplasmic NapAB complex that is likely to be reduced by the membrane-bound menaquinol dehydrogenase complex NapGH thereby rendering the Nap system to catalyse electroneutral nitrate reduction by menaquinol (cf. Fig. 3.2C). The molybdopterin guanine dinucleotide cofactor (Mo) and the iron–sulfur centre of NapA are attached in the cytoplasm by specific maturation systems prior to Holo-NapA export by the twin-arginine translocation (Tat) system. The NapD chaperone is required to coordinate maturation and export of NapA. The question mark denotes speculative interactions between NapF and NapA. There are several variations known in proteobacterial Nap systems, many of which involve the alternative quinol dehydrogenase NapC, a member of the NapC/NrfH/Cyt. c_{m552} family that is supposed to transfer electrons to NapB. Fe/S, iron–sulfur centre.

might enable the iron–sulfur centres of NapG and NapH to come sufficiently close to allow electron transport between them. These assumptions would suggest a Q-cycle-like mechanism for the NapGH complex in which two molecules of menaquinol are oxidized at the periplasmic side and one molecule of menaquinone is reduced at the cytoplasmic side of the membrane giving net translocation of two positive charges per menaquinol oxidized.

Recently, *W. succinogenes* NapH was shown to interact with the cytoplasmic iron–sulfur protein NapF which is encoded in many, though not all, *nap* loci (Kern and Simon, 2009b). This finding suggests the formation of a membrane-bound NapGHF complex (Fig. 3.4). As a *napF* deletion mutant was apparently retarded in NapA maturation, it is possible that an electron transfer pathway from NapH via NapF to NapA exists that serves in NapA maturation, for example in reductive activation of newly inserted cofactors. Supporting this hypothesis, NapF from *E. coli* was reported to interact with NapA in the cytoplasm and *Rhodobacter sphaeroides* NapF was proposed to play a role in incorporating the iron–sulfur centre into NapA (Olmo-Mira *et al.*, 2004; Nilavongse *et al.*, 2006).

Despite being absent from *W. succinogenes* and other Epsilonproteobacteria, the NapC protein is an essential component of the Nap system in *E. coli* and other Proteobacteria and was shown to be required for menaquinol oxidation within the Nap system (Potter and Cole, 1999; Brondijk *et al.*, 2002). In addition, the putative iron–sulfur proteins NapG and NapH were postulated to form a ubiquinol dehydrogenase complex in *E. coli* that donates electrons via NapC to NapAB (Brondijk *et al.*, 2002, 2004). This model was deduced mainly from growth experiments performed with mutants that produced only one type of quinone in combination with either NapC or NapGH. It therefore seems that the quinol specificity of NapH is different in *W. succinogenes* and *E. coli* since the former organism does not synthesize ubiquinone. Furthermore, it provides a good example that respiratory systems, especially at the level of quinone/quinol-reactive proteins may differ considerably between bacterial species despite the fact that a similar terminal reductase (like NapA) is employed.

Respiratory nitrite reduction pathways

In the environment, many bacteria predominantly produce nitrite as the major product of nitrate reduction (Gregory *et al.*, 2003). There are, however, two widespread respiratory pathways (used mainly in organotrophic bacteria) that serve to reduce the bioenergetically favourable electron acceptor nitrite (Table 3.1). Depending on the enzymic inventories of such organisms, the corresponding nitrite reduction products are either NO, N_2O and/or N_2 (in denitrification) or ammonium (in respiratory nitrite ammonification).

Denitrification

The denitrification process appears to be mainly associated with free-living bacteria but it is also found in some clinically relevant species like *Pseudomonas aeruginosa*. One of the well-characterized model organisms of denitrification is *Paracoccus denitrificans* that synthesizes a periplasmic electron transfer network in order to reduce nitrite to N_2 in three consecutive steps (according to reactions 3.3–3.5) (Fig. 3.5) (van Spanning *et al.*, 2007).

Denitrifying nitrite reductase:

$$NO_2^- + 1e^- + 2H^+ \rightarrow NO + H_2O \qquad (3.3)$$

Nitric oxide reductase:

$$2NO + 2e^- + 2H^+ \rightarrow N_2O + H_2O \qquad (3.4)$$

Nitrous oxide reductase:

$$N_2O + 2e^- + 2H^+ \rightarrow N_2 + H_2O \qquad (3.5)$$

During heterotrophic growth ubiquinone is reduced using electrons donors like NADH or succinate. In turn, ubiquinol is oxidized by the electrogenic cytochrome bc_1 complex that feeds electrons via its cytochrome *c* subunit into a periplasmic network of small electron transfer proteins. In case of *P. denitrificans*, these are Cyt. c_{550} and a mono-domain cupredoxin called pseudoazurin (Fig. 3.5) (Pearson *et al.*, 2003). In their reduced

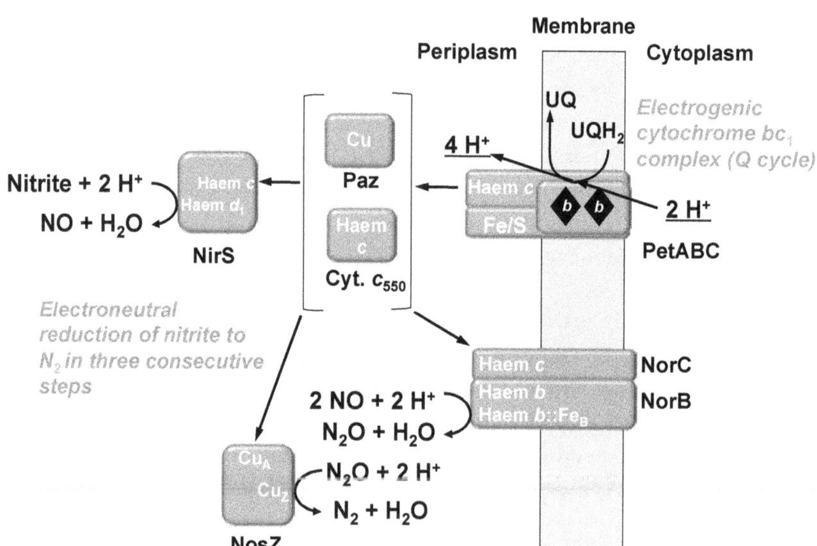

Figure 3.5 Architecture of the electron transport chain that catalyses denitrification of nitrite in *P. denitrificans*. Reactions driven by the oxidation of the ubiquinol pool are shown. The latter is kept reduced by NADH dehydrogenase, succinate dehydrogenase and/or acyl CoA dehydrogenase. Electrons are transferred via the electrogenic cytochrome bc_1 complex to cytochrome cd_1 nitrite reductase (NirS), nitric oxide reductase (NorBC) and nitrous oxide reductase (NosZ). The latter three enzymes catalyse the stepwise reduction of nitrite to N_2 in the periplasm. Note that the physiological form of NosZ is a homodimer. Cu_Z and Cu_A denote the binuclear Cu_A and tetranuclear Cu_Z centres of NosZ. Haem *b*::Fe_B refers to the dinuclear NO-reducing reaction centre of NorBC. Paz, pseudoazurin.

form, such proteins form an electron reservoir from which the denitrifying terminal reductases can draw electrons from. These reductases are periplasmic cytochrome cd_1 nitrite reductase [NirS; catalysing reaction 3.3], membrane-bound nitric oxide reductase [the NorBC complex or cNOR; catalysing reaction 3.4] and periplasmic nitrous oxide reductase [the homodimeric copper protein NosZ; catalysing reaction 3.5] (Fig. 3.5) (Zumft, 2005a,b; Rinaldo and Cutruzzolà, 2007; Zumft and Kroneck, 2007). The three reactions can be regarded as electron sinks as they are not coupled to *pmf* generation. However, it seems vital to reduce nitrite and NO quickly as these compounds have detrimental effects on various cell constituents in higher concentrations (note that *P. denitrificans* produces nitrite from nitrate by both the Nar and Nap systems). On the other hand, several denitrifying bacteria reduce nitrate or nitrite only to the level of N_2O as they do not have a *nosZ* gene. This genetic deficit was proposed to contribute significantly to the release of the potent greenhouse gas N_2O into the atmosphere (Richardson *et al.*, 2009).

There are many variations concerning the organization of denitrifying electron transport chains of different bacteria. Most notably, some bacteria (e.g. *Achromobacter cycloclastes, Rhodobacter sphaeroides, Bradyrhizobium japonicum* and *Hyphomicrobium denitrificans*) use a copper-dependent nitrite reductase (NirK) to catalyse reaction 3.3 (Rinaldo and Cutruzzolà, 2007). As with cytochrome cd_1 nitrite reductase, NirK is reduced either by small *c*-type cytochromes or cupredoxins. Recently, a structural model of the complex between *Achromobacter xylosoxidans* NirK and its cognate redox partner, monohaem cytochrome c_{551}, was presented demonstrating that a hydrophobic electron transport pathway is established at the protein docking interface (Nojiri *et al.*, 2009). Also, there is a widespread subclass of NirK proteins that possess an additional N-terminal copper-containing cupredoxin domain that serves as a fused electron donor (Ellis *et al.*, 2007; Nojiri *et al.*, 2007). Genomic analysis revealed the presence of another variation of this theme, namely a family of NirK proteins harbouring a C-terminal monohaem cytochrome *c* domain of about 160 to 190 amino acid residues which is present for example in *Burkholderia* species and *Bdellovibrio bacteriovorus* (Ellis *et al.*, 2007).

There are also various forms of NO reductases including single subunit (mena)quinol-oxidizing NO reductases (qNOR) known from different organisms including denitrifying and even some ammonifying Epsilonproteobacteria (de Vries *et al.*, 2007; Kern and Simon, 2009a). *P. stutzeri* uses a tetrahaem cytochrome *c* of the NapC/NrfH/Cyt. c_{m552} family (named NirT) as a redox partner for cytochrome cd_1 nitrite reductase (Jüngst *et al.*, 1991). In *Thermus thermophilus* (phylum *Deinococcus–Thermus*) a novel Nar complex was described that, in addition to conventional NarG, -H and -I subunits, uses an additional dihaem cytochrome *c* subunit (NarC) that apparently transfers electrons to the terminal reductases of nitrite denitrification, thereby bypassing the cytochrome bc_1 complex whose synthesis is even repressed under denitrifying conditions (Cava *et al.*, 2008, 2009). Furthermore, a complex between NADH dehydrogenase and this special Nar complex was demonstrated that might play a crucial role in *pmf* generation in *T. thermophilus* (Cava *et al.*, 2008).

Concerning electron transport to the active site of N_2O reductase, an interesting variation was reported to be present in Epsilonproteobacteria which comprise both denitrifying and ammonifying species (Kern and Simon, 2009a). Reminiscent of the situation described above for NirK-type nitrite reductase, the N_2O reductase from *W. succinogenes* was found to contain a C-terminal monohaem cytochrome *c* domain that appears to be a fused redox partner to the Cu_A centre of this enzyme, hence the designation cNosZ (Teraguchi and Hollocher, 1989; Simon *et al.*, 2004). Enzymes of the cNosZ-type are encoded in the genomes of several Epsilonproteobacteria as part of extensive *nos* gene clusters that predict four other proteins (the two monohaem *c*-type cytochromes NosC1 and NosC2 as well as the above mentioned NosH and NosG proteins) that are absent from other proteobacterial *nos* gene clusters (Kern and Simon, 2009a). Fig. 3.6 illustrates the deduced model of what appears to be a typical epsilonproteobacterial electron transport chain connecting the menaquinol pool

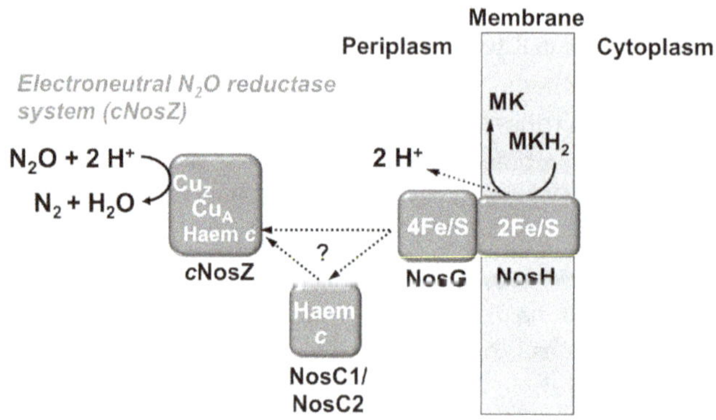

Figure 3.6 Model of the respiratory cNos system that catalyses the reduction of nitrous oxide by menaquinol in *W. succinogenes*. The function of the menaquinol-oxidizing NosGH complex and the electron transfer pathway to cytochrome c nitrous oxide reductase (cNosZ) via NosC1 and NosC2 is speculative. See legends to Figs 3.2 and 3.5 for further explanations.

with periplasmic *c*NosZ. NosG and NosH are expected to form a NosGH complex that is envisaged to be functionally equivalent to NapGH (see above). NosC1 and NosC2 are located either in the periplasm or attached to the membrane via an N-terminal helix (which is hard to distinguish from a Sec-dependent signal peptide by standard bioinformatic protocols). Recently, chimeric *nap/nos* operons were expressed in *W. succinogenes* indicating that a NapG–NosH complex is functional in nitrate respiration whereas NosG was apparently not capable to interact with the NapAB system when expressed in either a *nosGH* or *nosG-napH* arrangement (Kern and Simon, 2008). The NapG-NosH assembly was found to be less stable than the NapGH complex since NapG could be easily washed off the membrane by applying low concentrations of sodium chloride. These results suggest that NapG and NosG represent adaptor proteins that determine the specificity of periplasmic electron transport to either the Nap or the Nos system as shown in Figs 3.4 and 3.6. Remarkably, the polycysteine motif arrangements of NapH/NosH proteins are also present in NosR, a protein encoded in *nos* gene clusters of Proteobacteria like *P. stutzeri* and *P. denitrificans* (Zumft and Kroneck, 2007). *P. stutzeri* NosR was shown to contain an additional N-terminal periplasmic flavin-binding domain that proved essential for N_2O respiration (Wunsch and Zumft, 2005). It cannot be excluded that NosR facilitates electron transport to NosZ in these organisms.

Respiratory nitrite ammonification

Many Bacteria including pathogenic proteobacterial species of the genera *Escherichia, Haemophilus, Salmonella, Pasteurella* and *Campylobacter* catalyse the respiratory reduction of nitrite to ammonium (Simon, 2002; Mohan and Cole, 2007). Increasing interest has been attracted recently to this metabolic pathway since this mode of respiration might play an important role in host invasion and colonization as it enables ATP synthesis in micro-oxic environments such as the intestinal tract (see, for example, Weingarten *et al.*, 2008). In addition, the respiratory nitrite ammonification system is involved in the production or reductive detoxification of reactive or mutagenic nitrogen species such as NO and hydroxylamine (Poock *et al.*, 2002; Rudolf *et al.*, 2002; Poole, 2005; Mills *et al.*, 2008) The respiratory system that catalyses nitrite reduction to ammonium is called the Nrf system (nitrite reduction by formate) and the corresponding terminal reductase is the pentahaem cytochrome *c* nitrite reductase (NrfA) that catalyses reaction 3.6 (Einsle *et al.*, 2002). This enzyme is an exceptional cytochrome *c* whose active site contains a special haem *c* group that is covalently bound via a CX_2CK haem *c* binding motif, as opposed to the standard CX_2CH motif (Einsle *et al.*, 1999, 2000; Clarke *et al.*, 2008).

Ammonifying nitrite reductase:

$$NO_2^- + 6\,e^- + 8H^+ \rightarrow NH_4^+ + 2H_2O \qquad (3.6)$$

Similar to the Nap system described above, two different electron transport pathways have been identified that connect the quinol pool with NrfA. These are either NrfH (in Delta- and Epsilonproteobacteria, Firmicutes and Bacteroidetes) or the NrfD/NrfC proteins (in Gammaproteobacteria) (Fig. 3.2B and D). The electron pathway from NrfD/NrfC to NrfA is mediated by a pentahaem cytochrome *c* (NrfB) which is not related to NrfH (Clarke *et al.*, 2007).

In the following the Nrf system of *W. succinogenes* is described in more detail (Fig. 3.7). The NrfH protein from *W. succinogenes* was shown to form a stable membrane-bound complex with NrfA which is in striking contrast to other members of the NrfH/NapC/Cyt. c_{m552} family that apparently do not form stable membrane-bound complexes with their redox partner proteins (Simon *et al.*, 2000, 2001). The NrfHA complex catalyses menaquinol oxidation by nitrite and reconstitution of the purified enzyme in proteoliposomes confirmed that this reaction is electroneutral (Simon *et al.*, 2000). Non-polar inactivation of the *nrfH* gene in *W. succinogenes* abolished growth by nitrite respiration and resulted in NrfA being located exclusively in the periplasmic space of the mutant strain (Simon *et al.*, 2001). Primary structure alignments and site-directed modification experiments suggested different axial haem *c* iron ligation patterns between the related NapC- and NrfH-type proteins (Gross *et al.*, 2005; Simon and Kern, 2008). The only known high-resolution structure of a NapC/NrfH-type cytochrome *c* is that of *Desulfovibrio vulgaris* NrfH, a subunit of the $(NrfHA_2)_2$ crystal complex (Rodrigues *et al.*, 2006, 2008). *D. vulgaris* NrfH forms an N-terminal (putatively transmembrane) helix and contains a menaquinone-binding site situated close to the first of the four haem *c* groups. These haem *c* groups form a linear electron transfer wire that feeds electrons into the NrfA subunit(s) suggesting that each haem *c* group is functionally essential. This assumption is supported by the finding that individual replacement of each cysteine and histidine residue within the four CX_2CH haem *c* binding motifs of *W. succinogenes* NrfH abolished electron transport from formate to nitrite (Simon *et al.*, 2002; Kern *et al.*, 2008). Notably, two of the four haem *c* groups in *D. vulgaris* NrfH were found to deviate from the standard *bis*-histidine haem *c* iron ligation proposed for NapC-like proteins. Haem 1 has an unusual proximal methionine ligand that derives from a conserved CX_2CHXM sequence and the corresponding methionine residue was shown to be functionally essential in *W. succinogenes* NrfH (Gross *et al.*, 2005). The distal ligand position is occupied by an aspartate residue in *D. vulgaris* NrfH which is not coordinated to the haem 1 iron atom in the structural model (Rodrigues *et al.*, 2006). It is conceivable that such a high-spin

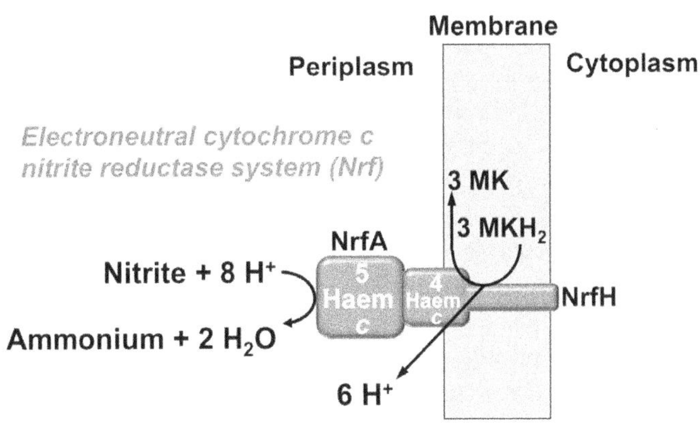

Figure 3.7 The respiratory Nrf system of *W. succinogenes*. The NrfHA complex comprises the tetrahaem cytochrome *c* membrane anchor NrfH and cytochrome *c* nitrite reductase NrfA. Together, they form a membrane-bound menaquinol:nitrite oxidoreductase complex. Protons generated during menaquinol oxidation are liberated on the periplasmic side of the membrane thus making the overall reaction electroneutral.

methionine-coordinated haem has evolved to establish a suitable (i.e. rather positive) midpoint potential of haem 1 in order to favour menaquinol oxidation. Haem 4 of NrfH is histidine/lysine-coordinated, with the distal lysine ligand being provided by an NrfA subunit in the *D. vulgaris* $(NrfHA_2)_2$ crystal structure but it is unknown whether this represents a general feature of NrfHA complexes. Several conserved residues in close vicinity of the menaquinol-binding site of *D. vulgaris* NrfH were replaced in *W. succinogenes* NrfH and the phenotypes of the mutants indicated that the quinone-binding site is conserved in NrfH-type cytochromes (Kern *et al.*, 2008). This view is also supported by a homology model of *W. succinogenes* NrfH that underlines the overall conservation of tertiary structure in spite of a low sequence homology (the NrfH proteins from *W. succinogenes* and *D. vulgaris* share only 23% identical residues) (Kern *et al.*, 2008).

There are some Epsilonproteobacteria that lack the Nrf system despite the fact that they were described as nitrite ammonifying organisms (Kern and Simon, 2009a; Campbell *et al.*, 2009). Intriguingly, the genomes of such organisms (e.g. *Campylobacter concisus, Campylobacter curvus, Caminibacter mediatlanticus* and *Nautilia profundicola*) encode a hydroxylamine oxidoreductase-type enzyme, here called Epsilon-Hao. Although biochemical evidence is not yet available, it is quite likely that this enzyme works *in vivo* as a nitrite reductase that might produce ammonium in the periplasm that subsequently can be taken up by the cells via an ammonium transport system. In this context, Epsilon-Hao enzymes were suggested to represent an evolutionary link between NrfA and Hao enzymes in the phylogenetic tree of multihaem *c*-type cytochromes (Klotz *et al.*, 2008). Notably, each Epsilon-Hao lacks the tyrosine residue that cross-links the monomers in Hao enzymes in nitrifying bacteria (see below). This strengthens the hypothesis that the tyrosine–haem cross-link is a prerequisite to enable oxidative, rather than reductive, chemistry. Interestingly, some, but not all, Epsilon-Hao proteins are encoded within gene arrangements that also comprise a gene encoding a member of the NrfH/NapC/Cyt. c_{m552} family.

Electron transport processes in nitrification

Nitrification comprises the aerobic oxidations of ammonium to nitrite (via hydroxylamine) and of nitrite to nitrate (Ferguson *et al.*, 2007). These reactions are catalysed by chemolithotrophic as well as by heterotrophic bacteria but most model bacteria are actually chemolithoautotrophs belonging to the genera *Nitrosomonas* or *Nitrobacter*. As these organisms grow to rather low cell densities in pure cultures, the biochemistry of electron transport enzymes and the organization of respiratory chains is less well understood when compared to the different forms of anaerobic respiration discussed in the previous sections. In the light of the positive standard potentials of the redox pairs nitrite/ammonium and nitrate/nitrite (Table 3.1), nitrifying autotrophic bacteria also face the problem to produce the NADH (which is needed for CO_2 assimilation) by reversed electron transport. This implies that endergonic NAD^+ reduction by e.g. ubiquinol or cytochrome *c* has to be driven at the expense of the *pmf*. It is assumed, however, that less than 10% of the electrons derived from ammonium oxidation are used in NAD^+ reduction.

Aerobic oxidation of ammonium to nitrite

Nitrosomonas europaea is the best-known bacterial model organism for ammonium oxidation to nitrite (Hooper *et al.*, 2004 and references therein). In the first step of this conversion, ammonium is oxidized to hydroxylamine by ammonium monooxygenase (Amo) that catalyses reaction 3.7. Amo is a membrane-bound copper enzyme (probably a heterotrimeric complex encoded by the *amoCAB* gene cluster) that requires molecular oxygen and two electrons for ammonium oxidation. The electrons are thought to be provided upon ubiquinol oxidation, either directly or via an additional redox mediator yet to be discovered. The enzyme releases hydroxylamine into the periplasmic space where it is further oxidized by hydroxylamine oxidoreductase (Hao) according to reaction 3.8.

Ammonium monooxygenase:

$$NH_4^+ + O_2 + 2e^- + H^+ \rightarrow NH_2OH + H_2O \quad (3.7)$$

Hydroxylamine oxidoreductase:

$$NH_2OH + H_2O \rightarrow NO_2^- + 4\,e^- + 5H^+ \quad (3.8)$$

Hao is a homotrimeric octahaem cytochrome *c* complex that catalyses the four-electron oxidation of hydroxylamine to nitrite (Hooper *et al.*, 2004). Redox partner of Hao is the periplasmic tetrahaem cytochrome c_{554} that probably donates electrons to cytochrome c_{m552}, a membrane-bound ubiquinone reductase of the NapC/NrfH/Cyt. c_{m552} family (Fig. 3.8) (Kim *et al.*, 2008). Thus, hydroxylamine oxidation by ubiquinone is envisaged as an electroneutral electron transport process catalysed by three different multihaem *c*-type cytochromes whose genes (*haoA, cycA* and *cycB*) belong to the same gene cluster. To underline its modular composition, the oxidative part of the electron transport chain is referred to as the hydroxylamine ubiquinone redox module (HURM) (Klotz and Stein, 2008). However, the interaction between cytochrome c_{554} and cytochrome c_{m552} has not been experimentally confirmed. Two molecules of ubiquinol are produced per hydroxylamine oxidized. One of these is needed for ammonium oxidation by Amo while the other enters a *pmf*-generating respiratory electron transport chain that ultimately serves in oxygen reduction catalysed by a haem copper oxidase. Presumably, other constituents of this chain are the cytochrome bc_1 complex and a monohaem cytochrome c_{552} (Fig. 3.8).

Interestingly, *N. europaea* and other nitrifiers are also capable to produce N_2O from nitrite under oxic and hypoxic conditions in a process known as nitrifier denitrification. Under oxic conditions, the release of NO and N_2O is most likely catalysed by Hao while under hypoxic conditions, a NirK-type copper nitrite reductase and a *c*NOR-type nitric oxide reductase are involved. These enzymes might be important both in nitrite and nitric oxide detoxification as well as in redox balancing processes during the transition from oxic to anoxic environments. NirK is thought to use cytochrome c_{552} as electron donor that was previously reduced by the cytochrome bc_1 complex. Thus, in analogy to denitrification, ubiquinol oxidation by nitrite or nitric oxide in nitrifiers would be electrogenic due to the Q-cycle mechanism of the cytochrome bc_1 complex.

Aerobic nitrite oxidation to nitrate

Model bacteria for this type of chemolithotrophic lifestyle are species of the genera *Nitrobacter* (phylum Proteobacteria) and *Nitrospira* (phylum Nitrospirae). In both cases, our picture of the organization of the short electron transport chain from nitrite to oxygen is far from complete. Given the positive standard potentials of the

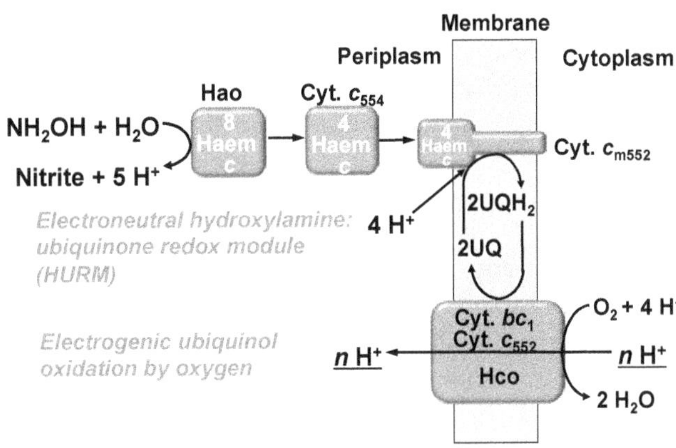

Figure 3.8 Model of the respiratory electron transport chain from hydroxylamine to oxygen in *N. europaea*. Hydroxylamine oxidoreductase is a periplasmic homotrimer whose subunits are covalently linked between a tyrosine residue and the active site haem *c* group of the neighbouring monomer. Electrons derived from hydroxylamine oxidation are transferred via cytochrome c_{554} to the ubiquinone-reducing membrane-bound cytochrome c_{m552}, a member of the NapC/NrfH/Cyt. c_{m552} family. Half of the generated ubiquinol molecules are oxidized by an electrogenic aerobic respiratory chain comprising the cytochrome bc_1 complex, one or more cytochromes *c* and a terminal haem copper oxidase (Hco) while the other half is probably oxidized by Amo (not shown).

involved redox pairs (Table 3.1), it is unlikely that a quinone is involved. Instead, in *Nitrobacter winogradsky*, nitrite is assumed to be oxidized by a membrane-associated nitrite oxidoreductase complex (NxrAB, formerly abbreviated NorAB) that catalyses reaction 3.9 and transfers electrons to one of several periplasmic high-potential cytochromes c_{550} that probably act as electron donor to a *pmf*-generating terminal haem copper oxidase of the aa_3-type (Nomoto *et al.*, 1993).

Nitrite oxidoreductase:

$$NO_2^- + H_2O \rightarrow NO_3^- + 2e^- + 2H^+ \quad (3.9)$$

The NxrAB complex contains molybdenum and iron and, strikingly, the primary structure of the NxrA and NxrB proteins is related to the NarG and NarH subunits of the membrane-bound nitrate reductase (Starkenburg *et al.*, 2006). Most likely, NxrA and NxrB are located at the cytoplasmic surface of the membrane but it is unknown how electrons are delivered across the membrane in order to reduce cytochrome c_{550}. This pathway might involve a NarI-like membrane-bound dihaem cytochrome *b* and/or another *c*-type cytochrome (Starkenburg *et al.*, 2006). Furthermore, cytoplasmic nitrite oxidation necessitates nitrite import and nitrate export which might be catalysed by specific transporters of the proteobacterial NarK-type that are encoded in the genome of *N. winogradsky*. The genome also encodes a pathway for reversed electron transport from cytochrome c_{550} to NADH via the cytochrome bc_1 complex and a NADH:ubiquinone oxidoreductase.

Conclusions and perspectives

It is astonishing that the large variety of highly optimized electron transport enzymes discussed in this article is electronically connected by only a few types of soluble electron transfer proteins (mainly small *c*-type cytochromes and cupredoxins) and a limited number of quinone/quinol-reactive electron transport modules. This raises the question as to the specificity of such proteins in terms of their participation in distinct electron transport chains. In this respect, quinone/quinol-reactive proteins seem to interact with cognate redox partners with whom they might form stable or loosely associated complexes. On the other hand, small *c*-type cytochromes and cupredoxins are apparently rather promiscuous in electron transfer to different redox partners, thus forming the central part of what seems to be a periplasmic electron exchange network from which different enzymes draw electrons to enable efficient turnover of available substrates. Apparently, this strategy is particularly successful in denitrification which relies on concomitant electron supply to up to three periplasmic enzymes. However, one has to keep in mind that many molecular interactions discussed in this chapter have not been demonstrated biochemically but rather were predicted as a result of the individual characterization of potentially interacting enzymes or from the composition of gene loci that encode them (derived from either complete genomes or from metagenomic data). Especially in the light of the emerging wealth of genomic data, it is critical that modelling electron transfer chains or networks is not exclusively based on genetic information alone (despite the fact that many models established in this way proved to be correct later on). One also has to realize that only a very small proportion of the known (and cultivable) bacteria has been investigated in any depth and surely we can await many surprises in the future in terms of the biochemical characterization of respiratory electron transport chains in bacteria like e.g. *Nitrospira* spp. or members of the NC10 division (e.g. *Candidatus Methylomirabilis oxyfera*) that have been described more recently (Spieck *et al.*, 2006; Ettwig *et al.*, 2010, and references therein). The latter organism was shown to couple methane oxidation to denitrification and apparently employs a hitherto completely unknown oxygenic pathway that involves the reaction of two molecules of nitric oxide to yield both N_2 and O_2 (an NO dismutase reaction). Furthermore, crystal structures of many key enzymes discussed in this article are still lacking, e.g. those of Amo, Nxr, Epsilon-Hao or NapC.

Looking at the bioenergetic principles discussed above, it is surprising that many highly exergonic reactions are not directly connected to *pmf* generation but rather serve in replenishment of the quinone/quinol pool (e.g. quinol oxidation by the periplasmic Nrf system where nitrite merely serves as electron sink). By using such electron transfer systems, a substantial amount of free energy is not conserved in ATP synthesis but, obviously, microorganisms tolerate this inefficiency in favour of fast electron transport from reductant to oxidant. Another advantage of periplasmic respiratory systems is that dissolved substrates do not have to be taken up into the cytoplasmic space by potentially energy-consuming transport processes. Therefore, when new models of electron transport chains are proposed, one has to keep in mind that bioenergetic systems are obviously not optimized to obtain maximal H^+/e^- ratios.

Finally, one of the most discussed challenges in modern microbiology is the rational design of synthetic microorganisms intended to provide potential benefits in e.g. healthcare and environmental microbiology. Creating such organisms clearly needs a proper design of their energy metabolism (especially when using respiratory rather than fermentative pathways) and to this end a good starting point could be to test appropriate electron transport modules for their capacity to exchange electrons with suitable substrate-converting enzymes. Such a scenario, however, needs more basic research dealing with the organization of respiratory electron transport chains in a diverse range of microorganisms. Furthermore, a more detailed knowledge on enzyme maturation processes, synthesis of co-factors and their insertion into proteins as well as of transport processes of apo- or holoproteins across membranes is required. At present, our picture of these fields is still rather sketchy which also hampers to establish efficient heterologous production systems for metalloenzymes or enzyme complexes in suitable host bacteria. Ultimately, the combined knowledge may lead to the *de novo* construction of organisms whose energy metabolism meets favoured requirements in industrial and environmental applications.

References

Baar, C., Eppinger, M., Raddatz, G., Simon, J., Lanz, C., Klimmek, O., Nandakumar, R., Gross, R., Rosinus, A., Keller, H., *et al.* (2003). Complete genome sequence and analysis of *Wolinella succinogenes*. Proc. Natl. Acad. Sci. U.S.A. *100*, 11690–11695.

Bothe, H., Ferguson, S.J., and Newton, W.E. (2007). Biology of the Nitrogen Cycle (Amsterdam: Elsevier).

Brondijk, T.H.C., Fiegen, D., Richardson, D.J., and Cole, J.A. (2002). Roles of NapF, NapG and NapH, subunits of the *Escherichia coli* periplasmic nitrate reductase, in ubiquinol oxidation. Mol. Microbiol. *44*, 245–255.

Brondijk, T.H.C., Nilavongse, A., Filenko, N., Richardson, D.J., and Cole, J.A. (2004). NapGH components of the periplasmic nitrate reductase of *Escherichia coli* K-12: location, topology and physiological roles in quinol oxidation and redox balancing. Biochem. J. *379*, 47–55.

Campbell, B.J., Smith, J.L., Hanson, T.E., Klotz, M.G., Stein, L.Y., Lee, C.K., Wu, D., Robinson, J.M., Khouri, H.M., Eisen, J.A., and Cary, S.C. (2009). Adaptations to submarine hydrothermal environments exemplified by the genome of *Nautilia profundicola*. PLoS Genet. *5*, e1000362

Cartron, M.L., Roldán, M.D., Ferguson, S.J., Berks, B.C., and Richardson, D.J. (2002). Identification of two domains and distal histidine ligands to the four haems in the bacterial *c*-type cytochrome NapC: the prototype connector between quinol/quinone and periplasmic oxido-reductases. Biochem. J. *368*, 425–432.

Cava, F., Zafra, O., and Berenguer, J. (2008). A cytochrome *c* containing nitrate reductase plays a role in electron transport for denitrification in *Thermus thermophilus* without involvement of the respiratory *bc* respiratory complex. Mol. Microbiol. *70*, 507–518.

Cava, F., Hidalgo, A., and Berenguer, J. (2009). *Thermus thermophilus* as biological model. Extremophiles *13*, 213–231.

Clarke, T.A., Cole, J.A., Richardson, D.J., and Hemmings, A.M. (2007). The crystal structure of the pentahaem *c*-type cytochrome NrfB and characterization of its solution–state interaction with the pentahaem nitrite reductase NrfA. Biochem. J. *406*, 19–30.

Clarke, T.A., Mills, P.C., Poock, S.R., Butt, J.N., Cheesman, M.R., Cole, J.A., Hinton, J.C., Hemmings, A.M., Kemp, G., Söderberg, C.A., Spiro, S., van Wonderen, J., and Richardson D.J. (2008). *Escherichia coli* cytochrome *c* nitrite reductase NrfA, Meth. Enzymol. 437, 63–77.

de Vries, S., Suharti, and Pouvreau, L.A.M. (2007). Nitric oxide reductase: Structural variations and catalytic mechanism. In Biology of the Nitrogen Cycle, Bothe, H., Ferguson, S.J., and Newton, W.E., eds. (Amsterdam: Elsevier), pp. 57–66.

Einsle, O., Messerschmidt, A., Stach, P., Bourenkov, G.P., Bartunik, H.D., Huber, R., and Kroneck P.M.H. (1999). Structure of cytochrome *c* nitrite reductase. Nature *400*, 476–480.

Einsle, O., Stach, P., Messerschmidt, A., Simon, J., Kröger, A., Huber, R., and Kroneck, P.M.H. (2000). Cytochrome *c* nitrite reductase from *Wolinella succinogenes*. Structure at 1.6 Å resolution, inhibitor binding, and heme-packing motifs. J. Biol. Chem. *275*, 39608–39616

Einsle, O., Messerschmidt, A., Huber, R., Kroneck, P.M.H., and Neese F. (2002). Mechanism of the six-electron reduction of nitrite to ammonia by cytochrome *c* nitrite reductase. J. Am. Chem. Soc. *124*, 11737–11745.

Einsle, O., and Kroneck, P.M.H. (2004). Structural basis of denitrification. Biol. Chem. *385*, 875–883.

Ellis, M.J., Grossmann, J.G., Eady, R.R., and Hasnain, S.S. (2007). Genomic analysis reveals widespread occurrence of new classes of copper nitrite reductases. J. Biol. Inorg. Chem. *12*, 1119–1127.

Ettwig, K.F., Butler, M.K., Le Paslier, D., Pelletier, E., Mangenot, S., Kuypers, M.M., Schreiber, F., Dutilh, B.E., Zedelius, J., de Beer, D., *et al.* (2010). Nitrite-driven anaerobic methane oxidation by oxygenic bacteria. Nature *464*, 543–548.

Ferguson, S.J., Richardson, D.J., and van Spanning, R.J.M. (2007). Biochemistry and molecular biology of nitrification. In Biology of the Nitrogen Cycle, Bothe, H., Ferguson, S.J., and Newton, W.E., eds. (Amsterdam: Elsevier), pp. 209–222.

Fritz, G., Einsle, O., Rudolf, M., Schiffer, A., and Kroneck P.M.H. (2005). Key bacterial multi-centered metal enzymes involved in nitrate and sulfate respiration. J. Mol. Microbiol. Biotechnol. *10*, 223–233.

Gregory, L.G., Bond, P.L., Richardson, D.J., and Spiro, S. (2003). Characterization of a nitrate-respiring bacterial community using the nitrate reductase gene (*narG*) as a functional marker. Microbiology *149*, 229–237.

Gross, R., Pisa, R., Sänger, M., Lancaster, C.R.D., and Simon, J. (2004). Characterization of the menaquinone reduction site in the diheme cytochrome *b* membrane anchor of *Wolinella succinogenes* NiFe-hydrogenase. J. Biol. Chem. *279*, 274–281.

Gross, R., Eichler, R., and Simon J. (2005). Site-directed modifications indicate differences in axial haem *c* iron ligation between the related NrfH and NapC families of multihaem *c*-type cytochromes. Biochem. J. *390*, 689–693.

Hooper, A.B., Arciero, D., Bergmann, D., and Hendrich, M.P. (2004). The oxidation of ammonia as an energy source in Bacteria. In: Respiration in Archaea and Bacteria. Vol. 2: Diversity of prokaryotic respiratory systems, Zannoni, D., ed. (Dordrecht: Springer), pp. 121–147.

Hussain, H., Grove, J., Griffiths, L., Busby, S., and Cole, J. (1994). A seven-gene operon essential for formate-dependent nitrite reduction to ammonia by enteric bacteria. Mol. Microbiol. *12*, 153–163.

Igarashi, N., Moriyama, H., Fujiwara, T., Fukumori, Y., and Tanaka, N. (1997). The 2.8 Å structure of hydroxylamine oxidoreductase from a nitrifying chemolithotrophic bacterium, *Nitrosomonas europaea*. Nat. Struct. Biol. *4*, 276–284.

Jormakka, M., Yokoyama, K., Yano, T., Tamakoshi, M., Akimoto, S., Shimamura, T., Curmi, P., and Iwata, S. (2008). Molecular mechanism of energy conversion in polysulfide respiration. Nat. Struct. Mol. Biol. *15*, 730–737.

Jüngst, A., Wakabayashi, S., Matsubara, H., and Zumft, W.G. (1991). The *nirSTBM* region coding for cytochrome cd_1-dependent nitrite respiration of *Pseudomonas stutzeri* consists of a cluster of mono-, di-, and tetraheme proteins. FEBS Lett. *279*, 205–209.

Kern, M., Mager, A.M., and Simon, J. (2007). Role of individual *nap* gene cluster products in NapC-independent nitrate respiration of *Wolinella succinogenes*. Microbiology *153*, 3739–3747.

Kern, M., and Simon, J. (2008). Characterization of the NapGH quinol dehydrogenase complex involved in *Wolinella succinogenes* nitrate respiration. Mol. Microbiol. *69*, 1137–1152.

Kern, M., Einsle, O., and Simon J. (2008). Variants of the tetrahaem cytochrome *c* quinol dehydrogenase NrfH characterize the menaquinol binding site, the haem *c* binding motifs and the transmembrane segment. Biochem. J. *414*, 73–79.

Kern, M., and Simon, J. (2009a). Electron transport chains and bioenergetics of respiratory nitrogen metabolism in *Wolinella succinogenes* and other Epsilonproteobacteria. Biochim. Biophys. Acta *1787*, 646–656.

Kern, M., and Simon, J. (2009b). Periplasmic nitrate reduction in *Wolinella succinogenes*: cytoplasmic NapF facilitates NapA maturation and requires the menaquinol dehydrogenase NapH for membrane attachment. Microbiology *155*, 2784–2794.

Kim, H.J., Zatsman, A., Upadhyay, A.K., Whittaker, M., Bergmann, D., Hendrich, M.P., and Hooper, A.B. (2008). Membrane tetraheme cytochrome c_{m552} of the ammonia-oxidizing *Nitrosomonas europaea*: a ubiquinone reductase. Biochemistry *47*, 6539–6551.

Klimmek, O., Dietrich, W., Dancea, F., Lin, Y.-J., Pfeiffer, S., Löhr, F., Rüterjans, H., Gross, R., Simon, J., and Kröger A. (2004). Sulfur respiration. In: Respiration in Archaea and Bacteria. Vol. 2: Diversity of prokaryotic respiratory systems, Zannoni, D., ed. (Dordrecht: Springer), pp. 217–232.

Klotz, M.G., and Stein, L.Y. (2008). Nitrifier genomics and evolution of the nitrogen cycle. FEMS Microbiol. Lett. *278*, 146–156.

Klotz, M.G., Schmid, M.C., Strous, M., op den Camp, H.J.M., Jetten, M.S.M., and Hooper A.B. (2008). Evolution of an octahaem cytochrome *c* protein family that is key to aerobic and anaerobic ammonia oxidation by bacteria. Env. Microbiol. *10*, 3150–3158.

Kuenen, J.G. (2008). Anammox bacteria: from discovery to application. Nat. Rev. Microbiol. *6*, 320–326.

Marietou, A., Richardson, D., Cole, J., and Mohan, S. (2005). Nitrate reduction by *Desulfovibrio desulfuricans*: a periplasmic nitrate reductase system that lacks NapB, but includes a unique tetraheme *c*-type cytochrome, NapM. FEMS Microbiol. Lett. *248*, 217–225.

Maillard, J., Spronck, C.A., Buchanan, G., Lyall, V., Richardson, D.J., Palmer, T., Vuister, G.W., and Sargent F. (2007). Structural diversity in twin-arginine signalpeptide-binding proteins. Proc. Natl. Acad. Sci. U.S.A. *104*, 15641–15646.

Mills, P.C., Rowley, G., Spiro, S., Hinton, J.C.D., and Richardson D.J. (2008). A combination of cytochrome *c* nitrite reductase (NrfA) and flavorubredoxin (NorV) protects *Salmonella enterica* serovar Typhimurium against killing by NO in anoxic environments. Microbiology *154*, 1218–1228.

Mohan, S.B., and Cole, J.A. (2007). The dissimilatory reduction of nitrate to ammonia by anaerobic bacteria. In Biology of the Nitrogen Cycle, Bothe, H., Ferguson, S.J., and Newton, W.E., eds. (Amsterdam: Elsevier), pp. 93–106.

Nilavongse, A., Brondijk, T.H.C., Overton, T.W., Richardson, D.J., Leach, E.R., and Cole, J.A. (2006). The NapF protein of the *Escherichia coli* periplasmic nitrate reductase system: demonstration of a cytoplasmic location and interaction with the catalytic subunit, NapA. Microbiology *152*, 3227–3237.

Nojiri, M., Xie, Y., Inoue, T., Yamamoto, T., Matsumura, H., Kataoka, K., Deligeer, Yamaguchi, K., Kai, Y., and Suzuki, S. (2007). Structure and function of a hexameric copper-containing nitrite reductase. Proc. Natl. Acad. Sci. U.S.A. *104*, 4315–4320.

Nojiri, M., Koteishi, H., Nakagami, T., Kobayashi, K., Inoue, T., Yamaguchi, K., and Suzuki, S. (2009). Structural basis of inter-protein electron transfer for nitrite reduction in denitrification. Nature *462*, 117–120.

Nomoto, T., Fukumori, Y., and Yamanaka, T. (1993). Membrane-bound cytochrome *c* is an alternative electron donor for cytochrome aa_3 in *Nitrobacter winogradsky*. J. Bacteriol. *175*, 4400–4404.

Olmo-Mira, M.F., Gavira, M., Richardson, D.J., Castillo, F., Moreno-Vivián, C., and Roldán, M.D. (2004). NapF is a cytoplasmic iron–sulfur protein required for Fe–S cluster assembly in the periplasmic nitrate reductase. J. Biol. Chem. *279*, 49727–49735.

Pearson, I.V., Page, M.D., van Spanning, R.J.M., and Ferguson, S.J. (2003). A mutant of *Paracoccus denitrificans* with disrupted genes coding for cytochrome c_{550} and pseudoazurin establishes these two proteins as the *in vivo* electron donors to cytochrome cd_1 nitrite reductase. J. Biol. Chem. *185*, 6308–6315.

Poock, S.R., Leach, E.R., Moir, J.W.B., Cole, J.A., and Richardson, D.J. (2002). Respiratory detoxification of nitric oxide by the cytochrome *c* nitrite reductase of *Escherichia coli*. J. Biol. Chem. *277*, 23664–23669.

Poole, R.K. (2005). Nitric oxide and nitrosative stress tolerance in bacteria. Biochem. Soc. Trans. *33*, 176–180.

Potter, L.C., and Cole J.A. (1999). Essential roles for the products of the *napABCD* genes, but not *napFGH*, in periplasmic nitrate reduction by *Escherichia coli* K-12. Biochem. J. *344*, 69–76.

Potter, L., Angove, H., Richardson, D., and Cole, J. (2001). Nitrate reduction in the periplasm of gram-negative bacteria. Adv. Microb. Physiol. *45*, 51–112.

Richardson, D.J. (2000). Bacterial respiration: a flexible process for a changing environment. Microbiology *146*, 551–571.

Richardson, D.J., Berks, B.C., Russell, D.A., Spiro, S., and Taylor, C.J. (2001). Functional, biochemical and genetic diversity of prokaryotic nitrate reductases. Cell. Mol. Life Sci. *58*, 165–178.

Richardson, D., and Sawers G. (2002). PMF through the redox loop. Science *295*, 1842–1843.

Richardson, D.J., Felgate, H., Watmough, N., Thomson, A., and Baggs, E. (2009). Mitigating release of the potent greenhouse gas N_2O from the nitrogen cycle – could enzymic regulation hold the key? Trends Biotechnol. *27*, 388–397.

Rinaldo, S., and Cutruzzolà, F. (2007). Nitrite reductases in denitrification. In Biology of the Nitrogen Cycle, Bothe, H., Ferguson, S.J., and Newton, W.E., eds. (Amsterdam: Elsevier), pp. 37–55.

Rodrigues, M.L., Oliveira, T.F., Pereira, I.A.C., and Archer, M. (2006). X-ray structure of the membrane-bound cytochrome *c* quinol dehydrogenase NrfH reveals novel haem coordination. EMBO J. *25*, 5951–5960.

Rodrigues, M.L., Scott, K.A., Sansom, M.S., Pereira, I.A.C., and Archer M. (2008). Quinol oxidation by *c*-type cytochromes: structural characterization of the menaquinol binding site of NrfHA. J. Mol. Biol. *381*, 341–350.

Rudolf, M., Einsle, O., Neese, F., and Kroneck, P.M.H. (2002). Pentahaem cytochrome *c* nitrite reductase: reaction with hydroxylamine, a potential reaction intermediate and substrate. Biochem. Soc. Trans. *30*, 649–653.

Rudolf, M., and Kroneck, P.M.H. (2005). The nitrogen cycle: its biology. Met. Ions Biol. Syst. *43*, 75–103.

Sargent, F. (2007). The twin-arginine transport system: moving folded proteins across membranes. Biochem. Soc. Trans. *35*, 835–847.

Simon, J., Gross, R., Einsle, O., Kroneck, P.M.H., Kröger, A., and Klimmek O. (2000). A NapC/NirT-type cytochrome *c* (NrfH) is the mediator between the quinone pool and the cytochrome *c* nitrite reductase of *Wolinella succinogenes*. Mol. Microbiol. *35*, 686–696.

Simon, J., Pisa, R., Stein, T., Eichler, R., Klimmek, O., and Gross, R. (2001). The tetraheme cytochrome *c* NrfH is required to anchor the cytochrome *c* nitrite reductase (NrfA) in the membrane of *Wolinella succinogenes*. Eur. J. Biochem. *268*, 5776–5782.

Simon, J. (2002). Enzymology and bioenergetics of respiratory nitrite ammonification. FEMS Microbiol. Rev. *26*, 285–309.

Simon, J., Eichler, R., Pisa, R., Biel, S., and Gross, R. (2002). Modification of haem *c* binding motifs in the small subunit (NrfH) of the *Wolinella succinogenes* cytochrome *c* nitrite reductase complex. FEBS Lett. *522*, 83–87.

Simon, J., Sänger, M., Schuster, S.C., and Gross R. (2003). Electron transport to periplasmic nitrate reductase (NapA) of *Wolinella succinogenes* is independent of a NapC protein. Mol. Microbiol. *49*, 69–79.

Simon, J., Einsle, O., Kroneck, P.M.H., and Zumft W.G. (2004). The unprecedented *nos* gene cluster of *Wolinella succinogenes* encodes a novel respiratory electron transfer pathway to cytochrome *c* nitrous oxide reductase. FEBS Lett. *569*, 7–12.

Simon, J., and Kern M. (2008). Quinone-reactive proteins devoid of haem *b* form widespread membrane-bound electron transport modules in bacterial anaerobic respiration. Biochem. Soc. Trans. *36*, 1011–1016.

Simon, J., van Spanning, R.J.M., and Richardson D.J. (2008). The organisation of proton motive and non-proton motive redox loops in prokaryotic respiratory systems. Biochim. Biophys. Acta *1777*, 1480–1490.

Spieck, E., Hartwig, C., McCormack, I., Maixner, F., Wagner, M., Lipski, A., and Daims, H. (2006). Selective enrichment and molecular characterization of a previously uncultured *Nitrospira*-like bacterium from activated sludge. Environ. Microbiol. *8*, 405–415.

Starkenburg, S.R., Chain, P.S.G., Sayavedra-Soto, L.A., Hauser, L., Land, M.L., Larimer, F.W., Malfatti, S.A., Klotz, M.G., Bottomly, P.J., Arp, D.J., and Hickey, W.J. (2006). Genome sequence of the chemolithotrophic nitrite-oxidizing bacterium *Nitrobacter winogradsky* Nb-255. Appl. Environ. Microbiol. 72, 2050–2063.

Tavares, P., Pereira, A.S., Moura, J.J.G., and Moura, I. (2006). Metalloenzymes of the denitrification pathway. J. Inorg. Biochem. *100*, 2087–2100.

Teraguchi, S., and Hollocher T.C. (1989). Purification and some characteristics of a cytochrome *c*-containing nitrous oxide reductase from *Wolinella succinogenes*. J. Biol. Chem. *264*, 1972–1979.

van Spanning, R.J.M., Richardson, D.J., and Ferguson, S.J. (2007). Introduction to the biochemistry and molecular biology of denitrification. In Biology of the Nitrogen Cycle, Bothe, H., Ferguson, S.J., and Newton, W.E., eds. (Amsterdam: Elsevier), pp. 3–20.

Weingarten, R.A., Grimes, J.L., and Olson J.W. (2008). Role of *Campylobacter jejuni* oxidases and reductases in host colonization. Appl. Environ. Microbiol. *74*, 1367–1375.

Wunsch, P., and Zumft W.G. (2005). Functional domains of NosR, a novel transmembrane iron–sulfur flavoprotein necessary for nitrous oxide respiration. J. Bacteriol. *187*, 1992–2001.

Zumft, W.G. (1997). Cell biology and molecular basis of denitrification. Microbiol. Mol. Biol. Rev. *61*, 533–616.

Zumft, W.G. (2005a). Nitric oxide reductases of prokaryotes with emphasis on the respiratory, heme-copper oxidase type. J. Inorg. Chem. *99*, 194–215.

Zumft, W.G. (2005b). Biogenesis of the bacterial respiratory Cu_A, Cu-S enzyme nitrous oxide reductase. J. Mol. Microbiol. Biotechnol. *10*, 154–166.

Zumft, W.G., and Kroneck P.M.H. (2007). Respiratory transformation of nitrous oxide (N_2O) to dinitrogen by *Bacteria* and *Archaea*. Adv. Microb. Physiol. *52*, 107–227.

Biochemistry of Mo-Nitrogenase

4

John W. Peters, Eric S. Boyd, Trinity Hamilton and Luis M. Rubio

Abstract

The large majority of biological nitrogen fixation occurs by the activity of Mo-nitrogenase. Mo-nitrogenase is found in a wide variety of bacteria and some archaea and is a complex two component enzyme that contains multiple metal-containing prosthetic groups. The biochemistry of nitrogenase has a rich history and the enzyme is a model system for examining more general processes in biology such as electron transfer, metal-cofactor assembly, and even nucleotide dependent signal transduction. In addition, studies examining nitrogenase have pushed the envelope in terms of the practical application of various spectroscopic methods. This chapter treats the historical perspective and development of key advances in our understanding of the biochemistry of Mo-nitrogenase from its infancy and early beginnings in the 1960s to the state of the field today.

Introduction

The largest reservoir of nitrogen on the planet is in the form of dinitrogen gas (N_2), although this form is bio-unavailable. The nitrogenase enzyme evolved to convert bio-unavailable N_2 gas to bio-available NH_3, and is considered to be one of the most fundamental biochemical innovations in the evolution of life on earth. Detailed investigations into the biochemistry of nitrogenase began to be reported frequently in the late 1960s and early 1970s. However, understanding nitrogenase biochemistry has been a challenge given the sensitivity of the nitrogenase complex to oxygen and the fact that the complex exists as two easily separable components which have no measurable enzymatic activity on their own. Although biological nitrogen fixation is commonly associated by many as being a property of species of legume symbionts such as *Rhizobium* spp. and *Bradyrhizobium* spp., the process is significantly more widely distributed in nature and is found in environments ranging from the marine water column (Zehr and Capone, 2002) to freshwater environments (Affourtit *et al.*, 2001). Restricted to prokaryotes, nitrogen fixation is a property of certain obligate aerobes (e.g. *Azotobacter vinelandii*), obligate anaerobes (e.g. *Clostridium pasteurianum*), facultative anaerobes (e.g. *Klebsiella pneumoniae*), as well as some microaerophiles (e.g. *Rhizobium meliloti*). In addition, nitrogen-fixation is a common process among anoxygenic and oxygenic photosynthetic organisms such as *Rhodopseudomanas* spp. and *Cyanothece* spp., respectively. As evidence of the commonality of *nif* in oxygenic phototrophs, *nif* genes have been identified in 21 of the 44 sequenced cyanobacterial genomes, including those that inhabit terrestrial environments (e.g. *Cyanothece* and *Synechococcus* strains) and those that inhabit the marine water column (*Crocosphaera watsonii*). In addition, *nif* regulons are common in the *Firmicutes, Chloroflexi, Chlorobi*, and *Bacteroidetes* as well as in several lineages of *Actinobacteria* and *Proteobacteria*. In contrast with the widespread distribution of *nif* among bacteria, the ability to fix nitrogen in archaea is highly restricted with demonstrated evidence for *nif* catalysed N_2 fixation only in the methanogenic archaeal orders including the *Methanococcales* and *Methanosarcinales* (Leigh, 2000; Raymond *et al.*,

2004). Interestingly, homologues of nitrogenase-encoding genes have yet to be identified in the genomes of the Crenarchaeota.

The majority of present-day N_2 reduction is catalysed by the molybdenum-dependent nitrogenase (Mo-nitrogenase or *nif*) (Burgess and Lowe, 1996). Alternative (Mo-independent) forms of nitrogenase, which contain vanadium and iron (V-nitrogenase or *vnf*) or iron alone (Fe-nitrogenase or *anf*) in the active site, are found in a limited subset of diazotrophs. These systems were discovered in the mid to late 1980s on investigating nitrogen fixation of several microbial species (*A. chroococcum, A. vinelandii,* and *Rhodobacter capsulatus*) under conditions of Mo starvation (Chisnell *et al.*, 1988; Hales *et al.*, 1986; Robson *et al.*, 1986; Schneider *et al.*, 1991). Sequences of alternative forms of nitrogenase (*anf*/*vnf*) are found in both bacteria and archaea, but like *nif* have yet to be identified in eukaryotes. Both *anf* and *vnf* nitrogenase are present in the genomes of several Methanosarcina strains, but have yet to be identified in a genome from a member of the *Methanococcales* or *Methanobacteriales*. *anf* regulons are also present in the several phylogenetically disparate bacterial genomes including a single cyanobacterial genome (*Anabaena variabilis*), a single chlorobi genome (*Chloroherpeton thalassium*), several firmicutes genomes, several proteobacterial genomes (*Rhodospirillum* spp. and *Rhodopseudomonas* spp.), as well as an uncharacterized Opitutae strain. The distribution of *vnf* regulons in bacteria is constrained to several proteobacterial and firmicute genomes, and *A. variabilis*. In addition to *nif*, *vnf*, and *anf* forms of nitrogenase, several nitrogenase homologues with uncharacterized active site cluster metal content have been identified in the genomes of the firmicutes *Candidatus Desulforudis audaxviator* MP104C and *Caldicellulosiruptor saccharolyticus*, several unclassified methanogen genomes, and two strains of *Roseiflexus* (*Chloroflexi*) (Soboh *et al.*, 2010). The nitrogenases with uncharacterized active site metal contents from the methanogen strains have recently been shown to actively catalyse nitrogen fixation. One of the strains inhabits marine sediments and grows in syntrophy with a sulfate reducing bacterium (Dekas *et al.*, 2009) and the other strain inhabits deep sea hydrothermal vents (Mehta and Baross, 2006).

Nitrogenase exists as two separable iron–sulfur cluster-containing protein components that act in concert to catalyse the multiple electron reduction of dinitrogen gas (N_2) to ammonia (previously reviewed in Burgess and Lowe, 1996; Hoffman *et al.*, 2009; Igarashi and Seefeldt, 2003; Lee and Holm, 2004; Peters *et al.*, 1995a; Peters and Szilagyi, 2006; Rees and Howard, 2000; Seefeldt *et al.*, 2009). All forms of nitrogenase (Nif, Anf, Vnf) contain oxygen sensitive iron–sulfur clusters and assays for enzymatic activity require strict anaerobic conditions (Chisnell *et al.*, 1988; Eady *et al.*, 1972, 1987; Hales *et al.*, 1986; Schneider *et al.*, 1994, 1997). Interestingly, as mentioned above, there are several examples of organisms that require oxygen for growth, concomitant with nitrogen-fixation activity. *A. vinelandii* uses a strategy for protecting nitrogenase from oxygen inactivation during aerobic growth while actively fixing nitrogen termed respiratory protection that involves active consumption of oxygen by a specific *bd*-type cytochrome oxidase (Poole and Hill, 1997). In addition, *A. vinelandii* has conformational protection of nitrogenase by a specific iron–sulfur protein called the Shethna protein (Maier and Moshiri, 2000) and there is evidence that this mode of oxygen protection may be present in other organisms (Ureta and Nordlund, 2002). In the case of the symbiotic nitrogen fixers (Vance, 2001) that grow in the nodules of leguminous plants, the plant nodule itself is an essentially anaerobic environment and although the microaerophilic nitrogen fixers need oxygen for growth the concentration of free oxygen is kept at a very low level. This is achieved in plant nodules through the production of a protein called leghaemoglobin by the plant, a protein that binds oxygen in much the same manner as haemoglobin (Downie, 2005). Spatial decoupling is a strategy utilized in many filamentous cyanobacteria that form differentiated cells called heterocysts (Fay, 1992). The heterocysts are anaerobic compartments that utilize carbon produced by the vegetative cells for the energetic needs of nitrogen fixation and in turn the heterocyst cells supply vegetative cells with fixed nitrogen. In facultative anaerobes, the strategy of temporal separation of nitrogenase expression and activity often is utilized to decouple aerobic respiration and nitrogen fixation activity through regulation in response to oxygen (Dixon *et al.*, 1980). Perhaps the most interesting mechanisms that enable nitrogen fixation to occur in aerobic organisms occurs among the oxygenic Cyanobacteria. For nitrogen-fixing unicellular cyanobacteria living in microbial communities characterized by dense biomass (e.g. microbial mats), nitrogen fixation is temporally separated and occurs only at night when oxygenic photosynthesis is inactive and the immense respiratory capacity of the biomass renders the mats anaerobic (Stal and Krumbien, 1985).

Basic biochemistry

Although from an agricultural standpoint symbiotic nitrogen fixers are of more interest, they are not the most ideal model organisms for the study of nitrogenase biochemistry. This is to a great extent a result of difficulties in culturing these organisms at large scale. The most extensive biochemical studies have been conducted on nitrogenases from three free-living organisms: *A. vinelandii*, *Clostridium pasteurianum*, and *Klebsiella pneumoniae*.

Results of the initial biochemical characterization of Mo-nitrogenase first appeared in the literature in the late 1960s (Bui and Mortenson, 1969; Bulen and LeComte, 1966; Carnahan *et al.*, 1960; Jeng *et al.*, 1969; Kelly *et al.*, 1967; Mortenson, 1966; Mortenson *et al.*, 1967; Moustafa and Mortenson, 1969; Stewart *et al.*, 1970; Vandecasteele and Burris, 1970; Winter and Burris, 1968). Early studies revealed that Mo-nitrogenase consisted of two separable iron–sulfur proteins now termed the Fe protein and MoFe protein, reflecting the proteins relative metal cluster composition (Vandecasteele and Burris, 1970). In this early work, the two nitrogenase components were referred as molybdoferredoxin and azoferredoxin also highlighting their relative metal content and their relationship between the metalloclusters of the nitrogenase components and basic ferredoxins (Mortenson *et al.*, 1967). The proteins have also been termed dinitrogenase reductase and dinitrogenase reflecting their enzymatic activities (Emerich *et al.*, 1981). During nitrogenase catalysis the [4Fe–4S] cluster containing Fe protein associates with and dissociates from the MoFe protein in a manner that couples the hydrolysis of MgATP to electron transfer to the MoFe protein (Fig. 4.1). Since single electrons are transferred in each cycle, at least 6 cycles are required for substrate (N_2) reduction.

Assaying nitrogenase activity is complicated and requires multiple protein components and strict anaerobic conditions. Both Fe protein and MoFe protein are required in the assay mixture and optimal activities are achieved under conditions of high Fe protein to MoFe protein ratios and excess reductant (Dilworth *et al.*, 1965). Under most conditions, at least two MgATP molecules are hydrolysed for each electron transferred to substrate. Since the nitrogenase is very sensitive to inhibition by MgADP, assaying steady state activity requires an active ATP regenerating system in the assay mixture. Typically what has been used in nitrogenase assays is a mixture of creatine phosphate and the enzyme creatine phosphokinase. In the presence of excess creatine phosphate, ADP is rapidly converted back into ATP such that ADP does not accumulate to inhibitory levels in the nitrogenase assay. Assays are typically conducted in sealed vials as a function of time, with assay termination achieved through the introduction of concentrated acid or EDTA.

Ammonia formation is typically measured colorimetrically (Dilworth *et al.*, 1965) and is very time intensive. Fortunately, nitrogenase is a promiscuous enzyme capable of reducing other substrates and thus substrate reduction that offers a more facile detection can and is commonly used as a proxy for enzyme activity. The most common proxy is the reduction of acetylene to ethylene that can be easily monitored by analysing the gas head space using gas chromatography equipped with a flame ionization detector (Stewart *et al.*, 1968, 1970). The analysis of the substrate reduction properties of alternative substrates and inhibitors has been fruitful in terms of providing insights into the nitrogenase mechanism and these details will be discussed later in this chapter. Interestingly, nitrogenase is also an MgATP dependent hydrogenase (Bulen and LeComte, 1966). In the presence of MgATP and reductant and in the absence of added substrate, nitrogenase

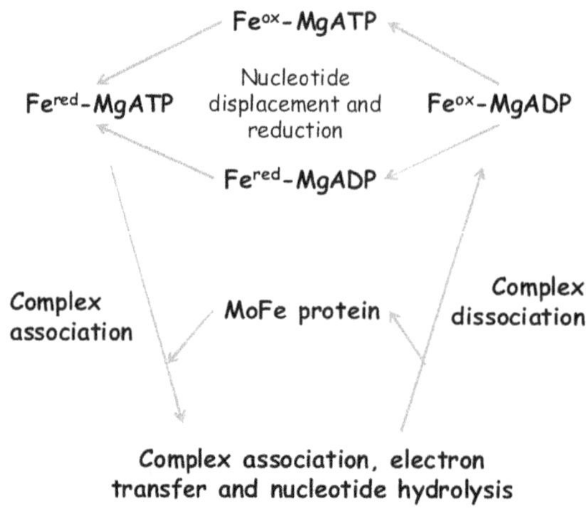

Figure 4.1 The three state cycle of a single nitrogenase Fe protein – MoFe protein complex association and dissociation and electron transfer. Fe protein in the MgADP bound oxidized state (Fe^{ox}-MgADP – upper right) is reduced and undergoes a nucleotide exchange in which MgADP is replaced by MgATP (Fe^{red}-MgATP – upper left). The reduced MgATP bound Fe protein (Fe^{red}-MgATP) then associates with the MoFe protein to form a complex triggering MgATP hydrolysis, electron transfer, and finally dissociation of the oxidized MgADP bound state of the Fe protein (Fe^{ox}-MgADP).

will direct all electron flux to the reduction of protons to molecular hydrogen (H_2). Proton reduction has always been a fascination biochemically and recently has gained additional attention due to our increasing awareness of the potential for hydrogen as an alternative energy source. In the early days of nitrogenase biochemistry, it was thought that hydrogen production by nitrogenase was the result of the inability to maintain high enough partial pressures of N_2 to ensure that 100% of the electron flux could be directed towards N_2 reduction. In essence, H_2 production was thought not necessarily to be part of the nitrogenase mechanism. This idea was somewhat supported by observations of acetylene reduction in that the addition of acetylene resulted in the diversion of the majority of electron flux towards acetylene reduction, although a small number of electrons were always directed at H_2 evolution (Hageman and Burris, 1980). During the same period prevailing mechanistic proposals placed H_2 evolution/proton reduction to be integral to the N_2 reduction mechanism (Chatt, 1980; Lowe and Thorneley, 1984a). This was addressed in the mid-1980s when Robert Burris performed an experiment in which nitrogenase was assayed under fifty atmospheres pressure N_2 gas, demonstrating that even under these conditions H_2 was evolved from nitrogenase at a stoichiometry of one mole of H_2 produced for every 1 mole of N_2 reduced to 2 moles of ammonia (Simpson and Burris, 1984). These considerations support to the generally accepted stoichiometry for the reaction shown below.

$$N_2 + 16MgATP + 8e^- + 8H^+ \rightarrow 2NH_3 + H_2 + 16MgADP + 16Pi \quad (4.1)$$

In the late 1980s, biochemical experimentation progressed to the point where rates of individual electron transfer events and ATP hydrolysis could be monitored in addition to basic substrate reduction rates, enabling more detailed kinetic schemes for catalysis. In a series of papers published in the *Biochemical Journal*, Roger Thorneley and David Lowe of the nitrogen fixation research unit at Sussex in the UK laid out a rational kinetic scheme for nitrogenase catalysis, for which the salient points still hold true today (Lowe and Thorneley, 1984a,b; Thorneley and Lowe, 1984a,b). Kinetic data from studies on nitrogenase from *K. pneumoniae* were used to determine many of the rate constants in this model. Given that we are talking about a mechanism that involves eight individual reduction steps and coordinated proton and nitrogen reduction steps, the Lowe-Thorneley mechanism (Lowe and Thorneley, 1984a) is quite complicated but more than half of the 17 rate constants that define the mechanism were determined (Lowe and Thorneley, 1984b). For nitrogenase the rate limiting step (~5/s) (Thorneley and Lowe, 1983) is about two orders of magnitude slower than intermolecular electron transfer (~150/s) (Thorneley, 1975). Furthermore, H_2 evolution or proton reduction precedes N_2 binding and this only occurs when the enzyme is reduced by at least three or four electrons (Fig. 4.2). The rate-limiting step was determined to be complex dissociation first by examining the inhibition of dithionite mediated Fe protein reduction in the presence of the MoFe protein (Thorneley and Lowe, 1983). This can also be estimated by observing the rate of oxidation of reduced flavodoxin (a terminal electron donor to the Fe protein) in the presteady state reactions initiated with reduced Fe protein (Peters *et al.*, 1995b). In this experiment oxidized Fe protein as a substrate for reduced flavodoxin only becomes available after dissociation of the Fe protein from the MoFe protein. Nitrogenase is a slow enzyme given that the enzyme has to conduct a minimum of eight cycles of electron transfer each occurring at ~5/s.

The nitrogenase Fe protein

The nitrogenase Fe protein is encoded by *nifH* and is responsible for transferring electrons derived from central metabolism in the form of reduced ferredoxin or flavodoxin to the MoFe

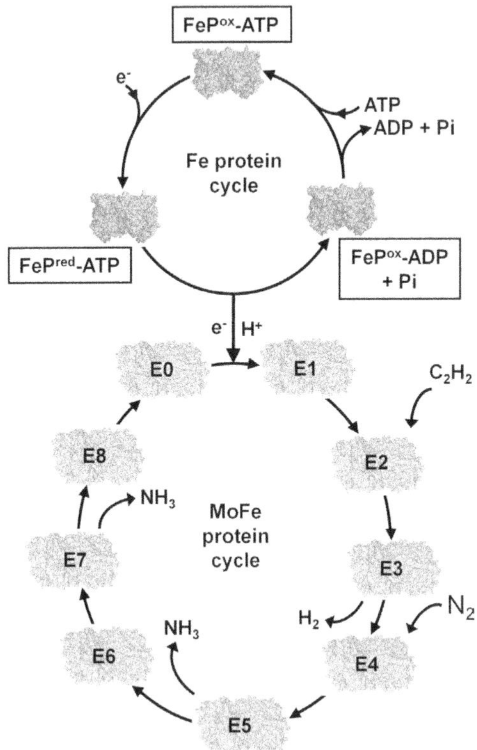

Figure 4.2 The three state cycle of the Fe protein shown in figure one (top) superimposed on the 8 state MoFe protein cycle (bottom). The MoFe protein is shown to be successively reduced by one additional electron with each step indicated as E_n where *n* indicates the total number of electrons transferred to the Fe protein and accumulated on the MoFe protein within the metal clusters or reduced products or intermediates. The binding of substrates and release of products as indicated through previous kinetic studies with N_2 binding occurring subsequently to hydrogen release in a proposed displacement mechanism with ammonia released as a product later in the cycle (Seefeldt *et al.*, 2009).

protein during nitrogenase catalysis (Burgess and Lowe, 1996; Lanzilotta and Seefeldt, 1997; Peters and Szilagyi, 2006). The Fe protein is the only electron donor to the MoFe protein and the requirement for ATP binding and hydrolysis in intramolecular electron transfer in nitrogenase catalysis has always been of significant interest. There are many electron transfer events in biology and most do not require ATP hydrolysis and thus for nitrogenase the requirement of ATP must be much more complicated than merely the delivery of electrons to the MoFe protein.

Prior to obtaining any structural information, the Fe protein had been classified as a member of a large family of proteins that couple nucleotide binding and hydrolysis and that modulate protein conformational change and the formation of macromolecular complexes (Koonin, 1993; Schulz, 1992). This family of so-called nucleotide-dependent signal transduction proteins and enzymes includes transducin (Sondek *et al.*, 1994), ras p21 (Pai *et al.*, 1989, 1990; Tong *et al.*, 1989a, 1991), elongation factor Tu (Jurnak, 1985; Kjeldgaard and Nyborg, 1992), RecA (Story *et al.*, 1993; Story and Steitz, 1992; Story *et al.*, 1992), and Gα of heterotrimeric G proteins (Sprang, 1997a,b). Of course, unlike the nitrogenase Fe protein, these aforementioned proteins do not participate in redox reactions and thus their biochemical roles and the role of nucleotide binding and hydrolysis in their function is somewhat different. For the heterotrimeric G proteins, nucleotide displacement from bound MgGDP to bound MgGTP signalled by the binding of a small molecule to the G protein αβ integral membrane subunits on the outside of the cell, results in dissociation of Gα enabling it to migrate and interact to stimulate the activity of an enzyme within the cell. The classic example of this phenomenon is the transduction of the signal from the compound adrenaline to activate the production of the second messenger cyclic AMP. G proteins are involved in the regulation of metabolic enzymes, ion channels and transporters, as well as having roles in controlling transcription, motility, contractility, secretion, embryonic development, and homeostasis.

Overall structure

The structure of the Fe protein was first determined by Rees and coworkers in the early 1990s where the structure was highlighted in the journal *Science* as the 'Iron Butterfly' (Georgiadis *et al.*, 1992; Orme-Johnson, 1992). The first structure of the Fe protein was that isolated from *A. vinelandii* (Georgiadis *et al.*, 1992) which was later refined at higher resolution (Schlessman *et al.*, 1998) and determined with bound MgADP (Jang *et al.*, 2000). The structure of the Fe protein from *C. pasteurianum* has also been determined (Schlessman *et al.*, 1998). The Fe protein exists as a dimer of identical subunits bridged by a single [4Fe–4S] cluster that is coordinated by two Cys residues (Cys^{97} and Cys^{132} in the *A. vinelandii* Fe protein) from each subunit (Fig. 4.3). Each Fe protein subunit harbours a MgATP binding site located approximately ~15 Å away from the [4Fe–4S] cluster. Each subunit consists of a conserved twisted β-sheet core surrounded by α-helices and was predicted by primary sequence similarity to have a protein fold similar to that of other structurally characterized members of the class (Howard and Rees, 1994). In the aforementioned G protein family, nucleotide-dependent 'Switch Regions' define pathways by which the nucleotide bound state of the G protein (GDP – 'off state' or GTP 'on state') modulates protein–protein interactions with either the G protein α- and β- subunits in the GDP bound 'off state' or effecter targets liked adenylate cyclase in the GTP bound 'on state'. The analogous 'switch regions' of nitrogenase discussed in detail later provide pathways for communication between the sites for MgATP and MgADP binding and the sites for MoFe protein interactions and complex formation 'Switch I' and the [4Fe–4S] 'Switch II'.

Each αβ subunit of the Fe protein consists of a core eight-stranded β-sheet surrounded by nine α-helices. The core twisted β-sheet resembles a propeller and consists of seven parallel strands and a single short antiparallel strand. Primary sequence comparisons indicate the highest levels of residue conservation occur within this β-core. The nine α-helices surround the core β-sheet with the largest helix spanning the long axis of each subunit. Another pair of α-helices are oriented

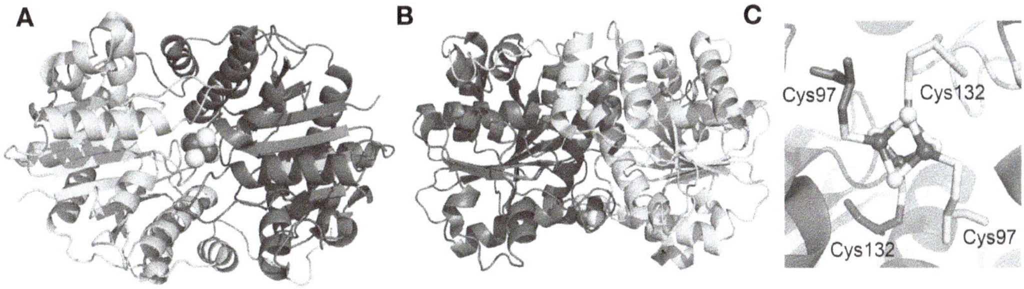

Figure 4.3 (A) Ribbons representation of the overall fold of the *A. vinelandii* Fe protein dimer viewed from two orientations (A and B) down the two fold axis of symmetry. (C) Expanded view of the cluster symmetrically coordinated with two analogous Cys ligands from each subunit (amino acid residue numbers are from *A. vinelandii* Fe protein).

such that the long axis and helix dipoles point to the [4Fe–4S] cluster, and thus potentially contribute to modulating the redox properties of the [4Fe–4S] cluster. In comparison to other [4Fe–4S] cluster containing proteins, the Fe protein's cluster is located closer to the enzyme surface, a property that has been suggested to potentially contribute to the oxygen sensitivity of this enzyme.

The [4Fe–4S] cluster

The Fe protein's [4Fe–4S] cluster is symmetrically coordinated by two Cys residues from each subunit (Cys^{97} and Cys^{132} of the *A. vinelandii* Fe protein). This symmetric bridging cluster coordination is rare in biology but has also been observed in the structures of 2-hydroxyglutaryl-CoA dehydratase (Locher *et al.*, 2001) and the L protein of protochlorophyllide reductase (BchL) (Sarma *et al.*, 2008). The latter protein is evolutionarily related to the Fe protein (Burke *et al.*, 1993) and its structure has just recently been determined and is strikingly similar to the nitrogenase Fe protein (Sarma *et al.*, 2008). Despite the uniqueness of the coordinating ligand arrangement of the Fe protein [4Fe–4S] cluster, the cluster itself is analogous to [4Fe–4S] clusters that are essentially ubiquitous in biology and most often involved in electron transfer reactions. These [4Fe–4S] cluster-containing proteins typically exist in two stable redox states, a state having two ferrous and two ferric ions commonly termed the $[4Fe–4S]^{+2}$ or oxidized state and a state having three ferrous ions and one ferric termed the $[4Fe–4S]^{+1}$ state of the reduced state (Beinert *et al.*, 1997; Holm *et al.*, 1996). The oxidation–reduction potentials of these simple regular cubane clusters in different proteins are very sensitive to their respective protein environments and have been shown to have midpoint potentials that span greater than 1 V. Factors influencing the midpoint potentials include (1) the direct interactions of dipoles (e.g. hydrogen bonding of amide dipoles to the cluster S), (2) solvent accessibility, and (3) bond strain induced by the conformation of the clusters covalent Cys thiolate ligands.

The Fe protein $[4Fe–4S]^{+2}$ oxidized state can be reduced by one electron to yield the $[4Fe–4S]^{+1}$ state using a variety of reducing agents including dithionite, reduced ferredoxin, or reduced flavodoxin. In the presence of a very strong reducing agent such as the methyl viologen, Ti(III)Citrate, and the hydroquinone state of flavodoxin, the $[4Fe–4S]^{+1}$ state of the Fe protein can be reduced further to a stable all ferrous $[4Fe–4S]^{0}$ state (Angove *et al.*, 1998; Nyborg *et al.*, 2000a,b). Stabilization of [4Fe–4S] clusters in an all ferrous state is rare for biological [4Fe–4S] clusters. The structure of the Fe protein poised in the all ferrous state has been determined and although the overall conformation of the protein is analogous to other observed Fe protein structures (Strop *et al.*, 2001), parallel EXAFS studies have implicated differences manifested in asymmetry and shorter Fe-Fe distances when compared to other observed redox states of the protein (Musgrave *et al.*, 2000). The $[4Fe–4S]^{+1/+2}$ redox couple has been estimated in the range of approximately

–200 to –300 mV depending on the microbial source of the protein in question. In addition, the $[4Fe–4S]^{0/1+}$ couple has been estimated to be approximately –450 to 500 mV but has only been characterized for the Fe protein from *A. vinelandii* (Angove *et al.*, 1998; Nyborg *et al.*, 2000a,b). The different oxidation states of the Fe protein have been interrogated by a variety of spectroscopic methods including electron paramagnetic resonance (EPR), nuclear magnetic resonance (NMR), circular dichroism (CD), and Mössbauer spectroscopies (Burgess and Lowe, 1996; Chan *et al.*, 2000; Hagen *et al.*, 1985a,b; Lanzilotta and Seefeldt, 1997; Lindahl *et al.*, 1985, 1987; Morgan *et al.*, 1990; Orme-Johnson *et al.*, 1972; Ryle *et al.*, 1995, 1996a,b; Stephens *et al.*, 1979). EPR is used to probe the electron structure of paramagnetic species probing the transition in energy states of an unpaired electron induced by the interaction of a magnetic field. The majority of EPR studies on the Fe protein are of the $[4Fe–4S]^{+1}$ paramagnetic state. The EPR spectra resembles that of a $[4Fe–4S]^{+1}$ cluster-containing ferredoxin with the predominant group state sublevel of $S = 1/2$ ground state ($g = 1.94$) with some contribution of $S = 3/2$ ground state ($g = 5$). These spin systems can be interrogated with standard perpendicular mode EPR.

Under optimal conditions (maximal activity and high Fe protein to MoFe protein ratios) using standard reducing agents such as dithionite approximately 4–5 MgATP molecules are hydrolysed per electron pair that ends up in reduced products (Ljones and Burris, 1972; Watt *et al.*, 1975). Using the more reduced electron donors such as Ti(III) citrate or flavodoxin hydroquinone to reduce the Fe protein to the all ferrous state, the ratio can be decreased to values approaching two MgATP hydrolysed per electron pair suggesting that all the $[4Fe–4S]^0$ are capable of transferring an electron pair in one cycle of complex association and dissociation (Angove *et al.*, 1998; Nyborg *et al.*, 2000a; Nyborg *et al.*, 2000b). Although the aforementioned midpoint potential for the $[4Fe–4S]^{0/1+}$ couple of approximately –450 to –500 mV is low but within the range to be considered physiologically relevant, it has yet to be determined if the $[4Fe–4S]^0$ is relevant *in vivo*.

MgATP binding and hydrolysis

The MgATP hydrolysis activity of the Fe protein is dependent on the presence of the MoFe protein and is presumed to occur during catalysis only when the Fe protein and MoFe protein are complexed. The MgATP binding sites resemble that of nucleotide binding sites in other members of the family (Howard and Rees, 1994; Koonin, 1993; Schulz, 1992) with the key regions of protein being conserved in structure across the family (Fig. 4.4). Two nucleotide-binding motifs termed the Walker motifs are involved in directly interacting with the Mg^{2+} and phosphate groups of the nucleotide (Howard and Rees, 1994; Seefeldt and Dean, 1997). The consensus Walker A motif (Gly-X-X-X-X-Gly-Lys/Ser) and the consensus Walker B motif (Asp-X-X-Gly) correspond to positions 9–16 and 125–128 of the *A. vinelandii* Fe protein, respectively. Key residues in these motifs are likely involved in the conserved interactions with the γ-phosphate of nucleoside triphosphate and Mg^{2+}. A third conserved region of the class, the phosphate binding loop or 'P loop' (consensus sequence: Gly-X-X-Gly-X-Gly) provides direct hydrogen bonding interactions with the nucleotide phosphate groups via hydrogen bonding. The P loop is adjacent to the Walker A motif and consists of residues 10 through 17 of the *A. vinelandii* Fe protein.

Switch regions couple the protein conformational changes in the Fe protein to signal MoFe protein docking and complex formation and also impose changes on the properties of the bridging [4Fe–4S] cluster. The switch regions detect changes in the nucleotide bound states via the direct and indirect interactions of amino acid side chains of the Fe protein and Mg^{2+} that is in turn coordinated by phosphates of the nucleotide. The hydrolysis of MgATP results in a change in the phosphate composition and charge modulating Mg^{2+} coordination. In the *A. vindelandii* Fe protein, 'Switch I' residues Lys^{41} and Asp^{43} interact with and detect changes in the phosphate composition and Mg^{2+} coordination, respectively, and function to transduce these conformational changes to modulate change at the Fe protein – MoFe protein docking interface. 'Switch II' communicates signals from the Walker B nucleotide binding motif to the Fe protein's

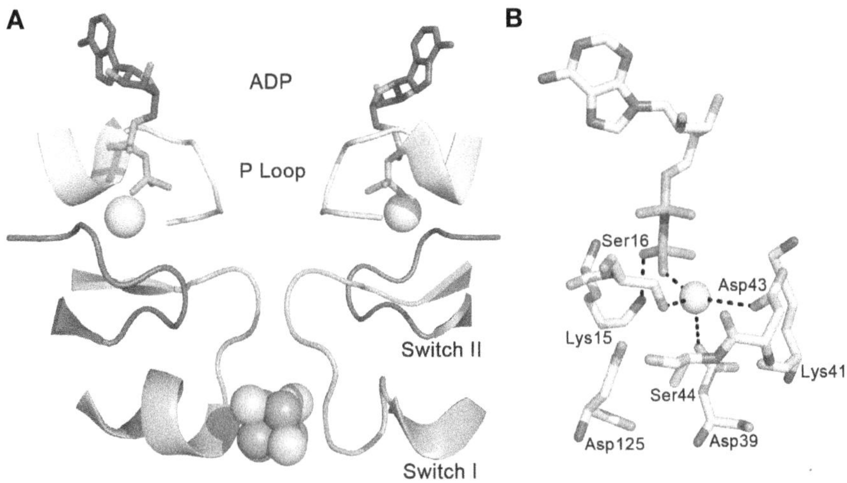

Figure 4.4 (A) Ribbons representation of the two nucleotide dependent Switch regions and the phosphate binding 'P loop' region of the Fe protein shown relative to the location of the nucleotide (stick representation) in the structure of the MgADP state of the Fe protein. (B) MgADP and associated key residues that interact with phosphates and magnesium and have been implicated in Fe protein nucleotide dependent signal transduction.

[4Fe–4S] cluster. Asp^{125} of the Walker B motif in the *A. vinelandii* Fe protein is linked through a short region of the polypeptide to Cys^{132} to the [4Fe–4S] cluster.

Although the first structure of the native Fe protein revealed what was modelled as a single partial occupancy MgADP spanning the nucleotide binding sites (Georgiadis *et al.*, 1992), subsequent structures of the Fe protein with bound MgADP and nucleotide bound complex structures revealed what was modelled as full occupancy nucleotide bound to the nucleotide binding sites of each subunit. The structure of the MgADP state of the Fe protein (Jang *et al.*, 2000) together with the structure of the Fe protein within various stabilized nitrogenase complexes (Chiu *et al.*, 2001; Schindelin *et al.*, 1997; Tezcan *et al.*, 2005) confirmed the detailed nucleotide interactions that were inferred by the sequence conservation and comparison to structures of other members of the protein family. In addition these structures provide detailed information on the interaction of the adenine base by hydrogen bonding to the side chains of Asp^{185}, Gln^{218}, and Gln^{236}, features that could account for the specificity of the protein for MgADP and MgATP over other nucleotides. The structures reveal that nucleotide binding occurs through a channel that is located on the opposite face of the protein from where the [4Fe–4S] cluster approaches the protein surface.

In early work, it was observed that the presence of nucleotide influenced Fe protein conformation and the spectroscopic properties of the Fe protein (Burgess and Lowe, 1996). Distinct differences in nucleotide bound states could be observed through changes in EPR line shape and intensity (Lindahl *et al.*, 1987; Morgan *et al.*, 1990) as well as changes in the paramagnetic NMR (Lanzilotta *et al.*, 1995) and CD spectra (Chan *et al.*, 2000; Lanzilotta *et al.*, 1997; Ryle *et al.*, 1995,1996a,b; Stephens *et al.*, 1979). Changes in the EPR spectra of the Fe protein that occur upon MgATP binding are manifested mainly in relative distribution of ground state spin states. The evidence for spin conversion was also supported by magnetic circular dichroism (MCD) studies (Onate *et al.*, 1993). Interestingly a prior study also indicated that different solvents such as ethylene glycol can also affect the relative distribution of spin states (Lindahl *et al.*, 1985). Structural changes in the cluster upon MgATP binding cannot be detected by using Fe X-ray absorption spectroscopy Fe XAS (Lindahl *et al.*, 1987) and results of another study probing the Fe protein using a combination of magnetic susceptibility and NMR suggest that the changes in spin states may be a result of freezing (Meyer *et al.*, 1988).

In addition, it was observed that upon binding of MgATP, the [4Fe–4S] cluster became more susceptible to metal chelators (Ljones and Burris, 1978; Walker and Mortenson, 1974) and chelation behaviour is dependent on oxidation state (Anderson and Howard, 1984). Moreover, nucleotide binding resulted in changes in the [4Fe–4S] cluster midpoint potential (Morgan *et al.*, 1986; Watt *et al.*, 1986; Zumft *et al.*, 1974). Changes in the susceptibility of the cluster were interpreted to indicate that upon the addition of MgATP the cluster becomes more surface exposed. The binding of MgATP has also been shown to cause an approximately 100mV shift towards a more negative midpoint potential. A negative shift in the midpoint potential during MgATP binding would result in the Fe protein becoming a better electron donor in catalysis; however, similar shifts towards a more negative potential are observed with the binding of MgADP. Relative affinities have been estimated for MgATP and MgADP using isothermal titration calorimetry (ITC) and have revealed that the Fe protein has a higher affinity for MgADP than MgATP in the oxidized state (Lanzilotta *et al.*, 1999) consistent with the aforementioned need to regenerate ATP during nitrogenase assays. This difference in affinity may also reflect an adaptation to cease this energy and reductant demanding process during periods of energy limitation and low MgATP/MgADP ratios.

Mutagenesis probing MgATP binding and hydrolysis

A number of different mutagenesis studies in the 1990's probed the direct role of residues implicated in nucleotide binding and hydrolysis as well as nucleotide-dependent signal transduction during nitrogenase catalysis. The ability to evaluate the effects of nucleotide binding, especially in the context of communication between nucleotide binding and the [4Fe–4S] cluster, was made possible by the aforementioned nucleotide dependent changes in the spectroscopic properties of the cluster. From examinations of the first crystal structure (Georgiadis *et al.*, 1992), it was observed that Lys^{15} of the Walker A motif formed a salt bridge with Asp^{125} of the Walker B motif. It was hypothesized that the salt bridge would be disrupted upon nucleotide binding and this event could be integral to nucleotide dependent signal transduction. In the MgADP structure of the Fe protein (Jang *et al.*, 2000), it was observed that the salt bridge was supplanted by interactions between Lys^{15} and the ADP β-phosphate and that Asp^{125} was hydrogen bonded to a water molecule in the first coordination sphere of the Mg^{2+}. Substitution of Lys^{15} by Gln results in a Fe protein that can bind MgATP but which does not communicate with the [4Fe–4S] cluster as assessed by spectroscopy (Seefeldt *et al.*, 1992). The substitution of the Asp^{125} by Glu resulted in a protein that in the presence of MgADP, resembled some of the biophysical properties of the MgATP bound form of the Fe protein (Wolle *et al.*, 1992). This could be rationalized given its interaction with the Mg^{2+} and the introduction of the additional methylene group. This displacement of the Mg^{2+} in this altered protein was later confirmed in the Glu^{125} variant structure (Jang *et al.*, 2004). Substitution of Ser^{16}, which coordinates directly with Mg^{2+} in the MgADP bound structure of the Fe protein, with Thr resulted in an active enzyme but higher activities were observed in the presence of Mn^{2+} than Mg^{2+} (Seefeldt and Mortenson, 1993). Other substitution studies had direct effects on MoFe protein interaction and nitrogenase complex formation. These included substitutions at Asp^{39} and Asp^{129} and the deletion of Leu^{127} (Lanzilotta *et al.*, 1997; Ryle and Seefeldt, 1996). The latter of which (ΔLeu^{127}) resulted in a Fe protein that mimicked a number of spectroscopic properties similar to the MgATP-bound state of the Fe protein in absence of bound nucleotides (Lanzilotta *et al.*, 1997). The protein was presumed to be captured in a conformation that approximated the MgATP bound state; however this conformation to date has not been structurally characterized. Although this variant was inactive, it was observed to form a very stable complex with the MoFe protein (Lanzilotta *et al.*, 1996). The subsequent characterization of the ΔLeu^{127} resulted in an unanticipated conformation of the Fe protein manifested in large scale reorientation of the Fe protein subunits (Sen *et al.*, 2004). The Fe protein subunits are rotated away from one another with the pivot point being the [4Fe–4S] cluster, which is the site of bridging covalent attachment. There are a number of attractive qualities of this conformation in the context of rationalizing a role for MgATP binding and hydrolysis in nitrogenase catalysis (Peters and Szilagyi, 2006). In this conformation the sites for the MgATP binding and hydrolysis are separated and the docking interface adopts a conformation with structural and chemical complementarity to the MoFe protein docking surface. It has been suggested that intersubunit interactions between the side chain amine group of Lys^{41} stabilize the charge that develops during ATP hydrolysis (Schindelin *et al.*, 1997). The control of MgATP hydrolysis and the gating of electron transfer could then be presumably controlled by conformational changes imposed on the Fe protein by the MoFe protein. These ideas will be discussed more thoroughly later, in the section on complex formation.

The structure of the variant (ΔLeu^{127}) was of significant interest in light of its detailed biochemical characterization which indicated that it had properties similar to MgATP bound state and the difficulties in structurally characterizing the true MgATP bound state. Interestingly, the results were not consistent with previous solution small angle X-ray scattering (SAXS) studies indicating that the nucleotide bound states of the Fe protein exhibited a similar radius of gyration values (~25 Å) to the nucleotide-free states for Fe proteins from both *A. vinelandii* and *K. pneumoniae* (Chen *et al.*, 1994; Grossmann *et al.*, 2001). Further SAXS experiments were able to show that the ΔLeu^{127} Fe protein indeed adopted the unique open elongated conformation observed in the X-ray structure in solution evinced by an increase in radius of gyration values (~27 Å) and reasonable shape reconstructions (Sarma *et al.*, 2007). Together the results indicate that the ΔLeu^{127} state of the Fe protein is not a good structural mimic for the MgATP bound conformation. Thus, the MgATP-bound conformation still remains a mystery as does defining the true structural determinants that result in the changes in biophysical properties that are observed upon MgATP binding to the Fe protein.

The MoFe protein

Overall structure

The *nifD*- and *nifK*-encoded polypeptides share significant primary sequence identity and are the product of an ancient gene duplication event (Fani *et al.*, 2000; Soboh *et al.*, 2010). Multiple sequence alignments of deduced amino acid sequences of NifD and NifK from multiple sources provided the basis for the first generation of mutagenesis studies probing mainly potential P cluster and FeMo-co ligands (Brigle *et al.*, 1985; Dean *et al.*, 1990; Kent *et al.*, 1989, 1990; May *et al.*, 1991; Scott *et al.*, 1990, 1992). These studies as well as the aforementioned mutagenesis on the Fe protein were really made possible by the established genetic tools and methods for facile gene replacement in the model organisms, *A. vinelandii* and *K. pneumoniae* (Brigle *et al.*, 1987a; Kent *et al.*, 1989; Page and von Tigerstrom, 1979; Robinson *et al.*, 1986). The structure of the MoFe protein has been determined from three microbial sources *A. vinelandii, K. pneumoniae,* and *C. pasterianum* (Bolin *et al.*, 1993; Chan *et al.*, 1993; Einsle *et al.*, 2002; Kim and Rees, 1992; Kim *et al.*, 1993; Mayer *et al.*, 1999; Peters *et al.*, 1997). As anticipated based on primary sequence alignments, the overall folds of NifD (α-subunit) and NifK (β-subunit) are similar and each αβ dimer or symmetric half of the MoFe protein exists as two interdigitated shaped subunits (Fig. 4.5). The α- and β-subunits are of similar size and the structure of each exists as three domains each having a single parallel β-sheet and associated α-helices.

The MoFe protein P clusters are located at the pseudosymmetric interface of the α- and β-subunits and the FeMo-cofactors are located solely within the α-subunit. The α- and β-subunit interface is quite extensive and each αβ dimer unit of the tetramer interact solely by contacts between each β-subunit. The MoFe protein is evolutionarily related to dark protochlorophyllide reductase (DPOR), an enzyme complex that is involved in chlorophyll biosynthesis (Burke *et al.*, 1993). Like the MoFe protein, DPOR is a multicomponent/multiprotein enzyme but in contrast to nitrogenase acts to catalyse a double bond reduction of protochlorophyllide (Brocker *et al.*, 2010; Brocker *et al.*, 2008a,b; Fujita, 1996;

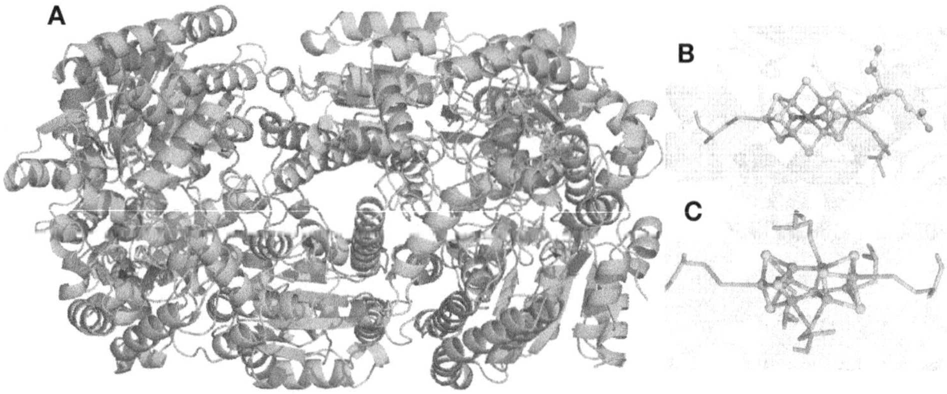

Figure 4.5 (A) Ribbons representation of the nitrogen MoFe protein from *A. vinelandii* viewed down the twofold axis of symmetry that relates each equivalent dimer of the $\alpha_2\beta_2$ heterotetramer. (B) Two FeMo-cofactors are located solely within the two α-subunits. Each α- and β-subunits are similar in structure and are related by pseudo-twofold symmetry and the MoFe protein P clusters in the P^{ox} state (C) are coordinated symmetrically by three Cys ligands of each subunit in a similar fashion to the [4Fe–4S] cluster of the Fe protein which is coordinated symmetrically by equivalent Cys residues of two separate protein chains.

Fujita and Bauer, 2000; Fujita *et al.*, 1998; Muraki *et al.*, 2010; Sarma *et al.*, 2008). Interestingly, at the P cluster binding site, DPOR contains a simple [4Fe–4S] cluster and at the FeMo-co binding site retains a cavity for substrate (protochlorophyllide) binding (Brocker *et al.*, 2010; Muraki *et al.*, 2010).

The expression of active MoFe protein requires the expression of numerous gene products that are involved in synthesis and maturation of the P cluster and FeMo-cofactor (Dos Santos *et al.*, 2004a; Rubio and Ludden, 2008). During nitrogenase catalysis the electrons are transferred intermolecularly from the [4Fe–4S] cluster of the Fe protein to the MoFe protein P clusters and then subsequently are transferred intramolecularly to the FeMo-cofactor site where substrates are bound and reduced.

FeMo-cofactor

There are several lines of experimental evidence that indicate that the FeMo-cos are located at the sites for substrate binding and reduction. Mutant strains of nitrogen-fixing bacteria deficient in FeMo-co biosynthesis produce inactive nitrogenases that can be activated by the addition of isolated FeMo-co (Christiansen *et al.*, 1998; Ribbe *et al.*, 2002). MoFe proteins with compositionally altered FeMo-co, for example MoFe protein with citrate rather than homocitrate as an organic constituent (Imperial *et al.*, 1989; Madden *et al.*, 1990, 1991) or V in the place of Mo (Eady, 1996; Eady, 2003; Hales and Blanchard, 1996; Hales *et al.*, 1986; Lee *et al.*, 2009b, 2010; Morningstar *et al.*, 1987), possess altered substrate reduction properties relative to native MoFe protein.

The FeMo-co can be best described as a 'complex bridged metal assembly' made up of two modified [4Fe–4S] cluster cubanes (Fig. 4.6). The cluster consists of a [4Fe–3S] partial cubane bridged to a Mo substituted [Mo–3Fe–3S] partial cubane through three bridging sulfides. The FeMo-co [Mo–7Fe–9S] exists with limited coordination to the MoFe protein, with just a single Cys ligand at the terminal Fe of the cluster and a single His δN coordination at the Mo (For *A. vinelandii*, α-Cys^{275} at the cluster terminal unique Fe and at the Mo by the Nδ of α-His^{442}). The Mo is in an octahedral coordination environment which, in addition to the three sulfides and the His N ligand, is ligated by two oxygens from the cofactor organic constituent homocitrate. The [Mo–7Fe–9S] unit has three-fold symmetry although is covalently coordinated by only two amino acid residues. Several other amino acid side chains interact with the FeMo-co though non-covalent interactions. The side chains of *A. vinelandii* MoFe protein

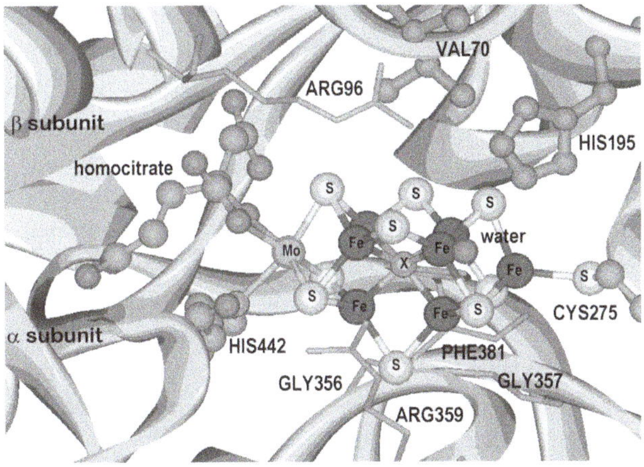

Figure 4.6 The FeMo-co structure exists as a [4Fe–3S] partial cubane bridged to a [Mo–3Fe–4S] partial cubane via three sulfides. The two partial cubanes are additional linked by a hexacoordinated light atom X within the FeMo-cofactor interior. For the *A. vinelandii* MoFe protein the FeMo-co is covalently bound by the thiolate sulfur of α-Cys^{275} to a terminal Fe and the δ-nitrogen of α-His^{442} to the Mo. Additional residues highlighted include α-His^{195} which hydrogen bonds to one of the bridging sulfides of the FeMo-co and α-Val^{70} which resides in the vicinity of the four Fe face of the FeMo-co implicated in substrate interactions.

-Arg^{96}, α-His^{195}, and α-Arg^{359} as well as the peptide bond amide groups of α-Gly^{356} and α-Gly^{357}. The side chains of α-Gln^{191}, α-Glu^{427}, and α-Glu^{440} provide interactions with the carboxylates of homocitrate. Other residues that do not provide direct interactions and have a more nonpolar character include α-Val^{70}, α-Tyr^{229}, α-Ile^{231}, α-Ile^{355}, α-Leu^{358}, and α-Phe^{381}.

Spectroscopy on the MoFe protein was much more complicated than that of the Fe protein due to the unique nature of the clusters and the lack of appropriate precedents. Early studies involving Mössbauer spectroscopy analysing MoFe protein from *A. vinelandii, K. pneumoniae,* and *C. pasteurianum* identified 30 atoms divided into two different types of clusters (Huynh *et al.*, 1980; Huynh *et al.*, 1979; Munck *et al.*, 1975; Rawlings *et al.*, 1978; Smith and Lang, 1974; Smith *et al.*, 1980; Yoo *et al.*, 2000; Zimmermann *et al.*, 1978). The FeMo-co is unique and has been characterized in both the MoFe protein and isolated intact and has a characteristic $S=3/2$ EPR signal in the reduced resting state (Lowe and Smith, 1985; Morgan *et al.*, 1988; Munck *et al.*, 1975; O'Donnell and Smith, 1978; Orme-Johnson *et al.*, 1972; Watt *et al.*, 1980). Although $S=3/2$ states have been observed in other metalloprotein systems the low rhombicity and the sharp inflections ($g=2.0$, 3.7, and 4.3) of the resting state $S=3/2$ EPR signal of the MoFe protein is rare. The inflections in the $g=3$–4 range occur in a region of the spectra where metal centre EPR signals are not typically observed and this allows for the signal to be visible in even whole cells (Davis *et al.*, 1972).

The most recent crystallographic study on the MoFe protein provides evidence for a light atom within the interior of the FeMo-co coordinated by six Fe ions (Einsle *et al.*, 2002). The study carefully refined C, N, and O in the appropriate ionization states at this site and the analysis indicated that N was the best fit to the crystallographic data. Subsequent studies using electron nuclear double resonance (ENDOR) spectroscopy using preparations of the MoFe protein enriched with ^{15}N were not consistent with either an exchangeable (Lee *et al.*, 2003) or nonexchangeable N (Yang *et al.*, 2005) at this site. However, parallel computational studies indicate that the low spin density of this atom may confound detection (Lukoyanov *et al.*, 2007b). Generating corroborating biochemical and chemical data has been difficult and the composition of the interstitial light atom of FeMo-co is still an outstanding question and one that may have significant mechanistic implications.

Probing the interactions of substrates at the FeMo-co has been difficult because the interactions with substrates or changes in the spectroscopic properties are not observed in the resting state. A number of spectroscopic approaches have been used to turnover states of the MoFe protein using EPR, ENDOR, EXAFS, Mössbauer, and FTIR spectroscopy under a variety of different conditions and with different substrates and inhibitors (Cameron and Hales, 1998; Christiansen *et al.*, 1995; Christie *et al.*, 1996; Davis *et al.*, 1979; George *et al.*, 1997; Ibrahim *et al.*, 1999; Lee *et al.*, 1997a,b; Lowe *et al.*, 1978; Munck *et al.*, 1975; Pollock *et al.*, 1995; Rasche and Seefeldt, 1997; Ryle *et al.*, 2000; Seefeldt *et al.*, 1995; Yates and Lowe, 1976; Yoo *et al.*, 2000). These studies have revealed that under turnover conditions changes do occur in FeMo-co. In these earlier studies, however, although different states were observed, specifics concerning binding and mechanism could not be inferred. Recent work has revealed that under conditions of lower electron flux (nitrogenase assays conducted under low Fe protein to MoFe protein ratios) intermediates in the reduction of N_2 are observed by EPR and ENDOR spectroscopy (Barney *et al.*, 2009; Barney *et al.*, 2005b; Lukoyanov *et al.*, 2007a). These new results provide a framework for examining the mechanism of nitrogen reduction at FeMo-co and even effectively connecting experimentally observed intermediates with elements of the aforementioned Lowe–Thorneley kinetic scheme (Hoffman *et al.*, 2009; Lukoyanov *et al.*, 2007a; Seefeldt *et al.*, 2009).

Active site mutagenesis and substrate reduction mechanism

Initial mutagenesis, in advance of the structure, implicated residues involved in cluster coordination, residues that have important roles in substrate reduction, and highlighted the key importance of not just the coordinating ligands but the cluster environment in the functioning of nitrogenase (Brigle *et al.*, 1987a; Kent *et al.*, 1989, 1990; Scott *et al.*, 1990, 1992). Although many substitutions of residues in the immediate cofactor environment could support some level of proton and acetylene reduction activity, nitrogen reduction in even the most conservative substitutions was often completely abolished. The first two mutants that were extensively characterized were *A. vinelandii* MoFe protein α-His[195] and α-Lys[191] (Kim *et al.*, 1995; Scott *et al.*, 1990, 1992; Sorlie *et al.*, 2001). Interestingly these residues were chosen prior to the determination of an X-ray crystal structure of the MoFe protein and were based on multiple sequence comparison of diverse MoFe proteins and the comparison of MoFe protein α- and β-subunits (Brigle *et al.*, 1987a; Kent *et al.*, 1990; Kent *et al.*, 1989; Peters *et al.*, 1995a; Scott *et al.*, 1990, 1992). These residues are strictly conserved and their substitution led to interesting and unique substrate reduction properties.

The substitution of α-Gln[191] for Lys results in a MoFe protein that is unable to reduce N_2 to NH_3 but will still catalyse low levels of acetylene and proton reduction (Scott *et al.*, 1992). This altered nitrogenase was observed to produce, in addition to the two electron reduction product of acetylene, ethylene, a small percentage of the four electron reduction product, ethane. In addition, proton reduction was observed to be sensitive to CO inhibition (Scott *et al.*, 1992) which is not observed in wild-type nitrogenase but observed in mutants of the MoFe protein defective in homocitrate synthesis (Hawkes *et al.*, 1984). These mutants defective in *nifV*, which encodes homocitrate synthase, have been shown to produce MoFe proteins having citrate in the place of homocitrate in FeMoco (Hoover *et al.*, 1988a,b, 1989). Later it was observed in the structure of nitrogenase that the role of α-Gln[191] is in directly interacting with homocitrate carboxylate group (Kim and Rees, 1992).

The substitution of α-His[195] to Gln or Asn results in elimination of N_2 reduction activities but retains H_2 evolution activity and acetylene reduction. Interestingly, α-Gln[195] MoFe protein exhibits acetylene reduction activity that is sensitive to inhibition by N_2 but for the substitution to Asn (α-Asn[195]), N_2 is not inhibitory indicating that the latter substitution results in a MoFe protein that no longer binds N_2 (Kim *et al.*, 1995). Later work involving structural characterization of the α-Gln[195] MoFe protein indicated that Gln provided an analogous hydrogen bonding interaction as the εN group of His does in the native enzyme (Sorlie *et al.*, 2001) and that Asn being a carbon unit shorter would not be able to support the same interaction. Early results of electron spin echo envelope modulation (ESEEM) studies probing nitrogen modulation in wild-type and MoFe proteins with substitution at this site were interpreted to implicate *A. vinelandii* α-His^{195} as a direct FeMo-co ligand (Thomann *et al.*, 1991; Thomann *et al.*, 1987). After the structure was determined however it was revealed that the His ligand was α-His^{442} (Kim and Rees, 1992). The inability to generate mutants that have altered N_2 reduction activity but do not eliminate N_2 reduction activity is among the many challenges facing biochemists interested in probing the mechanism of catalysis of this multifaceted enzyme.

As mentioned above acetylene has been used as a surrogate substrate for nitrogenase activity (Stewart *et al.*, 1968, 1970). Although mutations in the FeMo-co environment have resulted in eliminating N_2 reduction but retain acetylene reduction activity, substrate reduction is believed to occur at a common site. This can be rationalized given that N_2 reduction requires the FeMo-co to achieve more reduced states for substrate binding and reduction and the activation barrier that must overcome for N_2 reduction must be much higher. This is supported by the observation that acetylene is an inhibitor of N_2 reduction (Davis and Wang, 1980; Rivera-Ortiz and Burris, 1975). The inhibition of N_2 reduction by acetylene was exploited to gain insights into the nature of substrate interactions (Christiansen *et al.*, 2000; Dos Santos *et al.*, 2005). An elegant genetic strategy was developed to identify amino acid residues in the substrate binding site by challenging *A. vinelandii* cells transformed with mutagenized DNA to grow in the absence of fixed nitrogen but in the presence of inhibitory levels of acetylene. DNA sequence analysis of the nitrogenase acetylene resistant mutants revealed that a substitution at a single site (α-Gly^{69}) resulted in the acetylene resistant phenotype restricting access to proximal 4Fe face of the central 6Fe prism of the FeMo-co. This presumably allowing the active site to discriminate between acetylene and nitrogen substrates. Subsequent substitution studies of the adjacent amino acid residue (*A. vinelandii* MoFe protein α-Val^{70} substituted by Ala and Gly) resulted in an extended substrate size range whereas substitution by Ile resulted in a MoFe protein that is hampered in its ability to reduce a range of substrates including acetylene and N_2, yet retains normal proton reduction activity (Barney *et al.*, 2004; Benton *et al.*, 2003; Dos Santos *et al.*, 2007b; Igarashi *et al.*, 2004; Lee *et al.*, 2004; Mayer *et al.*, 2002). Subsequent structural characterization of the α-Ile^{70} (Fig. 4.7) indicates that the additional carbon unit places a methyl group in close proximity to the 4Fe face of the FeMo-co that is believed to be involved in substrate binding (Sarma *et al.*, 2010).

Figure 4.7 (A) Stereoview of the 2*Fo*-*Fc* electron density maps (1.0σ of the region surrounding the FeMo-co and the adjacent α–70^{Ile} residue. The protein backbone is shown in ribbons with residues in the FeMo-co environment from the α-subunit are shown in line angle representations. (B) FeMo-co binding site in the α–70^{Ile} MoFe protein structure showing the location of the surrounding water molecules represented as dark spheres. The residues in the FeMo-co and surrounding protein environment shown with line angle representations.

Biochemical and spectroscopic characterization of the α-Ala^{70} MoFe protein revealed that propargyl alcohol and propargyl amine were substrates for the variant and bound substrate or intermediates could be observed by EPR spectroscopy (Barney *et al.*, 2004). ENDOR studies on this MoFe protein variant involving isotopically labelled alkyne and nitrogenous substrates have revealed potential binding modes for intermediates at the FeMo-co (Barney *et al.*, 2005a). The ability to capture and characterize nitrogenous substrates and intermediates at FeMo-co by freeze quench ENDOR spectroscopy in the wild-type and amino acid substituted variants of the MoFe protein has provided significant insight into substrate interaction and the mechanism of substrate reduction (Barney *et al.*, 2004, 2005a,b, 2006a,b, 2007, 2009; Lukoyanov *et al.*, 2007a). These studies suggest a symmetric binding of nitrogen limiting the number of potential binding modes and favour an alternating mechanism in which individual reduction steps alternate between the two nitrogen atoms of N_2 (Hoffman *et al.*, 2009; Seefeldt *et al.*, 2009).

P cluster

Several lines of evidence support that the P cluster has a role in mediating electron transfer from the Fe protein's [4Fe–4S] cluster and the FeMo-co substrate reduction sites. Firstly, the structures of nitrogenase complexes and mutagenesis studies of the MoFe protein at the component protein docking interface indicate the P cluster resides in an intermediate position between the [4Fe–4S] cluster of the Fe protein and the FeMo-co in the nitrogenase complex during catalysis (Bolin *et al.*, 1993; Chan *et al.*, 1993; Einsle *et al.*, 2002; Kim and Rees, 1992; Kim *et al.*, 1993; Mayer *et al.*, 1999; Peters *et al.*, 1997). In addition, the distance between the Fe protein [4Fe–4S] cluster and the FeMo-co in the nitrogenase complex structures (Chiu *et al.*, 2001; Schindelin *et al.*, 1997; Schmid *et al.*, 2002; Tezcan *et al.*, 2005) are too long for typical biological electron transfer reactions. Secondly, amino acid substitutions in the protein environment between the P clusters and FeMo-cos perturb intramolecular electron transfer (Peters *et al.*, 1995b), and thirdly stopped flow UV/Vis experiments and freeze quench EPR experiments of turnover conditions indicate that the P cluster is oxidized during substrate reduction and the oxidized P clusters can be reduced by the Fe protein (Lowe *et al.*, 1993).

Initially, the P clusters were a challenge to characterize spectroscopically since the reduced resting or as isolated state of the cluster, as determined later by Mössbauer studies, exists in an all ferrous oxidation state (termed P^N) and is EPR silent (Lindahl *et al.*, 1988; Yoo *et al.*, 2000). Oxidation of the P cluster yields, however, two stable states that can be probed spectroscopically by EPR (P^{1+} and the P^{2+} or P^{OX} states) (Hagen *et al.*, 1987; Lindahl *et al.*, 1988; Pierik *et al.*, 1993). The P^{1+} state exists as a mixed spin system with $S = 1/2$ and $S = 5/2$ components and P^{2+} is an $S \geq 3$ system which can be probed via parallel mode EPR (Tittsworth and Hales, 1993). Oxidation of the P cluster beyond the P^{2+} is possible but not reversible and thus not thought to be physiologically relevant. The midpoint potential of the P^N/P^{1+} and P^{1+}/P^{2+} couples are ~ −300mV and since the midpoint potential of the Fe protein is ~−420 mV when bound to MgATP(Lanzilotta and Seefeldt, 1997) the P cluster can presumably transfer one or two electrons but interestingly not until the resting state of the P cluster (P^N) is oxidized.

The P cluster is a rare example of a cluster that contains only Fe and S at a ratio other than one-to-one and that are coordinated by bridging thiolates. In some of initial reports of the MoFe protein published in the early 1990s, the structure of the P cluster was proposed to consist of two regular [4Fe–4S] cluster cubanes bridged by a shared sulfide or by a disulfide bond (Chan *et al.*, 1993; Kim *et al.*, 1993). However in the late 1990s, structural studies of the MoFe protein poised in both the P^N and P^{OX} states indicated that the P^N existed as the formerly undescribed structure with a single hexacoordinated shared sulfide (Peters *et al.*, 1997). In the oxidized P^{OX} state one nearly standard [4Fe–4S] cubane located mainly in the α-subunit is coordinated through two terminal Cys ligands from the α-subunit (α-Cys^{62} and α-Cys^{154} in *A. vinelandii*) and one bridging thiolate from each the α- and β-subunit (α-Cys^{88} and β-Cys^{95} in *A. vinelandii*). The second partial cubane located mainly in the β-subunit exists as a distorted [4Fe–3S] unit with terminal Cys coordination from two β-subunit Cys residues (β-Cys^{70} and β-Cys^{153} in *A. vinelandii*) and the same bridging thiolate coordination (α-Cys^{88} and β-Cys^{95} in *A. vinelandii*) and with additional coordination to supplant the missing S of the subcluster via the amide nitrogen of α-Cys^{88} and the side chain oxygen of β-Ser^{188}. Upon reduction and conversion of the P^{OX} state to the P^{N} state the P cluster undergoes a redox dependent structural rearrangement in which the unique ligands (the amide nitrogen of α-Cys^{88} and the side chain oxygen of β-Ser^{188}) are lost and replaced by bonds to a sulfide of the more regular α-subunit localized cubane resulting in a cluster that can be best described as two regular [4Fe–4S] clusters that share a corner sulfide (Peters *et al.*, 1997) (Fig. 4.8).

Oxidation state-dependent structural rearrangements are uncommon for FeS clusters in biology and for the P cluster it was suggested that in addition to its role as an intermediate in electron transfer from the Fe protein to the FeMo-co, the cluster may serve in the coupling of electron transfer events to proton uptake and proton transfer (Peters *et al.*, 1997). One of the hallmarks of proton coupled electron transfer is that the oxidation-reduction potential of the cluster is sensitive to pH. Redox titrations in which the signal intensity of the P cluster signals were monitored in parallel mode EPR clearly show pH dependence, supporting proton coupled electron transfer (Lanzilotta *et al.*, 1998). However, substitution of the P^{OX} coordinating Ser residue by Cys resulted in a P cluster with a pH-dependent midpoint potential (Chan *et al.*, 1999) so the role of redox-dependent structural changes observed for the P cluster is still not resolved. A number of amino acid substitution studies were conducted on the P cluster coordinating ligands and substitution of the ligands by either Ser or Ala typically resulted in abolishing nitrogenase activity (Dean *et al.*, 1990; Kent *et al.*, 1990; Kent *et al.*, 1989; May *et al.*, 1991) and even in cases failure to form the tetrameric MoFe protein structure suggesting that the P clusters are involved in MoFe protein assembly (Govezensky and Zamir, 1989; Kent *et al.*, 1990). Interestingly, substitution of either the α- or β-subunit bridging Cys ligands eliminates activity, however when both are substituted by Ala in tandem some activity is retained (Kent *et al.*, 1990). Biochemical characterization of the MoFe protein expressed in a background devoid of the Fe protein indicate that the P clusters are assembled stepwise via an intermediate that resembles closely associated regular [4Fe–4S] clusters (Broach *et al.*, 2006; Corbett *et al.*, 2004). Coupling of the clusters and maturation of the P cluster can be accomplished by addition of the Fe protein, dithionite, and MgATP (Hu *et al.*, 2005a; Lee *et al.*, 2009a).

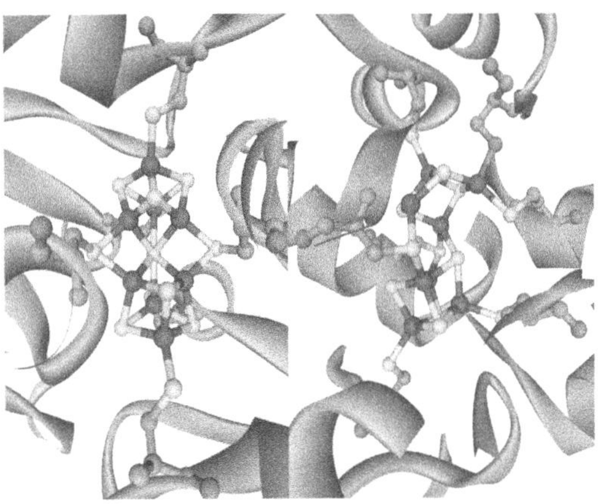

Figure 4.8 Ball-and-stick representations of the (A) P^{N} and (B) P^{OX} states of the *A. vinelandii* MoFe protein P cluster. The interconversion between the all ferrous or P^{N} state (left) and the P^{ox} state or P cluster two electron oxidized state (right) is accompanied by a redox dependent structural rearrangement in which the Fe ligands supplied by the central hexacoordinated sulfide in the P^{N} state are replaced by protein ligands supplied by the amide nitrogen of α-Cys^{88} and the side chain oxygen of β-Ser^{188} in the *A. vinelandii* structures.

Nitrogenase complex formation

The structures of the individual components as well as biochemical work involving regulation of nitrogenase by ADP-ribosylation (Pope *et al.*, 1985) and chemical cross-linking (Willing *et al.*, 1989) laid the groundwork for the initial component protein docking model (Howard and Rees, 1994). This model was based on pairing the two-fold symmetric axis of the homodimeric Fe protein with the pseudosymmetrical axis that relates the structurally similar α- and β-subunits of the MoFe protein. The generalities of the docking models were confirmed with the elucidation of the first structure of a stable nitrogenase Fe protein – MoFe protein complex (Schindelin *et al.*, 1997) providing significant insights into the role of MgATP binding and hydrolysis in nitrogenase catalysis. The first structure of a nitrogenase complex was the Fe protein – MoFe protein complex stabilized in the presence of MgADP and tetrafluoroaluminate (Schindelin *et al.*, 1997). The tetrafluoroaluminate-containing nitrogenase complex has been characterized biochemically (Duyvis *et al.*, 1996; Renner and Howard, 1996) and has been proposed to correspond to the hydrolytic transition state by analogy to the equivalent complex of myosin (Fisher *et al.*, 1995). Probably the most striking observation revealed in the structure was the distinct conformational differences between the Fe protein of the complex and that of the previously characterized states of the Fe protein. The conformational changes implied by the aforementioned comparison consist, to the greatest extent, of rigid body reorientations of the Fe protein subunits with respect to one another. Shortly after the tetrafluoroaluminate-containing nitrogenase complex was determined (Fig. 4.9A), the structure of a stable nitrogenase complex of the ΔLeu^{127} Fe protein and the MoFe protein was published and a very similar structure was revealed (Chiu *et al.*, 2001). Both structures revealed conformational changes within the Fe protein manifested in rigid body reorientations of the Fe protein subunits and translation of the Fe protein's [4Fe–4S] cluster towards the docking interface affording a close approach to the MoFe protein P cluster in order to affect intermolecular electron transfer. The observations provided in this work could be interpreted to imply that Fe protein – MoFe protein complex formation occurs through the concerted and symmetrical interaction of the Fe protein and MoFe protein and the trigger for ATP hydrolysis and Fe protein conformational change promotes intermolecular electron transfer and transition to the MgADP-bound state of the Fe protein, promoting symmetric dissociation (Fig. 4.10A).

More recently however, the elucidation of additional complex structures (Fig. 4.9B–D) affords an alternative interpretation of the cycle of interaction that occurs within the nitrogenase complex (Schmid *et al.*, 2002; Tezcan *et al.*, 2005). The nitrogenase complex determined with the Fe protein in nucleotide-free, MgADP-bound, and MgAMPPCP-bound states (Tezcan *et al.*, 2005) as well as the structure of a chemically cross-linked complex (Schmid *et al.*, 2002) reveals that the Fe protein can interact with the MoFe protein in several different bound conformations leading to an alternative model in which the MgATP-bound state of the Fe protein initially interacts with one of the MoFe protein subunits and in a rolling motion moves first into more symmetric interactions which in the transition promotes MgATP and electron transfer (Fig. 4.10B). In this model, dissociation is mainly being promoted through interactions between the Fe protein and the opposite MoFe protein subunit. These studies and

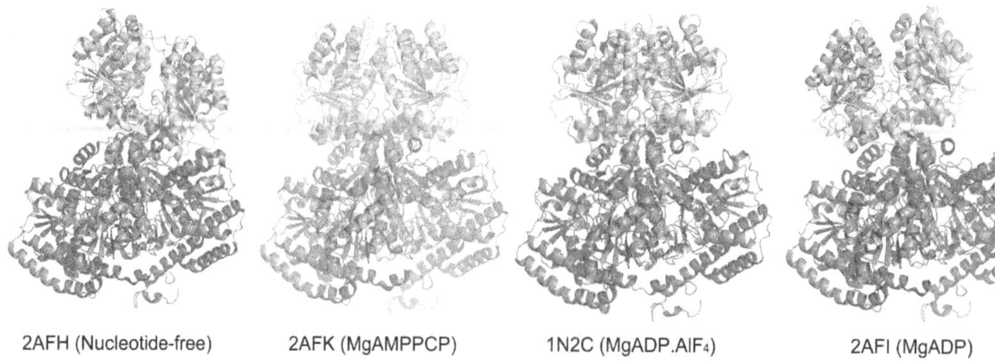

Figure 4.9 Ribbon representation of one symmetric half of the (A) tetraflouroaluminate stabilized, (B) nucleotide free, (C) MgAMPPCP and (D) MgADP bound nitrogenase complexes.

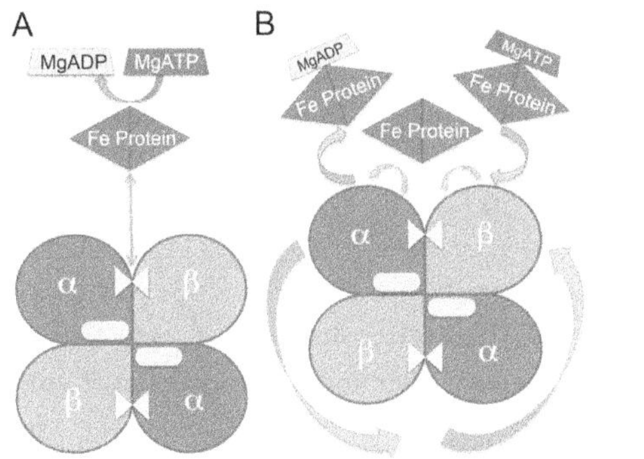

Figure 4.10 Model for docking interactions involving (A) reciprocal symmetric interactions only in nitrogenase catalysis versus (B) the rolling interaction model suggested by Rees and coworkers involving the initial interactions through largely β–subunit interactions with dissociation occurring through largely α-subunit interactions.

previous studies involving chemical cross-linking of the Fe protein and MoFe protein have led to the proposal that initial docking occurs primarily through interactions of the Fe protein with the MoFe protein β-subunit with dissociation being promoted primarily through interactions of the α-subunit. These two distinctly different models for nitrogenase complex interactions: (1) direct symmetric interaction versus (2) asymmetric rolling interaction have yet to be experimentally discriminated.

Regardless of how this complex interacts it is presumed that nitrogenase complex, MgATP hydrolysis, and electron transfer are conformationally gated to maintain electron transfer towards substrate reduction even in the higher Lowe-Thorneley E states required for N_2 binding and reduction (Fig. 4.2). Recent studies examining the oxidation of the Fe protein in the presteady state in the presence of solutions of varying viscosity and osmotic pressure are consistent with an overall model of conformational gating within the complex consistent with either of the aforementioned docking mechanisms (Danyal *et al.*, 2010). In this work estimates of the uptake of water during complex interactions and nitrogen catalysis can be estimated. This has been used as a proxy to estimate the level of docking interface remodelling that occurs within the complex during the cycle of Fe protein MoFe protein association, MgATP hydrolysis and electron transfer, and dissociation. Water uptake clearly indicates docking interface remodelling in this work but suggests that it may not occur to the extent required for the asymmetric rolling interaction model.

From the structural work, complementary chemical and structural interactions were realized between the Fe protein and MoFe protein that contributed to complex formation. Intermolecular interactions involved both amino side chain and main chain components and the salient side chain interactions have been summarized in a number of reviews (Seefeldt *et al.*, 2009). Key interactions are evident between the Arg^{100} residues of the Fe protein and the pseudo symmetric α-Glu^{120} and β-Glu^{120} side chains and non-symmetrical interaction with α-Glu^{184} and β-Glu^{156} of the MoFe protein. Substitution of Arg^{100} has been studied in detail and it has been shown that substitution by Tyr, Phe, Leu, Thr, His, and Lys all affected diazotrophic growth of mutants expressing these altered proteins. It was also shown that substitution of bulky groups such as Tyr or Phe still were capable of some level of diazotrophic growth indicating that the positive charge at this position was not essential for nitrogenase activity. In other previous work, substitution of α-Glu^{120} and β-Glu^{120} by Gln was shown to partially complement the effects of substitution of the Fe protein Arg^{100} to His. Fe protein Arg^{140} residues interact with backbone carbonyls of α-Gly^{160} and β-Gly^{159} and substitution studies at the Arg^{140} position have directly implicated these residues as having a role in component protein interactions. Additional side chain interactions are evident in the nitrogenase complex occurring between pseudo symmetric hydrophobic side chains of α-Phe^{125} and β-Phe^{125} in the MoFe protein and Met^{58} of the Fe protein. Substitution of these Phe residues either singly or in tandem by Ala was shown to affect complex formation and it was also shown that the single substitutions (either α- and β-subunits) did not have equivalent biochemical phenotypes. Mutagenesis studies adjacent to Met^{58} of the Fe protein were conducted in attempt to probe the observation that the *C. pasteurianum* Fe protein is not compatible with the *A. vinelandii* MoFe protein. In this work substitution of the region of the Fe protein between residues 59 to 67 of the *A. vinelandii* Fe protein with the corresponding regions of the *C. pasteurianum* Fe protein was found to affect component protein interactions.

MoFe protein maturation

The synthesis of the FeMo-co and the maturation of the MoFe protein requires a suite of *nif* regulated genes that encode various enzymes, carrier proteins, scaffolds, and maturases (Fig. 4.11). The products of the *nifD* and *nifK* genes are not required for the biosynthesis of FeMo-co (Ugalde *et al.*, 1984). Rather, FeMo-co is assembled elsewhere in the cell and is then incorporated into a FeMo-co deficient apo-MoFe protein to generate active holo- MoFe protein. The presence of P-clusters in the apo- MoFe protein is a pre-requisite to incorporate FeMo-co (White *et al.*, 1992). Thus, a pathway to mature the NifD and NifK polypeptides into an apo- MoFe protein that can readily incorporate FeMo-co also exists.

Aside from the FeMo-co biosynthetic requirements, maturation of MoFe protein involves the functions of NifU, NifS, NifH, NifZ, and NafY (Allen *et al.*, 1993; Hu *et al.*, 2007; Jacobson *et al.*, 1989b; Rubio *et al.*, 2002) The involvement of the general GroEL chaperone in this process has also been postulated (Govezensky *et al.*, 1991; Ribbe and Burgess, 2001). Fig. 4.12 depicts a model for the maturation of the MoFe protein from the nascent NifD and K polypeptides to the active holo- MoFe protein nitrogenase component. This model starts with the loading of a nascent NifDK polypeptide tetramer with [Fe–S] clusters (step 1). NifU and NifS are responsible for this initial step.

NifU and NifS are the major providers of Fe–S clusters for the nitrogenase component proteins (Johnson *et al.*, 2005a; Johnson *et al.*, 2005b). NifS is a cysteine desulfurase that provides S for the assembly of the clusters while NifU serves as molecular scaffold to hold the nascent [2Fe–2S] and [4Fe–4S] clusters (Smith *et al.*, 2005; Zheng *et al.*, 1993). [Fe–S] clusters synthesized on NifU are transferred to target apo-proteins by direct protein–protein interactions. Transfer of [4Fe–4S] clusters from NifU to apo-Fe protein has been demonstrated (Dos Santos *et al.*, 2004a). Cell-free extracts of *A. vinelandii nifUS* double mutants exhibited a 10-fold decrease in MoFe protein activity that could not be cured by FeMo-co addition (Jacobson *et al.*, 1989b), strongly suggesting this protein lacks P-clusters. Although there is no direct experimental evidence obtained from purified protein preparations, it is likely that the Δ*nifUS* apo-MoFe protein completely lacks [Fe–S] clusters. The assembly of P-clusters by the Fe protein and MgATP requires the presence of [4Fe–4S] cluster substrates (see below). As mentioned above, NifU and NifS are also involved in the synthesis of the [4Fe–4S] cluster of the Fe protein, and it is tempting to speculate that the lack of P-clusters in Δ*nifUS* apo- MoFe protein is indirectly due to Fe protein cluster deficiency. This scenario is, however, unlikely because there is evidence showing that [4Fe–4S] cluster-deficient Fe protein functions in apo-MoFe protein maturation (Curatti *et al.*, 2005; Rangaraj *et al.*, 1997)

The Δ*nifH* apo-MoFe protein represents the maturation stage prior to P-cluster formation; it carries two pairs of [4Fe–4S] clusters located at the interfaces of the α and β subunits, which electronic and structural properties have been well established (Broach *et al.*, 2006; Corbett *et al.*, 2004; Ribbe *et al.*, 2002). The FeMo-co sites in Δ*nifH* apo-MoFe protein are inaccessible, and this protein cannot be activated simply by FeMo-co addition. The exact role of the Fe

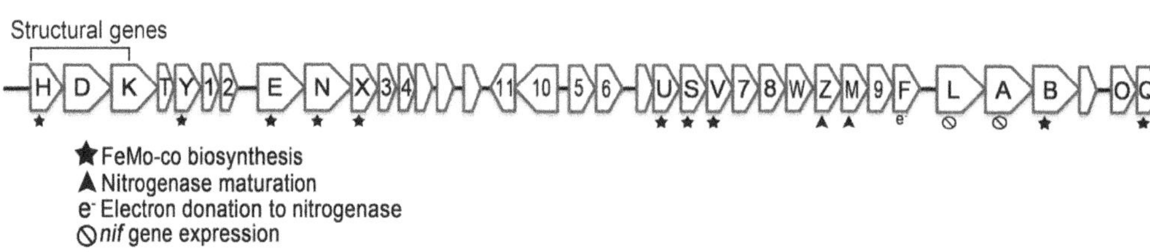

Figure 4.11 Organization of nitrogen fixation (*nif*) genes in *A. vinelandii*. Shown are all those genes given the designation *nif* and associated open reading frames that are part of *nif* specific transcriptional units.

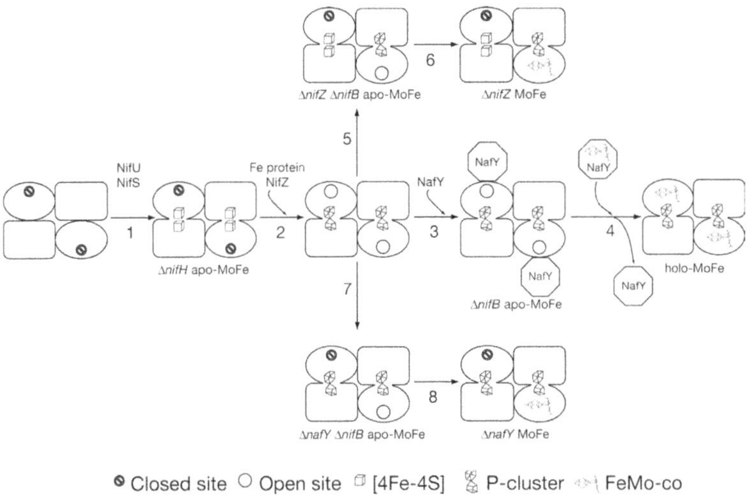

Figure 4.12 Model for the maturation of the MoFe protein. The α (NifD), β (NifK) and γ (NafY) subunits of the apo-MoFe protein are represented by oval, rectangular and octagonal cartoons, respectively. The complete apo-MoFe protein maturation pathway leading to the formation of active MoFe protein is depicted in the centre of the figure (reactions 1–4). The different forms of apo-MoFe protein characterized to date are labelled. The defective apo-MoFe protein maturation pathway in a *nifZ* mutant is shown in the upper part of the figure (reactions 5 and 6). In the absence of NifZ and FeMo-co, 50% of the FeMo-co sites remain closed. The defective apo-MoFe protein maturation pathway in a *nafY* mutant is shown in the lower part of the figure (reactions 7 and 8). In the absence of NafY/NifY and FeMo-co, a significant percentage of the open FeMo-co sites is damaged (50% in *A. vinelandii* and 95% in *K. pneumoniae*).

protein in P-cluster assembly is not well understood. Although it must be able to bind MgATP and to complex with apo-MoFe protein, MgATP hydrolysis, electron transfer or even the presence of the [4Fe–4S] cluster in the Fe protein are not requirements for P-cluster formation (Rangaraj *et al.*, 1997, 1999). Thus, there is not an obligate redox role by the Fe protein during P-cluster assembly. P-cluster formation appears to trigger major conformational changes in the apo-MoFe protein by bringing the α and β subunit together (Corbett *et al.*, 2007), which ultimately lead to a rearrangement of the region around residue α-Cys^{275} and the opening of the FeMo-co biding site (Magnuson *et al.*, 1997) (Fig. 4.12, step 2). The Fe protein-matured apo-MoFe protein contains P-clusters and is readily activated by FeMo-co.

Within each MoFe protein tetramer, the assembly of the two P-clusters appears to occur stepwise (Hu *et al.*, 2004, 2007). The first P-cluster is assembled by Fe protein/MgATP, whereas the assembly of the second P-cluster within each tetramer requires the additional presence of NifZ (Fig. 4.12, step 5). Consistently, the *A. vinelandii* Δ*nifZ* mutant synthesizes NifDK proteins having a P-cluster and FeMo-co in one of the α/β dimers while having only the P-cluster precursors on the other α/β dimer (step 6) (Hu *et al.*, 2004). The exact role of NifZ is unknown, but it has been proposed to serve as chaperone (Hu *et al.*, 2008).

Another chaperone protein, denominated as the γ subunit binds and stabilizes the P-cluster containing apo-MoFe protein in the absence of FeMo-co (step 3). Thus, strains that are solely defective in FeMo-co synthesis, such as Δ*nifB* or Δ*nifEN*, accumulate an apo-MoFe protein with $\alpha_2\beta_2$ subunit composition (Homer *et al.*, 1995). The γ subunit has been shown to be the product of the *nafY* gene in *A. vinelandii* (Rubio *et al.*, 2002) or the product of the *nifY* gene in *K. pneumoniae* (Homer *et al.*, 1993). Although the *nifB* apo-MoFe protein can be readily activated by FeMo-co *in vitro* (Paustian *et al.*, 1990) or *in vivo* (Rubio *et al.*, 2002), the capability of NafY to bind either apo-MoFe protein or FeMo-co led to the proposal that NafY plays an additional role in delivering FeMo-co into apo-MoFe protein (Homer *et al.*, 1995) (step 4).

NafY binding stabilizes the Δ*nifB* apo-MoFe protein in a conformation that is fully activatable by FeMo-co (Rubio *et al.*, 2002). The exact mechanism underlying NafY stabilization is not known, but it has been proposed that the presence of NafY stabilizes an open conformation with the FeMo-co binding residue α-Cys^{275} exposed to solvent (Magnuson *et al.*, 1997). In the absence of NafY/NifY and FeMo-co a significant percentage of the open FeMo-co sites become damaged. Thus, the Δ*nafY nifB* apo-MoFe protein can only be partially activated by FeMo-co (Homer *et al.*, 1993; Rubio *et al.*, 2002).

Overview of FeMo-co biosynthesis

FeMo-co biosynthesis proceeds outside of the MoFe protein in a complex pathway that involves the activities of a number of *nif* proteins (Fig. 4.13). Elucidation of this pathway resulted from systematic genetic and biochemical analyses in *A. vinelandii* and *K. pneumoniae* (reviewed in Hu *et al.*, 2008; Rubio and Ludden, 2008). A functional categorization of proteins involved in the biosynthesis of FeMo-co reveals that the pathway involves molecular scaffolds (NifU, NifB, NifQ, and NifEN), metallocluster carrier proteins (NifX and NafY) and proteins with enzymatic activity (NifS, NifB, and NifV). Although essential to FeMo-co biosynthesis, the exact role of the Fe protein remains undefined and it has not yet been assigned to any of the above categories.

The FeMo-co biosynthetic pathway comprises three clearly independent branches providing biosynthetic 'blocks' (Fe–S core, Mo and homocitrate) that converge onto NifEN, a MoFe protein homologue proposed to function as a central node in which FeMo-co assembly is completed (Brigle *et al.*, 1987a; Corbett *et al.*, 2006; Soboh *et al.*, 2006) (Fig. 4.13A). NifS, NifU and NifB are required to assemble the Fe–S core. NifUS provide NifB with simple [Fe–S] clusters (Zhao *et al.*, 2007) that are used by NifB to synthesize

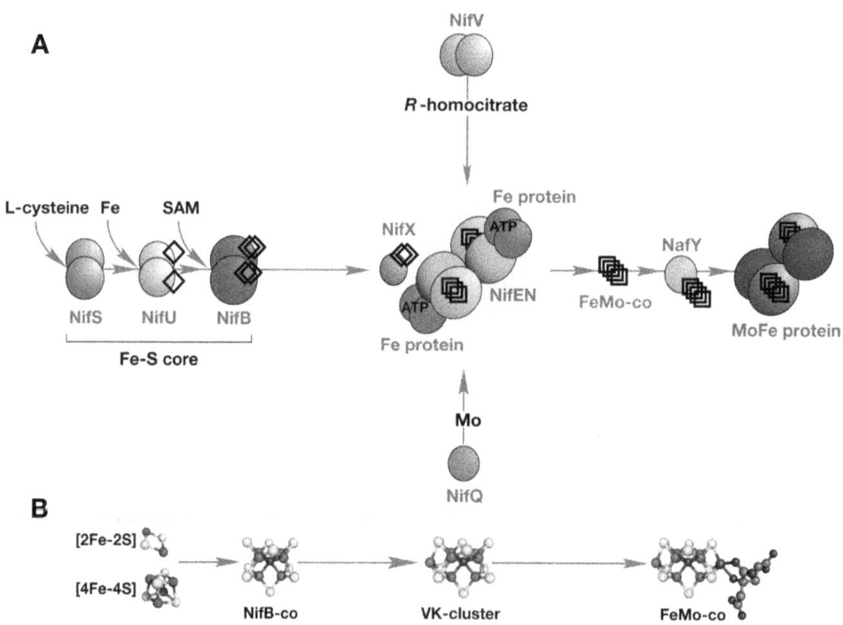

Figure 4.13 Model for the biosynthesis of FeMo-co. Panel A, scheme illustrating the convergence of FeMo-co 'building blocks' into a central node composed of the NifEN and the Fe protein. The metallocluster carrier proteins, NifX and NafY, serve as links between the Fe–S core biosynthetic branch, the NifEN/NifH central node, and the apo-MoFe protein, respectively. Panel B, proposed [Fe–S] cluster conversions during FeMo-co biosynthesis. The biosynthetic intermediates are aligned beneath the panel A proteins that assemble or carry them: NifU with the [2Fe–2S] and [4Fe–4S] clusters, NifB with NifB-co, NifEN with the VK-cluster, and NafY with FeMo-co.

NifB-co in a reaction dependent on S-adenosyl methionine (SAM) radical chemistry (Curatti *et al.*, 2006). NifX has the ability to carry NifB-co from NifB to NifEN (Hernandez *et al.*, 2007; Zhao *et al.*, 2007) where it is converted into the VK-cluster. The molybdenum branch involves the activity of NifQ, which has been shown to serve as Mo donor to a putative NifEN/Fe protein complex (Hernandez *et al.*, 2008). Finally, homocitrate is synthesized by the *nifV* gene product (Zheng *et al.*, 1997). Whether homocitrate would bind to isolated NifEN or to a NifEN/Fe protein complex has not been established. Assembled FeMo-co can be transferred from NifEN to apo-MoFe protein directly or via NafY, a FeMo-co binding protein that stabilizes the cofactor and increases the yield of reconstituted MoFe protein (Curatti *et al.*, 2007; Rubio *et al.*, 2004).

Fig. 4.13B depicts the current model for [Fe–S] cluster transformations ultimately leading to assembled FeMo-co. The positions of [Fe–S] cluster biosynthetic intermediates are lined up with those of the proteins involved in their synthesis (Fig. 4.13A). It is important to note that the structures shown for NifB-co and the VK-cluster are models proposed on the basis of EXAFS and NRVS analysis (Corbett *et al.*, 2006; George *et al.*, 2008). The reaction catalysed by NifB is perhaps the most relevant one from an [Fe–S] cluster biosynthetic perspective. It has been proposed that NifB-co comprises the central [6Fe–9S–X] symmetric core of FeMo-co (George *et al.*, 2008). Thus, NifB activity links the world of simple [Fe–S] clusters, which are ubiquitous, to the large and complex cofactors found in the active sites of the *nif*, *vnf* and *anf* nitrogenases. The VK-cluster is the FeMo-co biosynthetic intermediate accumulated on NifEN when the Fe protein is lacking (Corbett *et al.*, 2006; Hernandez *et al.*, 2007). The VK-cluster presumably contains all of FeMo-co Fe atoms but lack Mo and homocitrate.

Formation of the Fe–S core of FeMo-co

The products of the *nifU* and *nifS* genes are required to achieve full activity of both nitrogenase component proteins. *A. vinelandii nifU* or *nifS* deletion mutants exhibited 5% Fe protein activity and 20% MoFe protein activity compared to wild type (Jacobson *et al.*, 1989b). Similarly, *K. pneumoniae nifS* mutants of showed 4% MoFe protein activity and completely lacked Fe protein activity (Roberts *et al.*, 1978).

NifU and NifS are specialized *nif*-regulated homologues of the Isc (iron–sulfur cluster) housekeeping counterparts, IscU and IscS, respectively. IscU and IscS proteins are components of a cellular machinery dedicated to the assembly of [Fe–S] clusters that participate in general metabolic processes, whereas NifU and NifS provide [Fe–S] clusters for the maturation of the nitrogenase component proteins (Johnson *et al.*, 2005b). The boost in [Fe–S] cluster synthesis provided by the activities of NifU and NifS is necessary to cope with the high levels of nitrogenase protein required for diazotrophic growth. Limited replacement of functions between the Nif and the Isc systems is possible by artificially elevating the expression of the Isc system in the absence of the Nif system and vice versa (Dos Santos *et al.*, 2007a).

NifS is a pyridoxal phosphate (PLP)-containing enzyme that catalyses the desulfurization of L-cysteine to provide sulfur for [Fe–S] cluster formation (Zheng *et al.*, 1994; Zheng *et al.*, 1993). NifU is a homodimeric protein composed of three highly conserved domains (Agar *et al.*, 2000), all of which have the capacity to coordinate [Fe–S] clusters. Both, the N-terminal and the C-terminal domains of NifU can serve as molecular scaffolds to assemble transient [Fe–S] cluster, whereas the central domain of NifU binds a permanent [2Fe–2S] cluster that might have a redox role in the delivery of the transient [Fe–S] clusters (Fu *et al.*, 1994). Upon formation of a NifUS heterotetramer (Yuvaniyama *et al.*, 2000), the S provided by NifS directs the assembly of the transient [Fe–S] clusters on NifU (Smith *et al.*, 2005). Newly synthesized clusters are transferred to target apo-proteins, e.g. transfer of [4Fe–4S] clusters from NifU to apo-Fe protein and the reconstitution of functional Fe protein has been demonstrated (Dos Santos *et al.*, 2004b).

Participation of NifU and NifS in FeMo-co synthesis was demonstrated by showing the requirement of *nifU* and *nifS* genes to achieve wild-type levels of NifB-co synthesis (Zhao *et al.*, 2007). Since the NifUS machinery generates simple clusters with [2Fe–2S] and [4Fe–4S] composition, we hypothesize that NifB utilizes these types of clusters as substrates for NifB-co biosynthesis.

The activity of the *nifB* gene product is essential to the biosynthesis of FeMo-co. *A. vinelandii* and *K. pneumoniae nifB* mutants are impaired in FeMo-co biosynthesis (Joerger and Bishop, 1988; MacNeil *et al.*, 1978) and accumulate a P-cluster-containing FeMo-co-deficient apo-MoFe protein (see Fig. 4.12 above) (Christiansen *et al.*, 1998; Paustian *et al.*, 1990). The *nifB* gene is also required for the activities of the alternative V- and Fe-only nitrogenases (Joerger and Bishop, 1988; Rodríguez-Quiñones *et al.*, 1993), which implies that the metabolic product of NifB is a common intermediate for the biosynthetic pathways of FeMo-co, FeV-co and FeFe-co (the analogous active site cofactors of the V- and Fe-only nitrogenases). Consistently, *nifB* expression is regulated not only by the NifA transcriptional activator, but also by the activators of the alternative nitrogenases, VnfA and AnfA (Drummond *et al.*, 1996).

NifB catalyses the first committed step in FeMo-co biosynthesis. NifB uses simple [Fe–S] clusters as substrates to generate the complex [6Fe-9S-X] NifB-co cluster, which serves as specific Fe and S donor to FeMo-co biosynthesis. The incorporation of Fe and S atoms from NifB-co into FeMo-co has been shown by carrying out radiolabelled *in vitro* FeMo-co biosynthesis experiments (Allen *et al.*, 1995; Curatti *et al.*, 2006). In addition, purified NifB has been shown to substitute for NifB-co in the *in vitro* FeMo-co synthesis assay (Curatti *et al.*, 2006, 2007; Zhao *et al.*, 2007).

Amino acid sequence analysis of NifB indicates that it is composed of at least two different domains: a S-adenosyl methionine (SAM)-radical domain at the N-terminal region of NifB (Sofia *et al.*, 2001), and a NifX-like domain at the C-terminal region of the polypeptide (Moreno-Vivian *et al.*, 1989). SAM-radical proteins contain a [4Fe–4S] cluster coordinated by three cysteine residues and a molecule of SAM. The dependence of *in vitro* NifB-co synthesis on SAM strongly suggests that NifB catalytic mechanism involves radical chemistry. It has been suggested that very low potential radical chemistry is needed to incorporate the central atom X in NifB-co (Curatti *et al.*, 2006).

The NifX-like domain of NifB is found in a family of small proteins, such as NifX, NafY, and VnfX, that are able to bind either FeMo-co, NifB-co or the VK-cluster (Rubio *et al.*, 2002). Although there are no studies on the role of the C-terminal region of NifB, we propose that it could be a site that binds NifB-co precursors and serves as scaffold for the formation of NifB-co or a site that binds preassembled NifB-co for subsequent delivery to NifX. At present, it is not known whether both domains of NifB are necessary for NifB-co synthesis.

Incorporation of molybdenum into FeMo-co

NifQ is a iron–sulfur protein known by genetic and biochemical evidence to be involved in the incorporation of molybdenum into FeMo-co. *A. vinelandii* and *K. pneumoniae nifQ* mutants are impaired in FeMo-co biosynthesis and therefore exhibit a nitrogen fixation deficiency (Joerger and Bishop, 1988; MacNeil *et al.*, 1978; Rodríguez-Quiñones *et al.*, 1993). Although not affected in Mo transport, *nifQ* mutants accumulate low levels of intracellular Mo. The *nifQ* mutant phenotype can be reverted by adding high concentrations of molybdate or cysteine to the medium (Imperial *et al.*, 1984; Ugalde *et al.*, 1985), which led to the suggestion that NifQ catalysed a reaction between Mo and a sulfur donor and that such reaction could occur non-enzymatically at high concentrations of the reactants.

NifQ is a 20-kDa monomeric protein carrying a redox-responsive $[3Fe–4S]^{+1/0}$ cluster that is probably coordinated by the cysteine residues of a conserved $Cx_4Cx_2Cx_5C$ motif located at the C-terminal end of the polypeptide. As isolated NifQ preparations exhibited a mixture of [3Fe–4S] and [Mo–3Fe–4S] clusters. Metal analysis and EPR experiments suggested that each NifQ molecule carries either a [3Fe–4S] cluster or a [Mo–3Fe–4S] cluster (Hernandez *et al.*, 2008). Incubation of purified NifQ with Mo and sulfide under reducing conditions boosted the [Mo–3Fe–4S] cluster EPR signals, suggesting intramolecular cluster conversion from [3Fe–4S] to [Mo–3Fe–4S]. This conversion is in line with previous reports showing that proteins containing reduced $[3Fe–4S]^0$ clusters were able to coordinate heterometals in reversible equilibria to generate [Mo–3Fe–4S] clusters (Butt *et al.*, 1991).

The presence in NifQ of the [Mo–3Fe–4S] cluster correlates with its ability to serve as molybdenum donor for *in vitro* FeMo-co synthesis. Interestingly, Mo mobilization from NifQ in the *in vitro* FeMo-co synthesis assay requires the simultaneous presence of the Fe protein and NifEN (Hernandez *et al.*, 2008).

It is important to note that molybdate must undergo at least three chemical transformations during its incorporation into FeMo-co: replacement of oxo-ligands by sulfur ligands, reduction of Mo, and insertion of Mo into an [Fe–S] cluster environment (Hernandez *et al.*, 2009). The [3Fe–4S] cluster observed in purified NifQ is an obvious site for Mo binding, reduction and subsequent transfer (Fig. 4.14). Mo EXAFS analysis suggests the presence of a second molybdenum site in NifQ that would include short 1.73 Å Mo–O bonds and, possibly, Mo–S bonds at 2.23 Å (George, Hernandez and Rubio, unpublished results). It follows that the two Mo environments could represent two different stages in the process of Mo incorporation into FeMo-co.

The role of NifV and the incorporation of homocitrate into FeMo-co

The *nifV* gene product is a homocitrate synthase that catalyses the condensation of acetyl coenzyme A and α-ketoglutarate to form *R*-homocitrate (Zheng *et al.*, 1997). *A. vinelandii* and *K. pneumoniae nifV* mutants are defective in FeMo-co biosynthesis (Hoover *et al.*, 1986) and synthesize a nitrogenase enzyme in which homocitrate is replaced either by citrate (in the case of *K. pneumoniae* Liang *et al.*, 1990) or by a mixture of isocitrate and citrate (in the case of *A. vinelandii* Ludden *et al.*, 2004). The altered *nifV* nitrogenases are unable to effectively reduce N_2 and, therefore, *nifV* strains exhibit slow diazotrophic growth rates (Jacobson *et al.*, 1989b; MacNeil *et al.*, 1978; McLean *et al.*, 1983). Addition of homocitrate to the growth medium of *nifV* mutant strains during nitrogenase derepression results in a phenotypic reversion (Hoover *et al.*, 1988b).

In FeMo-co, the *R*-homocitrate coordinates the Mo atom through its C-2 carboxyl and hydroxyl groups. It is still not known how Mo becomes coordinated to *R*-homocitrate or why homocitrate and not another organic acid is incorporated into FeMo-co *in vivo*. Analogues of homocitrate

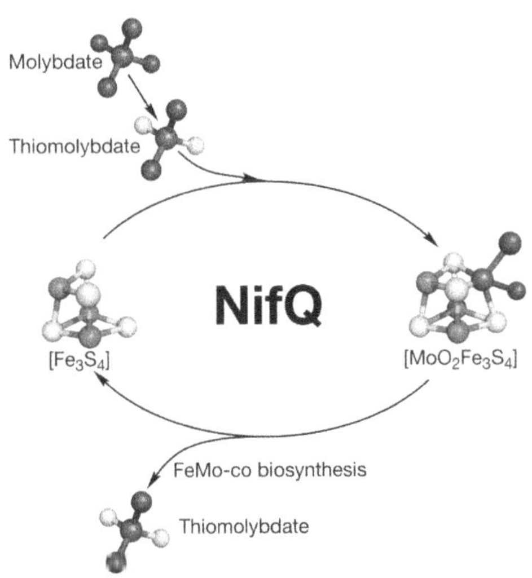

Figure 4.14 Biochemical transformations of Mo thought to occur on the NifQ protein. Binding of Mo by the [3Fe–4S] cluster generates the [Mo- 3Fe–4S]. A sulfur donor might be required to release Mo from the [Mo–3Fe–4S] cluster.

can successfully replace homocitrate in the *in vitro* FeMo-co synthesis system, and the resulting MoFe proteins exhibit altered substrate specificities and inhibitor susceptibilities (Imperial *et al.*, 1989). Thus, a mechanism that discriminates against organic acids similar to homocitrate must exist *in vivo*. It is also possible that the NifV activity produces such a large intracellular concentration of homocitrate that it avoids the incorporation of other analogues into FeMo-co.

Homocitrate incorporation into FeMo-co seems to follow Mo incorporation because ^{99}Mo can be incorporated into the VK-cluster in the absence of any organic acid (Rangaraj and Ludden, 2002). Citrate is an important analogue for the investigation of homocitrate incorporation into FeMo-co synthesis because it shifts the accumulation of ^{99}Mo FeMo-co precursors towards NifX instead of NifEN and the Fe protein (Rangaraj and Ludden, 2002). Although it is still a matter of debate, we propose that homocitrate incorporation into FeMo-co occurs on NifEN while bound to the Fe protein (Rubio and Ludden, 2008).

The central node in the FeMo-co biosynthetic pathway and the putative NifEN/Fe protein complex

NifEN is a 200-kDa $\alpha_2\beta_2$ tetramer with amino acid sequence similarity to the MoFe protein. Based on this similarity, and in the fact that the MoFe protein is not required for FeMo-co synthesis (Ugalde *et al.*, 1984), it was early proposed that NifEN could act as molecular scaffold for cofactor biosynthesis (Brigle *et al.*, 1987b). Indeed, NifEN function has been shown to be essential to FeMo-co biosynthesis *in vivo* (Jacobson *et al.*, 1989a; Roberts *et al.*, 1978) and *in vitro* (Curatti *et al.*, 2007).

NifEN is an [Fe–S] cluster-containing protein (Paustian *et al.*, 1989) containing two identical [4Fe–4S] clusters, which have been proposed to have an all-cysteinyl coordination at the α/β subunit interface analogously to that of the P-clusters in the MoFe protein (Goodwin *et al.*, 1998). In addition, on certain genetic backgrounds, NifEN contains other [Fe–S] clusters with the ability to serve as Fe and S donors to FeMo-co (Goodwin *et al.*, 1998; Hu *et al.*, 2005b; Soboh *et al.*, 2006).

NifEN is able to bind NifB-co (Allen *et al.*, 1995) and to transform it into the VK-cluster independently of the Fe protein (Hernandez *et al.*, 2007). NifB-co can be transferred directly by NifB or through the metallocluster carrier NifX, which is also able to bind VK-cluster and has been proposed to buffer the flux of FeMo-co precursors towards NifEN (Hernandez *et al.*, 2007). NifB-co and the VK-cluster are complex [Fe–S] cluster FeMo-co precursors that lack Mo and homocitrate (George *et al.*, 2008; Hernandez *et al.*, 2007; Hu *et al.*, 2005b; Shah *et al.*, 1994). They are, however, electronically different since NifB-co is EPR silent whereas the VK-cluster exhibits isotropic $S = 1/2$ and axial $S = 1/2$ EPR signals in the reduced and the oxidized states, respectively. ^{55}Fe labelling experiments suggest the incorporation of additional Fe into the cofactor precursor after the stage of NifB-co (Curatti *et al.*, 2006), which would probably occur at the VK-cluster stage. In addition, EXAFS analysis suggests that NifB-co comprises the [6Fe-9S-X] core of FeMo-co whereas the best EXAFS fits for the VK-cluster included one or two extra Fe atoms at the capping positions of the cofactor (Corbett *et al.*, 2006).

The Δ*nifH* NifEN protein also accumulates substoichiometric amounts of a [Mo–3Fe–(3 + *n*)–S] cluster that resembles the cluster found in the NifQ protein (George *et al.*, 2007). This Mo cluster within NifEN serves as Mo donor for FeMo-co biosynthesis (Soboh *et al.*, 2006) and our proposal is that this cluster would represent an intermediate step before the Fe protein-dependent incorporation of Mo into the VK-cluster that has been observed by others (Hu *et al.*, 2006a; Rangaraj and Ludden, 2002). The origin of the NifEN [Mo–3Fe–(3 + *n*)–S] cluster is unclear. It is tempting to speculate that NifQ would be the direct donor of Mo or even the entire cluster, but Mo transfer from NifQ to NifEN has been shown not to be efficient in the absence of the Fe protein (Hernandez *et al.*, 2008), at least under the conditions of the *in vitro* FeMo-co biosynthesis assay. Another possibility is that the formation of this [Mo–3Fe–(3 + *n*)–S] cluster is independent of NifQ, which could provide an explanation to the observed reversal of the

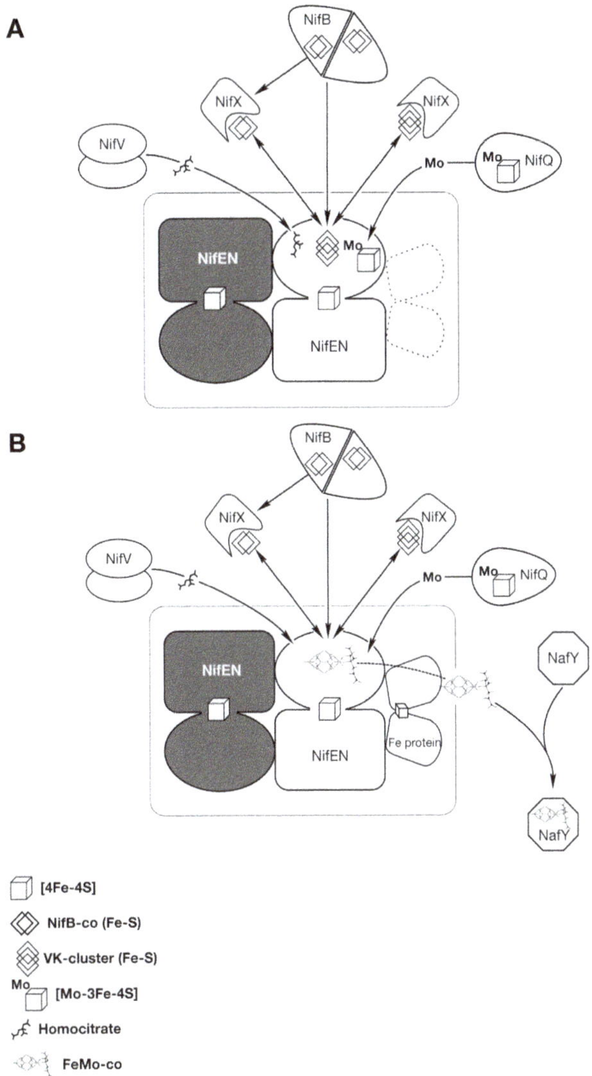

Figure 4.15 FeMo-co biosynthetic precursors in the NifEN scaffold protein. Panel A, accumulation of FeMo-co precursors within NifEN in the *A. vinelandii nifH* mutant. Panel B, flux of FeMo-co biosynthetic precursor through a putative NifEN/Fe protein complex in a nitrogenase derepressed wild-type *A. vinelandii* strain. NafY binds newly synthesized FeMo-co and delivers it to the apo-MoFe protein.

nifQ mutant phenotype by the presence of high levels of Mo and cysteine in the growth medium (Imperial *et al.*, 1984; Ugalde *et al.*, 1985).

Finally, the Δ*nifH* NifEN protein could carry very low levels of homocitrate since purified Δ*nifH* NifEN slightly activated apo-MoFe protein in FeMo-co biosynthesis assays lacking externally added homocitrate (Soboh *et al.*, 2006). *In vitro* experiments of homocitrate binding to NifEN would be required to confirm this proposal. Fig. 4.15A illustrates the accumulation of FeMo-co precursors observed to occur within the NifEN protein in a Δ*nifH A. vinelandii* mutant strain, which is completely deficient in FeMo-co biosynthesis *in vivo* (Robinson *et al.*, 1987).

Fig. 4.15B illustrates the Fe protein requirement in FeMo-co biosynthesis by showing the flux of FeMo-co biosynthetic precursors through NifEN in the presence of Fe protein. At this step of the pathway, the outcome of the joint participation of NifEN and the Fe protein together with the presence MgATP and reducing conditions is the synthesis of FeMo-co. The Fe protein requirement has been unequivocally demonstrated using the defined *in vitro* FeMo-co synthesis assay (Allen *et al.*, 1993; Curatti *et al.*, 2007).

The exact role of the Fe protein in FeMo-co synthesis is unclear. This is in part due to contradictory reports found in the literature. On one hand, neither the presence of the [4Fe–4S] cluster (Rangaraj *et al.*, 1997) nor the hydrolysis of MgATP (Gavini and Burgess, 1992; Rangaraj *et al.*, 1999) were shown to be necessary for the function of the Fe protein in FeMo-co biosynthesis. Moreover, *nifM* mutants that accumulate an immature form of the Fe protein that is catalytically inactive as dinitrogenase reductase remained competent for FeMo-co biosynthesis (Jacobson *et al.*, 1989b; Roberts *et al.*, 1978). On the other hand, there are *in vitro* experiments in which FeMo-co synthesis correlates with the presence of the [4Fe–4S] cluster of the Fe protein and with its ability to hydrolyse ATP (Hu *et al.*, 2006b).

The other puzzling observation is, that although the Fe protein has been repeatedly proposed as having a role in specifying molybdenum as the heterometal to be incorporated into FeMo-co (e.g. Hu *et al.*, 2006b; Rangaraj and Ludden, 2002), it has been demonstrated that the Fe protein homologue from the vanadium system (VnfH) can replace its function in the *in vitro* biosynthesis of FeMo-co (Chatterjee *et al.*, 1997). We have recently proposed that the role of the Fe protein could be to dock with NifEN and promote some sort of conformational change in the scaffold protein that would result in the incorporation of Mo into the VK-cluster (Rubio and Ludden, 2008).

The NifX and NafY metallocluster carrier proteins

NifX and NafY are members of a family of homologous proteins capable of binding FeMo-co or the structurally related FeMo-co precursors, NifB-co and the VK cluster (Table 4.1). The domain composition of members of the NifX family of proteins that have been biochemically characterized is shown in Fig. 4.16. Proteins of the NifX family bind the metalloclusters through a conserved domain denominated as the core domain. The NifX family includes proteins involved in FeMo-co biosynthesis (NifX, NafY, NifY and NifB) as well as proteins involved in the biosynthesis of the analogous iron-vanadium cofactor (FeV-co) located at the active site of the vanadium nitrogenase (VnfX, VnfY) (Rubio *et al.*, 2002; Ruttimann-Johnson *et al.*, 2003). NifX, VnfX and VnfY are the smallest members of the family and appear to be composed only by the core domain. NifY and NafY share amino acid sequence similarity through the entire polypeptide sequences and are composed of an N-terminal domain, with no known homologues in protein databases, and a C-terminal core domain. The NMR structure of N-terminal domain of NafY has revealed that it belongs to the sterile alpha motif family of domains which are frequently involved in protein–protein interactions (Hernandez *et al.*, unpublished results). Finally, two domains have been identified in NifB proteins: an N-terminal AdoMet domain (SAM radical) and a C-terminal core domain (Sofia *et al.*, 2001).

Neither NifX nor NafY are essential proteins for FeMo-co biosynthesis *in vivo* (Rubio *et al.*, 2002) or *in vitro* (Curatti *et al.*, 2007). However, they do play a role in cofactor biosynthesis or insertion because their presence in the *in vitro*

Table 4.1 Cluster binding capability of 'core-domain'-containing proteins

Protein	NifB-co	VK-cluster	FeMo-co
NafY	Yes	n.d.	Yes
NifY	n.d	n.d.	n.d.
NifX	Yes	Yes	Yes
VnfX	Yes	n.d.	Yes
NifB	Yes	n.d.	n.d.

n.d., binding capability not determined.

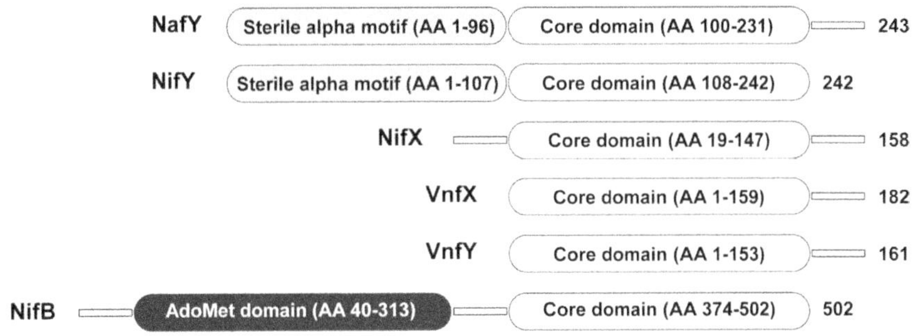

Figure 4.16 Conservation of the 'core domain' among members of the NafY/NifY/NifX/VnfX/VnfY/NifB family of proteins. NafY and NifY share an N-terminal sterile alpha motif and a C-terminal core domain. NifB carries an N-terminal AdoMet domain and a C-terminal core domain. Polypeptide lengths are not drawn to scale.

FeMo-co synthesis assay increases the yield of nitrogenase reconstitution (Curatti *et al.*, 2007). The capability of NifX to bind NifB-co from purified NifB protein (Zhao *et al.*, 2007) or to deliver NifB-co to the NifEN protein (Hernandez *et al.*, 2007; Rangaraj *et al.*, 2001) strongly suggests that the role of NifX is to carry NifB-co from NifB to NifEN. In addition, based on the exchange of VK-cluster between NifEN and NifX, it was suggested that NifX could also serve as storage of FeMo-co precursors to help control their flux during FeMo-co synthesis (Hernandez *et al.*, 2007).

NafY is a monomeric protein capable of binding either apo-MoFe protein or FeMo-co with high affinity (Rubio *et al.*, 2004). A role for NafY in the stabilization of the FeMo-co binding sites in the Δ*nifB* apo-MoFe protein has been established (see above) (Rubio *et al.*, 2002). A second role for NafY in FeMo-co insertion into apo-MoFe protein was proposed based on the bifunctional properties of NafY (Homer *et al.*, 1995). Indeed, an *in vitro*-generated NafY:FeMo-co complex is capable of activating Δ*nifB* apo-MoFe protein with high efficiency (Rubio, unpublished results). NafY relies on its N-terminal domain to bind tightly to apo-MoFe protein. We propose that, through this interaction, the FeMo-co binding module of NafY is recruited to the apo-MoFe protein thus facilitating cofactor delivery (Hernandez *et al.*, 2011).

Future directions

The field of nitrogenase structure and function has evolved dramatically. However, some of the most important questions have eluded biochemists to date, largely due to the complexity of the nitrogenase enzyme. The questions that remain unanswered to a large extent are (1) the mechanism of substrate (N_2) binding and reduction at the FeMo-co active site, (2) the role of MgATP binding and hydrolysis, and (3) the remaining ambiguities concerning cluster biosyntheses. With regard to these questions the difficulties have been largely a result of the inability to capture intermediates that can be characterized by spectroscopic or structural approaches. The main problem here is the inability to synchronize the system over the course of the multiple electron transfer and metal cluster reduction steps required to bind and reduce N_2. Creative genetic approaches have been implemented to alter substrate binding and identify the region of the cofactor that substrates approach, and recent spectroscopic studies of wild-type MoFe proteins and MoFe proteins altered by site directed mutagenesis have implicated N_2 binding at an Fe site on the FeMo-co. The individual reduction steps and the precise binding mode of N_2 and intermediates along the pathway still remain unclear. Certainly the approach of combining site directed mutagenesis and cutting edge spectroscopic methods has helped provide the tools necessary to advance the problem and it is anticipated that the characterization of additional altered MoFe proteins and captured intermediates will allow us to continue along the path of solving this important and difficult problem. With regard to the role of MgATP the challenges are similar in terms of the difficulties of capturing and characterizing the appropriate intermediates to tell the complete story. Clearly, as in the case of substrate binding and reduction, advances are being made in terms of the determination of nitrogenase complex structures and recent spectroscopic work.

Acknowledgements

Research in the Peters Laboratory is supported by the NASA Astrobiology Institute (NAI), the Department of Energy, and the AirForce Office of Scientific Research. T.L.H. was supported by an NSF-Integrated Graduate Educational Research and Training fellowship grant. E.S.B. graciously acknowledges support from the NAI Postdoctoral Program. Research in the Rubio lab is supported by an ERC Starting Grant 205442 and by MICINN BIO2009-12661 Grant.

References

Affourtit, J., Zehr, J.P., and Paerl, H.W. (2001). Distribution of nitrogen-fixing microorganisms along the Neuse River Estuary, North Carolina. Microb. Ecol. *41*, 114–123.

Agar, J.N., Yuvaniyama, P., Jack, R.F., Cash, V.L., Smith, A.D., Dean, D.R., and Johnson, M.K. (2000). Modular organization and identification of a mononuclear iron-binding site within the NifU protein. J. Biol. Inorg. Chem. *5*, 167–177.

Allen, R.M., Chatterjee, R., Ludden, P.W., and Shah, V.K. (1995). Incorporation of iron and sulfur from NifB cofactor into the iron-molybdenum cofactor of dinitrogenase. J. Biol. Chem. *270*, 26890–26896.

Allen, R.M., Homer, M.J., Chatterjee, R., Ludden, P.W., Roberts, G.P., and Shah, V.K. (1993). Dinitrogenase reductase- and MgATP-dependent maturation of apodinitrogenase from *Azotobacter vinelandii*. J. Biol. Chem. *268*, 23670–23674.

Anderson, G.L., and Howard, J.B. (1984). Reactions with the oxidized iron protein of *Azotobacter vinelandii* nitrogenase: formation of a 2Fe center. Biochemistry *23*, 2118–2122.

Angove, H.C., Yoo, S.J., Munck, E., and Burgess, B.K. (1998). An all-ferrous state of the Fe protein of nitrogenase. Interaction with nucleotides and electron transfer to the MoFe protein. J. Biol. Chem. *273*, 26330–26337.

Barney, B.M., Igarashi, R.Y., Dos Santos, P.C., Dean, D.R., and Seefeldt, L.C. (2004). Substrate interaction at an iron–sulfur face of the FeMo-cofactor during nitrogenase catalysis. J. Biol. Chem. *279*, 53621–53624.

Barney, B.M., Laryukhin, M., Igarashi, R.Y., Lee, H.I., Dos Santos, P.C., Yang, T.C., Hoffman, B.M., Dean, D.R., and Seefeldt, L.C. (2005a). Trapping a hydrazine reduction intermediate on the nitrogenase active site. Biochemistry *44*, 8030–8037.

Barney, B.M., Lee, H.I., Dos Santos, P.C., Hoffman, B.M., Dean, D.R., and Seefeldt, L.C. (2006a). Breaking the N_2 triple bond: insights into the nitrogenase mechanism. Dalton Trans., 2277–2284.

Barney, B.M., Lukoyanov, D., Igarashi, R.Y., Laryukhin, M., Yang, T.C., Dean, D.R., Hoffman, B.M., and Seefeldt, L.C. (2009). Trapping an intermediate of dinitrogen (N_2) reduction on nitrogenase. Biochemistry *48*, 9094–9102.

Barney, B.M., Lukoyanov, D., Yang, T.C., Dean, D.R., Hoffman, B.M., and Seefeldt, L.C. (2006b). A methyldiazene ($HN{=}N\text{-}CH_3$)-derived species bound to the nitrogenase active-site FeMo cofactor: Implications for mechanism. Proc. Natl. Acad. Sci. U.S.A. *103*, 17113–17118.

Barney, B.M., McClead, J., Lukoyanov, D., Laryukhin, M., Yang, T.C., Dean, D.R., Hoffman, B.M., and Seefeldt, L.C. (2007). Diazene (HN=NH) is a substrate for nitrogenase: insights into the pathway of N_2 reduction. Biochemistry *46*, 6784–6794.

Barney, B.M., Yang, T.C., Igarashi, R.Y., Dos Santos, P.C., Laryukhin, M., Lee, H.I., Hoffman, B.M., Dean, D.R., and Seefeldt, L.C. (2005b). Intermediates trapped during nitrogenase reduction of N triple bond N, $CH_3\text{-}N{=}NH$, and $H_2N\text{-}NH_2$. J. Am. Chem. Soc. *127*, 14960–14961.

Beinert, H., Holm, R.H., and Munck, E. (1997). Iron–sulfur clusters: nature's modular, multipurpose structures. Science *277*, 653–659.

Benton, P.M., Laryukhin, M., Mayer, S.M., Hoffman, B.M., Dean, D.R., and Seefeldt, L.C. (2003). Localization of a substrate binding site on the FeMo-cofactor in nitrogenase: trapping propargyl alcohol with an α-70-substituted MoFe protein. Biochemistry *42*, 9102–9109.

Bolin, J.T., Ronco, A.E., Morgan, T.V., Mortenson, L.E., and Xuong, N.H. (1993). The unusual metal clusters of nitrogenase: structural features revealed by x-ray anomalous diffraction studies of the MoFe protein from *Clostridium pasteurianum*. Proc. Natl. Acad. Sci. U.S.A. *90*, 1078–1082.

Brigle, K.E., Newton, W.E., and Dean, D.R. (1985). Complete nucleotide sequence of the *Azotobacter vinelandii* nitrogenase structural gene cluster. Gene 37, 37–44.

Brigle, K.E., Setterquist, R.A., Dean, D.R., Cantwell, J.S., Weiss, M.C., and Newton, W.E. (1987a). Site-directed mutagenesis of the nitrogenase MoFe protein of *Azotobacter vinelandii*. Proc. Natl. Acad. Sci. U.S.A. *84*, 7066–7069.

Brigle, K.E., Weiss, M.C., Newton, W.E., and Dean, D.R. (1987b). Products of the iron-molybdenum cofactor-specific biosynthetic genes, *nifE* and *nifN*, are structurally homologous to the products of the nitrogenase molybdenum-iron protein genes, *nifD* and *nifK*. J. Bacteriol. *169*, 1547–1553.

Broach, R.B., Rupnik, K., Hu, Y., Fay, A.W., Cotton, M., Ribbe, M.W., and Hales, B.J. (2006). Variable-temperature, variable-field magnetic circular dichroism spectroscopic study of the metal clusters in the Δ*nifB* and Δ*nifH* MoFe proteins of nitrogenase from *Azotobacter vinelandii*. Biochemistry 45, 15039–15048.

Brocker, M.J., Schomburg, S., Heinz, D.W., Jahn, D., Schubert, W.D., and Moser, J. (2010). Crystal structure of the nitrogenase-like dark operative protochlorophyllide oxidoreductase catalytic complex (ChlN/ChlB)2. J. Biol. Chem. *285*, 27336–27345.

Brocker, M.J., Virus, S., Ganskow, S., Heathcote, P., Heinz, D.W., Schubert, W.D., Jahn, D., and Moser, J. (2008a). ATP-driven reduction by dark-operative protochlorophyllide oxidoreductase from *Chlorobium tepidum* mechanistically resembles nitrogenase catalysis. J. Biol. Chem. *283*, 10559–10567.

Brocker, M.J., Watzlich, D., Uliczka, F., Virus, S., Saggu, M., Lendzian, F., Scheer, H., Rudiger, W., Moser, J., and Jahn, D. (2008b). Substrate recognition of nitrogenase-like dark operative protochlorophyllide oxidoreductase from *Prochlorococcus marinus*. J. Biol. Chem. *283*, 29873–29881.

Bui, P.T., and Mortenson, L.E. (1969). The hydrolysis of adenosine triphosphate by purified components of nitrogenase. Biochemistry *8*, 2462–2465.

Bulen, W.A., and LeComte, J.R. (1966). The nitrogenase system from *Azotobacter*: two-enzyme requirement for N_2 reduction, ATP-dependent H_2 evolution, and ATP hydrolysis. Proc. Natl. Acad. Sci. U.S.A. *56*, 979–986.

Burgess, B.K., and Lowe, D.J. (1996). Mechanism of molybdenum nitrogenase. Chem. Rev. *96*, 2983–3012.

Burke, D.H., Hearst, J.E., and Sidow, A. (1993). Early evolution of photosynthesis: clues from nitrogenase and chlorophyll iron proteins. Proc. Natl. Acad. Sci. U.S.A. *90*, 7134–7138.

Butt, J.N., Armstrong, F.A., Breton, J., George, S.J., Thomson, A.J., and Hatchikian, E.C. (1991). Investigation of metal ion uptake reactivities of [3Fe-4S] clusters in proteins: voltammetry of co-adsorbed ferredoxin-aminocyclitol films at graphite electrodes and spectroscopic identification of transformed clusters. J. Am. Chem. Soc. *113*, 6663–6670.

Cameron, L.M., and Hales, B.J. (1998). Investigation of CO binding and release from Mo-nitrogenase during catalytic turnover. Biochemistry *37*, 9449–9456.

Carnahan, J.E., Mortenson, L.E., Mower, H.F., and Castle, J.E. (1960). Nitrogen fixation in cell-free extracts of *Clostridium pasteurianum*. Biochim. Biophys. Acta *44*, 520–535.

Chan, J.M., Christiansen, J., Dean, D.R., and Seefeldt, L.C. (1999). Spectroscopic evidence for changes in the redox state of the nitrogenase P-cluster during turnover. Biochemistry *38*, 5779–5785.

Chan, J.M., Wu, W., Dean, D.R., and Seefeldt, L.C. (2000). Construction and characterization of a heterodimeric iron protein: defining roles for adenosine triphosphate in nitrogenase catalysis. Biochemistry *39*, 7221–7228.

Chan, M.K., Kim, J., and Rees, D.C. (1993). The nitrogenase FeMo-cofactor and P-cluster pair: 2.2 Å resolution structures. Science *260*, 792–794.

Chatt, J. (1980). Chemistry relevant to the biological fixation of nitrogen. In Nitrogen Fixation, W.D.P. Stewart, and J.R. Gallon, eds. (London, Academic Press), pp. 1–18.

Chatterjee, R., Allen, R.M., Ludden, P.W., and Shah, V.K. (1997). *In vitro* synthesis of the iron-molybdenum cofactor and maturation of the *nif*-encoded apodinitrogenase. Effect of substitution of VnfH for NifH. J. Biol. Chem. *272*, 21604–21608.

Chen, L., Gavini, N., Tsuruta, H., Eliezer, D., Burgess, B.K., Doniach, S., and Hodgson, K.O. (1994). MgATP-induced conformational changes in the iron protein from *Azotobacter vinelandii*, as studied by small-angle x-ray scattering. J. Biol. Chem. *269*, 3290–3294.

Chisnell, J.R., Premakumar, R., and Bishop, P.E. (1988). Purification of a second alternative nitrogenase from a *nifHDK* deletion strain of *Azotobacter vinelandii*. J. Bacteriol. *170*, 27–33.

Chiu, H., Peters, J.W., Lanzilotta, W.N., Ryle, M.J., Seefeldt, L.C., Howard, J.B., and Rees, D.C. (2001). MgATP-bound and nucleotide-free structures of a nitrogenase protein complex between the Leu 127Δ-Fe-protein and the MoFe-protein. Biochemistry *40*, 641–650.

Christiansen, J., Cash, V.L., Seefeldt, L.C., and Dean, D.R. (2000). Isolation and characterization of an acetylene-resistant nitrogenase. J. Biol. Chem. *275*, 11459–11464.

Christiansen, J., Goodwin, P.J., Lanzilotta, W.N., Seefeldt, L.C., and Dean, D.R. (1998). Catalytic and biophysical properties of a nitrogenase Apo-MoFe protein produced by a *nifB*-deletion mutant of *Azotobacter vinelandii*. Biochemistry 37, 12611–12623.

Christiansen, J., Tittsworth, R.C., Hales, B.J., and Cramer, S.P. (1995). Fe and Mo EXAFS of *Azotobacter vinelandii* nitrogenase in partially oxidized and singly reduced Forms. J. Am. Chem. Soc. *117*, 10017–10024..

Christie, P.D., Lee, H.-I., Cameron, L.M., Hales, B.J., Orme-Johnson, W.H., and Hoffman, B.M. (1996). Identification of the CO-binding cluster in nitrogenase MoFe protein by ENDOR of ^{57}Fe isotopomers. J. Am. Chem. Soc. *118*, 8707–8709.

Corbett, M.C., Hu, Y., Fay, A.W., Ribbe, M.W., Hedman, B., and Hodgson, K.O. (2006). Structural insights into a protein-bound iron-molybdenum cofactor precursor. Proc. Natl. Acad. Sci. U.S.A. *103*, 1238–1243.

Corbett, M.C., Hu, Y., Fay, A.W., Tsuruta, H., Ribbe, M.W., Hodgson, K.O., and Hedman, B. (2007). Conformational differences between *Azotobacter vinelandii* nitrogenase MoFe proteins as studied by small-angle X-ray scattering. Biochemistry *46*, 8066–8074.

Corbett, M.C., Hu, Y., Naderi, F., Ribbe, M.W., Hedman, B., and Hodgson, K.O. (2004). Comparison of iron-molybdenum cofactor-deficient nitrogenase MoFe proteins by X-ray absorption spectroscopy: implications for P-cluster biosynthesis. J. Biol. Chem. *279*, 28276–28282.

Curatti, L., Brown, C.S., Ludden, P.W., and Rubio, L.M. (2005). Genes required for rapid expression of nitrogenase activity in *Azotobacter vinelandii*. Proc. Natl. Acad. Sci. U.S.A. *102*, 6291–6296.

Curatti, L., Hernandez, J.A., Igarashi, R.Y., Soboh, B., Zhao, D., and Rubio, L.M. (2007). *In vitro* synthesis of the iron-molybdenum cofactor of nitrogenase from iron, sulfur, molybdenum and homocitrate using purified proteins. Proc. Natl. Acad. Sci. U.S.A. *104*, 17626–17631.

Curatti, L., Ludden, P.W., and Rubio, L.M. (2006). NifB-dependent *in vitro* synthesis of the iron-molybdenum cofactor of nitrogenase. Proc .Natl. Acad. Sci. U.S.A. *103*, 5297–5301.

Danyal, K., Mayweather, D., Dean, D.R., Seefeldt, L.C., and Hoffman, B.M. (2010). Conformational gating of electron transfer from the nitrogenase Fe protein to MoFe protein. J. Am. Chem. Soc. *132*, 6894–6895.

Davis, L.C., Henzl, M.T., Burris, R.H., and Orme-Johnson, W.H. (1979). Iron–sulfur clusters in the molybdenum-iron protein component of nitrogenase. Electron paramagnetic resonance of the carbon monoxide inhibited state. Biochemistry *18*, 4860–4869.

Davis, L.C., Shah, V.K., Brill, W.J., and Orme-Johnson, W.H. (1972). Nitrogenase. II. Changes in the EPR signal of component I (iron-molybdenum protein) of *Azotobacter vinelandii* nitrogenase during repression and derepression. Biochim. Biophys. Acta *256*, 512–523.

Davis, L.C., and Wang, Y.L. (1980). *In vivo* and *in vitro* kinetics of nitrogenase. J. Bacteriol. *141*, 1230–1238.

Dean, D.R., Setterquist, R.A., Brigle, K.E., Scott, D.J., Laird, N.F., and Newton, W.E. (1990). Evidence that conserved residues Cys-62 and Cys-154 within the *Azotobacter vinelandii* nitrogenase MoFe protein α-subunit are essential for nitrogenase activity but conserved residues His-83 and Cys-88 are not. Mol. Microbiol. *4*, 1505–1512.

Dekas, A.E., Poretsky, R.S., and Orphan, V.J. (2009). Deep-sea archaea fix and share nitrogen in methane-consuming microbial consortia. Science *326*, 422–426.

Dilworth, M.J., Subramanian, D., Munson, T.O., and Burris, R.H. (1965). The adenosine triphosphate requirement for nitrogen fixation in cell-free extracts of *Clostridium pasteurianum*. Biochim. Biophys. Acta *99*, 486–503.

Dixon, R., Eady, R.R., Espin, G., Hill, S., Iaccarino, M., Kahn, D., and Merrick, M. (1980). Analysis of regulation of *Klebsiella pneumoniae* nitrogen fixation (*nif*) gene cluster with gene fusions. Nature *286*, 128–132.

Dos Santos, P.C., Dean, D.R., Hu, Y., and Ribbe, M.W. (2004a). Formation and insertion of the nitrogenase iron-molybdenum cofactor. Chem. Rev. *104*, 1159–1173.

Dos Santos, P.C., Igarashi, R.Y., Lee, H.I., Hoffman, B.M., Seefeldt, L.C., and Dean, D.R. (2005). Substrate interactions with the nitrogenase active site. Acc. Chem. Res. *38*, 208–214.

Dos Santos, P.C., Johnson, D.C., Ragle, B.E., Unciuleac, M.C., and Dean, D.R. (2007a). Controlled expression of *nif* and *isc* iron–sulfur protein maturation components reveals target specificity and limited functional replacement between the two systems. J. Bacteriol. *189*, 2854–2862.

Dos Santos, P.C., Mayer, S.M., Barney, B.M., Seefeldt, L.C., and Dean, D.R. (2007b). Alkyne substrate interaction within the nitrogenase MoFe protein. J. Inorg. Biochem. *101*, 1642–1648.

Dos Santos, P.C., Smith, A.D., Frazzon, J., Cash, V.L., Johnson, M.K., and Dean, D.R. (2004b). Iron–sulfur cluster assembly: NifU-directed activation of the nitrogenase Fe protein. J. Biol. Chem. *279*, 19705–19711.

Downie, J.A. (2005). Legume haemoglobins: symbiotic nitrogen fixation needs bloody nodules. Curr. Biol. *15*, R196–198.

Drummond, M., Walmsley, J., and Kennedy, C. (1996). Expression from the *nifB* promoter of *Azotobacter vinelandii* can be activated by NifA, VnfA, or AnfA transcriptional activators. J. Bacteriol. *178*, 788–792.

Duyvis, M.G., Wassink, H., and Haaker, H. (1996). Formation and characterization of a transition state complex of *Azotobacter vinelandii* nitrogenase. FEBS Lett. *380*, 233–236.

Eady, R.R. (1996). Structure–function relationships of alternative nitrogenases. Chem. Rev *96*, 3013–3030.

Eady, R.R. (2003). Current status of structure function relationships of vanadium nitrogenase. Coordin. Chem. Rev. *237*, 23–30.

Eady, R.R., Robson, R.L., Richardson, T.H., Miller, R.W., and Hawkins, M. (1987). The vanadium nitrogenase of *Azotobacter chroococcum*. Purification and properties of the VFe protein. Biochem. J. *244*, 197–207.

Eady, R.R., Smith, B.E., Cook, K.A., and Postgate, J.R. (1972). Nitrogenase of *Klebsiella pneumoniae*. Purification and properties of the component proteins. Biochem. J. *128*, 655–675.

Einsle, O., Tezcan, F.A., Andrade, S.L., Schmid, B., Yoshida, M., Howard, J.B., and Rees, D.C. (2002). Nitrogenase MoFe-protein at 1.16 Å resolution: a central ligand in the FeMo-cofactor. Science *297*, 1696–1700.

Emerich, D.W., Hageman, R.V., and Burris, R.H. (1981). Interactions of dinitrogenase and dinitrogenase reductase. Adv. Enzymol. Relat. Areas Mol. Biol. *52*, 1–22.

Fani, R., Gallo, R., and Lio, P. (2000). Molecular evolution of nitrogen fixation: the evolutionary history of the *nifD*, *nifK*, *nifE*, and *nifN* genes. J. Mol. Evol. *51*, 1–11.

Fay, P. (1992). Oxygen relations of nitrogen fixation in cyanobacteria. Microbiol. Rev. *56*, 340–373.

Fisher, A.J., Smith, C.A., Thoden, J.B., Smith, R., Sutoh, K., Holden, H.M., and Rayment, I. (1995). X-ray structures of the myosin motor domain of *Dictyostelium discoideum* complexed with MgADP•BeF_x and MgADP•AlF_4. Biochemistry *34*, 8960–8972.

Fu, W., Jack, R.F., Morgan, T.V., Dean, D.R., and Johnson, M.K. (1994). *nifU* gene product from *Azotobacter vinelandii* is a homodimer that contains two identical [2Fe–2S] clusters. Biochemistry *33*, 13455–13463.

Fujita, Y. (1996). Protochlorophyllide reduction: a key step in the greening of plants. Plant Cell .Physiol. *37*, 411–421.

Fujita, Y., and Bauer, C.E. (2000). Reconstitution of light-independent protochlorophyllide reductase from purified bchl and BchN-BchB subunits. *In vitro* confirmation of nitrogenase-like features of a bacteriochlorophyll biosynthesis enzyme. J. Biol. Chem. *275*, 23583–23588.

Fujita, Y., Takagi, H., and Hase, T. (1998). Cloning of the gene encoding a protochlorophyllide reductase: the physiological significance of the co-existence of light-dependent and -independent protochlorophyllide reduction systems in the cyanobacterium *Plectonema boryanum*. Plant Cell. Physiol. *39*, 177–185.

Gavini, N., and Burgess, B.K. (1992). FeMo cofactor synthesis by a *nifH* mutant with altered MgATP reactivity. J. Biol. Chem. *267*, 21179–21186.

George, S.J., Ashby, G., Wharton, C.W., and Thorneley, R.N.F. (1997). Time-resolved binding of carbon monoxide to nitrogenase monitored by stopped-flow infrared spectroscopy. J. Am. Chem. Soc. *119*, 6450–6451.

George, S.J., Igarashi, R.Y., Piamonteze, C., Soboh, B., Cramer, S.P., and Rubio, L.M. (2007). Identification of a Mo-Fe–S cluster on NifEN by Mo K-edge extended X-ray absorption fine structure. J. Am. Chem. Soc. *129*, 3060–3061.

George, S.J., Igarashi, R.Y., Xiao, Y., Hernandez, J.A., Demuez, M., Zhao, D., Yoda, Y., Ludden, P.W., Rubio, L.M., and Cramer, S.P. (2008). Extended X-ray absorption fine structure and nuclear resonance vibrational spectroscopy reveal that NifB-co, a FeMo-co precursor, comprises a 6Fe core with an interstitial light atom. J. Am. Chem. Soc. *130*, 5673–5680.

Georgiadis, M.M., Komiya, H., Chakrabarti, P., Woo, D., Kornuc, J.J., and Rees, D.C. (1992). Crystallographic structure of the nitrogenase iron protein from *Azotobacter vinelandii*. Science *257*, 1653–1659.

Goodwin, P.J., Agar, J.N., Roll, J.T., Roberts, G.P., Johnson, M.K., and Dean, D.R. (1998). The *Azotobacter vinelandii* NifEN complex contains two identical [4Fe–4S] clusters. Biochemistry *37*, 10420–10428.

Govezensky, D., Greener, T., Segal, G., and Zamir, A. (1991). Involvement of GroEL in *nif* gene regulation and nitrogenase assembly. J. Bacteriol. *173*, 6339–6346.

Govezensky, D., and Zamir, A. (1989). Structure–function relationships in the alpha subunit of *Klebsiella pneumoniae* nitrogenase MoFe protein from analysis of *nifD* mutants. J. Bacteriol. *171*, 5729–5735.

Grossmann, J.G., Hasnain, S.S., Yousafzai, F.K., and Eady, R.R. (2001). Evidence for the selective population of FeMo-cofactor sites in MoFe protein and its molecular recognition by the Fe protein in transition state complex analogues of nitrogenase. J. Biol. Chem. *276*, 6582–6590.

Hageman, R.V., and Burris, R.H. (1980). Electron allocation to alternative substrates of *Azotobacter* nitrogenase is controlled by the electron flux through dinitrogenase. Biochim. Biophys. Acta *591*, 63–75.

Hagen, W.R., Dunham, W.R., Braaksma, A., and Haaker, H. (1985a). On the prosthetic group(s) of component II from nitrogenase. EPR of the Fe-protein from *Azotobacter vinelandii*. FEBS Lett *187*, 146–150.

Hagen, W.R., Eady, R.R., Dunham, W.R., and Haaker, H. (1985b). A novel S = 3/2 EPR signal associated with native Fe proteins of nitrogenase. FEBS Lett. *189*, 250–254.

Hagen, W.R., Wassink, H., Eady, R.R., Smith, B.E., and Haaker, H. (1987). Quantitative EPR of an S = 7/2 system in thionine-oxidized MoFe proteins of nitrogenase. A redefinition of the P-cluster concept. Eur. J. Biochem. *169*, 457–465.

Hales, B.J., and Blanchard, C.Z. (1996). Isolation of two forms of the nitrogenase VFe protein from *Azotobacter vinelandii*. Biochemistry *35*, 472–478.

Hales, B.J., Case, E.E., Morningstar, J.E., Dzeda, M.F., and Mauterer, L.A. (1986). Isolation of a new vanadium-containing nitrogenase from *Azotobacter vinelandii*. Biochemistry *25*, 7251–7255.

Hawkes, T.R., McLean, P.A., and Smith, B.E. (1984). Nitrogenase from *nifV* mutants of *Klebsiella pneumoniae* contains an altered form of the iron–molybdenum cofactor. Biochem. J. *217*, 317–321.

Hernandez, J.A., Igarashi, R.Y., Soboh, B., Curatti, L., Dean, D.R., Ludden, P.W., and Rubio, L.M. (2007). NifX and NifEN exchange NifB cofactor and the VK-cluster, a newly isolated intermediate of the iron-molybdenum cofactor biosynthetic pathway. Mol. Microbiol. *63*, 177–192.

Hernandez, J.A., Curatti, L., Aznar, C.P., Perova, Z., Britt, R.D., and Rubio, L.M. (2008). Metal trafficking for nitrogen fixation: NifQ donates molybdenum to NifEN/NifH for the biosynthesis of the nitrogenase FeMo-cofactor. Proc. Natl. Acad. Sci. U.S.A. *105*, 11679–11684.

Hernandez, J.A., George, S.J., and Rubio, L.M. (2009). Molybdenum trafficking for nitrogen fixation. Biochemistry *48*, 9711–9721.

Hernandez, J.A., Phillips, A.H., Erbil, W.K., Zhao, D., Demuez, M., Zeymer, C., Pelton, J.G., Wemmer, D.E., and Rubio, L.M. (2011). A sterile alpha-motif domain in NafY targets apo-NifDK for iron-molybdenum cofactor delivery via a tethered domain. J. Biol. Chem. *286*, 6321–6328.

Hoffman, B.M., Dean, D.R., and Seefeldt, L.C. (2009). Climbing nitrogenase: toward a mechanism of enzymatic nitrogen fixation. Acc. Chem Res. *42*, 609–619.

Holm, R.H., Kennepohl, P., and Solomon, E.I. (1996). Structural and functional aspects of metal sites in biology. Chem Rev. *96*, 2239–2314.

Homer, M.J., Dean, D.R., and Roberts, G.P. (1995). Characterization of the γ protein and its involvement in the metallocluster assembly and maturation of dinitrogenase from *Azotobacter vinelandii*. J. Biol. Chem. *270*, 24745–24752.

Homer, M.J., Paustian, T.D., Shah, V.K., and Roberts, G.P. (1993). The *nifY* product of *Klebsiella pneumoniae* is associated with apodinitrogenase and dissociates upon activation with the iron-molybdenum cofactor. J. Bacteriol. *175*, 4907–4910.

Hoover, T.R., Imperial, J., Ludden, P.W., and Shah, V.K. (1988a). Biosynthesis of the iron-molybdenum cofactor of nitrogenase. Biofactors *1*, 199–205.

Hoover, T.R., Imperial, J., Ludden, P.W., and Shah, V.K. (1988b). Homocitrate cures the NifV- phenotype in *Klebsiella pneumoniae*. J. Bacteriol. *170*, 1978–1979.

Hoover, T.R., Imperial, J., Ludden, P.W., and Shah, V.K. (1989). Homocitrate is a component of the iron-molybdenum cofactor of nitrogenase. Biochemistry *28*, 2768–2771.

Hoover, T.R., Shah, V.K., Roberts, G.P., and Ludden, P.W. (1986). *nifV*-dependent, low-molecular-weight factor required for *in vitro* synthesis of iron-molybdenum cofactor of nitrogenase. J. Bacteriol. *167*, 999–1003.

Howard, J.B., and Rees, D.C. (1994). Nitrogenase: a nucleotide-dependent molecular switch. Annu. Rev. Biochem. *63*, 235–264.

Hu, Y., Corbett, M.C., Fay, A.W., Webber, J.A., Hedman, B., Hodgson, K.O., and Ribbe, M.W. (2005a). Nitrogenase reactivity with P-cluster variants. Proc. Natl. Acad. Sci. U.S.A. *102*, 13825–13830.

Hu, Y., Corbett, M.C., Fay, A.W., Webber, J.A., Hodgson, K.O., Hedman, B., and Ribbe, M.W. (2006a). FeMo cofactor maturation on NifEN. Proc. Natl. Acad. Sci. U.S.A. *103*, 17119–17124.

Hu, Y., Fay, A.W., Dos Santos, P.C., Naderi, F., and Ribbe, M.W. (2004). Characterization of *Azotobacter vinelandii nifZ* deletion strains. Indication of stepwise MoFe protein assembly. J. Biol. Chem. *279*, 54963–54971.

Hu, Y., Fay, A.W., Lee, C.C., and Ribbe, M.W. (2007). P-cluster maturation on nitrogenase MoFe protein. Proc. Natl. Acad. Sci. U.S.A. *104*, 10424–10429.

Hu, Y., Fay, A.W., Lee, C.C., Yoshizawa, J., and Ribbe, M.W. (2008). Assembly of nitrogenase MoFe protein. Biochemistry *47*, 3973–3981.

Hu, Y., Fay, A.W., and Ribbe, M.W. (2005b). Identification of a nitrogenase FeMo cofactor precursor on NifEN complex. Proc. Natl. Acad. Sci. U.S.A. *102*, 3236–3241.

Hu, Y.L., Corbettt, M.C., Fay, A.W., Webber, J.A., Hodgson, K.O., Hedman, B., and Ribbe, M.W. (2006b). Nitrogenase Fe protein: A molybdate/homocitrate insertase. Proc. Natl. Acad. Sci. U.S.A. *103*, 17125–17130.

Huynh, B.H., Henzl, M.T., Christner, J.A., Zimmermann, R., Orme-Johnson, W.H., and Munck, E. (1980). Nitrogenase XII. Mössbauer studies of the MoFe protein from *Clostridium pasteurianum* W5. Biochim. Biophys. Acta *623*, 124–138.

Huynh, B.H., Munck, E., and Orme-Johnson, W.H. (1979). Nitrogenase XI: Mössbauer studies on the cofactor centers of the MoFe protein from *Azotobacter vinelandii* OP. Biochim. Biophys. Acta *576*, 192–203.

Ibrahim, S.K., Gormal, C., Smith, B.E., Pickett, C.J., Vincent, K., and Best, S.P. (1999). The isolated iron–molybdenum cofactor of nitrogenase binds carbon monoxide upon electrochemically accessing reduced states. Chem. Commun. *11*, 1019–1020.

Igarashi, R.Y., Dos Santos, P.C., Niehaus, W.G., Dance, I.G., Dean, D.R., and Seefeldt, L.C. (2004). Localization of a catalytic intermediate bound to the FeMo-cofactor of nitrogenase. J. Biol. Chem. *279*, 34770–34775.

Igarashi, R.Y., and Seefeldt, L.C. (2003). Nitrogen fixation: the mechanism of the Mo-dependent nitrogenase. Crit. Rev. Biochem. Mol. Biol. *38*, 351–384.

Imperial, J., Hoover, T.R., Madden, M.S., Ludden, P.W., and Shah, V.K. (1989). Substrate reduction properties of dinitrogenase activated *in vitro* are dependent upon the presence of homocitrate or its analogues during iron-molybdenum cofactor synthesis. Biochemistry *28*, 7796–7799.

Imperial, J., Ugalde, R.A., Shah, V.K., and Brill, W.J. (1984). Role of the *nifQ* gene product in the incorporation of molybdenum into nitrogenase in *Klebsiella pneumoniae*. J. Bacteriol. *158*, 187–194.

Jacobson, M.R., Brigle, K.E., Bennett, L.T., Setterquist, R.A., Wilson, M.S., Cash, V.L., Beynon, J., Newton, W.E., and Dean, D.R. (1989a). Physical and genetic map of the major *nif* gene cluster from *Azotobacter vinelandii*. J. Bacteriol. *171*, 1017–1027.

Jacobson, M.R., Cash, V.L., Weiss, M.C., Laird, N.F., Newton, W.E., and Dean, D.R. (1989b). Biochemical and genetic analysis of the *nifUSVWZM* cluster from *Azotobacter vinelandii*. Mol. Gen. Genet. *219*, 49–57.

Jang, S.B., Jeong, M.S., Seefeldt, L.C., and Peters, J.W. (2004). Structural and biochemical implications of single amino acid substitutions in the nucleotide-dependent switch regions of the nitrogenase Fe protein from *Azotobacter vinelandii*. J. Biol. Inorg. Chem. *9*, 1028–1033.

Jang, S.B., Seefeldt, L.C., and Peters, J.W. (2000). Insights into nucleotide signal transduction in nitrogenase: structure of an iron protein with MgADP bound. Biochemistry *39*, 14745–14752.

Jeng, D.Y., Devanathan, T., and Mortenson, L.E. (1969). Components of cell-free extracts of *Clostridium pasteurianum* W5 required for acetylene reduction and N_2 fixation. Biochem Biophys Res Commun *35*, 625–633.

Joerger, R.D., and Bishop, P.E. (1988). Nucleotide sequence and genetic analysis of the *nifB-nifQ* region from *Azotobacter vinelandii*. J. Bacteriol. *170*, 1475–1487.

Johnson, D.C., Dean, D.R., Smith, A.D., and Johnson, M.K. (2005a). Structure, function, and formation of biological iron–sulfur clusters. Annu. Rev. Biochem. *74*, 247–281.

Johnson, D.C., Dos Santos, P.C., and Dean, D.R. (2005b). NifU and NifS are required for the maturation of nitrogenase and cannot replace the function of isc-gene products in *Azotobacter vinelandii*. Biochem. Soc. Trans. *33*, 90–93.

Jurnak, F. (1985). Structure of the GDP domain of EF-Tu and location of the amino acids homologous to ras oncogene proteins. Science *230*, 32–36.

Kelly, M., Klucas, R.V., and Burris, R.H. (1967). Fractionation and storage of nitrogenase from *Azotobacter vinelandii*. Biochem. J. *105*, 3C–5C.

Kent, H.M., Baines, M., Gormal, C., Smith, B.E., and Buck, M. (1990). Analysis of site-directed mutations in the alpha- and beta-subunits of *Klebsiella pneumoniae* nitrogenase. Mol. Microbiol. *4*, 1497–1504.

Kent, H.M., Ioannidis, I., Gormal, C., Smith, B.E., and Buck, M. (1989). Site-directed mutagenesis of the *Klebsiella pneumoniae* nitrogenase. Effects of modifying conserved cysteine residues in the alpha- and beta-subunits. Biochem. J. *264*, 257–264.

Kim, C.H., Newton, W.E., and Dean, D.R. (1995). Role of the MoFe protein alpha-subunit histidine-195 residue in FeMo-cofactor binding and nitrogenase catalysis. Biochemistry *34*, 2798–2808.

Kim, J., and Rees, D.C. (1992). Structural models for the metal centers in the nitrogenase molybdenum-iron protein. Science *257*, 1677–1682.

Kim, J., Woo, D., and Rees, D.C. (1993). X-ray crystal structure of the nitrogenase molybdenum-iron protein from *Clostridium pasteurianum* at 3.0-Å resolution. Biochemistry *32*, 7104–7115.

Kjeldgaard, M., and Nyborg, J. (1992). Refined structure of elongation factor EF-Tu from *Escherichia coli*. J. Mol. Biol. *223*, 721–742.

Koonin, E.V. (1993). A superfamily of ATPases with diverse functions containing either classical or deviant ATP-binding motif. J. Mol. Biol. *229*, 1165–1174.

Lanzilotta, W.N., Christiansen, J., Dean, D.R., and Seefeldt, L.C. (1998). Evidence for coupled electron and proton transfer in the [8Fe-7S] cluster of nitrogenase. Biochemistry *37*, 11376–11384.

Lanzilotta, W.N., Fisher, K., and Seefeldt, L.C. (1996). Evidence for electron transfer from the nitrogenase iron protein to the molybdenum-iron protein without MgATP hydrolysis: characterization of a tight protein–protein complex. Biochemistry *35*, 7188–7196.

Lanzilotta, W.N., Fisher, K., and Seefeldt, L.C. (1997). Evidence for electron transfer-dependent formation of a nitrogenase iron protein-molybdenum-iron protein tight complex. The role of aspartate 39. J. Biol. Chem. *272*, 4157–4165.

Lanzilotta, W.N., Holz, R.C., and Seefeldt, L.C. (1995). Proton NMR investigation of the $[4Fe–4S]^{1+}$ cluster environment of nitrogenase iron protein from *Azotobacter vinelandii*: defining nucleotide-induced conformational changes. Biochemistry *34*, 15646–15653.

Lanzilotta, W.N., Parker, V.D., and Seefeldt, L.C. (1999). Thermodynamics of nucleotide interactions with the *Azotobacter vinelandii* nitrogenase iron protein. Biochim. Biophys. Acta *1429*, 411–421.

Lanzilotta, W.N., and Seefeldt, L.C. (1997). Changes in the midpoint potentials of the nitrogenase metal centers as a result of iron protein-molybdenum-iron protein complex formation. Biochemistry *36*, 12976–12983.

Lee, C.C., Blank, M.A., Fay, A.W., Yoshizawa, J.M., Hu, Y., Hodgson, K.O., Hedman, B., and Ribbe, M.W. (2009a). Stepwise formation of P-cluster in nitrogenase MoFe protein. Proc. Natl. Acad. Sci. U.S.A. *106*, 18474–18478.

Lee, C.C., Hu, Y., and Ribbe, M.W. (2009b). Unique features of the nitrogenase VFe protein from *Azotobacter vinelandii*. Proc. Natl. Acad. Sci. U.S.A. *106*, 9209–9214.

Lee, C.C., Hu, Y., and Ribbe, M.W. (2010). Vanadium nitrogenase reduces CO. Science *329*, 642.

Lee, H.-I., Cameron, L.M., Hales, B.J., and Hoffman, B.M. (1997a). Binding to the FeMo cofactor of CO-inhibited nitrogenase: ^{13}CO and ^{1}H Q-band ENDOR investigation. J. Am. Chem. Soc. *119*, 10121–10126.

Lee, H.-I., Hales, B.J., and Hoffman, B.M. (1997b). Metal-ion valencies of the FeMo-cofactor in CO-inhibited and resting state nitrogenase by ^{57}Fe Q-band ENDOR. J. Am. Chem. Soc. *119*, 11395–11400.

Lee, H.I., Benton, P.M., Laryukhin, M., Igarashi, R.Y., Dean, D.R., Seefeldt, L.C., and Hoffman, B.M. (2003). The interstitial atom of the nitrogenase FeMo-cofactor: ENDOR and ESEEM show it is not an exchangeable nitrogen. J. Am. Chem. Soc. *125*, 5604–5605.

Lee, H.I., Igarashi, R.Y., Laryukhin, M., Doan, P.E., Dos Santos, P.C., Dean, D.R., Seefeldt, L.C., and Hoffman, B.M. (2004). An organometallic intermediate during alkyne reduction by nitrogenase. J. Am. Chem. Soc. *126*, 9563–9569.

Lee, S.C., and Holm, R.H. (2004). The clusters of nitrogenase: synthetic methodology in the construction of weak-field clusters. Chem. Rev. *104*, 1135–1158.

Leigh, J.A. (2000). Nitrogen fixation in methanogens: the archaeal perspective. Curr. Issues Mol. Biol. *2*, 125–131.

Liang, J., Madden, M., Shah, V.K., and Burris, R.H. (1990). Citrate substitutes for homocitrate in nitrogenase of a *nifV* mutant of *Klebsiella pneumoniae*. Biochemistry *29*, 8577–8581.

Lindahl, P.A., Day, E.P., Kent, T.A., Orme-Johnson, W.H., and Munck, E. (1985). Mössbauer, EPR, and magnetization studies of the *Azotobacter vinelandii* Fe protein. Evidence for a $[4Fe–4S]^{1+}$ cluster with spin $S = 3/2$. J. Biol. Chem. *260*, 11160–11173.

Lindahl, P.A., Gorelick, N.J., Munck, E., and Orme-Johnson, W.H. (1987). EPR and Mössbauer studies of nucleotide-bound nitrogenase iron protein from *Azotobacter vinelandii*. J. Biol. Chem. *262*, 14945–14953.

Lindahl, P.A., Papaefthymiou, V., Orme-Johnson, W.H., and Munck, E. (1988). Mössbauer studies of solid thionin-oxidized MoFe protein of nitrogenase. J. Biol. Chem. *263*, 19412–19418.

Ljones, T., and Burris, R.H. (1972). ATP hydrolysis and electron transfer in the nitrogenase reaction with different combinations of the iron protein and the molybdenum-iron protein. Biochim. Biophys. Acta *275*, 93–101.

Ljones, T., and Burris, R.H. (1978). Nitrogenase: the reaction between the Fe protein and bathophenanthrolinedisulfonate as a probe for interactions with MgATP. Biochemistry *17*, 1866–1872.

Locher, K.P., Hans, M., Yeh, A.P., Schmid, B., Buckel, W., and Rees, D.C. (2001). Crystal structure of the *Acidaminococcus fermentans* 2-hydroxyglutaryl-CoA dehydratase component A. J. Mol. Biol. *307*, 297–308.

Lowe, D.J., Eady, R.R., and Thorneley, N.F. (1978). Electron-paramagnetic-resonance studies on nitrogenase of *Klebsiella pneumoniae*. Evidence for acetylene- and ethylene-nitrogenase transient complexes. Biochem. J. *173*, 277–290.

Lowe, D.J., Fisher, K., and Thorneley, R.N. (1993). *Klebsiella pneumoniae* nitrogenase: pre-steady-state absorbance changes show that redox changes occur in the MoFe protein that depend on substrate and component protein ratio; a role for P-centres in reducing dinitrogen? Biochem. J. *292*, 93–98.

Lowe, D.J., and Smith, B.E. (1985). Electron-paramagnetic-resonance spectroscopy and related techniques in the study of nitrogenase. Biochem. Soc. Trans. *13*, 579–581.

Lowe, D.J., and Thorneley, R.N. (1984a). The mechanism of *Klebsiella pneumoniae* nitrogenase action. Pre-steady-state kinetics of H_2 formation. Biochem. J. *224*, 877–886.

Lowe, D.J., and Thorneley, R.N. (1984b). The mechanism of *Klebsiella pneumoniae* nitrogenase action. The determination of rate constants required for the simulation of the kinetics of N_2 reduction and H_2 evolution. Biochem. J. *224*, 895–901.

Ludden, P.W., Rangaraj, P., and Rubio, L.M. (2004). Biosynthesis of the iron-molybdenum and iron-vanadium cofactors of the *nif*- and *vnf*- encoded nitrogenases. In Catalysts for nitrogen fixation: nitrogenases, relevant chemical models, and commercial processes, B.E. Smith, R.L. Richards, and W.E. Newton, eds. (Dordretch, Kluwer Academic Publishers), pp. 219–253.

Lukoyanov, D., Barney, B.M., Dean, D.R., Seefeldt, L.C., and Hoffman, B.M. (2007a). Connecting nitrogenase intermediates with the kinetic scheme for N_2 reduction by a relaxation protocol and identification of the N_2 binding state. Proc. Natl. Acad. Sci. U.S.A. *104*, 1451–1455.

Lukoyanov, D., Pelmenschikov, V., Maeser, N., Laryukhin, M., Yang, T.C., Noodleman, L., Dean, D.R., Case, D.A., Seefeldt, L.C., and Hoffman, B.M. (2007b). Testing if the interstitial atom, X, of the nitrogenase molybdenum-iron cofactor is N or C: ENDOR, ESEEM, and DFT studies of the S = 3/2 resting state in multiple environments. Inorg. Chem. *46*, 11437–11449.

MacNeil, T., MacNeil, D., Roberts, G.P., Supiano, M.A., and Brill, W.J. (1978). Fine-structure mapping and complementation analysis of *nif* (nitrogen fixation) genes in *Klebsiella pneumoniae*. J. Bacteriol. *136*, 253–266.

Madden, M.S., Kindon, N.D., Ludden, P.W., and Shah, V.K. (1990). Diastereomer-dependent substrate reduction properties of a dinitrogenase containing 1-fluorohomocitrate in the iron-molybdenum cofactor. Proc. Natl. Acad. Sci. U.S.A. *87*, 6517–6521.

Madden, M.S., Paustian, T.D., Ludden, P.W., and Shah, V.K. (1991). Effects of homocitrate, homocitrate lactone, and fluorohomocitrate on nitrogenase in NifV- mutants of *Azotobacter vinelandii*. J. Bacteriol. *173*, 5403–5405.

Magnuson, J.K., Paustian, T.D., Shah, V.K., Dean, D.R., Roberts, G.P., Rees, D.C., and Howard, J.B. (1997). Nitrogenase iron-molybdenum cofactor binding site: Protein conformational changes associated with cofactor binding. Tetrahedron 53, 11971–11984.

Maier, R.J., and Moshiri, F. (2000). Role of the *Azotobacter vinelandii* nitrogenase-protective shethna protein in preventing oxygen-mediated cell death. J. Bacteriol. *182*, 3854–3857.

May, H.D., Dean, D.R., and Newton, W.E. (1991). Altered nitrogenase MoFe proteins from *Azotobacter vinelandii*. Analysis of MoFe proteins having amino acid substitutions for the conserved cysteine residues within the beta-subunit. Biochem. J. *277*, 457–464.

Mayer, S.M., Lawson, D.M., Gormal, C.A., Roe, S.M., and Smith, B.E. (1999). New insights into structure–function relationships in nitrogenase: A 1.6 A resolution X-ray crystallographic study of *Klebsiella pneumoniae* MoFe-protein. J. Mol. Biol. *292*, 871–891.

Mayer, S.M., Niehaus, W.G., and Dean, D.R. (2002). Reduction of short chain alkynes by an u70ala-substituted MoFe protein. Dalton Trans. *5*, 802–807.

McLean, P.A., Smith, B.E., and Dixon, R.A. (1983). Nitrogenase of *Klebsiella pneumoniae nifV* mutants. Biochem. J. *211*, 589–597.

Mehta, M.P., and Baross, J.A. (2006). Nitrogen fixation at 92°C by a hydrothermal vent archaeon. Science *314*, 1783–1786.

Meyer, J., Gaillard, J., and Moulis, J.M. (1988). Hydrogen-1 nuclear magnetic resonance of the nitrogenase iron protein (Cp2) from *Clostridium pasteurianum*. Biochemistry *27*, 6150–6156.

Moreno-Vivian, C., Schmehl, M., Masepohl, B., Arnold, W., and Klipp, W. (1989). DNA sequence and genetic analysis of the *Rhodobacter capsulatus nifENX* gene region: homology between NifX and NifB suggests involvement of NifX in processing of the iron-molybdenum cofactor. Mol. Gen. Genet. *216*, 353–363.

Morgan, T.V., McCracken, J., Orme-Johnson, W.H., Mims, W.B., Mortenson, L.E., and Peisach, J. (1990). Pulsed electron paramagnetic resonance studies of the interaction of Mg-ATP and D_2O with the iron protein of nitrogenase. Biochemistry *29*, 3077–3082.

Morgan, T.V., Mortenson, L.E., McDonald, J.W., and Watt, G.D. (1988). Comparison of redox and EPR properties of the molybdenum iron proteins of *Clostridium pasteurianum* and *Azotobacter vinelandii* nitrogenases. J. Inorg. Biochem. 33, 111–120.

Morgan, T.V., Prince, R.C., and Mortenson, L.E. (1986). Electrochemical titration of the S = Image and S = Image states of the iron protein of nitrogenase FEBS Lett. *206*, 4–8.

Morningstar, J.E., Johnson, M.K., Case, E.E., and Hales, B.J. (1987). Characterization of the metal clusters in the nitrogenase molybdenum-iron and vanadium-iron proteins of *Azotobacter vinelandii* using magnetic circular dichroism spectroscopy. Biochemistry *26*, 1795–1800.

Mortenson, L.E. (1966). Components of cell-free extracts of *Clostridium pasteurianum* required for ATP-dependent H_2 evolution from dithionite and for N_2 fixation. Biochim. Biophys. Acta *127*, 18–25.

Mortenson, L.E., Morris, J.A., and Jeng, D.Y. (1967). Purification, metal composition and properties of molybdoferredoxin and azoferredoxin, two of the components of the nitrogen-fixing system of *Clostridium pasteurianum*. Biochim. Biophys. Acta *141*, 516–522.

Moustafa, E., and Mortenson, L.E. (1969). Properties of azoferredoxin purified from nitrogen-fixing extracts of

Clostridium pasteurianum. Biochim. Biophys. Acta *172*, 106–115.

Munck, E., Rhodes, H., Orme-Johnson, W.H., Davis, L.C., Brill, W.J., and Shah, V.K. (1975). Nitrogenase. VIII. Mössbauer and EPR spectroscopy. The MoFe protein component from *Azotobacter vinelandii* OP. Biochim. Biophys. Acta *400*, 32–53.

Muraki, N., Nomata, J., Ebata, K., Mizoguchi, T., Shiba, T., Tamiaki, H., Kurisu, G., and Fujita, Y. (2010). X-ray crystal structure of the light-independent protochlorophyllide reductase. Nature *465*, 110–114.

Musgrave, K.B., Lim, B.S., Sung, K.M., Holm, R.H., Hedman, B., and Hodgson, K.O. (2000). X-ray spectroscopy of enzyme active site analogues and related molecules: bis(dithiolene)molybdenum(IV) and -tungsten(IV,VI) complexes with variant terminal ligands. Inorg Chem *39*, 5238–5247.

Nyborg, A.C., Erickson, J.A., Johnson, J.L., Gunn, A., Truscott, S.M., and Watt, G.D. (2000a). Reactions of *Azotobacter vinelandii* nitrogenase using Ti(III) as reductant. J. Inorg. Biochem. *78*, 371–381.

Nyborg, A.C., Johnson, J.L., Gunn, A., and Watt, G.D. (2000b). Evidence for a two-electron transfer using the all-ferrous Fe protein during nitrogenase catalysis. J. Biol. Chem. *275*, 39307–39312.

O'Donnell, M.J., and Smith, B.E. (1978). Electron-paramagnetic-resonance studies on the redox properties of the molybdenum-iron protein of nitrogenase between +50 and -450 mV. Biochem. J. *173*, 831–838.

Onate, Y.A., Finnegan, M.G., Hales, B.J., and Johnson, M.K. (1993). Variable temperature magnetic circular dichroism studies of reduced nitrogenase iron proteins and [4Fe–4S]+ synthetic analog clusters. Biochim. Biophys. Acta *1164*, 113–123.

Orme-Johnson, W.H. (1992). Nitrogenase structure: where to now? Science *257*, 1639–1640.

Orme-Johnson, W.H., Hamilton, W.D., Jones, T.L., Tso, M.Y., Burris, R.H., Shah, V.K., and Brill, W.J. (1972). Electron paramagnetic resonance of nitrogenase and nitrogenase components from *Clostridium pasteurianum* W5 and *Azotobacter vinelandii* OP. Proc. Natl. Acad. Sci. U.S.A. *69*, 3142–3145.

Page, W.J., and von Tigerstrom, M. (1979). Optimal conditions for transformation of *Azotobacter vinelandii*. J. Bacteriol. *139*, 1058–1061.

Pai, E.F., Kabsch, W., Krengel, U., Holmes, K.C., John, J., and Wittinghofer, A. (1989). Structure of the guanine-nucleotide-binding domain of the Ha-ras oncogene product p21 in the triphosphate conformation. Nature *341*, 209–214.

Pai, E.F., Krengel, U., Petsko, G.A., Goody, R.S., Kabsch, W., and Wittinghofer, A. (1990). Refined crystal structure of the triphosphate conformation of H-ras p21 at 1.35 Å resolution: implications for the mechanism of GTP hydrolysis. EMBO J. *9*, 2351–2359.

Paustian, T.D., Shah, V.K., and Roberts, G.P. (1989). Purification and characterization of the *nifN* and *nifE* gene products from *Azotobacter vinelandii* mutant UW45. Proc. Natl. Acad. Sci. U.S.A. *86*, 6082–6086.

Paustian, T.D., Shah, V.K., and Roberts, G.P. (1990). Apodinitrogenase: purification, association with a 20-kilodalton protein, and activation by the iron-molybdenum cofactor in the absence of dinitrogenase reductase. Biochemistry *29*, 3515–3522.

Peters, J.W., Fisher, K., and Dean, D.R. (1995a). Nitrogenase structure and function: a biochemical-genetic perspective. Annu. Rev. Microbiol. *49*, 335–366.

Peters, J.W., Fisher, K., Newton, W.E., and Dean, D.R. (1995b). Involvement of the P cluster in intramolecular electron transfer within the nitrogenase MoFe protein. J. Biol. Chem. *270*, 27007–27013.

Peters, J.W., Stowell, M.H., Soltis, S.M., Finnegan, M.G., Johnson, M.K., and Rees, D.C. (1997). Redox-dependent structural changes in the nitrogenase P-cluster. Biochemistry *36*, 1181–1187.

Peters, J.W., and Szilagyi, R.K. (2006). Exploring new frontiers of nitrogenase structure and mechanism. Curr. Opin. Chem. Biol. *10*, 101–108.

Pierik, A.J., Wassink, H., Haaker, H., and Hagen, W.R. (1993). Redox properties and EPR spectroscopy of the P cluster of *Azotobacter vinelandii* MoFe protein. Eur. J. Biochem. *169*, 51–61.

Pollock, R.C., Lee, H.-I., Cameron, L.M., DeRose, V.J., Hales, B.J., Orme-Johnson, W.H., and Hoffman, B.M. (1995). Investigation of CO bound to inhibited forms of nitrogenase MoFe protein by ^{13}C ENDOR. J. Am. Chem. Soc. *117*, 8686–8687.

Poole, R.K., and Hill, S. (1997). Respiratory protection of nitrogenase activity in *Azotobacter vinelandii* – roles of the terminal oxidases. Biosci Rep. *17*, 303–317.

Pope, M.R., Murrell, S.A., and Ludden, P.W. (1985). Covalent modification of the iron protein of nitrogenase from *Rhodospirillum rubrum* by adenosine diphosphoribosylation of a specific arginine residue. Proc. Natl. Acad. Sci. U.S.A. *82*, 3173–3177.

Rangaraj, P., and Ludden, P.W. (2002). Accumulation of ^{99}Mo-containing iron-molybdenum cofactor precursors of nitrogenase on NifNE, NifH, and NifX of *Azotobacter vinelandii*. J. Biol. Chem. *277*, 40106–40111.

Rangaraj, P., Ruttimann-Johnson, C., Shah, V.K., and Ludden, P.W. (2001). Accumulation of ^{55}Fe-labeled precursors of the iron–molybdenum cofactor of nitrogenase on NifH and NifX of *Azotobacter vinelandii*. J. Biol. Chem. *276*, 15968–15974.

Rangaraj, P., Ryle, M.J., Lanzilotta, W.N., Ludden, P.W., and Shah, V.K. (1999). *In vitro* biosynthesis of iron–molybdenum cofactor and maturation of the *nif*-encoded apodinitrogenase. Effect of substitution for NifH with site-specifically altered forms of NifH. J. Biol. Chem. *274*, 19778–19784.

Rangaraj, P., Shah, V.K., and Ludden, P.W. (1997). ApoNifH functions in iron-molybdenum cofactor synthesis and apodinitrogenase maturation. Proc. Natl. Acad. Sci. U.S.A. *94*, 11250–11255.

Rasche, M.E., and Seefeldt, L.C. (1997). Reduction of thiocyanate, cyanate, and carbon disulfide by nitrogenase: kinetic characterization and EPR spectroscopic analysis. Biochemistry *36*, 8574–8585.

Rawlings, J., Shah, V.K., Chisnell, J.R., Brill, W.J., Zimmermann, R., Munck, E., and Orme-Johnson, W.H. (1978). Novel metal cluster in the iron-molybdenum cofactor of nitrogenase. Spectroscopic evidence. J. Biol. Chem. *253*, 1001–1004.

Raymond, J., Siefert, J.L., Staples, C.R., and Blankenship, R.E. (2004). The natural history of nitrogen fixation. Mol. Biol. Evol. *21*, 541–554.

Rees, D.C., and Howard, J.B. (2000). Nitrogenase: standing at the crossroads. Curr. Opin. Chem. Biol. *4*, 559–566.

Renner, K.A., and Howard, J.B. (1996). Aluminum fluoride inhibition of nitrogenase: stabilization of a nucleotide·Fe-protein·MoFe–protein complex. Biochemistry *35*, 5353–5358.

Ribbe, M.W., and Burgess, B.K. (2001). The chaperone GroEL is required for the final assembly of the molybdenum-iron protein of nitrogenase. Proc. Natl. Acad. Sci. U.S.A. *98*, 5521–5525.

Ribbe, M.W., Hu, Y., Guo, M., Schmid, B., and Burgess, B.K. (2002). The FeMoco-deficient MoFe protein produced by a *nifH* deletion strain of *Azotobacter vinelandii* shows unusual P-cluster features. J. Biol. Chem. *277*, 23469–23476.

Rivera-Ortiz, J.M., and Burris, R.H. (1975). Interactions among substrates and inhibitors of nitrogenase. J. Bacteriol. *123*, 537–545.

Roberts, G.P., MacNeil, T., MacNeil, D., and Brill, W.J. (1978). Regulation and characterization of protein products coded by the *nif* (nitrogen fixation) genes of *Klebsiella pneumoniae*. J. Bacteriol. *136*, 267–279.

Robinson, A.C., Burgess, B.K., and Dean, D.R. (1986). Activity, reconstitution, and accumulation of nitrogenase components in *Azotobacter vinelandii* mutant strains containing defined deletions within the nitrogenase structural gene cluster. J. Bacteriol. *166*, 180–186.

Robinson, A.C., Dean, D.R., and Burgess, B.K. (1987). Iron-molybdenum cofactor biosynthesis in *Azotobacter vinelandii* requires the iron protein of nitrogenase. J. Biol. Chem. *262*, 14327–14332.

Robson, A.L., Eady, R.R., Richardson, T.H., Miller, R.W., Hawkins, M., and Postgate, J.R. (1986). The alternative nitrogenase from *Azotobacter chroococcum* is a vanadium enzyme. Nature *322*, 388–390.

Rodríguez-Quiñones, F., Bosch, R., and Imperial, J. (1993). Expression of the *nifBfdxNnifOQ* region of *Azotobacter vinelandii* and its role in nitrogenase activity. J. Bacteriol. *175*, 2926–2935.

Rubio, L.M., and Ludden, P.W. (2008). Biosynthesis of the iron-molybdenum cofactor of nitrogenase. Annu. Rev. Microbiol. *62*, 93–111.

Rubio, L.M., Rangaraj, P., Homer, M.J., Roberts, G.P., and Ludden, P.W. (2002). Cloning and mutational analysis of the γ gene from *Azotobacter vinelandii* defines a new family of proteins capable of metallocluster-binding and protein stabilization. J. Biol. Chem. *277*, 14299–14305.

Rubio, L.M., Singer, S.W., and Ludden, P.W. (2004). Purification and characterization of NafY (apodinitrogenase γ subunit) from *Azotobacter vinelandii*. J. Biol. Chem. *279*, 19739–19746.

Ruttimann-Johnson, C., Rubio, L.M., Dean, D.R., and Ludden, P.W. (2003). VnfY is required for full activity of the vanadium-containing dinitrogenase in *Azotobacter vinelandii*. J. Bacteriol. *185*, 2383–2386.

Ryle, M.J., Lanzilotta, W.N., Mortenson, L.E., Watt, G.D., and Seefeldt, L.C. (1995). Evidence for a central role of lysine 15 of *Azotobacter vinelandii* nitrogenase iron protein in nucleotide binding and protein conformational changes. J. Biol. Chem. *270*, 13112–13117.

Ryle, M.J., Lanzilotta, W.N., and Seefeldt, L.C. (1996a). Elucidating the mechanism of nucleotide-dependent changes in the redox potential of the [4Fe–4S] cluster in nitrogenase iron protein: the role of phenylalanine 135. Biochemistry *35*, 9424–9434.

Ryle, M.J., Lanzilotta, W.N., Seefeldt, L.C., Scarrow, R.C., and Jensen, G.M. (1996b). Circular dichroism and x-ray spectroscopies of *Azotobacter vinelandii* nitrogenase iron protein. MgATP and MgADP induced protein conformational changes affecting the [4Fe–4S] cluster and characterization of a [2Fe–2S] form. J. Biol. Chem. *271*, 1551–1557.

Ryle, M.J., Lee, H.I., Seefeldt, L.C., and Hoffman, B.M. (2000). Nitrogenase reduction of carbon disulfide: freeze-quench EPR and ENDOR evidence for three sequential intermediates with cluster-bound carbon moieties. Biochemistry 39, 1114–1119.

Ryle, M.J., and Seefeldt, L.C. (1996). Elucidation of a MgATP signal transduction pathway in the nitrogenase iron protein: formation of a conformation resembling the MgATP-bound state by protein engineering. Biochemistry *35*, 4766–4775.

Sarma, R., Barney, B.M., Hamilton, T.L., Jones, A., Seefeldt, L.C., and Peters, J.W. (2008). Crystal structure of the L protein of *Rhodobacter sphaeroides* light-independent protochlorophyllide reductase with MgADP bound: a homologue of the nitrogenase Fe protein. Biochemistry *47*, 13004–13015.

Sarma, R., Barney, B.M., Keable, S., Dean, D.R., Seefeldt, L.C., and Peters, J.W. (2010). Insights into substrate binding at FeMo-cofactor in nitrogenase from the structure of an α-70^{Ile} MoFe protein variant. J. Inorg. Biochem. *104*, 385–389.

Sarma, R., Mulder, D.W., Brecht, E., Szilagyi, R.K., Seefeldt, L.C., Tsuruta, H., and Peters, J.W. (2007). Probing the MgATP-bound conformation of the nitrogenase Fe protein by solution small-angle X-ray scattering. Biochemistry *46*, 14058–14066.

Schindelin, H., Kisker, C., Schlessman, J.L., Howard, J.B., and Rees, D.C. (1997). Structure of ADP × AIF_4^- stabilized nitrogenase complex and its implications for signal transduction. Nature *387*, 370–376.

Schlessman, J.L., Woo, D., Joshua-Tor, L., Howard, J.B., and Rees, D.C. (1998). Conformational variability in structures of the nitrogenase iron proteins from *Azotobacter vinelandii* and *Clostridium pasteurianum*. J. Mol. Biol. *280*, 669–685.

Schmid, B., Einsle, O., Chiu, H.J., Willing, A., Yoshida, M., Howard, J.B., and Rees, D.C. (2002). Biochemical and structural characterization of the cross-linked complex of nitrogenase: comparison to the ADP-AlF_4^--stabilized structure. Biochemistry *41*, 15557–15565.

Schneider, K., Gollan, U., Drottboom, M., Selsemeier-Voigt, S., and Muller, A. (1997). Comparative biochemical characterization of the iron-only nitrogenase and the molybdenum nitrogenase from *Rhodobacter capsulatus*. Eur. J. Biochem. *244*, 789–800.

Schneider, K., Gollan, U., Selsemeier-Voigt, S., Plass, W., and Muller, A. (1994). Rapid purification of the protein components of a highly active 'iron only' nitrogenase. Naturwissenschaften *81*, 405–408.

Schneider, K., Muller, A., Schramm, U., and Klipp, W. (1991). Demonstration of a molybdenum- and vanadium-independent nitrogenase in a *nifHDK*-deletion mutant of *Rhodobacter capsulatus*. Eur. J. Biochem. *195*, 653–661.

Schulz, G.E. (1992). Binding of nucleotides by proteins. Curr. Opin. Struct. Biol. *2*, 61–67.

Scott, D.J., Dean, D.R., and Newton, W.E. (1992). Nitrogenase-catalyzed ethane production and CO-sensitive hydrogen evolution from MoFe proteins having amino acid substitutions in an alpha-subunit FeMo cofactor-binding domain. J. Biol. Chem. *267*, 20002–20010.

Scott, D.J., May, H.D., Newton, W.E., Brigle, K.E., and Dean, D.R. (1990). Role for the nitrogenase MoFe protein alpha-subunit in FeMo-cofactor binding and catalysis. Nature *343*, 188–190.

Seefeldt, L.C., and Mortenson, L.E. (1993). Increasing nitrogenase catalytic efficiency for MgATP by changing serine 16 of its Fe protein to threonine: use of Mn^{2+} to show interaction of serine 16 with Mg^{2+}. Protein Sci. *2*, 93–102.

Seefeldt, L.C., and Dean, D.R. (1997). Role of nucleotides in nitrogenase catalysis. Acc. Chem. Res. *30*, 260–266.

Seefeldt, L.C., Morgan, T.V., Dean, D.R., and Mortenson, L.E. (1992). Mapping the site(s) of MgATP and MgADP interaction with the nitrogenase of *Azotobacter vinelandii*. Lysine 15 of the iron protein plays a major role in MgATP interaction. J. Biol. Chem. *267*, 6680–6688.

Seefeldt, L.C., Rasche, M.E., and Ensign, S.A. (1995). Carbonyl sulfide and carbon dioxide as new substrates, and carbon disulfide as a new inhibitor, of nitrogenase. Biochemistry *34*, 5382–5389.

Seefeldt, L.C., Hoffman, B.M., and Dean, D.R. (2009). Mechanism of Mo-dependent nitrogenase. Annu. Rev. Biochem. *78*, 701–722.

Sen, S., Igarashi, R., Smith, A., Johnson, M.K., Seefeldt, L.C., and Peters, J.W. (2004). A conformational mimic of the MgATP-bound 'on state' of the nitrogenase iron protein. Biochemistry *43*, 1787–1797.

Shah, V.K., Allen, J.R., Spangler, N.J., and Ludden, P.W. (1994). *In vitro* synthesis of the iron-molybdenum cofactor of nitrogenase. Purification and characterization of NifB cofactor, the product of NifB protein. J. Biol. Chem. *269*, 1154–1158.

Simpson, F.B., and Burris, R.H. (1984). A nitrogen pressure of 50 atmospheres does not prevent evolution of hydrogen by nitrogenase. Science *224*, 1095–1097.

Smith, A.D., Jameson, G.N., Dos Santos, P.C., Agar, J.N., Naik, S., Krebs, C., Frazzon, J., Dean, D.R., Huynh, B.H., and Johnson, M.K. (2005). NifS-mediated assembly of [4Fe–4S] clusters in the N- and C-terminal domains of the NifU scaffold protein. Biochemistry *44*, 12955–12969.

Smith, B.E., and Lang, G. (1974). Mössbauer spectroscopy of the nitrogenase proteins from *Klebsiella pneumoniae*. Structural assignments and mechanistic conclusions. Biochem. J. *137*, 169–180.

Smith, B.E., O'Donnell, M.J., Lang, G., and Spartalian, K. (1980). A Mössbauer spectroscopic investigation of the redox behaviour of the molybdenum-iron protein from *Klebsiella pneumoniae* nitrogenase. Mechanistic and structural implications. Biochem. J. *191*, 449–455.

Soboh, B., Boyd, E.S., Zhao, D., Peters, J.W., and Luis M. Rubio, L.M. (2010). Substrate specificity and evolutionary implications of a NifDK enzyme carrying NifB-co at its active site. FEBS Lett *584*, 1487–1492.

Soboh, B., Igarashi, R.Y., Hernandez, J.A., and Rubio, L.M. (2006). Purification of a NifEN protein complex that contains bound Mo and a FeMo-co precursor from an *Azotobacter vinelandii* Δ*nifHDK* strain. J. Biol. Chem. *281*, 36701–36709.

Sofia, H.J., Chen, G., Hetzler, B.G., Reyes-Spindola, J.F., and Miller, N.E. (2001). Radical SAM, a novel protein superfamily linking unresolved steps in familiar biosynthetic pathways with radical mechanisms: functional characterization using new analysis and information visualization methods. Nucleic Acids Res. *29*, 1097–1106.

Sondek, J., Lambright, D.G., Noel, J.P., Hamm, H.E., and Sigler, P.B. (1994). GTPase mechanism of G proteins from the 1.7-Å crystal structure of transducin alpha-GDP-AlF_4^-. Nature *372*, 276–279.

Sorlie, M., Christiansen, J., Lemon, B.J., Peters, J.W., Dean, D.R., and Hales, B.J. (2001). Mechanistic features and structure of the nitrogenase α-Gln^{195} MoFe protein. Biochemistry *40*, 1540–1549.

Sprang, S.R. (1997a). G protein mechanisms: insights from structural analysis. Annu. Rev. Biochem. *66*, 639–678.

Sprang, S.R. (1997b). G proteins, effectors and GAPs: structure and mechanism. Curr. Opin. Struct. Biol. *7*, 849–856.

Stal, L.J., and Krumbien, W.E. (1985). Nitrogenase activity in the non-heterocystous cyanobacterium *Oscillatoria sp*. grown under alternating light-dark cycles. Arch Microbiol *143*, 67–71.

Stephens, P.J., McKenna, C.E., Smith, B.E., Nguyen, H.T., McKenna, M.C., Thomson, A.J., Devlin, F., and Jones, J.B. (1979). Circular dichroism and magnetic circular dichroism of nitrogenase proteins. Proc. Natl. Acad. Sci. U.S.A. *76*, 2585–2589.

Stewart, W.D., Fitzgerald, G.P., and Burris, R.H. (1968). Acetylene reduction by nitrogen-fixing blue-green algae. Arch. Mikrobiol. *62*, 336–348.

Stewart, W.D., Fitzgerald, G.P., and Burris, R.H. (1970). Acetylene reduction assay for determination of phosphorus availability in Wisconsin lakes. Proc. Natl. Acad. Sci. U.S.A. *66*, 1104–1111.

Story, R.M., Bishop, D.K., Kleckner, N., and Steitz, T.A. (1993). Structural relationship of bacterial RecA proteins to recombination proteins from bacteriophage T4 and yeast. Science *259*, 1892–1896.

Story, R.M., and Steitz, T.A. (1992). Structure of the recA protein–ADP complex. Nature 355, 374–376.

Story, R.M., Weber, I.T., and Steitz, T.A. (1992). The structure of the *E. coli* recA protein monomer and polymer. Nature 355, 318–325.

Strop, P., Takahara, P.M., Chiu, H., Angove, H.C., Burgess, B.K., and Rees, D.C. (2001). Crystal structure of the all-ferrous $[4Fe–4S]^0$ form of the nitrogenase iron protein from *Azotobacter vinelandii*. Biochemistry *40*, 651–656.

Tezcan, F.A., Kaiser, J.T., Mustafi, D., Walton, M.Y., Howard, J.B., and Rees, D.C. (2005). Nitrogenase complexes: multiple docking sites for a nucleotide switch protein. Science *309*, 1377–1380.

Thomann, H., Bernardo, M., Newton, W.E., and Dean, D.R. (1991). N coordination of FeMo cofactor requires His-195 of the MoFe protein alpha subunit and is essential for biological nitrogen fixation. Proc. Natl. Acad. Sci. U.S.A. *88*, 6620–6623.

Thomann, H., Morgan, T.V., Jin, H., Burgmayer, S.J.N., Bare, R.E., and Stiefel, E.I. (1987). Protein nitrogen coordination to the iron-molybdenum center of nitrogenase from *Clostridium pasteurianum*. J. Am. Chem. Soc. *109*, 7913–7914.

Thorneley, R.N. (1975). Nitrogenase of *Klebsiella pneumoniae*. A stopped-flow study of magnesium-adenosine triphosphate-induce electron transfer between the compeonent proteins. Biochem. J. *145*, 391–396.

Thorneley, R.N., and Lowe, D.J. (1983). Nitrogenase of *Klebsiella pneumoniae*. Kinetics of the dissociation of oxidized iron protein from molybdenum-iron protein: identification of the rate-limiting step for substrate reduction. Biochem. J. *215*, 393–403.

Thorneley, R.N., and Lowe, D.J. (1984a). The mechanism of *Klebsiella pneumoniae* nitrogenase action. Pre-steady-state kinetics of an enzyme-bound intermediate in N_2 reduction and of NH_3 formation. Biochem. J. *224*, 887–894.

Thorneley, R.N., and Lowe, D.J. (1984b). The mechanism of *Klebsiella pneumoniae* nitrogenase action. Simulation of the dependences of H_2-evolution rate on component-protein concentration and ratio and sodium dithionite concentration. Biochem. J. *224*, 903–909.

Tittsworth, R.C., and Hales, B.J. (1993). Detection of EPR signals assigned to the 1-equiv-oxidized P-clusters of the nitrogenase MoFe-protein from *Azotobacter vinelandii*. J. Am. Chem. Soc. *116*, 9763–9767.

Tong, L., Milburn, M.V., de Vos, A.M., and Kim, S.H. (1989a). Structure of ras proteins. Science *245*, 244.

Tong, L.A., de Vos, A.M., Milburn, M.V., Jancarik, J., Noguchi, S., Nishimura, S., Miura, K., Ohtsuka, E., and Kim, S.H. (1989b). Structural differences between a ras oncogene protein and the normal protein. Nature *337*, 90–93.

Tong, L.A., de Vos, A.M., Milburn, M.V., and Kim, S.H. (1991). Crystal structures at 2.2 Å resolution of the catalytic domains of normal ras protein and an oncogenic mutant complexed with GDP. J. Mol. Biol. *217*, 503–516.

Ugalde, R.A., Imperial, J., Shah, V.K., and Brill, W.J. (1984). Biosynthesis of iron-molybdenum cofactor in the absence of nitrogenase. J. Bacteriol. *159*, 888–893.

Ugalde, R.A., Imperial, J., Shah, V.K., and Brill, W.J. (1985). Biosynthesis of the iron molybdenum cofactor and the molybdenum cofactor in *Klebsiella pneumoniae*: effect of sulfur source. J. Bacteriol. *164*, 1081–1087.

Ureta, A., and Nordlund, S. (2002). Evidence for conformational protection of nitrogenase against oxygen in *Gluconacetobacter diazotrophicus* by a putative FeSII protein. J. Bacteriol. *184*, 5805–5809.

Vance, C.P. (2001). Symbiotic nitrogen fixation and phosphorus acquisition. Plant nutrition in a world of declining renewable resources. Plant Physiol *127*, 390–397.

Vandecasteele, J.P., and Burris, R.H. (1970). Purification and properties of the constituents of the nitrogenase complex from *Clostridium pasteurianum*. J. Bacteriol. *101*, 794–801.

Walker, G.A., and Mortenson, L.E. (1974). Effect of magnesium adenosine 5'-triphosphate on the accessibility of the iron of clostridial azoferredoxin, a component of nitrogenase. Biochemistry *13*, 2382–2388.

Watt, G.D., Bulen, W.A., Burns, A., and Hadfield, K.L. (1975). Stoichiometry, ATP/2e values, and energy requirements for reactions catalyzed by nitrogenase from *Azotobacter vinelandii*. Biochemistry *14*, 4266–4272.

Watt, G.D., Burns, A., Lough, S., and Tennent, D.L. (1980). Redox and spectroscopic properties of oxidized MoFe protein from *Azotobacter vinelandii*. Biochemistry *19*, 4926–4932.

Watt, G.D., Wang, Z.C., and Knotts, R.R. (1986). Redox reactions of and nucleotide binding to the iron protein of *Azotobacter vinelandii*. Biochemistry 25, 8156–8162.

White, T.C., Harris, G.S., and Orme-Johnson, W.H. (1992). Electrophoretic studies on the assembly of the nitrogenase molybdenum-iron protein from the *Klebsiella pneumoniae nifD* and *nifK* gene products. J. Biol. Chem. *267*, 24007–24016.

Willing, A.H., Georgiadis, M.M., Rees, D.C., and Howard, J.B. (1989). Cross-linking of nitrogenase components. Structure and activity of the covalent complex. J. Biol. Chem. *264*, 8499–8503.

Winter, H.C., and Burris, R.H. (1968). Stoichiometry of the adenosine triphosphate requirement for N_2 fixation and H_2 evolution by a partially purified preparation of *Clostridium pasteurianum*. J. Biol. Chem. *243*, 940–944.

Wolle, D., Dean, D.R., and Howard, J.B. (1992). Nucleotide-iron–sulfur cluster signal transduction in the nitrogenase iron-protein: the role of Asp125. Science 258, 992–995.

Yang, T.C., Maeser, N.K., Laryukhin, M., Lee, H.I., Dean, D.R., Seefeldt, L.C., and Hoffman, B.M. (2005). The interstitial atom of the nitrogenase FeMo-cofactor:

ENDOR and ESEEM evidence that it is not a nitrogen. J. Am. Chem. Soc. *127*, 12804–12805.

Yates, M.G., and Lowe, D.J. (1976). Nitrogenase of *Azotobacter chroococcum*: a new electron-paramagnetic-resonance signal associated with a transient species of the Mo-Fe protein during catalysis. FEBS Lett 72, 121–126.

Yoo, J.Y., Angove, H.C., Papaefthymiou, V., Burgess, B.K., and Munck, E. (2000). Mössbauer study of the MoFe protein of nitrogenase from *Azotobacter vinelandii* using selective ^{57}Fe enrichment of the M-centers. J. Am. Chem. Soc. *122*, 4926–4936.

Yuvaniyama, P., Agar, J.N., Cash, V.L., Johnson, M.K., and Dean, D.R. (2000). NifS-directed assembly of a transient [2Fe–2S] cluster within the NifU protein. Proc. Natl. Acad. Sci. U.S.A. *97*, 599–604.

Zehr, J.P., and Capone, D.G., eds. (2002). N_2 Fixation in the Marine Environment (John Wiley & Sons.).

Zhao, D., Curatti, L., and Rubio, L.M. (2007). Evidence for *nifU* and *nifS* participation in the biosynthesis of the iron-molybdenum cofactor of nitrogenase. J. Biol. Chem. *282*, 37016–37025.

Zheng, L., White, R.H., Cash, V.L., and Dean, D.R. (1994). Mechanism for the desulfurization of L-cysteine catalyzed by the *nifS* gene product. Biochemistry *33*, 4714–4720.

Zheng, L., White, R.H., Cash, V.L., Jack, R.F., and Dean, D.R. (1993). Cysteine desulfurase activity indicates a role for NifS in metallocluster biosynthesis. Proc. Natl. Acad. Sci. U.S.A. *90*, 2754–2758.

Zheng, L., White, R.H., and Dean, D.R. (1997). Purification of the *Azotobacter vinelandii nifV*-encoded homocitrate synthase. J. Bacteriol. *179*, 5963–5966.

Zimmermann, R., Munck, E., Brill, W.J., Shah, V.K., Henzl, M.T., Rawlings, J., and Orme-Johnson, W.H. (1978). Nitrogenase X: Mössbauer and EPR studies on reversibly oxidized MoFe protein from *Azotobacter vinelandii* OP. Nature of the iron centers. Biochim. Biophys. Acta *537*, 185–207.

Zumft, W.G., Mortenson, L.E., and Palmer, G. (1974). Electron-paramagnetic-resonance studies on nitrogenase. Investigation of the oxidation-reduction behaviour of azoferredoxin and molybdoferredoxin with potentiometric and rapid-freeze techniques. Eur. J. Biochem. *46*, 525–535.

Transport and Assimilation of Inorganic Nitrogen in Bacteria

5

Conrado Moreno-Vivián, Víctor M. Luque-Almagro, Purificación Cabello, M. Dolores Roldán and Francisco Castillo

Abstract

The incorporation of inorganic nitrogen into cell material is known as nitrogen assimilation. Usually, ammonium is the preferred inorganic nitrogen source for microorganisms. Ammonium assimilation requires the transport of this ion into the cells and its further incorporation into carbon skeletons, mainly through the glutamine synthetase-glutamate synthase pathway. Alternatively, glutamate dehydrogenase may also contribute to ammonium assimilation under certain conditions. Glutamine synthetase is the key enzyme for the regulation of ammonium assimilation; its activity is usually controlled by reversible covalent modification or feedback mechanisms and, at the gene expression level, transcription is often controlled by general nitrogen regulatory systems that vary depending on the organisms. In addition, ammonium transport is also subjected to regulation by carbon and nitrogen availability. Oxidized nitrogen compounds like nitrate and nitrite may be also used as nitrogen sources by many bacteria and archaea. Nitrate assimilation requires nitrate transport into the cells and two enzymes, nitrate and nitrite reductases, which catalyse the two-electron reduction of nitrate to nitrite and the six-electron reduction of nitrite to ammonium, respectively. These assimilatory enzymes are structural and functionally different to respiratory nitrate and nitrite reductases. Control of nitrate assimilation in different organisms may involve distinct regulatory proteins and mechanisms, but usually the process is regulated by nitrate and/or nitrite induction (pathway-specific control) and by ammonium repression (general nitrogen control).

Introduction

Nitrogen is an essential element for life that constitutes, on average, about 6% dry weight of organisms. It is present in proteins, nucleic acids, many carbohydrate and lipid derivatives, and a wide variety of important metabolites. Therefore, all organisms need an adequate nitrogen source for growth, and nitrogen availability limits microbial and plant growth in many environments. Nitrogen is distributed between three major pools, atmosphere, soil/groundwater and biomass, with complex exchange processes among them that constitutes the so-called nitrogen cycle (Cabello *et al.*, 2009). Nitrogen assimilation consists in the incorporation of inorganic nitrogen into cell material. Although inorganic nitrogen may be found in different oxidation states, from +5 in the most oxidized compound (nitrate) to –3 in the most reduced form (ammonium), in the organic compounds nitrogen is usually found in the fully reduced state, ammonium being the preferred nitrogen source for most microorganisms. Ammonium assimilation requires the transport of this ion into the cells and its subsequent incorporation into cellular biomass, mainly via the glutamine synthetase-glutamate synthase (GS-GOGAT) pathway. Alternatively, glutamate dehydrogenase (GDH) may also be involved in ammonium assimilation. Different cellular N-compounds are further synthesized by secondary amino or amido transfer reactions (Fig. 5.1). Other inorganic nitrogen sources, like molecular nitrogen or nitrate, need to be reduced to ammonium before their incorporation to C-skeletons. Nitrogen fixation consists

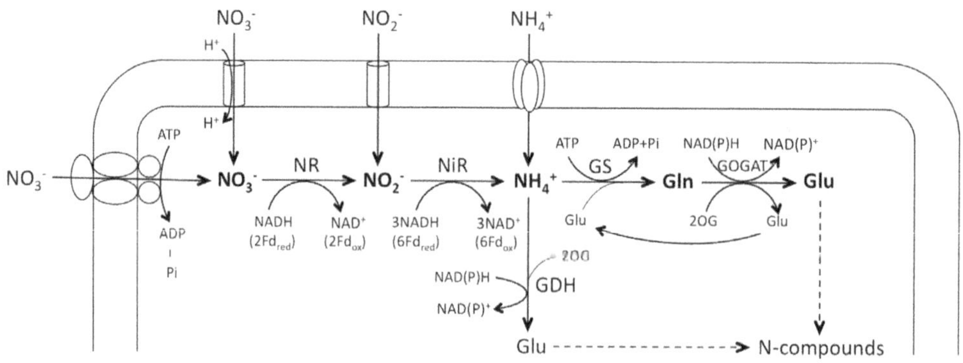

Figure 5.1 The nitrate and ammonium assimilation pathways. NR, assimilatory nitrate reductase; NiR, assimilatory nitrite reductase; GS, glutamine synthetase; GOGAT, glutamate synthase; GDH, glutamate dehydrogenase; Fd_{red}, reduced ferredoxin; Fd_{ox}, oxidized ferredoxin; 2OG, 2-oxoglutarate.

on the reduction of N_2 to ammonia, a process carried out by some free-living and symbiotic prokaryotes (see Chapters 4 and 10). Nitrate assimilation is based on reduction of nitrate to ammonium, via nitrite. This process requires an uptake system to import nitrate into the cells and the subsequent action of two enzymes: nitrate reductase, which catalyses the two-electron reduction of nitrate to nitrite, and nitrite reductase, which mediates the six-electron reduction of nitrite to ammonium (Fig. 5.1). This process is carried out by higher plants, fungi, algae and many bacteria and archaea and mobilizes more than 2×10^{13} kg of N per year (Moreno-Vivián and Flores, 2007). Although nitrate is found in the environment at relatively low concentrations (about 0.03–40 μM in oceanic seawaters), the excessive use of fertilizers in agricultural activities has led to nitrate accumulation up to mM concentrations in the groundwater of many areas, causing environmental and public health concerns (Moreno-Vivián *et al.*, 1999; Moreno-Vivián and Flores, 2007).

Inorganic nitrogen assimilation has been extensively studied in eukaryotic organisms. Comprehensive reviews covering nitrate and/or ammonium assimilation in higher plants, yeasts and algae have been published elsewhere (Siverio, 2002; Márquez *et al.*, 2005; Cánovas *et al.*, 2007; Fernández and Galván, 2007; Tischner and Kaiser, 2007; Fernández and Galván, 2008; Miller *et al.*, 2008; Bernard and Habash, 2009). In this chapter the current state of the studies on ammonium and nitrate assimilation in prokaryotes is reviewed.

Ammonium assimilation

The assimilation of ammonium requires its transport into the cells and its further incorporation into carbon skeletons. Although ammonia (NH_3) may enter the cells by membrane diffusion at high pH, ammonium (NH_4^+) transporters are ubiquitous in prokaryotic and eukaryotic organisms. Once in the cells, ammonium is mainly incorporated into carbon skeletons by the sequential action of glutamine synthetase (GS) and glutamate synthase (glutamine:2-oxoglutarate amidotransferase, GOGAT). GS is a ubiquitous enzyme that catalyses the ATP-dependent amidation of glutamate to generate glutamine, and GOGAT catalyses the reductive transfer of the amide group of glutamine to 2-oxoglutarate, yielding two molecules of glutamate. Therefore, the global GS-GOGAT pathway results in the incorporation of ammonium to 2-oxoglutarate to generate a molecule of glutamate with consumption of ATP, required for GS activity, and reducing power, required for GOGAT activity (Fig. 5.1). Then, other organic nitrogen compounds are produced by a set of secondary transfer reactions. The direct incorporation of ammonium to 2-oxoglutarate is catalysed by glutamate dehydrogenase (GDH), but this alternative pathway normally plays a minor role in ammonium assimilation. GDH usually acts as a deaminating enzyme for glutamate catabolism, showing a high K_M value for ammonium. Therefore, although this pathway does not consume ATP, is less effective at low ammonium concentrations than the GS-GOGAT route (Muro-Pastor *et al.*, 2005). For similar reasons, the incorporation of ammonium to pyruvate, catalysed by alanine dehydrogenase, only acts as an alternative ammonium assimilation pathway under certain specific conditions (Moreno-Vivián *et al.*, 1983).

The main carbon skeleton for ammonium assimilation, 2-oxoglutarate, represents the connection between carbon and nitrogen metabolism in most organisms. In this sense, 2-oxoglutarate is the main molecule used as a signal of nitrogen limitation in bacteria. On the other hand, glutamine, the first product of ammonium assimilation, is usually the signal of nitrogen sufficiency. The so-called P_{II} proteins (GlnB, GlnK) are ubiquitously distributed among prokaryotes as 2-oxoglutarate sensors. Also, in some bacteria, the P_{II} proteins are covalently modified by enzymes that respond to glutamine (for comprehensive reviews see Arcondéguy *et al.*, 2001; Reitzer, 2003; Forchhammer, 2004; Ninfa and Jiang, 2005; Forchhammer, 2007a,b; Leigh and Dodsworth, 2007).

Ammonium assimilation has also received special attention for its potential applied interest since the GS enzyme is a target for antimicrobial drugs and plant herbicides. Thus, it has been proposed that GS action allows pathogenic mycobacteria to avoid destruction by macrophages, probably preventing acidification of phagosomes and their fusion with lysosomes (Gordon *et al.*, 1980). The *glnA1* gene of *Mycobacterium tuberculosis* is essential for growth, and it has been suggested that the GS encoded by this gene plays a role in enabling the bacterium to survive during infection in the human host, so that GS inhibitors may be successfully used as effective anti-tuberculosis agents (Harth and Horwitz, 2003; Berlicki, 2008; Nilsson *et al.*, 2009).

Ammonium transport

Ammonium transporters (Amt) are homotrimeric integral membrane proteins, each 45–50 kDa subunit composed of 10–12 transmembrane helices. Many Amt proteins may also use the ammonium analogue methylammonium as substrate, providing a suitable mechanism to study the transport process using [^{14}C]methylammonium. Ammonium transporter genes (*amt*) are widely distributed among different organisms, including archaea, bacteria, yeast and plants. Also, there are homologous proteins in mammalian erythrocytes (Rhesus blood-type proteins) that could function as ammonium exporters (Hemker *et al.*, 2003). Most prokaryotic genomes contain several putative *amt* genes. Thus, in *Synechocystis* sp. PCC 6803 *amt1* codes for the main ammonium permease but additional *amt2* and *amt3* genes code for proteins that contribute little to ammonium uptake (Montesinos *et al.*, 1998). Some archaea, like *Methanosarcina acetivorans* and *Archaeoglobus fulgidus*, also contain three *amt* genes (Cabello *et al.*, 2004b; Andrade *et al.*, 2005). However, despite the wide distribution and the physiological relevance of Amt proteins, the nature of the transported substrate (ammonia or ammonium ion) and the transport mechanism (passive diffusion of NH_3, potential-driven uniport or proton-coupled symport of NH_4^+) still remain under discussion (Wang *et al.*, 1994; Ludewig *et al.*, 2002; Soupene *et al.*, 2002; Khademi *et al.*, 2004; Zheng *et al.*, 2004; Andrade *et al.*, 2005). The crystal structure of *E. coli* AmtB reveals that this protein is a stable trimer with 11 membrane-spanning helices per monomer, although an additional N-terminal helix acting as a leader sequence is cleaved upon insertion of AmtB into the cytoplasmic membrane (Khademi *et al.*, 2004; Zheng *et al.*, 2004). As substrate binding does not induce a conformational change in the Amt structure, it has been proposed that NH_3 rather than NH_4^+ is passively transported through the narrow hydrophobic pore that acts as a channel in the subunits (Khademi *et al.*, 2004). The crystal structure of the trimeric Amt1 protein of the hyperthermophilic archaeon *A. fulgidus* is also consistent with translocation of NH_3 (Andrade *et al.*, 2005). The unidirectional transport of NH_3 would require deprotonation of NH_4^+ outside the cells and reprotonation in the cytoplasm, resulting in a net electroneutral antiport of NH_4^+ versus H^+. Net uniport of NH_4^+ would be possible if NH_3 transport is linked to proton transport in the same direction. This secondary active transport driven by proton motive force could explain both the electrogenicity observed in the LeAMT1;1

protein of tomato (Ludewig *et al.*, 2002) and the apparent passive transport of NH_3 deduced from crystal structure (Andrade *et al.*, 2005).

In almost all organisms, with exception of cyanobacteria, *amt* genes are closely linked to the regulatory *glnK* gene, which codes for a P_{II} protein. As discussed below, both *amtB* and *glnK* gene products interact to regulate ammonium transport activity in response to nitrogen availability (Coutts *et al.*, 2002; Javelle *et al.*, 2004; Conroy *et al.*, 2007; Gruswitz *et al.*, 2007). Also, as methylammonium transport is usually inhibited by ammonium, it has been proposed that Amt permeases play a more relevant physiological role at low ammonium concentrations, the normal conditions found in natural environments (Montesisnos *et al.*, 1998; Vázquez-Bermúdez *et al.*, 2002b).

Ammonium assimilation by the glutamine synthetase–glutamate synthase pathway

GS catalyses the ATP-dependent synthesis of glutamine from glutamate and ammonium. This reaction requires the presence of divalent cations like Mg^{2+} or Mn^{2+}. There are three different types of GS (Eisenberg *et al.*, 2000). The GSI found in most prokaryotes is a dodecameric protein with identical subunits of about 50 kDa arranged in two hexagonal rings. Sequence analysis of the *glnA* gene encoding the GSI subunits reveals two main subgroups: the less common GSIα of some archaea and low G+C Gram-positive bacteria and the GSIβ found in all other bacteria. GSIα is regulated by feedback mechanisms whereas GSIβ is subjected to covalent modification by adenylylation of a conserved tyrosine residue (Reitzer, 2003). GSII is an octameric enzyme with identical 45–50 kDa subunits found in eukaryotic organisms (Llorca *et al.*, 2006), some bacterial strains, and in the haloarchaeon *Haloferax mediterranei* (Martínez-Espinosa *et al.*, 2006). Finally, the GSIII is a hexameric enzyme with 80-kDa identical subunits encoded by the *glnN* gene, which is found in some bacterial species as *Bacteroides fragilis*, *Butyrivibrio fibrisolvens* and the cyanobacterium *Pseudanabaena* sp. PCC 6903 (Crespo *et al.*, 1998). *Synechocystis* and *Synechococcus* strains contain both GSI and GSIII, although GSIII is only observed under nitrogen deficiency (Reyes and Florencio, 1994). Although both enzymes are different in their amino acid sequences, they share five conserved domains that include the residues involved in the catalytic site and the binding of divalent cations (Muro-Pastor *et al.*, 2005).

GOGAT catalyses the synthesis of two molecules of glutamate from glutamine and 2-oxoglutarate, in a two-electron reductive reaction. There are three different types of GOGAT, depending on the specificity for the electron donor. All of them are flavoproteins that also contain iron–sulfur clusters (van den Heuvel *et al.*, 2004; Muro-Pastor *et al.*, 2005; Suzuki and Knaff, 2005; Vanoni and Curti, 2008). The NADPH-dependent GOGAT, commonly found in heterotrophic bacteria, is composed of two different subunits arranged in $\alpha_4\beta_4$ structure. The large α subunit, encoded by the *gltB* gene, has a size of about 150 kDa and contains FMN and one [3Fe–4S] centre. The small β subunit, encoded by *gltA*, has a size of about 50 kDa and contains FAD. In the holoenzyme, two additional [4Fe–4S] centres are located in the interface of the subunits (van den Heuvel *et al.*, 2004; Vanoni and Curti, 2008). In the genomes of some archaeal species there are putative genes that could represent ancestral forms of the large or small subunits of the bacterial NADPH-GOGAT (Cabello *et al.*, 2004b). Two other types of GOGAT, the ferredoxin- and NADH-dependent enzymes, are usually present in cyanobacteria. Ferredoxin-GOGAT is a 170-kDa monomer encoded by the *glsF* (*glsS*) gene that binds FMN and one [3Fe–4S] centre, thus being similar both in size and cofactors to the large α subunit of NADPH-GOGAT. Some cyanobacteria also contain an NADH-dependent GOGAT, which is composed of two subunits of about 160 and 60 kDa encoded by the *gltB* and *gltD* genes, respectively. These subunits are homologous to the large and small subunits of NADPH-GOGAT (Muro-Pastor *et al.*, 2005; Suzuki and Knaff, 2005).

Alternative pathways for ammonium assimilation

The GDH is present in all living organisms and catalyses both oxidative glutamate deamination and reductive 2-oxoglutarate amination (Smith *et al.*, 1975; Hudson and Daniel, 1993; Smith and Stanley, 2008). However, GDH usually shows a high apparent K_m value for ammonium, in the 1–5 mM range, and it is believed that its main physiological role is the deamination of glutamate for catabolic purposes. Nevertheless, at high ammonium concentrations this enzyme may participate in ammonium assimilation avoiding the requirement of ATP in this process (Muro-Pastor *et al.*, 2005). In bacteria and lower eukaryotes, GDH shows preference for a particular coenzyme, either NAD^+ or $NADP^+$, but in most archaea and higher eukaryotes GDH presents dual specificity for both coenzymes (Smith *et al.*, 1975; Schinkinger *et al.*, 1991; Robb *et al.*, 1992; Peterson and Smith, 1999). Lateral gene transfer seems to have played an important role in the evolution of the GDH families (DiRuggiero *et al.*, 2000). The vast majority of GDH are hexamers of identical subunits of about 45–55 kDa, and the crystal structures of the bacterial, archaeal and mammalian enzymes reveal that the quaternary structure of the GDH can be considered as a dimer of trimers, with the catalytically important residues conserved during evolution (Baker *et al.*, 1992; Stillman *et al.*, 1993; Yip *et al.*, 1995; Knapp *et al.*, 1997; Smith *et al.*, 2002). However, while prokaryotic, fungal and plant GDH are not allosterically regulated, the enzyme of mammals is controlled by different ligands, and this regulation plays a key role in insulin homeostasis in humans (Smith and Stanley, 2008).

In some bacteria, the reductive amination of pyruvate to alanine catalysed by the alanine dehydrogenase may also constitute an alternative minor pathway for ammonium assimilation under certain conditions. In phototrophic bacteria, inhibition of GS by L-methionine-D,L-sulfoximine provokes induction of alanine dehydrogenase to generate alanine, which is further converted into glutamate by the alanine:2-oxoglutarate aminotransferase. As this transamination reaction regenerates pyruvate, the net result of the whole pathway is the synthesis of glutamate from 2-oxoglutarate and ammonium with consumption of two electrons from NAD(P)H (Moreno-Vivián *et al.*, 1983).

Regulation of ammonium assimilation

GS, the key enzyme for primary ammonium assimilation, is tightly regulated at both transcriptional and post-translational levels. Post-translational control of GS activity may operate through feedback mechanisms or reversible covalent modification (Leigh and Dodsworth, 2007). These regulatory mechanisms are complex and vary depending on the organism, but in general nitrogen limitation regulates positively both GS activity and *glnA* gene transcription whereas nitrogen excess causes a decrease in both enzyme activity and gene expression.

The regulation of GS is particularly well known in enterobacteria (Reitzer, 2003; Ninfa and Jiang, 2005). In *E. coli*, the activity of the dodecameric GSI-type enzyme is regulated by covalent modification through adenylylation/deadenylylation of each monomer. The GS adenylylation state is controlled by two effectors, 2-oxoglutarate and glutamine (Fig. 5.2). When the level of 2-oxoglutarate is low and the concentration of glutamine is high (nitrogen excess), GS is adenylylated and its activity is very low. On the contrary, under nitrogen starvation, when the level of 2-oxoglutarate is high and the amount of glutamine is low, GS is found in its deadenylylated state, which shows the maximal activity. Three proteins participate in this process: an adenylyltransferase (ATase or GlnD), the P_{II} protein GlnB, and an uridylyltransferase/uridylyl-removing enzyme (UTase/UR). The ATase is a bifunctional enzyme that catalyses both the ATP-dependent adenylylation of each GS subunit in a specific tyrosine residue and the removing of these AMP moieties by phosphorylytic cleavage (Jiang *et al.*, 2007). The GS adenylylation state is controlled by the trimeric P_{II} protein GlnB, a 2-oxoglutarate sensor that interacts with a variety of target proteins to promote the nitrogen response (Ninfa and Jiang, 2005). GlnB may be found in two forms, deuridylylated or uridylylated in all three subunits. In its deuridylylated state, GlnB stimulates GS inactivation by adenylylation, whereas the uridylylated form promotes GS deadenylylation, and consequently its activation. The activity of GlnB is modulated by a bifunctional UTase/UR that senses glutamine as signal of nitrogen excess to catalyse either the

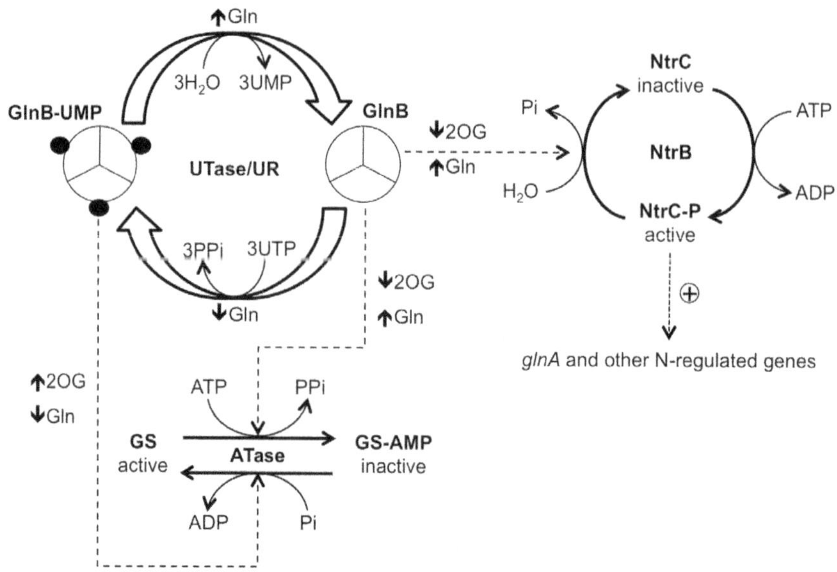

Figure 5.2 Control of glutamine synthetase activity and *glnA* gene expression in enterobacteria. Under nitrogen deficiency (high 2-oxoglutarate, low glutamine), the P_{II} protein GlnB is uridylylated by the UTase/UR enzyme and this leads to GS deadenylylation (activation). Phosphorylation of NtrC by NtrB also allows high expression of *glnA* and other N-regulated genes. Nitrogen sufficiency (low 2-oxoglutarate, high glutamine) causes P_{II} deuridylylation, which leads to GS adenylylation (inactivation) by the adenylyltransferase ATase and NtrC dephosphorylation by NtrB to decrease *glnA* gene expression. See text for details.

transference of UMP from UTP to a tyrosine residue of each GlnB subunit or the hydrolytic removal of the UMP groups (Fig. 5.2).

In *E. coli* and many other bacteria, the expression of the *glnA* gene is controlled by the so-called Ntr system, a two-component system for general nitrogen regulation. NtrB is dual quinase/phosphatase protein that, in response to the nitrogen status, modulates the activity of NtrC, a transcription regulator. Most nitrogen-regulated genes, including *glnA*, are recognized by the specific RNA polymerase σ^{54} factor and also contain specific target sequences for binding of the NtrC regulator, which is active in its phosphorylated form (Leigh and Dodsworth, 2007). Under nitrogen excess (low 2-oxoglutarate/glutamine ratio), deuridylylated GlnB stimulates NtrC dephosphorylation by NtrB. As a consequence of NtrC inactivation, transcription of *glnA* and other nitrogen-regulated genes decreases. On the contrary, under nitrogen deficiency (high 2-oxoglutarate/glutamine ratio), NtrB activates NtrC by phosphorylation of a specific aspartic residue, and this active NtrC protein binds to its target sites to allow the transcription of the nitrogen-regulated genes (Fig. 5.2).

In addition to the GS enzyme, the activity of the ammonium permease is also regulated by the 2-oxoglutarate sensing P_{II} protein GlnK, a GlnB paralogue present in many bacteria and archaea. As mentioned, the *glnK* and *amt* genes are usually clustered. In *E. coli*, it was first described that GlnK and AmtB interact to form a membrane complex that inhibits the ammonium transport (Coutts *et al.*, 2002; Javelle *et al.*, 2004). At high nitrogen, the GlnK protein becomes deuridylylated as a consequence of the low 2-oxoglutarate/glutamine ratio and it binds to AmtB forming an AmtB–GlnK complex that inhibits ammonium transport. Conversely, when nitrogen is limiting, GlnK is uridylylated and dissociates from the AmtB permease to facilitate transport activity (Fig. 5.3). In archaea and the Gram-positive bacterium *Bacillus subtilis*, the GlnK-type P_{II} protein is not modified by UMP binding and seems to be solely dependent on effector molecules. In *B. subtilis*, the Amt–GlnK complex dissociates in the presence of ATP and 2-oxoglutarate. The regulatory protein TnrA, a member of the MerR family (see below), also interacts with this Amt–GlnK complex, but not with free GlnK, and this interaction depends

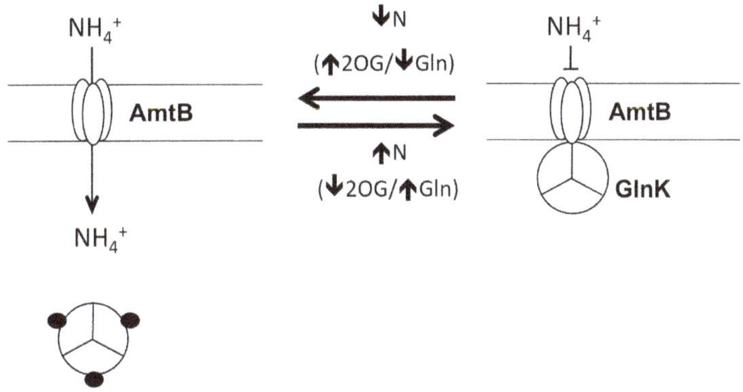

Figure 5.3 Control of ammonium transport activity of the AmtB permease by GlnK. In response to a nitrogen excess (low 2-oxoglutarate and high glutamine levels), the P_{II} protein GlnK becomes deuridylylated and binds to the AmtB permease in the cytoplasmic membrane to inhibit ammonium transport. Reversely, under nitrogen deficiency (high 2-oxoglutarate, low glutamine), GlnK is uridylylated and the GlnK–AmtB complex dissociates to allow ammonium transport. See text for details.

on ATP and 2-oxoglutarate levels (Heinrich *et al.*, 2006; Forchhammer, 2007a). TnrA may be inactivated by proteolysis in the cytoplasm, but it seems that the interaction with the Amt–GlnK complex in the membrane protects TnrA from degradation (Kayumov *et al.*, 2008). In the hyperthermophilic archaeon *A. fulgidus*, the GlnK-type GlnB-1 protein interacts with Amt-1 to form a tight, stable complex that inhibits ammonium transport simply through sterical blocking of the transport pathway, and dissociation of the complex occurs by binding of ATP or 2-oxoglutarate (Andrade *et al.*, 2005; Forchhammer, 2007a).

Although the mechanisms for regulation of ammonium assimilation are widely distributed among bacteria, there are significant differences depending on the species. In cyanobacteria GS activity is not covalently modified by adenylylation/deadenylylation (Fig. 5.4). Instead, it is controlled by two small polypeptides of 7 and 17 kDa (IF7 and IF17) encoded by the *gifA* and *gifB* genes, respectively, that inactivate GS under nitrogen excess (García-Domínguez *et al.*, 1999). In addition, cyanobacteria lack the Ntr system and the main nitrogen regulator is NtcA, a transcription factor of the cAMP receptor protein family (Vega-Palas *et al.*, 1992; Herrero *et al.*, 2001). The ammonium permease *amt1* gene and the GS *glnA* gene are activated by NtcA, but the *gifA* and *gifB* genes encoding the inactivating factors IF7 and IF17 are repressed by NtcA (Muro-Pastor *et al.*, 2005). Upon ammonium removal, GS is rapidly reactivated by proteolytic degradation of the inactivating factors (Galmozzi *et al.*, 2007). Furthermore, cyanobacteria only contain one P_{II} gene that is not clustered together *amt* genes. Rather than by uridylylation, the P_{II} protein encoded by this gene is modified by phosphorylation in response to 2-oxoglutarate (Forchhammer, 2007a). NtcA also binds to 2-oxoglutarate and it seems that the signalling and transcriptional activities of both NtcA and P_{II} are mutually dependent (Muro-Pastor *et al.*, 2005). A small protein called PipX interacts with both P_{II} and NtcA depending on the presence of 2-oxoglutarate (Fig. 5.4). When 2-oxoglutarate concentration is low (nitrogen excess), binding of PipX to the P_{II} protein is favoured, but at high 2-oxoglutarate levels (nitrogen starvation), PipX forms a stable complex with NtcA and enhances expression of NtcA-activated genes (Espinosa *et al.*, 2006; Forchhammer, 2007a).

The Gram-positive bacterium *Bacillus subtilis* also lacks the Ntr system, but two general transcription factors GlnR and TnrA regulate gene expression in response to nitrogen availability (Fisher, 1999; Wray and Fisher, 2006). The *tnrA* gene is monocistronic but the *glnR* gene is cotranscribed together *glnA*. The GlnR and TnrA regulators are active under different growth

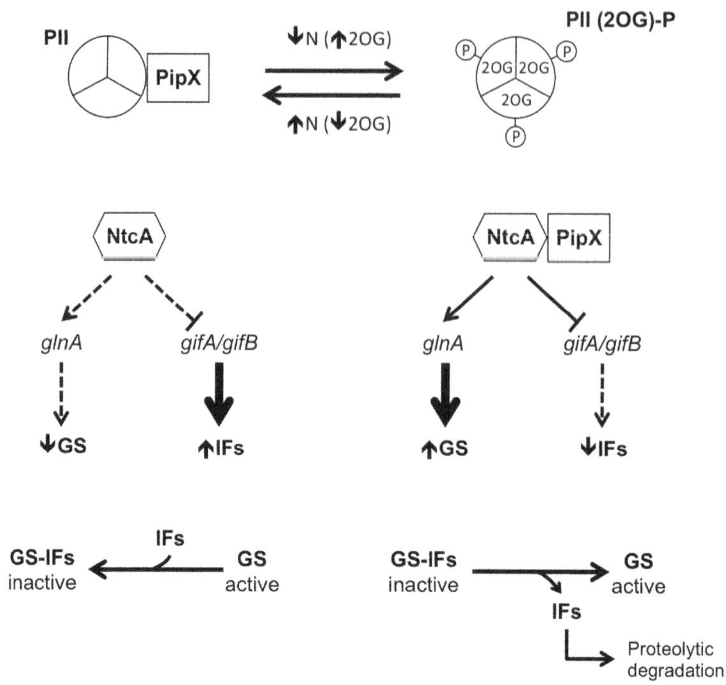

Figure 5.4 Control of glutamine synthetase activity and *glnA* gene expression in cyanobacteria. The P_{II} protein is phosphorylated at low nitrogen (high 2-oxoglutarate) and dephosphorylated at high nitrogen (low 2-oxoglutarate). The PipX protein may bind either to the dephosphorylated P_{II} protein (at low 2-oxoglutarate levels) or to the global nitrogen control factor NtcA (at high 2-oxoglutarate or low nitrogen). Binding of PipX to NtcA increases the expression of the glutamine synthetase *glnA* gene but decreases the expression of the *gifAB* genes coding for the GS inactivating factors IF7 and IF17, which control GS activity. See text for details.

conditions; while GlnR is active under nitrogen excess and represses *glnRA* genes, TnrA is active during nitrogen depletion. TnrA activates the expression of genes involved in assimilation of various nitrogen sources, like nitrate, purine and urea, but negatively regulates the *glnA* and *gltAB* genes coding for GS and GOGAT, respectively. In *B. subtilis*, the GS acts as a sensor of nitrogen availability. When nitrogen is in excess, glutamine and AMP are feedback inhibitors of GS and the feedback-inhibited enzyme forms a complex with TnrA, which inactivates the TnrA regulator. In contrast, feedback-inhibited GS acts as a chaperone that stabilizes the GlnR–DNA complexes under nitrogen excess. Therefore, feedback inhibition of GS not only controls GS activity but also acts as the nitrogen signal for the regulation of the activity of the main nitrogen regulators GlnR and TnrA (Wray *et al.*, 2001; Fisher *et al.*, 2002; Wray and Fisher, 2006; Fisher and Wray, 2008, 2009). Concerning other Gram-positive bacteria, in *Streptomyces coelicolor* GlnR regulates the expression of almost all important genes related to nitrogen metabolism (Tiffert *et al.*, 2008), whereas the TetR-type repressor AmtR is the main regulator of nitrogen genes in corynebacteria (Beckers *et al.*, 2005; Walter *et al.*, 2007; Buchinger *et al.*, 2009). In *Mycobacterium smegmatis* both GlnR and AmtR proteins are found, while in other mycobacteria AmtR is absent and GlnR is the only nitrogen regulator (Amon *et al.*, 2009).

Archaea may also use a different mechanism to control gene expression in response to the C/N balance. Thus, in the methanogenic euryarchaeon *Methanococcus maripaludis* the tetrameric protein NrpR is a repressor of *glnA* (GS) and *nif* (nitrogen fixation) genes (Lie and Leigh, 2003; Leigh and Dodsworth, 2007). The affinity of NrpR for its DNA-binding sequence decreases in the presence of 2-oxoglutarate, allowing *glnA* and *nif* gene expression under nitrogen limitation (Lie and Leigh, 2003).

Nitrate assimilation

The importance of nitrate as a key nutrient for autotrophic, oxygenic photosynthetic organisms has been long recognized in the classical models of the carbon and nitrogen cycles. On the contrary, nitrate utilization by heterotrophic bacteria has historically received less attention, and studies on nitrate assimilation have been largely restricted only to *Klebsiella oxytoca* (Lin *et al.*, 1994; Lin and Stewart, 1998), *Bacillus subtilis* (Ogawa *et al.*, 1995; Nakano *et al.*, 1998), *Azotobacter vinelandii* (Ramos *et al.*, 1993; Gutiérrez *et al.*, 1995) and *Rhodobacter capsulatus* (Blasco *et al.*, 1997; Cabello *et al.*, 2004a; Pino *et al.*, 2006). However, the analysis of emerging prokaryotic genome sequences reveals that nitrate assimilation systems are widespread in bacteria and archaea, and it is believed that nitrate utilization by heterotrophic prokaryotes is relevant in many environments (Allen *et al.*, 2001; 2005). The phylogenetic analyses, together with the biochemical studies on model organisms, reveal significant differences among prokaryotic nitrate assimilation systems. Thus, in cyanobacteria nitrate reduction is functionally linked to the photosynthetic process and both nitrate and nitrite reductases use photosynthetically reduced ferredoxin or flavodoxin as the physiological electron donor, whereas in most heterotrophic bacteria these reductases are dependent on NADH (Flores *et al.*, 2005; Moreno-Vivián and Flores, 2007). Also, depending on the organism, nitrate and nitrite uptake systems involved in nitrate assimilation are either ATP-binding cassette (ABC) transporters or major facilitator superfamily (MFS) permeases. Finally, although bacterial assimilatory nitrate reduction is usually controlled by nitrate induction (pathway-specific control) and ammonium repression (global nitrogen control), different regulatory patterns, mechanisms and proteins are involved in the regulation of nitrate assimilation in different bacteria.

Nitrate assimilation gene clusters

The genes coding for the regulatory and structural proteins required for transport and reduction of both nitrate and nitrite have been identified in a wide range of prokaryotes. The genetic nomenclature for bacterial nitrate assimilation genes is confusing, with different designations for homologous genes and, in some cases, with names as *nar*, *nir* or *nap* that are also used for respiratory or periplasmic nitrate and nitrite reductases. Here, in absence of a common nomenclature, nitrate assimilation genes will be termed *nas*, although for specific genes of different organisms the nomenclature of the original papers will be used.

The bacterial *nas* genes are usually clustered, but content and organization of the gene clusters vary depending on the organism, reflecting the presence of different transport systems (ABC- or MFS-transporters), nitrate and nitrite reductases (ferredoxin- or NADH-dependent enzymes) and regulatory proteins. Some *nas* clusters also include genes that code for putative detoxification systems for reactive nitrogen species or their regulators, like the *hcp* and *nsrR* genes of *Rhodobacter capsulatus* (Cabello *et al.*, 2004a) and the *hmp* gene of *Bradyrhizobium japonicum* (Meakin *et al.*, 2006). Hcp (hybrid cluster protein) and Hmp (flavohaemoglobin) are proposed to detoxify hydroxylamine and nitric oxide, respectively, and NsrR is a nitric oxide-sensor that negatively regulates both *hcp* and *hmp* genes (Efromovich *et al.*, 2008). As hydroxylamine and nitric oxide are potential toxic by-products generated by incomplete reduction of nitrite to ammonium, it makes sense the presence and co-regulation of these detoxification systems with the nitrite reductase.

The *nas* gene clusters of some relevant prokaryotic organisms are shown in Fig. 5.5. The gene structure *nirA* (nitrite reductase)-*nrtABCD* (nitrate-nitrite transporter)-*narB* (nitrate reductase) found in *Synechococcus elongatus* is conserved in various cyanobacteria, although in *Synechocystis* sp. strain PCC 6803 *nirA* is not part of the main gene cluster and in some *Synechococcus* marine strains the nitrate assimilation structural genes are clustered together cofactor biosynthesis genes (Flores *et al.*, 2005). In the photoheterotroph *Rhodobacter capsulatus* E1F1 the *nas* gene cluster includes the *hcp* gene encoding a hybrid cluster protein with hydroxylamine reductase activity, three regulatory genes (*nsrR* and *nasTS*), the *nasFED* genes coding for an ABC-type nitrate transporter, the *nasB* and *nirD* genes encoding the nitrite reductase subunits,

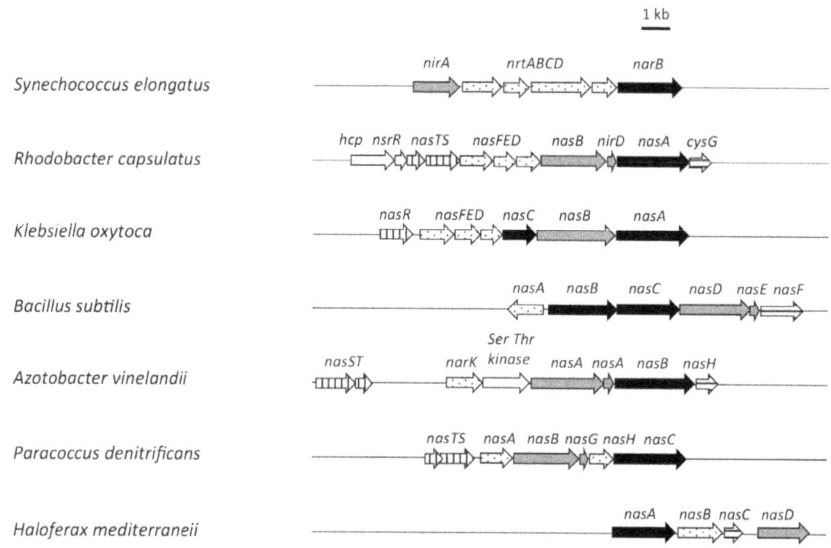

Figure 5.5 Organization of the nitrate assimilation (*nas*) gene clusters in some relevant prokaryotic organisms. The genes coding for the different components of the assimilatory nitrate reduction systems are indicated by arrows: black arrows, nitrate reductase genes; grey arrows, nitrite reductase genes; dotted arrows, nitrate/nitrite transport genes; vertical-dashed arrows, regulatory genes; horizontal-dashed arrows, cofactor biosynthesis genes; white arrows, other genes.

the gene encoding the catalytic subunit of nitrate reductase (*nasA*), and the *cysG* gene involved in biosynthesis of sirohaem, the nitrite reductase cofactor (Cabello *et al.*, 2004a; Pino *et al.*, 2006). In the γ-proteobacterium *Klebsiella oxytoca*, the genes coding for the ABC-type nitrate transporter (*nasFED*), the electron transfer subunit of nitrate reductase (*nasC*), the nitrite reductase (*nasB*) and the nitrate reductase catalytic subunit (*nasA*) are organized in an operon that is regulated by the product of *nasR*, a gene located upstream from *nasF* (Lin and Stewart, 1998). The *nas* cluster of the Gram-positive bacterium *Bacillus subtilis* includes a putative MFS-type permease gene (*nasA*), a gene encoding a flavoprotein (*nasB*) that may act as electron donor to the nitrate reductase and/or nitrite reductase, the genes coding for the catalytic subunits of the nitrate reductase (*nasC*) and nitrite reductase (*nasD*), a gene coding for a putative small subunit of the nitrite reductase (*nasE*), and the sirohaem biosynthesis *nasF* gene (Ogawa *et al.*, 1995). In the aerobic diazotroph *Azotobacter vinelandii*, two putative regulatory genes (*nasS* and *nasT*) are clustered together *nasBA* genes coding for the assimilatory nitrate reductase and nitrite reductase, respectively (Ramos *et al.*, 1993; Gutiérrez *et al.*, 1995). The *nas* gene cluster of the heterotrophic denitrifying bacterium *Paracoccus denitrificans* also includes the *nasST* regulatory genes, in addition to genes coding for nitrate and nitrite transporters (*nasA* and *nasH*), nitrate and nitrite reductases (*nasC* and *nasB*) and a gene (*nasG*) coding for a putative ferredoxin (GenBank accession number CP000490, Pden_4455–4449; D.J. Richardson, personal communication). Some archaea may also assimilate nitrate. The extreme halophilic archaeon *Haloferax mediterranei* contains a *nas* gene cluster comprising four genes (*nasABCD*) encoding, respectively, a ferredoxin-dependent nitrate reductase, a MFS-family NarK-type nitrate transporter, the MobA protein involved in nitrate reductase cofactor biosynthesis and a ferredoxin-dependent nitrite reductase. The presence of a NarK-type nitrate transporter in this haloarchaeon reveals that this type of transporter is not exclusive of bacteria and eukaryotes (Lledó *et al.*, 2005). The *nasABC* genes constitute a polycistronic unit, whereas the *nasD* gene encoding the nitrite reductase is monocistronic (Martínez-Espinosa *et al.*, 2009).

Nitrate and nitrite transport

For assimilatory purposes, nitrate or nitrite are incorporated into the cells by either ATP-dependent ABC-type transporters or MFS-type permeases. The fuelling mechanism for the MFS-type permeases is not clear, although it is believed that the transport of the negatively charged anions against the negatively inside membrane potential is driven by proton-motive force (Moir and Wood, 2001; Moreno-Vivián and Flores, 2007). The recent determination of the crystal structures of some ABC- and MFS-type transporters has represented an important step in the understanding of the biological function of these proteins (Abramson, 2004; Locher, 2004).

The ATP-driven, ABC-type transporters are members of the family 3.A.1.16.1 of the TC transporter classification (Busch and Saier, 2002; Saier *et al.*, 2009). These proteins are widely distributed among bacteria and archaea and usually accept nitrate and nitrite as substrates, showing high affinity for both compounds. ABC-type transporters are osmotic shock-sensitive since they require a periplasmic component for substrate binding (Muñoz-Centeno *et al.*, 1993). The basic genetic composition for these transporters is found in some bacteria like *Klebsiella oxytoca* and *Rhodobacter capsulatus*, in which three clustered genes, *nasFED*, encode the transporter (Wu and Stewart, 1998; Pino *et al.*, 2006). NasF is a periplasmic substrate-binding protein that is exported to the periplasm by the twin-arginine translocation (Tat) system, NasE is a homodimeric hydrophobic protein bearing six transmembrane helices, and NasD is a homodimeric cytoplasmic ATPase. The cyanobacterial ABC-type nitrate transporters are encoded by four genes, *nrtABCD*, and show some interesting variations over this transporter structure. The periplasmic substrate-binding protein NrtA contains a lipid moiety for membrane attachment, the transmembrane protein NrtB has only five transmembrane helices, and the ATPase component is a heterodimer composed of NrtD, the standard ATPase subunit, and NrtC, a protein with an additional C-terminal domain homologous to NrtA that has a regulatory function (Omata, 1991; Kobayashi *et al.*, 1997; Maeda and Omata, 1997; Flores *et al.*, 2005). In *Synechococcus elongatus* it has been recently reported that the cyanate transporter CynABD is also an active nitrite transporter that may contribute significantly to nitrite assimilation. This CynABD transporter is induced under nitrogen deficiency and shows high affinity for nitrite, with cyanate acting as a competitive inhibitor of nitrite transport (Maeda and Omata, 2009).

MFS-type nitrate permeases (family 2.A.1.8 of the TC transporter classification) are present in both bacteria and archaea, although they have received different names: NasA in *Bacillus subtilis* (Ogawa *et al.*, 1995), NasB in *Haloferax mediterranei* (Lledó *et al.*, 2005) and NrtP or NapA in some marine cyanobacterial strains (Sakamoto *et al.*, 1999; Wang *et al.*, 2000). A *nasA* gene coding for a putative MFS-type nitrate transporter is also found in the nitrate assimilation *nas* gene cluster of *Paracoccus denitrificans* (GenBank accession number CP000490; D.J. Richardson, personal communication). These MFS permeases usually consist of 12 transmembrane segments that are grouped into two domains, each with six-helices. As they are homologous to the well characterized NarK permeases (Goddard *et al.*, 2008), it may be assumed that they are H^+/NO_3^- symporters that use the proton electrochemical gradient as the driving force for transport (Fig. 5.1). The NrtP permease of *Nostoc punctiforme* transports nitrate preferentially over nitrite and, unlike the ABC-type transporters of other cyanobacteria, this transport activity is not regulated by ammonium (Aichi *et al.*, 2006).

Assimilatory nitrate reductases

Bacterial assimilatory nitrate reductases (Nas) are cytoplasmic molybdoenzymes that catalyse the two-electron reduction of nitrate to nitrite (Fig. 5.1). Some bacteria also contain other enzymes that reduce nitrate to nitrite during dissimilatory or respiratory processes, the dissimilatory periplasmic nitrate reductases (Nap) and the respiratory membrane-bound nitrate reductases (Nar). These three types of bacterial nitrate reductases (Nas, Nap and Nar) are structurally and functionally different, although all of them include a Mo-*bis*-molybdopterin guanine dinucleotide (Mo-*bis*-MGD) and at least one [4Fe–4S] cluster as cofactors (Moreno-Vivián *et al.*, 1999; Richardson *et al.*, 2001; see also Chapter 2).

Bacterial assimilatory nitrate reductases can be classified into two groups according to their electron donor specificity. The enzymes found in cyanobacteria, *Azotobacter vinelandii* and the archaeon *Haloferax mediterraneii* use ferredoxin or flavodoxin as reductant, whereas the enzymes present in most heterotrophic bacteria and in the photoheterotroph *Rhodobacter capsulatus* are NADH-dependent (Moreno-Vivián and Flores, 2007).

Cyanobacterial ferredoxin-dependent nitrate reductase is a monomeric protein encoded by the *narB* gene that is partially associated to the thylakoid membrane (Flores *et al.*, 2005). Especially well characterized is the enzyme of *Synechococcus elongatus*, a monomer of about 80 kDa that contains the Mo-*bis*-MGD cofactor and one [4Fe–4S] cluster (Butt *et al.*, 2002; Rubio *et al.*, 2002; Jepson *et al.*, 2004). The metal centres have midpoint redox potentials of –190 mV and -150 mV for $[4Fe–4S]^{1+/2+}$ and $Mo^{6+/5+}$, respectively. Electrons from very low potential electron donors, like [2Fe–2S] ferredoxin or flavin-containing flavodoxin, are transferred to the nitrate reductase [4Fe–4S] cluster and from this centre to the Mo-*bis*-MGD cofactor, which is the nitrate reduction site (Butt *et al.*, 2002; Jepson *et al.*, 2004). Although some cyanobacterial nitrate reductases show K_m values for nitrate of 1–10 mM, the *S. elongatus* enzyme exhibits values of 50–80 μM (Jepson *et al.*, 2004).

In *A. vinelandii*, the assimilatory nitrate reductase is a 105-kDa monomeric protein encoded by the *nasB* gene (Ramos *et al.*, 1993). The purified enzyme uses flavodoxin as physiological reductant and contains one Mo, four Fe and four acid-labile S atoms per molecule, according to the presence of Mo-*bis*-MGD and a [4Fe–4S] cluster (Gangeswaran *et al.*, 1993; Gangeswaran and Eady, 1996). The apparent K_m values for nitrate with reduced methyl viologen or bromophenol blue as artificial electron donor are 70 and 220 μM, respectively (Gangeswaran *et al.*, 1993).

The assimilatory nitrate reductase of the haloarchaeon *Haloferax mediterraneii* has also been purified. The enzyme uses ferredoxin as reductant, shows an apparent K_m for nitrate of 0.95 mM, and requires pH 9 and high temperatures (80°C) and salt concentrations (3.1 M NaCl) for optimal activity (Martínez-Espinosa *et al.*, 2001a). The sequence analysis of the *nas* gene cluster of this archaeon suggest that this ferredoxin-dependent nitrate reductase has an estimated molecular mass of about 76 kDa, similar to that described for the monomeric enzyme of cyanobacteria (Lledó *et al.*, 2005).

Putative NADH-nitrate reductase genes are present in many prokaryotic genomes, but interestingly, none of the predicted sequences of these enzymes include a typical NADH-binding site, thus making them different from the eukaryotic nitrate reductases, which contain an NAD(P) H-binding domain (Campbell, 2001). Therefore, in bacteria, an additional protein with an NADH-oxidizing domain (diaphorase) is required for the electron transfer from NADH to the nitrate reductase catalytic subunit. It is worth noting that ferredoxin/flavodoxin-dependent nitrate reductases do not contain flavin as prosthetic group but, on the contrary, bacterial NADH-dependent assimilatory nitrate reductases include FAD as a cofactor of the NADH oxidoreductase (diaphorase).

The properties of the NADH-dependent assimilatory nitrate reductases of *Klebsiella oxytoca* and *Bacillus subtilis* have been deduced from DNA sequence analysis. In *Klebsiella*, assimilatory nitrate reductase is composed of a 92-kDa catalytic subunit and a 43-kDa electron transfer subunit (Lin *et al.*, 1994; Lin and Stewart, 1998). The small subunit, encoded by the *nasC* gene, is predicted to be the FAD-containing NADH oxidoreductase that mediates electron transfer from NADH to the large catalytic subunit, the *nasA* gene product, which binds the Mo-*bis*-MGD cofactor, one [4Fe–4S] cluster in the N-terminal end and an additional [2Fe–2S] centre in an C-terminal extension that is similar to NifU and [2Fe–2S] cluster-containing proteins (Lin and Stewart, 1998). This C-terminal domain is usually absent in other bacterial Nas proteins, but it is also present in the assimilatory nitrate reductase of *Paracoccus denitrificans* (GenBank accession number CP000490; D.J. Richardson, personal communication). The predicted electron flow between redox cofactors in the *Klebsiella* NasCA nitrate reductase is NADH → FAD → [2Fe–2S] → [4Fe–4S] → Mo-*bis*-MGD → NO_3^-.

Genetic analysis of *B. subtilis* also suggests the presence of a dimeric NADH-dependent assimilatory nitrate reductase (Ogawa *et al.*, 1995). The large catalytic subunit (*nasC* gene product) contains the Mo-*bis*-MGD cofactor and one N-terminal [4Fe–4S] cluster, but it lacks the NifU-like C-terminal domain present in *Klebsiella* NasA protein. However, two putative [2Fe–2S] centres may be present in the electron transfer subunit, encoded by *nasB*, which also contains FAD (Ogawa *et al.*, 1995).

In the phototrophic bacterium *R. capsulatus*, the assimilatory NADH-dependent nitrate reductase is a heterodimer composed of a 45-kDa FAD-containing NADH oxidoreductase and a 95-kDa catalytic subunit with Mo-*bis*-MGD and one N-terminal [4Fe–4S] cluster, but without additional [2Fe–2S] centres. The purified enzyme shows an apparent K_m value for nitrate of 0.1 mM and is inhibited by NADH under aerobic conditions, probably by formation of superoxide anion at the diaphorase flavin cofactor (Blasco *et al.*, 1997). The large catalytic subunit is encoded by the *nasA* gene, which is located in a 17-kb *nas* gene region together other genes required for nitrate and nitrite assimilation. However, the putative gene coding for the small diaphorase subunit is not present in this *nas* gene cluster (Cabello *et al.*, 2004a; Pino *et al.*, 2006).

Assimilatory nitrite reductases

Assimilatory nitrite reductases are cytoplasmic enzymes that catalyse the six-electron reduction of nitrite to ammonium (Fig. 5.1). Two types of assimilatory nitrite reductases have been described according to their electron donor specificity: ferredoxin-dependent nitrite reductases are present in eukaryotic and prokaryotic photosynthetic organisms (plants, algae and cyanobacteria) and also in the haloarchaeon *H. mediterranei*, whereas NADH-dependent nitrite reductases are found in fungi and heterotrophic bacteria, including the photoheterotroph *R. capsulatus* (Moreno-Vivián and Flores, 2007). Both types of nitrite reductases contain one [4Fe–4S] centre and sirohaem, a special haem with eight carboxylic acid-containing side-chains that is also present in the sulfite reductase. NADH-nitrite reductases contain FAD, in addition to sirohaem and [4Fe–4S]. The FAD- and NADH-binding sites are located in an N-terminal extension of the enzyme that is not present in the ferredoxin-dependent nitrite reductases.

Cyanobacterial nitrite reductases are monomeric proteins of about 55 kDa (encoded by *nirA*) that receive electrons from photosynthetically reduced ferredoxin or flavodoxin (Flores *et al.*, 2005). The enzymes purified from different cyanobacteria show biochemical properties similar to those of plants and algae. Apparent K_m values of 5–20 μM for ferredoxin and 40–50 μM for nitrite have been reported (Flores *et al.*, 2005). The six electrons required for nitrite reduction probably flow from the [2Fe–2S] cluster of the reduced ferredoxin to the [4Fe–4S] centre of the nitrite reductase, and finally to the sirohaem group. NO and NH_2OH may be formed as enzyme-bound intermediates in the catalytic cycle, as described for plant nitrite reductases (Kuznetsova *et al.*, 2004). Curiously, the nitrite reductase from *Plectonema boryanum* contains a C-terminal extension that is similar to the plant-type [2Fe–2S] ferredoxins and probably mediates the electron flow between the iron–sulfur clusters of the electron donor (ferredoxin) and the enzyme (Suzuki *et al.*, 1995a).

A ferredoxin-dependent nitrite reductase has also been purified from the archaeon *H. mediterranei* (Martínez-Espinosa *et al.*, 2001b). The enzyme is a monomer of about 66 kDa with sirohaem and one [4Fe–4S] centre. Sequence analysis of the *nasD* gene encoding this ferredoxin-nitrite reductase confirms these enzymological studies (Lledó *et al.*, 2005). The enzyme exhibits a relatively high K_m for nitrite (8 mM), and shows the maximal activity at 60°C with 3.3 M NaCl (Martínez-Espinosa *et al.*, 2001b). The [2Fe–2S] ferredoxin that acts as the physiological electron donor for both nitrate and nitrite reductases in this haloarchaeon has also been purified and characterized (Martínez-Espinosa *et al.*, 2007b).

Biochemical studies on NADH-nitrite reductases of heterotrophic bacteria are scarce and most information has been obtained from DNA sequence analysis carried out in *Klebsiella* (Lin *et al.*, 1994) and *Bacillus* (Ogawa *et al.*, 1995), and from biochemical studies performed with the NADH-nitrite reductase of *E. coli*, although this nitrite reductase is not a proper assimilatory enzyme since it is only induced under anaerobic conditions and participates in detoxification of nitrite formed during nitrate respiration rather than in nitrate assimilation (Cole, 1996; Lin and Stewart, 1998). *E. coli* NADH-nitrite reductase consists of a large catalytic subunit (NirB) with sirohaem and one [4Fe–4S] cluster, and a small subunit (NirD) that is highly similar to the C-termini of fungal monomeric nitrite reductases (Harborne *et al.*, 1992). The nitrite reductase from *B. subtilis* is also composed of two subunits, NasDE, which are homologous to the *E. coli* NirBD proteins (Ogawa *et al.*, 1995). However, the assimilatory nitrite reductase of *Klebsiella* is a monomer encoded by the *nasB* gene (Lin *et al.*, 1994; Lin and Stewart, 1998). Two additional [2Fe–2S] centres may also be present in the NADH-nitrite reductases of *E. coli*, *B. subtilis* and *K. oxytoca*, as suggested by the presence of tandem NifU-like motifs (Lin and Stewart, 1998).

R. capsulatus is the only phototrophic organism described so far with an NADH-nitrite reductase (Olmo-Mira *et al.*, 2006). The nitrate assimilation *nas* gene region of this bacterium includes the *nasB* gene, which codes for a protein of about 87 kDa that shows high similarity to the catalytic subunit of the bacterial NADH-nitrite reductases, and the *nirD* gene, which codes for a 12-kDa protein homologous to *E. coli* NirD (Cabello *et al.*, 2004a; Pino *et al.*, 2006). The purified NasB contains FAD, sirohaem and labile iron and sulfur, like nitrite reductases from non-photosynthetic organisms (Olmo-Mira *et al.*, 2006). The heterotrophic denitrifying bacterium *Paracoccus denitrificans* also has an NADH-nitrite reductase, encoded by the *nasB* gene, which contains putative FAD- and NADH-binding domains in addition to the iron–sulfur cluster and the sirohaem prosthetic group (GenBank accession number CP000490; D.J. Richardson, personal communication).

Regulation of nitrate assimilation

Control of nitrate assimilation may involve distinct regulatory proteins and mechanisms in different bacteria, but usually this process is regulated at the transcriptional level by nitrate and/or nitrite induction (pathway-specific control) and by ammonium repression (general nitrogen control). Additionally, nitrate assimilation may be subjected to post-translational regulation and ammonium also causes a short-term inhibition of nitrate and nitrite uptake in many bacteria (Flores *et al.*, 1980, 1987; Caballero *et al.*, 1986; Muñoz-Centeno *et al.*, 1993).

Ammonium represses nitrate assimilation genes in almost all cyanobacteria. This global nitrogen control is exerted by NtcA, a transcription factor of the CAP family (Herrero *et al.*, 2001; Flores *et al.*, 2005). NtcA mediates nitrogen control activating transcription of nitrate assimilation genes at high C to N ratio, which is signalled by the 2-oxoglutarate levels (Herrero *et al.*, 2001; Tanigawa *et al.*, 2002; Vázquez-Bermúdez *et al.*, 2002a; Flores *et al.*, 2005). The P_{II} protein encoded by the *glnB* gene is a 2-oxoglutarate sensor that is required for the expression of NtcA-dependent genes (i.e., nitrate assimilation genes) under nitrogen deficiency (Aldehni *et al.*, 2003; Paz-Yepes *et al.*, 2003; Forchhammer, 2004). In the marine picocyanobacterium *Synechococcus* sp. strain WH8103 the expression of the genes *nrtP* (nitrate permease) and *narB* (nitrate reductase) is NtcA-dependent, but *nirA* (nitrite reductase) gene expression is not controlled by NtcA, so that this strain may coassimilate nitrite and ammonium (Wyman and Bird, 2007).

Nitrate control is observed in some cyanobacteria, like diazotrophic, heterocyst-forming *Anabaena* sp. strain PCC 7120, but not in unicellular cyanobacteria like *Synechococcus elongatus*. NtcB, a LysR family transcription regulator, and probably other regulators like CnaT and NirB, are involved in this pathway-specific positive control by nitrate/nitrite (Flores *et al.*, 2005). In *Anabaena* PCC 7120, NtcB binds to the *nir* operon promoter upstream from the NtcA-binding site to activate gene expression (Frías *et al.*, 2000). In other cyanobacteria NtcB also mediates nitrite induction of nitrate assimilation genes (Aichi *et al.*, 2001; Flores *et al.*, 2005). The CnaT protein, which is homologous to some glycosyl transferases, is also required for *nir* operon expression in *Anabaena* PCC 7120 (Frías *et al.*, 2003). Another protein, NirB, is also required for optimal expression of nitrite reductase in *S. elongatus* (Suzuki *et al.*, 1995b). Therefore, nitrate assimilation in cyanobacteria is subjected to a complex transcriptional regulation, with a strict requirement of NtcA for induction of gene expression in response to nitrogen availability but with a variable dependence on nitrate/nitrite through NtcB, CnaT and NirB in different cyanobacteria (Flores *et al.*, 2005).

Nitrate assimilation in cyanobacteria is also subjected to post-translational regulation. Nitrate uptake is rapidly inhibited by ammonium, and transport is restored after ammonium depletion (Flores *et al.*, 1980; 1987). This effect is mediated by the dephosphorylated form of the P_{II} protein, which acts on the NrtA-like C-terminal extension of NrtC to inhibit the nitrate/nitrite transport (Kobayashi *et al.*, 1997; Lee *et al.*, 1998; 2000; Forchhammer, 2004; Paz-Yepes *et al.*, 2009). Furthermore, nitrate reductase in *S. elongatus* may be directly regulated by the P_{II} protein (Takatani *et al.*, 2006). In *Nostoc punctiforme*, which has an NrtP nitrate permease that is insensitive to ammonium, the nitrate reductase activity is inhibited upon ammonium addition, and this confers the ammonium sensitivity at post-translational level (Aichi *et al.*, 2006).

In heterotrophic bacteria, the regulation of nitrate assimilation has been well characterized in *Klebsiella oxytoca*. In this bacterium, the expression of the *nas* operon is subjected to general nitrogen control by the Ntr system and to specific nitrate/nitrite induction by a transcription antitermination mechanism that involves the NasR protein (Lin and Stewart, 1996, 1998). The regulatory *nasR* gene is located upstream from the *nasFEDCBA* operon. A hairpin structure that causes early transcription termination has been identified in the mRNA upstream of the *nas* operon. NasR protein is proposed to be both a nitrate/nitrite sensor and a transcription antiterminator (Lin and Stewart, 1998; Chai and Stewart, 1999), and contains the so-called ANTAR motif for directly binding to mRNA (Shu and Zhulin, 2002). The *Klebsiella* Ntr system activates *nasF* operon expression in response to nitrogen limitation, but in absence of nitrate the transcription terminates very early by formation of a hairpin structure. However, when nitrate or nitrite is present, NasR binds to the mRNA sequence preventing hairpin formation and allowing the complete synthesis of the *nasF* transcript (Fig. 5.6a).

The dual nitrate sensor/transcription antiterminator NasR protein is replaced by a two-component system composed of a nitrate sensor (NasS) and a response regulator (NasT) in *Azotobacter vinelandii* (Gutiérrez *et al.*, 1995), *Rhodobacter capsulatus* (Pino *et al.*, 2006) and *Paracoccus denitrificans* (GenBank accession number CP000490; D.J. Richardson, personal communication). NasT proteins also contain the ANTAR domain for transcription antitermination, but unlike NasR, they do not sense nitrate or nitrite. This role seems to be carried out by NasS, which is predicted to be a soluble cytoplasmic protein sharing similarity to the periplasmic substrate-binding components of nitrate transporters. Therefore, NasS may bind nitrate/nitrite in the cytoplasm for sensing purpose rather than for transport. In organisms with this bicomponent system, the presence of nitrate/nitrite may allow the formation of an active NasT–NasS complex that could act as a transcription antiterminator, like the dual NasR protein. Alternatively, NasS may form a complex with NasT avoiding transcription antitermination activity of this protein in absence of nitrate, but in the presence of this compound NasS dissociates from NasT enabling this protein to bind mRNA for its transcription antitermination effect (Fig. 5.6b). Genes homologous to *nasT* and *nasS* are present in the genomes of many bacteria, indicating that this bicomponent system is widespread in bacteria. Although putative *nasS* genes are usually described as *ntrA*/*nasF* nitrate transport genes, it may be proposed as a general rule that products of these genes are involved in nitrate transport when they contain a signal peptide for periplasm translocation, but they are nitrate sensors if are devoid of this leader sequence (Moreno-Vivián and Flores, 2007).

Concerning to the general nitrogen regulation, in *A. vinelandii* the ammonium control is mediated by the Ntr system, like in *Klebsiella*, but in *R. capsulatus* a programmed translational frameshift −1 in the *nasS* transcript could allow the synthesis of a complete NasS protein able to sense nitrate only under nitrogen deficiency. The *R. capsulatus nasS* gene has a premature UGA stop codon

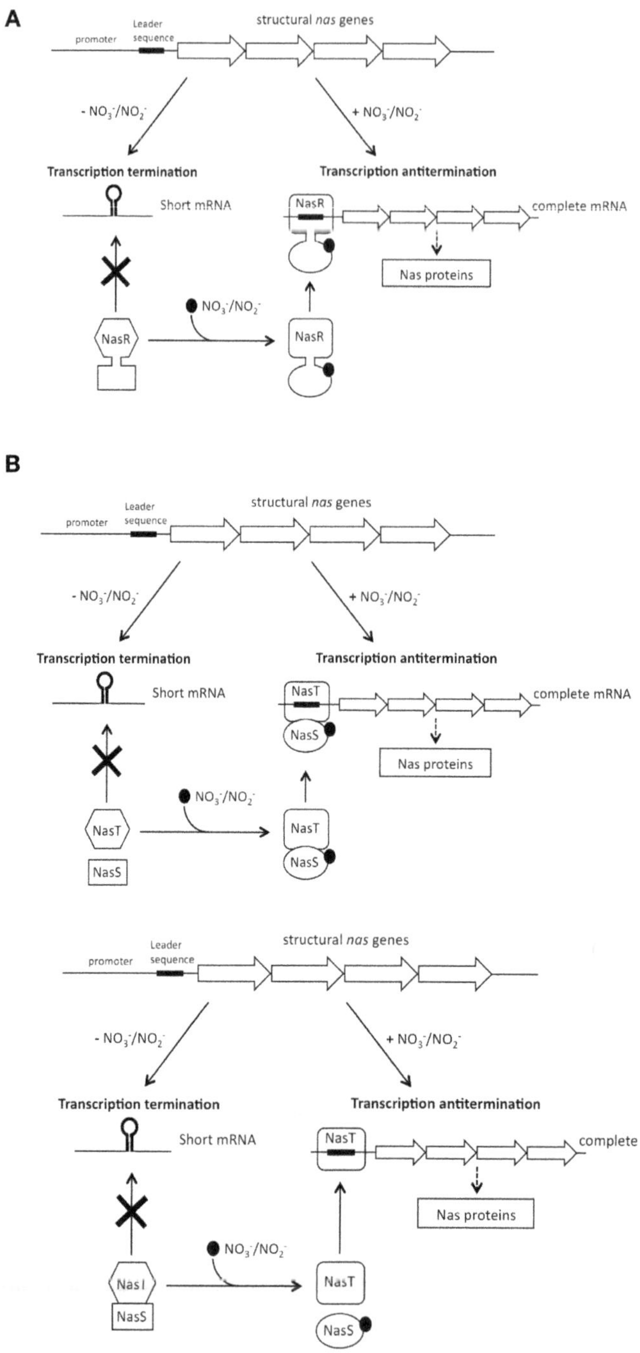

Figure 5.6 Control of the expression of nitrate assimilation genes by transcription antitermination. (A) In *Klebsiella*, the NasR protein acts as a dual nitrate sensor and transcription antiterminator. In absence of nitrate or nitrite, a hairpin structure is formed in the *nas* leader sequence and the transcription of the structural *nas* genes ceases. In the presence of nitrate/nitrite, NasR binds to this mRNA region avoiding transcription termination to allow *nas* gene expression. (B) In bacteria with the NasST bicomponent system, the NasS protein acts as a nitrate/nitrite sensor and the NasT protein acts as a transcription antiterminator. NasS may bind nitrate and then may form a complex with NasT to allow *nas* gene expression (up) or, alternatively, NasS may sequester NasT in absence of nitrate, but after binding nitrate, it may dissociate from NasT allowing this protein to act as a transcription antiterminator (down). See text for details.

downstream from a sequence coding for five almost consecutive Arg residues (RRRRKR), and this mRNA region may form a hairpin secondary structure. Therefore, a complete NasS protein may be only synthesized when a translational −1 frameshift takes place in this mRNA region. This frameshift may be induced under nitrogen limitation since the low Arg pool could favour ribosome pause and slippage at this five arginine coding sequence. Thus, in *R. capsulatus* the *nas* genes are induced by nitrate or nitrite only in the absence of ammonium, without the involvement of Ntr system, since the NasS protein (nitrate sensor) may be only synthesized under nitrogen deficiency, and if nitrate becomes available under these conditions, NasT acts as an antiterminator to allow *nas* gene expression (Pino *et al.*, 2006).

Significant differences also exist in the regulation of nitrate assimilation in *Bacillus subtilis*, which lacks the general Ntr system and controls nitrate assimilation through the general nitrogen regulator TnrA (Nakano *et al.*, 1998). This Gram-positive bacterium has two different nitrate reductases, the cytoplasmic assimilatory NasBC enzyme and the membrane-bound respiratory NarGHI complex, but only one nitrite reductase, the soluble NADH-dependent NasDE enzyme, which is required for both anaerobic respiration and aerobic assimilation processes. The *nasBC* and *nasDE* genes constitute an operon, but transcription of *nasDE* may occur from the *nas* promoter upstream of *nasB* or from an internal promoter between *nasC* and *nasD* (Nakano *et al.*, 1998). Under nitrogen limitation in aerobiosis, transcription from both promoters is activated by the nitrogen regulator TnrA, which also activates the expression of other nitrogen assimilation genes of this bacterium. However, under anaerobic conditions, expression of the *nasDE* genes from the internal promoter is induced when nitrite is present. The ResDE two-component regulatory system is also required for anaerobic induction of the *nasDE* nitrite reductase genes and the *narGHI* genes coding for the respiratory nitrate reductase (Nakano *et al.*, 1998). On the other hand, the Gram positive actinomycete *Streptomyces coelicolor* is also able to grow with nitrate as the sole nitrogen source, and the global regulator GlnR activates the expression of *nasA* and *nirB1B2C* genes that code for the assimilatory nitrate and nitrite reductases, respectively (Tiffert *et al.*, 2008; Wang and Zhao, 2009). The presence of a nitrate assimilation pathway has been also described in *M. smegmatis* (Khan *et al.*, 2008) and *M. tuberculosis* (MALM *et al.*, 2009). Curiously, in this last organism, the respiratory nitrate reductase encoded by the *narGHJI* genes is the enzyme involved in nitrate assimilation together with the nitrite reductase encoded by the *nirBD* genes. The GlnR regulator also activates the expression of the *nirBD* genes in this mycobacterium (Malm *et al.*, 2009).

Ammoniun also regulates nitrate assimilation in archaea. Thus, in *Haloferax mediterraneii* ammonium represses *nas* gene transcription (Lledó *et al.*, 2005) and also inhibits nitrate and nitrite reductase activities (Martínez-Espinosa *et al.*, 2007a).

Acknowledgements

We thank the financial support from Ministerio de Ciencia e Innovación (grants BIO2008-04542-CO2-01 and PET2008_0048, also supported with FEDER funds) and Junta de Andalucía (Proyecto de Excelencia CVI1728). Authors thank Dr D. J. Richardson for helpful advice and suggestions.

References

Abramson, J., Kaback, H.R., and Iwata, S. (2004). Structural comparison of lactose permease and the glycerol-3-phosphate antiporter: members of the major facilitator superfamily. Curr. Opin. Struc. Biol. *14*, 413–419.

Aichi, M., Takatani, N., and Omata, T. (2001). Role of NtcB in activation of nitrate assimilation genes in the cyanobacterium *Synechocystis* sp. strain PCC 6803. J. Bacteriol. *183*, 5840–584.

Aichi, M., Yoshihara, S., Yamashita, M., Nagai, K., and Omata, T. (2006). Characterization of the nitrate-nitrite transporter of the major facilitator superfamily (the *nrtP* gene product) from the cyanobacterium *Nostoc punctiforme* strain ATCC 29133. Biosci. Biotechnol. Biochem. *70*, 2682–2689.

Aldehni, M., Sauer, J., Spielhaupter, C., Schmid, R., and Forchhammer, K. (2003). Signal transduction protein P_{II} is required for NtcA-regulated gene expression during nitrogen deprivation in the cyanobacterium *Synechococcus elongatus* strain PCC 7942. J. Bacteriol. *185*, 2582–2591.

Allen, A.E., Booth, M.G., Frischer, M.E., Verity, P.G., Zehr, J.P., and Zani, S. (2001). Diversity and detection of nitrate assimilation genes in marine bacteria. Appl. Environ. Microbiol. *67*, 5343–5348.

Allen, A.E., Booth, M.G., Verity, P.G., and Frischer, M.E. (2005). Influence of nitrate availability on the distribution and abundance of heterotrophic bacterial nitrate assimilation genes in the Barents Sea during summer. Aquat. Microb. Ecol. *39*, 247–255.

Amon, J., Titgemeyer, F., and Burkovski, A. (2009). A genomic view on nitrogen metabolism and nitrogen control in mycobacteria. J. Mol. Microbiol. Biotechnol. *17*, 20–29.

Andrade, S.L.A., Dickmanns, A., Ficner, R., and Einsle, O. (2005). Crystal structure of the archaeal ammonium transporter Amt-1 from *Archaeoglobus fulgidus*. Proc. Natl. Acad. Sci. U.S.A. *102*, 14994–14999.

Arcondéguy, T., Jack, R., and Merrick, M. (2001). P(II) signal transduction proteins, pivotal players in microbial nitrogen control. Microbiol. Mol. Biol. Rev. *65*, 80–105.

Baker, P.J., Britton., K.L., Engel, P.C., Farrants, G.W., Lilley, K.S., Rice, D.W., and Stillman, T.J. (1992). Subunit assembly and active site location in the structure of glutamate dehydrogenase. Proteins Struct. Funct. Genet. *12*, 75–86.

Beckers, G., Strösser, J., Hildebrandt, U., Kalinowski, J., Farwick, M., Krämer, R., and Burkovski, A. (2005). Regulation of AmtR-controlled gene expression in *Corynebacterium glutamicum*: mechanism and characterization of the AmtR regulon. Mol. Microbiol. *58*, 580–595.

Berlicki, L. (2008). Inhibitors of glutamine synthetase and their potential application in medicine. Mini Rev. Med. Chem. *8*, 869–878.

Bernard, S.M., and Habash, D.Z. (2009). The importance of cytosolic glutamine synthetase in nitrogen assimilation and recycling. New Phytol. *182*, 608–620.

Blasco, R., Castillo, F., and Martínez-Luque, M.M. (1997). The assimilatory nitrate reductase from the phototrophic bacterium *Rhodobacter capsulatus* E1F1 is a flavoprotein. FEBS Lett. *414*, 45–49.

Buchinger, S., Strösser, J., Rehm, N., Hänbler, E., Hans, S., Bathe, B., Schomburg, D., Krämer, R., and Burkovski, A. (2009). A combination of metabolome and transcriptome analyses reveals new targets of the *Corynebacterium glutamicum* nitrogen regulator AmtR. J. Biotechnol. *140*, 68–74.

Busch, W., and Saier, M.H. (2002). The transporter classification (TC) system, 2002. Crit. Rev. Biochem. Mol. Biol. *37*, 287–337.

Butt, J.N., Anderson, L.J., Rubio, L.M., Richardson, D.J., Flores, E., and Herrero, A. (2002). Enzyme-catalysed nitrate reduction – themes and variations as revealed by protein film voltammetry. Bioelectrochemistry *56*, 17–18.

Caballero, F.J., Moreno-Vivián, C., Castillo, F., and Cárdenas, J. (1986). Nitrite uptake system in photosynthetic bacterium *Rhodopseudomanas capsulata* E1F1. Biochim. Biophys. Acta *848*, 16–23.

Cabello P., Pino, C., Olmo-Mira, M.F., Castillo, F., Roldán, M.D., and Moreno-Vivián, C. (2004a). Hydroxylamine assimilation by *Rhodobacter capsulatus* E1F1. Requirement of the *hcp* gene (hybrid cluster protein) located in the nitrate assimilation *nas* gene region for hydroxylamine reduction. J. Biol. Chem. *279*, 45485–45494.

Cabello, P., Roldán, M.D., and Moreno-Vivián, C. (2004b). Nitrate reduction and the nitrogen cycle in archaea. Microbiology *150*, 3527–3546.

Cabello, P., Roldán, M.D., Castillo, F., and Moreno-Vivián, C. (2009). Nitrogen cycle. In Encyclopedia of Microbiology, 3rd edn., Schaechter, M., ed. (Oxford, UK: Elsevier), pp. 299–321.

Campbell, W.H. (2001). Structure and function of eukaryotic NAD(P)H:nitrate reductase. Cell. Mol. Life Sci. *58*, 194–204.

Cánovas, F.M., Ávila, C., Cantón, F.R., Cañas, R.A., and de la Torre, F. (2007). Ammonium assimilation and amino acid metabolism in conifers. J. Exp. Bot. *58*, 2307–2318.

Chai, W., and Stewart, V. (1999). RNA sequence requirements for NasR-mediated, nitrate-responsive transcription antitermination of the *Klebsiella oxytoca* M5al *nasF* operon leader. J. Mol. Biol. *292*, 203–216.

Cole, J. (1996). Nitrate reduction to ammonia by enteric bacteria: redundancy or a strategy for survival during oxygen starvations? FEMS Microbiol. Lett. *136*, 1–11.

Conroy, M.J., Durand, A., Lupo, D., Li, X.D., Bullough, P.A., Winkler, F.K., and Merrick, M. (2007). The crystal structure of the *Escherichia coli* AmtB–GlnK complex reveals how GlnK regulates the ammonia channel. Proc. Natl. Acad. Sci. U.S.A. *104*, 1213–1218.

Coutts, G., Thomas, G., Blakey, D., and Merrick, M. (2002). Membrane sequestration of the signal transduction protein GlnK by the ammonium transporter AmtB. EMBO J. *21*, 536–545.

Crespo, J.L., García-Domínguez, M., and Florencio, F.J. (1998). Nitrogen control of the *glnN* gene that codes for GS type III, the only glutamine synthetase in the cyanobacterium *Pseudanabaena* sp. PCC 6903. Mol. Microbiol. *30*, 1101–1112.

DiRuggiero, J., Dunn, D., Maeder, D.L., Holley-Shanks, R., Chatard, J., Horlacher, R., Robb, F.T., Boos, W., and Weiss, R.B. (2000). Evidence of recent lateral gene transfer among hyperthermophilic archaea. Mol. Microbiol. *38*, 684–693.

Efromovich, S., Grainger, D., Bodenmiller, D., and Spiro, S. (2008). Genome-wide identification of binding sites for the nitric oxide-sensitive transcriptional regulator NsrR. Meth. Enzymol. *437*, 221–233.

Eisenberg, D., Gill, H.S., Pfluegl, G.M.U., and Rotstein, S.H. (2000). Structure–function relationships of glutamine synthetases. Biochim. Biophys. Acta *1477*, 122–145.

Espinosa, J., Forchhammer, K., Burillo, S., and Contreras, A. (2006). Interaction network in cyanobacterial nitrogen regulation: PipX, a protein that interacts in a 2-oxoglutarate dependent manner with P_{II} and NtcA. Mol. Microbiol. *61*, 457–469.

Fernández, E., and Galván, A. (2007). Inorganic nitrogen assimilation in *Chlamydomonas*. J. Exp. Bot. *58*, 2279–2287.

Fernández, E., and Galván, A. (2008). Nitrate assimilation in *Chlamydomonas*. Eukaryot. Cell. *7*, 555–559.

Fisher, S.H. (1999). Regulation of nitrogen metabolism in *Bacillus subtilis*: vive la difference! Mol. Microbiol. *32*, 223–232.

Fisher, S.H., and Wray, L.V. (2008). *Bacillus subtilis* glutamine synthetase regulates its own synthesis by acting as a chaperone to stabilize GlnR–DNA complexes. Proc. Natl. Acad. Sci. U.S.A. *105*, 1014–1019.

Fisher, S.H., and Wray, L.V. (2009). Novel *trans*-acting *Bacillus subtilis glnA* mutations that derepress *glnRA* expression. J. Bacteriol. *191*, 2485–2492.

Fisher, S.H., Brandenburg, J.L., and Wray, L.V. (2002). Mutations in *Bacillus subtilis* glutamine synthetase that block its interaction with transcription factor TnrA. Mol. Microbiol. *45*, 627–635.

Flores, E., Guerrero, M.G., and Losada, M. (1980). Short-term ammonium inhibition of nitrate utilization by *Anacystis nidulans* and other cyanobacteria. Arch. Microbiol. *128*, 137–144.

Flores, E., Herrero, A., and Guerrero, M.G. (1987). Nitrate uptake and its regulation in the cyanobacterium *Anacystis nidulans*. Biochim. Biophys. Acta *896*, 103–108.

Flores, E., Frías, J.E., Rubio, L.M., and Herrero, A. (2005). Photosynthetic nitrate assimilation in cyanobacteria. Photosynth. Res. *83*, 117–133.

Forchhammer, K. (2004). Global carbon/nitrogen control by P_{II} signal transduction in cyanobacteria: from signals to targets. FEMS Microbiol. Rev. *28*, 319–333.

Forchhammer, K. (2007a). P_{II} signal transducers: novel functional and structural insights. Trends Microbiol. *16*,65–72.

Forchhammer, K. (2007b). Glutamine signalling in bacteria. Front. Biosci. *12*, 358–370.

Frías, J.E., Flores, E., and Herrero, A. (2000). Activation of the *Anabaena nir* operon promoter requires both NtcA (CAP family) and NtcB (LysR family) transcription factors. Mol. Microbiol. *38*, 613–625.

Frías, J.E., Herrero, A., and Flores, E. (2003). Open reading frame *all0601* from *Anabaena* sp. strain PCC 7120 represents a novel gene, *cnaT*, required for expression of the nitrate assimilation *nir* operon. J. Bacteriol. *185*, 5037–5044.

Galmozzi, C.V., Fernández-Avila, M.J., Reyes, J.C., Florencio, F.J., and Muro-Pastor, M.I. (2007). The ammonium-inactivated cyanobacterial glutamine synthetase I is reactivated *in vivo* by a mechanism involving proteolytic removal of its inactivating factors. Mol. Microbiol. *65*, 166–179.

Gangeswaran, R., and Eady, R.A. (1996). Flavodoxin 1 of *Azotobacter vinelandii*: characterization and role in electron donation to purified assimilatory nitrate reductase. Biochem. J. *317*, 103–108.

Gangeswaran, R., Lowe, D.J., and Eady, R.A. (1993). Purification and characterization of the assimilatory nitrate reductase of *Azotobacter vinelandii*. Biochem. J. *289*, 335–342.

García-Domínguez, M., Reyes, J.C., and Florencio, F.J. (1999). Glutamine synthetase inactivation by protein–protein interaction. Proc. Natl. Acad. Sci. U.S.A. *96*, 7161–7166.

Goddard, A.D., Moir, J.W.B., Richardson, D.J., and Ferguson, S.J. (2008). Interdependence of two NarK domains in a fused nitrate/nitrite transporter. Mol. Microbiol. *70*, 667–681.

Gordon, A.H., Hart, P.D., and Young, M.R. (1980). Ammonia inhibits phagosome–lysosome fusion in macrophages. Nature *286*, 79–80.

Gruswitz, F., O'Connell, J., and Stroud, R.M. (2007). Inhibitory complex of the transmembrane ammonia channel, AmtB, and the cytosolic regulatory protein, GlnK, at 1.96 Å. Proc. Natl. Acad. Sci. U.S.A. *104*, 42–47.

Gutiérrez, J.C., Ramos, F., Ortner, L., and Tortolero, M. (1995). *nasST*, two genes involved in the induction of the assimilatory nitrite-nitrate reductase operon (*nasAB*) of *Azotobacter vinelandii*. Mol. Microbiol. *18*, 579–591.

Harborne, N.R., Griffiths, L., Busby, S.J., and Cole, J.A. (1992). Transcriptional control, translation and function of the products of the five open reading frames of the *Escherichia coli nir* operon. Mol. Microbiol. *6*, 2805–2813.

Harth, G., and Horwitz, M.A. (2003). Inhibition of *Mycobacterium tuberculosis* glutamine synthetase as a novel antibiotic strategy against tuberculosis: demonstration of efficacy *in vivo*. Infect. Immun. *71*, 456–464.

Heinrich, A., Woyda, K., Brauburger, K., Meiss, G., Detsch, C., Stülke, J., and Forchhammer, K. (2006). Interaction of the membrane-bound GlnK–AmtB complex with the master regulator of nitrogen metabolism TnrA in *Bacillus subtilis*. J. Biol. Chem. *281*, 34909–34917.

Hemker, B.M., Cheroutre, G., van Zwieten, R., Maaskent-van Wijk, P.A., Roos, D., Loos, J.A., van der Schoot, C.E., and von dem Borne, A.E. (2003). The Rh complex exports ammonium from human red blood cells. Br. J. Haematol. *122*, 333–340.

Herrero, A., Muro-Pastor, A.M., and Flores, E. (2001). Nitrogen control in cyanobacteria. J. Bacteriol. *183*, 411–425.

Hudson, R.C., and Daniel, R.M. (1993). L-glutamate dehydrogenases: distribution, properties and mechanism. Comp. Biochem. Physiol. *106*, 767–792.

Javelle, A., Severi, E., Thornton, J., and Merrick, M. (2004). Ammonium sensing in *Escherichia coli*. Role of the ammonium transporter AmtB and AmtB–GlnK complex formation. J. Biol. Chem. *279*, 8530–8538.

Jepson, B.J.N., Anderson, L.J., Rubio, L.M., Taylor, C.J., Butler, C.S., Flores, E., Herrero, A., Butt, J.N., and Richardson, D.J. (2004). Tuning a nitrate reductase for function: the first spectropotenciometric characterization of a bacterial assimilatory nitrate reductase reveals novel redox properties. J. Biol. Chem. *279*, 32212–32218.

Jiang, P., Pioszak, A.A., and Ninfa, A.J. (2007). Structure-function analysis of glutamine synthetase adenylyltransferase (ATase, EC 2.7.7.49) of *Escherichia coli*. Biochemistry *46*, 4117–4132.

Kayumov, A., Heinrich, A., Sharipova, M., Lljinskaya, O., and Forchhammer, K. (2008). Inactivation of the general transcription factor TnrA in *Bacillus subtilis* by proteolysis. Microbiology *154*, 2348–2355.

Khademi, S., O'Connell, J., Remis, J., Robles-Colmenares, Y., Miericke, L.J.W., and Stroud, R.M. (2004). Mechanism of ammonia transport by Amt/MEP/Rh: structure of AmtB at 1.35 Å. Science *305*, 1587–1594.

Khan, A., Akhtar, S., Ahmad, J.N., and Sarkar, D. (2008). Presence of a functional nitrate assimilation pathway in *Mycobacterium smegmatis*. Microb. Pathogenesis *44*, 71–77.

Knapp, S., de Vos, W.M., Rice, D., and Ladenstein, R. (1997). Crystal structure of glutamate dehydrogenase from hyperthermophilic eubacterium *Thermotoga maritima* at 3.0 Å resolution. J. Mol. Biol. *267*, 916–932.

Kobayashi, M., Rodríguez, R., Lara, C., and Omata, T. (1997). Involvement of the C-terminal domain of an ATP-binding subunit in the regulation of the ABC-type nitrate/nitrite transporter of the cyanobacterium *Synechococcus* sp. strain PCC 7942. J. Biol. Chem. *272*, 27197–27201.

Kuznetsova, S., Knaff, D.B., Hirasawa, M., Lagoutte, B., and Setif, P. (2004). Mechanism of spinach chloroplast ferredoxin-dependent nitrite reductase: spectroscopic evidence for intermediate states. Biochemistry *43*, 510–517.

Lee, H.M., Flores, E., Herrero, A., Houmard, J., and Tandeau de Marsac, N. (1998). A role for the signal transduction protein P_{II} in the control of nitrate/nitrite uptake in a cyanobacterium. FEBS Lett. *427*, 291–295.

Lee, H.M., Flores, E., Forchhammer, K., Herrero, A., and Tandeau de Marsac, N. (2000). Phosphorylation of the signal transducer P_{II} protein and an additional effector are required for P_{II}-mediated regulation of nitrate and nitrite uptake in the cyanobacterium *Synechococcus* sp. PCC 7942. Eur. J. Biochem. *267*, 591–600.

Leigh, J.A., and Dodsworth, J.A. (2007). Nitrogen regulation in bacteria and archaea. Annu. Rev. Microbiol. *61*, 349–377.

Lie, T.J., and Leigh, J.A. (2003). A novel repressor of *nif* and *glnA* expression in the methanogenic archaeon *Methanococcus maripaludis*. Mol. Microbiol. *47*, 235–246.

Lin J.T., and Stewart, V. (1996). Nitrate and nitrite-mediated transcription antitermination control of *nasF* (nitrate assimilation) operon expression in *Klebsiella pneumoniae* M5a1. J. Mol. Biol. *256*, 423–435.

Lin, J.T., and Stewart, V. (1998). Nitrate assimilation by bacteria. Adv. Microb. Physiol. *39*, 1–30.

Lin, J.T., Goldman, B.S., and Stewart, V. (1994). The *nasFEDCBA* operon for nitrate and nitrite assimilation in *Klebsiella pneumoniae* M5a1. J. Bacteriol. *176*, 2551–2559.

Lledó, B., Marhuenda-Egea, F.C., Martínez-Espinosa, R.M., and Bonete, M.J. (2005). Identification and transcriptional analysis of nitrate assimilation genes in the halophilic archaeon *Haloferax mediterranei*. Gene *361*, 80–88.

Llorca, O., Betti, M., González, J.M., Valencia, A., Márquez, A.J., and Valpuesta, J.M. (2006). The three-dimensional structure of an eukaryotic glutamine synthetase: functional implications of its oligomeric structure. J. Struct. Biol. *156*, 469–479.

Locher, K.P. (2004). Structure and mechanism of ABC transporters. Curr. Opin. Struc. Biol. *14*, 426–431.

Ludewig, U., von Wirén, N., and Frommer, W.B. (2002). Uniport of NH_4^+ by the root hair plasma membrane ammonium transporter LeAMT1;1. J. Biol. Chem. *277*, 13548–13555.

Maeda, S.I., and Omata, T. (1997). Substrate-binding lipoprotein of the cyanobacterium *Synechococcus* sp. strain PCC 7942 involved in the transport of nitrate and nitrite. J. Biol. Chem. *272*, 3036–3041.

Maeda, S., and Omata, T. (2009). Nitrite transport activity of the ABC-type cyanate transporter of the cyanobacterium *Synechococcus elongatus*. J. Bacteriol. *191*, 3265–3272.

Malm, S., Tiffert, Y., Micklinghoff, J., Schultze, S., Joost, I., Weber, I., Horst, S., Ackermann, B., Schmidt, M., Wohlleben, W., Ehlers, S., Geffers, R., Reuther, J., and Bange, F.C. (2009). The roles of the nitrate reductase NarGHJI, the nitrite reductase NirBD and the response regulator GlnR in nitrate assimilation of *Mycobacterium tuberculosis*. Microbiology *155*, 1332–1339.

Márquez, A.J., Betti, M., García-Calderón, M., Pal'ove-Balang, P., Díaz, P., and Monza, J. (2005). Nitrate assimilation in *Lotus japonicus*. J. Exp. Bot. *56*, 1741–1749.

Martínez-Espinosa, R.M., Marhuenda-Egea, F.C., and Bonete, M.J. (2001a). Assimilatory nitrate reductase from the haloarchaeon *Haloferax mediterranei*: purification and characterization. FEMS Microbiol. Lett. *204*, 381–385.

Martínez-Espinosa, R.M., Marhuenda-Egea, F.C., and Bonete, M.J. (2001b). Purification and characterization of a possible assimilatory nitrite reductase from the halophile archaeon *Haloferax mediterranei*. FEMS Microbiol. Lett. *196*, 113–118.

Martínez-Espinosa, Esclapez, J., Bautista, V., and Bonete, M.J. (2006). An octameric prokaryotic glutamine synthetase from the haloarchaeon *Haloferax mediterranei*. FEMS Microbiol. Lett. *264*, 110–116.

Martínez-Espinosa, R.M., Lledó, B., Marhuenda-Egea, F.C., and Bonete, M.J. (2007a). The effect of ammonium on assimilatory nitrate reduction in the haloarchaeon *Haloferax mediterranei*. Extremophiles *11*, 759–767.

Martínez-Espinosa, R.M., Richardson, D.J., Butt, J.N., and Bonete, M.J. (2007b). Spectropotentiometric properties and salt-dependent thermotolerance of a [2Fe–2S] ferredoxin-involved nitrate assimilation in *Haloferax mediterranei*. FEMS Microbiol. Lett. *277*, 50–55.

Martínez-Espinosa, R.M., Lledó, B., Marhuenda-Egea, F.C., Díaz, S., and Bonete, M.J. (2009). NO_3^-/NO_2^- assimilation in halophilic archaea: physiological analysis, *nasA* and *nasD* expressions. Extremophiles *13*, 785–792.

Meakin, G.E., Jepson, B.J.N., Richardson, D.J., Bedmar, E.J., and Delgado, M.J. (2006). The role of *Bradyrhizobium japonicum* nitric oxide reductase in nitric oxide detoxification in soya bean root nodules. Biochem. Soc. Trans. *34*, 195–196.

Miller, A.J., Fan, X., Shen, Q., and Smith, S.J. (2008). Amino acids and nitrate as signals for the regulation of nitrogen acquisition. J. Exp. Bot. *59*, 111–119.

Moir, J.W.B., and Wood, N.J. (2001). Nitrate and nitrite transport in bacteria. Cell. Mol. Life Sci. *58*, 215–224.

Montesinos, M.L., Muro-Pastor, A.M., Herrero, A., and Flores, E. (1998). Ammonium/methylammonium permeases of a cyanobacterium. Identification and analysis of three nitrogen-regulated *amt* genes in *Synechocystis* sp. PCC 6803. J. Biol. Chem. *273*, 31463–31470.

Moreno-Vivián, C., and Flores, E. (2007). Nitrate assimilation in bacteria. In Biology of the Nitrogen Cycle, Bothe, H., Ferguson, S.J., and Newton, W.E., eds. (Amsterdam, The Netherlands: Elsevier), pp. 263–282.

Moreno-Vivián, C., Cejudo, F.J., Cárdenas, J., and Castillo, F. (1983). Ammonia assimilation pathways in *Rhodopseudomonas capsulata* E1F1. Arch. Microbiol. *136*, 147–151.

Moreno-Vivián, C., Cabello, P., Martínez-Luque, M., Blasco, R., and Castillo, F. (1999). Prokaryotic nitrate reduction: molecular properties and functional distinction among bacterial nitrate reductases. J. Bacteriol. *181*, 6573–6584.

Muñoz-Centeno, M.C., Cejudo, F.J., Ruiz, M.T., and Paneque, A. (1993). The *Azotobacter chroococcum* nitrate permease is a multicomponent system. Biochim. Biophys. Acta *1141*, 75–80.

Muro-Pastor, M.I., Reyes, J.C., and Florencio, F.J. (2005). Ammonium assimilation in cyanobacteria. Photosynth. Res. *83*, 135–150.

Nakano, M.M., Hoffmann, T., Zhu, Y., and Jahn, D. (1998). Nitrogen and oxygen regulation of *Bacillus subtilis nasDEF* encoding NADH-dependent nitrite reductase by TnrA and ResDE. J. Bacteriol. *180*, 5344–5350.

Nilsson, M.T., Krajewski, W.W., Yellagunda, S., Prabhumurthy, S., Chamarahally, G.N., Siddamadappa, C., Srinivasa, B.R., Yahiaoui, S., Larhed, M., Karlén, A., Jones, T.A., and Mowbray, S.L. (2009). Structural basis for the inhibition of *Mycobacterium tuberculosis* glutamine synthetase by novel ATP-competitive inhibitors. J. Mol. Biol. *393*, 504–513.

Ninfa, A.J., and Jiang, P. (2005). P_{II} signal transduction proteins: sensors of alpha-ketoglutarate that regulate nitrogen metabolism. Curr. Opin. Microbiol. *8*, 168–173.

Ogawa, K.I., Akagawa, E., Yamane, K., Sun, Z.W., LaCelle, M., Zuber, P., and Nakano, M.M. (1995). The *nasB* operon and the *nasA* gene are required for nitrate/nitrite assimilation in *Bacillus subtilis*. J. Bacteriol. *177*, 1409–1413.

Olmo-Mira, M.F., Cabello, P., Pino, C., Martínez-Luque, M., Richardson, D.J., Castillo, F., Roldán, M.D., and Moreno-Vivián, C. (2006). Expression and characterization of the assimilatory NADH-nitrite reductase from the phototrophic bacterium *Rhodobacter capsulatus* E1F1. Arch. Microbiol. *186*, 339–344.

Omata, T. (1991). Cloning and characterization of the *nrtA* gene that encodes a 45-kDa protein involved in nitrate transport in the cyanobacterium *Synechococcus* PCC 7942. Plant Cell Physiol. *32*, 151–157.

Paz-Yepes, J., Flores, E., and Herrero, A. (2003). Transcriptional effects of the signal transduction protein P_{II} (*glnB* gene product) on NtcA-dependent genes in *Synechococcus* sp. PCC 7942. FEBS Lett. *543*, 42–46.

Paz-Yepes, J., Flores, E., and Herrero, A. (2009). Expression and mutational analysis of the *glnB* genomic region in the heterocyst-forming cyanobacterium *Anabaena* sp. strain PCC 7120. J. Bacteriol. *191*, 2353–2361.

Peterson, P.E., and Smith, T.J. (1999). The structure of bovine glutamate dehydrogenase provides insights into the mechanism of allostery. Structure *7*, 769–782.

Pino, C., Olmo-Mira, M.F., Cabello, P., Martínez-Luque, M., Castillo, F., Roldán, M.D., and Moreno-Vivián, C. (2006). The assimilatory nitrate reduction system of the phototrophic bacterium *Rhodobacter capsulatus* E1F1. Biochem. Soc. Trans. *34*, 127–129.

Ramos, F., Blanco, G., Gutiérrez, J.C., Luque, F., and Tortolero, M. (1993). Identification of an operon involved in the assimilatory nitrate-reducing system of *Azotobacter vinelandii*. Mol. Microbiol. *8*, 1145–1153.

Reitzer, L. (2003). Nitrogen assimilation and global regulation in *Escherichia coli*. Annu. Rev. Microbiol. *57*, 155–176.

Reyes, J.C., and Florencio, F.J. (1994). A new type of glutamine synthetase in cyanobacteria: the protein encoded by the *glnN* gene supports nitrogen assimilation in *Synechocystis* sp. strain PCC 6803. J. Bacteriol. *176*, 1260–1267.

Richardson, D.J., Berks, B.C., Russell, D.A., Spiro, S., and Taylor, C.J. (2001). Functional, biochemical and genetic diversity of prokaryotic nitrate reductases. Cell. Mol. Life Sci. *58*, 165–178.

Robb, F.T., Park, J.B., and Adams, M.W. (1992). Characterization of an extremely thermostable glutamate dehydrogenase: a key enzyme in the primary metabolism of the hyperthermophilic archaebacterium *Pyrococcus furiosus*. Biochim. Biophys. Acta *1120*, 267–272.

Rubio, L.M., Flores, E., and Herrero, A. (2002). Purification, cofactor analysis and site-directed mutagenesis of *Synechococcus* ferredoxin-nitrate reductase. Photosynth. Res. *72*, 13–26.

Saier, M.H., Yen, M.R., Noto, K., Tamang, D.G., and Elkan, C. (2009). The transporter classification database: recent advances. Nucleic Acid Res. *37*, 274–278.

Sakamoto, T., Inoue-Sakamoto, K., and Bryant, D.A. (1999). A novel nitrate/nitrite permease in the marine cyanobacterium *Synechococcus* sp. strain PCC 7002. J. Bacteriol. *181*, 7363–7372.

Schinkinger, M.F., Redl, B., and Stöffler, G. (1991). Purification and properties of an extreme thermostable glutamate dehydrogenase from the archaebacterium *Sulfolobus solfataricus*. Biochim. Biophys. Acta *1073*, 142–148.

Shu, C.J., and Zhulin, I.B. (2002). ANTAR: an RNA-binding domain in transcription antitermination regulatory proteins. Trends Biochem. Sci. *27*, 3–5.

Siverio, J.M. (2002). Assimilation of nitrate by yeast. FEMS Microbiol. Rev. *26*, 277–284.

Smith, T.J., and Stanley, C.A. (2008). Untangling the glutamate dehydrogenase allosteric nightmare. Trends Biochem. Sci. 33, 557–564.

Smith, E.L., Austen, B.M., Blumenthal, K.M., and Nyc, J.F. (1975). Glutamate dehydrogenase. In The Enzymes, vol. XI, P.D. Boyer, ed. (New York, USA:Academic Press), pp. 293–367.

Smith, T.J., Schmidt, T., Fang, J., Wu, J., Siuzdak, G., and Stanley, C.A. (2002). The structure of apo human glutamate dehydrogenase details subunit communication and allostery. J. Mol. Biol. *318*, 765–777.

Soupene, E., Lee, H., and Kustu, S. (2002). Ammonium/methylamonium transport (Amt) proteins facilitate diffusion of NH_3 bidirectionally. Proc. Natl. Acad. Sci. U.S.A. *99*, 3926–3931.

Stillman, T.J., Baker, P.J., Britton, K.L., and Rice, D.W. (1993). Conformational flexibility in glutamate dehydrogenase. Role of water in substrate recognition and catalysis. J. Mol. Biol. *234*, 1131–1139.

Suzuki, A., and Knaff, D.B. (2005). Glutamate synthase: structural, mechanistic and regulatory properties, and role in the amino acid metabolism. Photosynth. Res. *83*, 191–217.

Suzuki, I., Kikuchi, H., Nakanishi, S., Fujita., Y., Sugiyama, T., and Omata, T. (1995a). A novel nitrite reductase from the cyanobacterium *Plectonema boryanum*. J. Bacteriol. *177*, 6137–6143.

Suzuki, I., Horie, N., Sugiyama, T., and Omata, T. (1995b). Identification and characterization of two nitrogen-regulated genes of the cyanobacterium *Synechococcus* sp. strain PCC7942 required for maximum efficiency of nitrogen assimilation. J. Bacteriol. *177*, 290–296.

Takatani, N., Kobayashi, M., Maeda, S., and Omata, T. (2006). Regulation of nitrate reductase by non-modifiable derivatives of P_{II} in the cells of Synechococcus elongatus strain PCC 7942. Plant Cell Physiol. 47, 1182–1186.

Tanigawa, R., Shirokane, M., Maeda, S.I., Omata, T., Tanaka, K., and Takahashi, H. (2002). Transcriptional activation of NtcA-dependent promoters of *Synechococcus* sp. PCC 7942 by 2-oxoglutarate *in vitro*. Proc. Nat. Acad. Sci. U.S.A .99, 4251–4255.

Tiffert, Y., Supra, P., Wurm, R., Wohlleben, W., Wagner, R., and Reuther, J. (2008). The *Streptomyces coelicolor* GlnR regulon: identification of new GlnR targets and evidence for a central role of GlnR in nitrogen metabolism in actinomycetes. Mol. Microbiol. *67*, 861–880.

Tischner, R., and Kaiser, W. (2007). Nitrate assimilation in plants. In Biology of the Nitrogen Cycle, Bothe, H., Ferguson, S.J., and Newton, W.E., eds. (Amsterdam: Elsevier), pp. 283–302.

van den Heuvel, R.H., Curti, B., Vanoni, M.A., and Mattevi, A. (2004). Glutamate synthase: a fascinating pathway from L-glutamate to L-glutamine. Cell. Mol. Life Sci. *61*, 669–681.

Vanoni, M.A., and Curti, B. (2008). Structure-function studies of glutamate synthases: a class of self-regulated iron–sulfur flavoenzymes essential for nitrogen assimilation. IUBMB Life *60*, 287–300.

Vázquez-Bermúdez, M.F., Herrero, A., and Flores, E. (2002a). 2-Oxoglutarate increases the binding affinity of the NtcA (nitrogen control) transcription factor for the *Synechococcus glnA* promoter. FEBS Lett. *512*, 71–74.

Vázquez-Bermúdez, M.F., Paz-Yepes, J., Herrero, A., and Flores, E. (2002b). The NtcA-activated *amt1* gene encodes a permease required for uptake of low concentrations of ammonium in the cyanobacterium *Synechococcus* sp. PCC 7942. Microbiology *148*, 861–869.

Vega-Palas, M.A., Flores, E., and Herrero, A. (1992). NtcA, a global nitrogen regulator from the cyanobacterium *Synechococcus* that belongs to the Crp family of bacterial regulators. Mol. Microbiol. *6*, 1853–1859.

Walter, B., Hänssler, E., Kalinowski, J., and Burkovski, A. (2007). Nitrogen metabolism and nitrogen control in corynebacteria: variations of a common theme. J. Mol. Microbiol. Biotechnol. *12*, 131–138.

Wang, J., and Zhao, G.P. (2009). GlnR positively regulates *nasA* transcription in *Streptomyces coelicolor*. Biochem. Biophys. Res. Commu. *386*, 77–81.

Wang, M.Y., Glass, A., Shaff, J.E., and Kochian, L.V. (1994). Ammonium uptake by rice roots (III. Electrophysiology). Plant Physiol. *104*, 899–906.

Wang, Q., Li, H., and Post, A.F. (2000). Nitrate assimilation genes of the marine diazotrophic, filamentous cyanobacterium *Trichodesmium* sp. strain WH9601. J. Bacteriol. *182*, 1764–1767.

Wray, L.V., and Fisher, S.H. (2006). Functional analysis of the carboxy-terminal region of *Bacillus subtilis* TnrA, a MerR family protein. J. Bacteriol. *189*, 20–27.

Wray, L.V., Zalieckas, J.M., and Fisher, S.H. (2001). *Bacillus subtilis* glutamine synthetase controls gene expression through a protein–protein interaction with transcription factor TnrA. Cell *107*, 427–435.

Wu, Q., and Stewart, V. (1998). NasFED proteins mediate assimilatory nitrate and nitrite transport in *Klebsiella oxytoca (pneumoniae)* M5al. J. Bacteriol. *180*, 1311–1322.

Wyman, M., and Bird, C. (2007). Lack of control of nitrite assimilation by ammonium in an oceanic picocyanobacterium, *Synechococcus* sp. strain WH8103. Appl. Environ. Microbiol. 73, 3028–3033.

Yip, K.S.P., Stillman, T.J., Britton., K.L., Artymiuk, P.J., Baker, P.J., Sedelnikova, S.E., Engel, P.C., Pasquo, A., Chiraluce, R., Consalvi, V., Scandurra, R., and Rice, D.W. (1995). The structure of *Pyrococcus furiosus* glutamate dehydrogenase reveals a key role for ion–pair networks in maintaining enzyme stability at extreme temperatures. Structure 3, 1147–1158.

Zheng, L., Kostrewa, D., Bernèche, S., Winkler, F.K., and Li, X.D. (2004). The mechanism of ammonia transport based on the crystal structure of AmtB of *Escherichia coli*. Proc. Natl. Acad. Sci. U.S.A. *101*, 17090–17095.

Beyond Denitrification: Alternative Routes to Dinitrogen

6

Marc Strous

Abstract

Although nitrate is a powerful electron acceptor, it was generally believed that it could not be used to activate recalcitrant substrates such as ammonium and methane. Only in the past decades, bacteria were identified that could activate these compounds. These bacteria have become known as anaerobic ammonium oxidizing ('anammox') bacteria and 'denitrifying methanotrophs'. Each makes use of a different and so far unique pathway of nitrate reduction with dinitrogen gas as the end product. Anammox bacteria activate ammonia with nitric oxide, leading to the production of hydrazine (N_2H_4). Denitrifying methanotrophs dismutate two molecules of nitric oxide into molecular oxygen (O_2) and nitrogen (N_2). In this chapter bacteria, pathways, cell biology and environmental relevance are discussed.

Introduction

The nitrogen cycle has been investigated since the dawn of microbiology and most of the microbial processes have been known for more than a century. Best known are nitrification, denitrification and nitrogen fixation. In the course of the twentieth-century science grew more powerful experimentally and microbes performing each of these processes were isolated and studied. Subsequently also the enzymes involved were purified, characterized and crystalized. As more and more knowledge was gained the confidence grew and it appeared to many that microbial nitrogen cycling on earth was well understood. Engineers made use of this understanding to implement nitrogen removal in wastewater treatment – via nitrification and denitrification. Soil scientists and oceanographers used it to measure denitrification rates in nature – by acetylene inhibition. An important consequence of the conventional view was that recalcitrant compounds such as ammonia and methane could not be converted in the absence of oxygen (Table 6.1); it had been discovered that their activation by mono-oxygenase enzymes depended on molecular oxygen (O_2). Attempts to isolate microorganisms capable of anaerobic ammonium or methane oxidation coupled to denitrification (Eisentrager *et al.*, 2001) failed, supporting this notion.

Already in the 1960s and 1970s some individuals questioned the validity of the existing paradigms. Best known are the studies by Richards (1965) and Broda (1977), who addressed the possibility of anaerobic ammonium oxidation on experimental and theoretical grounds respectively. In the 1980s evidence for ammonia oxidation with nitrate was documented in a patent application (Mulder, 1989). Further hints that reality might be more complicated than anticipated came from the observation that methane was apparently converted with sulfate in marine sediments (e.g. Alperin and Reeburgh, 1985).

The view that a good understanding of element cycling had been achieved was further challenged with the development of molecular ecology in the 1990s. Only at that point, microbiologists realized that a large part of the microbial biodiversity had not been explored yet; this idea implied that also some microbial processes might have been overlooked.

Table 6.1 Thermodynamic properties of relevant compounds

Compound	Formula	$\Delta G_f^{0'}$
Ammonia	NH_4^+	–79.4
Hydrazine	N_2H_4	128.5
Hydroxylamine	NH_2OH	–22.9
Nitric oxide	NO	86.9
Nitrite	NO_2^-	–37.4
Nitrate	NO_3^-	–111.7
Dinitrogen	N_2	0
Oxygen	O_2	0
Methane	CH_4	–50.8
Carbon dioxide	CO_2	–394.4
Proton	H^+	–39.8
Water	H_2O	–237.2

In this context new experimental approaches to enrich for organisms performing anaerobic ammonia and methane oxidation were developed (Strous *et al.*, 2002). There were two basic assumptions behind these experiments. The first was that the four billion years of microbial evolution must have made possible the exploitation of any conceivable ecological niche; all reactions that are possible thermodynamically are actually used by microorganisms to gain energy for growth. The second idea was that the failure to enrich or isolate these associated microorganisms is caused by the experimental limitations of the conventional Petri dish microbiology. Most importantly, many organisms in nature are known to grow very slowly – at a doubling time in the order of weeks, months or even years (Jørgensen and Boetius, 2007). Such microorganisms are outside the scope of conventional cultivation approaches.

To accommodate slow-growing microorganisms, laboratory bioreactors were set up for continuous cultivation enrichment. Fresh medium was supplied to these bioreactors continuously and spent medium removed continuously. Freshly grown biomass was not removed but was retained inside the bioreactors by gravity (settling). This approach selected for a slowly growing microbial community that formed biofilm aggregates; suspended cells were washed out. The substrates were either ammonium and nitrite (Strous *et al.*, 1998, 1999) or methane and nitrite (Raghoebarsing *et al.*, 2006). Anoxic conditions were maintained by gentle sparging with a argon–carbon dioxide gas mixture (95:5). Although the nitrite concentration was sometimes very high in the fresh medium, the actual concentration experienced by the enriched microorganisms was always very low; the ratio of electron-donor (ammonia or methane) to nitrite was always chosen in such a way that nitrite was the growth limiting substrate. This was important because nitrite is a reactive, toxic compound.

Enrichment cultures like this were maintained for months and even years in the laboratory. Eventually, the persistently stable conditions favoured the selective growth of anaerobic methane and ammonia oxidizing communities. In all cases, these communities were dominated by a single population that constituted up to 80% of the total microbial community. Almost everything we know about the anaerobic methane and ammonia oxidizers has been learnt from the study of these communities. The research was less straightforward than with conventional pure cultures, because the presence of other community members always has to be taken into account when interpreting the results.

Anaerobic ammonium oxidation

Anaerobic ammonium oxidation (anammox) is performed by bacteria part of the Planctomycetes/Verrucomicrobia/Chlamydia superphylum. Within this superphylum anammox bacteria form a separate taxonomic order (the 'Brocadiales') within the Planctomycete line of descent (Jetten *et al.,* 2010). Four genera have been defined provisionally; they have the 'Candidatus' status because they have not been isolated in pure culture. The genus *Scalindua* is most abundant in the world's oceans. The other three genera (e.g. *Kuenenia*) are most abundant in wastewater treatment plants. The monophyly of anammox bacteria is exceptional, because almost all other bacterial lifestyles are polyphyletic. For example, nitrification is performed by Betaproteobacteria, Gammaproteobacteria and Archaea. Apparently, the anammox lifestyle is not easily acquired by lateral gene transfer and as will become clear, there are several good reasons why this may be the case.

A bacterial organelle, the intermediate hydrazine (N_2H_4) and ladderane lipids are the three most striking aspects of anammox bacteria that have attracted much attention. The facts that anammox bacteria can be applied in wastewater treatment to remove ammonia at lower cost than conventional nitrification/denitrification and that these bacteria contribute significantly to the global nitrogen cycle also make this process an interesting topic for scientific research. Their global significance is perhaps the biggest surprise because these bacteria have been overlooked for over a century.

Cell biology

Planctomycetes are best known for their internal membrane structures. These bacteria all have intracytoplasmic membrane-bound compartments. Such differentiation of the cytoplasm is rare among bacteria but other well known examples are magnetosomes and the vacuoles of the giant sulfur bacteria. Anammox bacteria have a single internal membrane-bound compartment that has become known as the 'anammoxosome' and qualifies as a true bacterial organelle (Van Niftrik *et al.,* 2009). It was shown experimentally to be the locus of anammox catabolism, so it has a well defined function (Van Niftrik *et al.,* 2008a,b). Each anammox cell contains a single anammoxosome; during cell division the anammoxosome is split into two halves and each daughter cell inherits one half. The anammoxosome has a strongly invaginated membrane with a large amount of membrane surface area, like mitochondria.

Anammox bacteria are not available in pure culture and conventional genetic experiments cannot be used. By performing metagenomic sequencing of enrichment cultures of Candidatus *'Kuenenia stuttgartiensis'* the genome of *Kuenenia* was almost completely (>98%) assembled (Strous *et al.,* 2006). By comparing the genes of *K. stuttgartiensis* to genes of well studied organisms predictions about the anammox cell biology and metabolism could still be made. In the meantime some of these predictions have been validated experimentally.

Like other planctomycetes, anammox cells lack the general bacterial cell division gene *ftsZ*. In other bacteria, the tubulin homologue FtsZ forms the division ring that constrains the cell during cell division. In anammox bacteria, a division ring was observed with electron tomography. This ring was shown to constrict and pinch off the anammoxosome during cell division. A large structural protein that was predicted to hydrolyse GTP (like FtsZ) was shown to be part of the anammox divisome and may perhaps function as a substitute for FtsZ (Van Niftrik *et al.,* 2009).

Anammox bacteria have a bright-red colour because they contain a large amount of cytochrome c, like the aerobic ammonia oxidizing Bacteria. In nitrifiers the octahaem protein hydroxylamine oxidoreductase (hao) is the main source of this colour. For anammox bacteria cytochemical staining showed that the cytochrome c was located exclusively inside the anammoxosome. The levels in the periplasm (in planctomcytes 'paryphoplasm') were too low to be detected. Immunolocalization showed that octahaem cytochrome c proteins homologous to hydroxylamine oxidoreductase were also located only inside the anammoxosome. Based on computational analyses of amino acid composition it was subsequently predicted that actually almost all genes involved in respiration were located inside the anammoxosome or on its membrane (Medema *et al.*, 2010). The anammoxosomal location of the ATP synthase was demonstrated experimentally with immunolocalization (Van Niftrik *et al.*, 2010) and an intracytoplasmic pH gradient was suggested by P31-NMR experiments (Van der Star *et al.*, 2010).

In conclusion, all current evidence suggests that anaerobic ammonium oxidation is performed inside the anammoxsome, the bacterial organelle of anammox bacteria. Although the topology of the anamoxosome is well resolved, that is not true for the layout of the cell envelope. It is unknown whether an additional cytoplasmic compartment exists on the outside, as was found for other planctomycetes or whether this fringing compartment should be considered as a regular Gram negative periplasm (Fig. 6.1).

The anammoxosomal membrane is composed of lipids that are so far unique in nature. These lipids contain concatenated cyclobutane residues that form a ladder-like structure (Sinninghe Damsté *et al.*, 2002). They had been synthesized by organic chemists before their discovery in anammox bacteria and were known as promising building blocks in opto-electronics. These molecules have very high ring strain and must be difficult to synthesize biochemically. Comparative genomics showed that it is most likely that polyunsaturated hydrocarbons are the biosynthetic precursor for the ladderane moieties (Rattray *et al.*, 2009a,b). The complete lipids molecules consist of several combinations of two types of ladderane moieties attached to the glycerol backbone via ether or ester bonds. The headgroup consists of phosphoethanolamine, phosphocholine or phosphoglycerol (Fig. 6.2).

Based on the analysis of core lipids it was initially proposed that ladderanes were only present in the anammoxosomal membrane. Later, analyses of the complete lipids (results presented in Fig. 6.2) showed that this could not be true – in

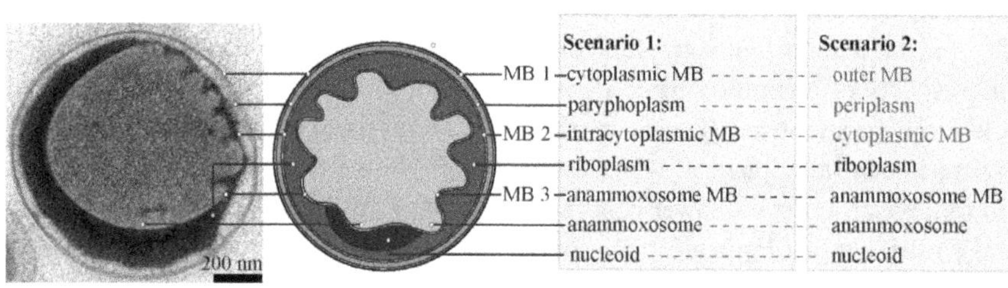

Figure 6.1 Electron micrograph (left) of an anammox bacterium and schematic overview (right) of the different scenarios regarding its cell plan (figure reproduced from Van Niftrik *et al.* (2010), Mol. Microbiol. *77*, 701–715).

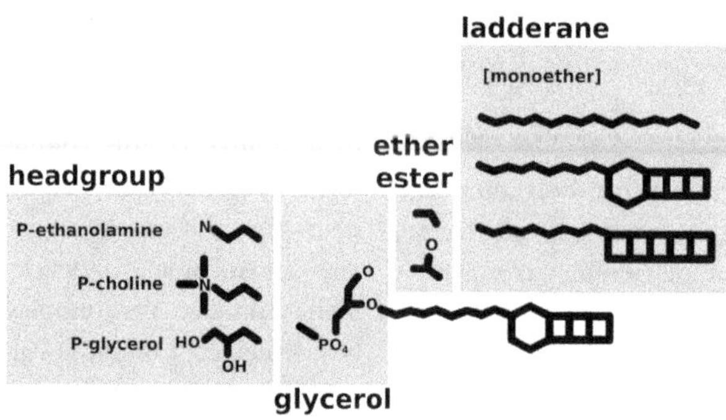

Figure 6.2 The complete lipids of the anammox bacteria are synthesized from the glycerol backbone with ether or ester bound combinations of ladderane moieties and one of three different headgroups (Boumann *et al.*, 2006; Rattray *et al.*, 2008).

some cases ladderane lipids were found to make up more than 90% of the extracted lipids (Rattray *et al.*, 2008). The biological function of the ladderanes might be the formation of a dense and highly viscous membrane that has reduced permeability to protons and the intermediates of anammox catabolism. High viscosity of ladderane membranes was demonstrated experimentally by fluorescence depolarization spectroscopy and other methods (Boumann *et al.*, 2009a,b). One can calculate that with the low growth rate of anammox bacteria (doubling time in the order of weeks) and with the permeability of a mitochondrial membrane it is impossible to create a proton motive force (Van Niftrik *et al.*, 2004). Protons would leak back through the membrane faster than they can be translocated by the respiratory chain. Prevention of the loss of energetic intermediates such as hydrazine might also be important. However, in physiological experiments, hydrazine could be measured in the extracellular medium, so in spite of the ladderane lipids, it still leaks out of the anammox cells.

Catabolism

Originally, it was believed that anammox catabolism proceeded by the four-electron reduction of nitrite to hydroxylamine (NH_2OH). Hydroxylamine could then condense with ammonia to form the hydrazine. This reaction would be performed by the hypothesized enzyme 'hydrazine hydrolase'. The four-electron oxidation of hydrazine to dinitrogen would complete the cycle. Although evidence for the last two reactions was established experimentally, the production of hydroxylamine from nitrite was never shown. Oxidation of hydrazine is most probably performed by hydrazine oxidoreductase enzymes, members of the octahaem hydroxylamine oxidoreductase family (Klotz *et al.*, 2008).

The genome of *Kuenenia stuttgartiensis* led to a revision of this pathway. A gene for cytochrome d nitrite reductase (nitric oxide forming) was found and it was proposed that nitric oxide (not hydroxylamine) might be the precursor for the hydrazine. It was also proposed that the electrons could flow from hydrazine oxidoreductase to nitrite reductase (one electron) and hydrazine hydrolase (three electrons) via complex III (the cytochrome bc1 complex) enabling the build up of a proton motive force (Fig. 6.3). Based on comparative genomics two gene clusters were identified as candidate genes for the hydrazine hydrolase. These predicted complexes were predicted to have both electron transfer and catalytic capacity and had no homologues in the databases.

Unfortunately the current idea is still problematic. Calculation of the Gibbs free energy changes and redox potentials of the half-reactions proposed yields the following:

$$NO_2^- + 1e^- + 2H^+ \rightarrow NO + H_2O \quad +0.35\,V \tag{6.1}$$

$$NO + NH_4^+ + 3\,e^- + 2H^+ \rightarrow N_2H_4 \quad + H_2O + 0.13\,V \tag{6.2}$$

$$N_2H_4 \rightarrow N_2 + 4e^- + 4H^+ \; -0.75\,V \tag{6.3}$$

$$QH_2 \rightarrow Q + 2e^- + 2H^+ + 0.11\,V \tag{6.4}$$

$$\text{cytochrome } bc_1^{RED} \rightarrow \text{Cytochrome } bc_1^{OX} + 2e^- + 0.25\,V \tag{6.5}$$

It is clear that under normal conditions it is impossible to condense hydrazine out of nitric oxide and ammonia with electrons originating from a normal cytochrome bc1 complex (reactions 6.2 and 6.5). Furthermore, proton translocation by the bc1 complex depends on the Q-loop. At a membrane potential of 150 mV, a ΔV of 150 mV between electron donor (ubiquinone, +0.11 V) and acceptor (hydrazine hydrolase, +0.13 V) would be required. With the proposed reactions only 20 mV is available. There is one theoretical possibility to overcome this problem, and that is related to anammox cell biology; the reactions are performed inside the anammoxosome, a small, membrane-bound compartment with an activated membrane. For this reason the activity of the protons may be higher than would be the case in a normal periplasm. If the full membrane potential adds to the activity of the protons, the potential of the electrons needed for reaction 6.2 increases to > 0.25 V, sufficient to drive the Q-loop. As a final remark, it would make much more sense thermodynamically to oxidize

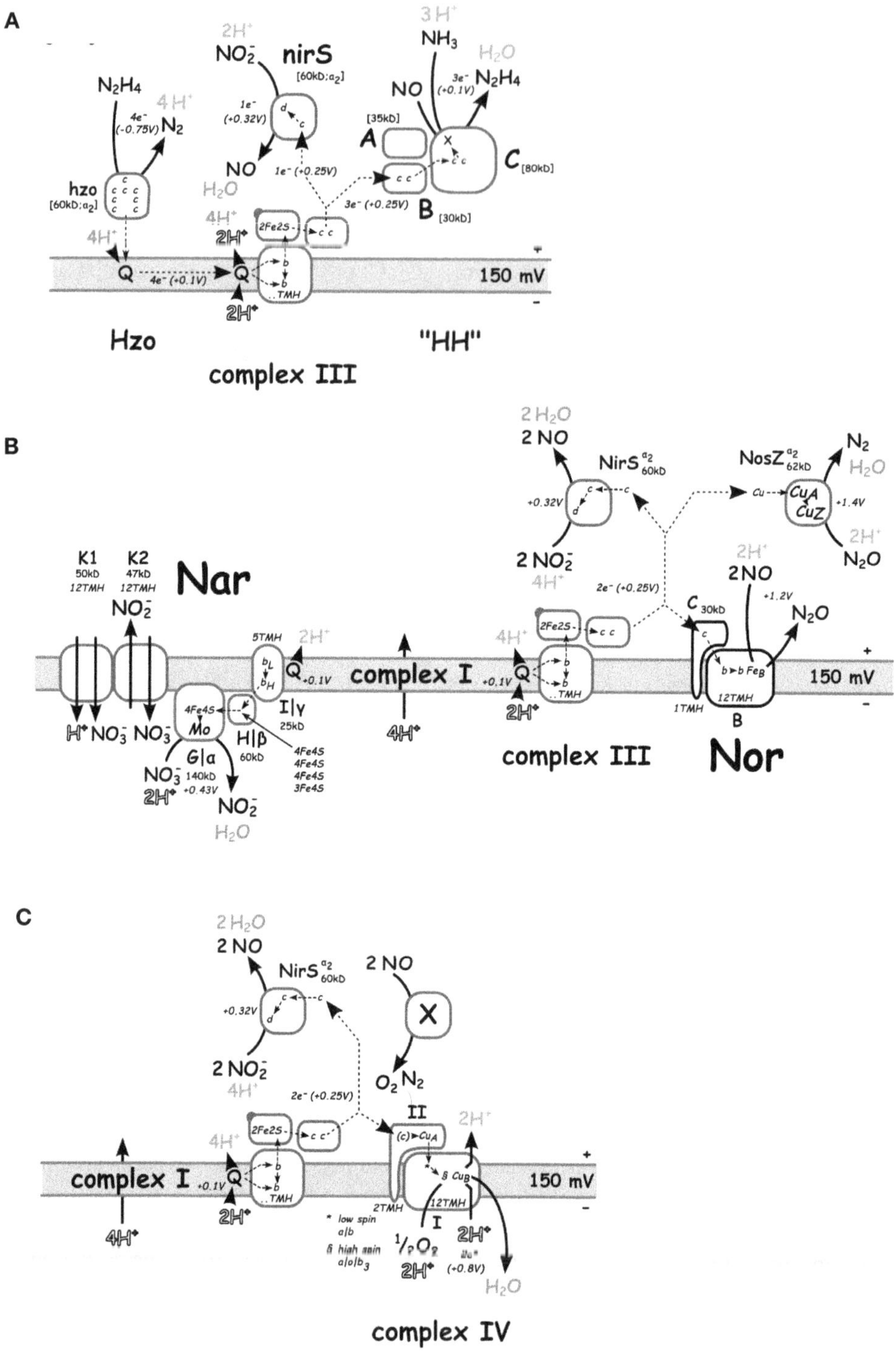

Figure 6.3 Current view of energy conservation via anaerobic ammonium oxidation (a), canonical denitrification (b) and 'intra-aerobic' denitrification (c). If combined with a proton pumping terminal oxidase, the latter pathway would be more energy-efficient than denitrification. For each complex reaction stoichiometry is shown for one electron pair. Outlined protons indicate conservation of energy.

hydrazine in the cytoplasm than inside the anammoxosome.

On chemical grounds, a three electron reduction is awkward in biochemistry. Usually, biochemical redox reactions deal with one, two, four or sometimes six electrons, but not three. One of the candidate complexes proposed contains a dihaem cytochrome c peroxidase that usually transfers two electrons. For this reason NO^- would be a more likely intermediate than nitric oxide itself. Because the electrons would then be split equally among the nitrite reductase and the hydrazine hydrolase, the thermodynamic constraints for the latter reaction would also be more relaxed.

In conclusion, the role of hydrazine in anammox catabolism is well supported by experiments; how the hydrazine is produced and more specifically which nitrogen species may be the precursor still remains shrouded in mystery. An intracytoplasmic compartment may be an essential requirement to increase proton activity and so host the reductive redox reactions of anammox catabolism that would be otherwise impossible. For this reason, it is perhaps not surprising that anaerobic ammonium oxidation is a trait that is difficult to share via lateral gene transfer.

Application in wastewater treatment

Anammox bacteria have been applied successfully in wastewater treatment. The advantage of anammox compared to conventional nitrification/denitrification is in the first place that anammox does not need an organic electron donor. Some wastewaters, for example effluents of anaerobic digestors, do not contain sufficient organic material to sustain denitrification (Hao and Van Loosdrecht, 2003). In those cases, an organic electron donor such as methanol is purchased and added to the wastewater. With anammox, the ammonia from the wastewater itself is used as electron donor leading to significant cost savings. An additional advantage of anammox is that only half of the wastewater ammonia needs to be nitrified to nitrite because the remainder is used as the electron donor. This leads to savings on aeration expenses. Anammox has been applied on full scale both in dedicated anaerobic bioreactors (Abma *et al.*, 2007) and in 'one-step' designs with a single oxygen-limited bioreactor where nitrification and anammox occur simultaneously.

Role in global nitrogen cycling

Anaerobic ammonium oxidizing bacteria and their activity have been detected mainly in the marine context (e.g. Kuypers *et al.*, 2003). Ocean minimum zones, anoxic basins, marine sediments and hydrothermal vents are well-known habitats of anammox bacteria. In such systems they can contribute up to 100% of the observed dinitrogen production. Because some of these ecosystems are important global sources of dinitrogen, it is not surprising that anammox sustains approximately half of the dinitrogen production globally, according to current estimates. Anaerobic ammonium oxidation in marine ecosystems is almost always performed by members of the genus *Scalindua* (Van de Vossenberg *et al.*, 2008). Dissimilatory nitrite reducers (producing ammonia) and aerobic nitrifiers (bacterial or archaeal) can provide substrates to anammox bacteria (Lam *et al.*, 2009).

Anammox has also been studied in wetlands, rivers, lakes and soils but in these habitats it is relatively unimportant compared to denitrification. This can be explained by the fact that in the ocean, the higher activity of sulfate reducers in the mineralization of organic matter leads to a much larger ammonia input to the chemocline. In freshwater systems the sulfate concentration is much lower, leading to a smaller flux of ammonia into the chemocline, favouring denitrification over anammox. In wastewater treatment plants much more ammonia is available and anaerobic ammonia oxidizers can be easily detected and enriched, even from the aerobic stages of the treatment plant.

Denitrifying methanotrophs

Denitrifying methanotrophs are a very recent discovery. In 2006 an enrichment culture was described that oxidized methane in the complete absence of oxygen with nitrite as the electron acceptor (Raghoebarsing *et al.*, 2006). The culture was dominated by a bacterium that represented the phylum 'NC10'. This phylum had been defined only based on 16S rDNA gene sequences. The enriched bacterium was later named Candidatus '*Methylomirabilis oxyfera*' and it is still the only cultivated representative of its phylum (Ettwig *et al.*, 2009). Together with this bacterium a population of Euryarchaea was coenriched. Based on 16S phylogeny these archaea were designated as methanogens or 'reversed' methanogens. Initially it was believed the methane was oxidized with the mechanism known from sulfate-dependent anaerobic methane oxidation. In this mechanism the bacterium (supposedly a denitrifier) would receive electrons from the archaeon (supposedly oxidizing the methane). The nature of the electron shuttling mechanism was unknown. Later, it was observed that the archaeal population declined and that the bacterium must also be capable of oxidizing methane by itself – without the archaea (Ettwig *et al.*, 2008). Candidatus '*Methylomirabilis oxyfera*' is characterized by very small cells that are rod-shaped. The rods are approximately 0.2 µm wide and 1 µm long. The overall reaction is as follows:

$$3CH_4 + 8NO_2^- + 8H^+ \rightarrow 3CO_2 + 4N_2 + 10H_2O - 928\,\text{kJ/mol}\,CH_4 \quad (6.7)$$

Catabolism

After metagenomic sequencing of the enrichment culture the genome of *M. oxyfera* could be closed (Ettwig *et al.*, 2010). The high quality genome sequence enabled a complete compilation of the genes involved in methane oxidation and nitrite reduction. Further, transcriptomic and proteomic approaches showed to what extent the relevant genes were also transcribed and expressed.

Interestingly, the genome clearly showed that *M. oxyfera* was genetically incapable of complete denitrification; no gene for nitrous oxide reductase was encoded in its genome. Otherwise the denitrification pathway was complete and consisted of both NarG and NapA nitrate reductases, a NirS nitrite reductase and a ubiquinol dependent nitric oxide reductase. Yet, experiments clearly showed that dinitrogen was the end product of denitrification in this culture. Consistent with the genetic evidence, experiments confirmed that nitrous oxide was not an intermediate in this case and that methane oxidation did not proceed with nitrous oxide as the electron acceptor. Also with nitrate no methane oxidation was observed. Methane oxidation was only observed when nitrite was supplied as the electron acceptor.

On the carbon side, the *M. oxyfera* genome encoded a complete aerobic methane oxidation pathway, consisting of a particulate (membrane-bound) methane monooxygenase, methanol dehydrogenase, genes for one-carbon metabolism via tetrahydromethanopterin, and formate dehydrogenase. These genes were all highly expressed including the methane monooxygenase.

This analysis led to the postulation of an alternative pathway for denitrification based on the dismutation of nitric oxide:

$$2NO \rightarrow O_2 + N_2 - 173.8\,\text{kJ/mol} \quad (6.8)$$

The oxygen produced could then be used by the methane monooxygenase. Because two molecules of nitric oxide are necessary to produce one molecule of oxygen, the reaction also nicely explained why nitrite, not nitrate was required for methane oxidation. With nitrite, the overall stoichiometry of nitrite to methane is 8:3. That means that four moles of oxygen are available to oxidize three moles of methane. Three are required to drive the monooxygenase. For nitrate the stoichiometry is 8:5. Thus, the 4 moles of oxygen are insufficient to oxidize the 5 moles of methane.

Experiments showed that *M. oxyfera* could indeed produce oxygen from nitrite. In these experiments, the culture was supplied with propylene and ^{18}O-labelled water. Propylene is a dead end substrate for the methane monooxygenase. It is oxidized to the propylene epoxide which can be analysed by gas chromatography/mass spectrometry. In the presence of nitrite reductase the ^{18}O from the water equilibrates with the nitrite leading to the production of ^{18}O labelled nitrite. In the presence of nitrite, the ^{18}O label was also recovered in propylene epoxide. Apparently, an oxygen atom of nitrite was used by the monooxygenase to oxidize propylene. Next it was shown that if the methane monooxygenase was (partially) inhibited by addition of acetylene, ^{18}O-labelled dioxygen (O_2) accumulated (Ettwig *et al.*, 2010). The pathway is compared to conventional denitrification in Fig. 6.3.

It is unknown what enzyme is responsible for the dismutation reaction. Multicopper oxidases and nitric oxide reductase were suggested as the most likely possibilities. Recently, the genome of the denitrifying alkane-degrading gammaproteobacterium HdN1 was sequenced (Zedelius *et al.*, 2010). This strain may use the same alkane activating strategy as *M. oxyfera*. Interestingly, it does not encode any multicopper oxidase similar to *M. oxyfera*.

It is also unknown why the archaea were initially coenriched with *M. oxyfera*. Their enrichment was not a coincidence as they were co-enriched in several independent trials (Hu *et al.*, 2009). Their numbers appear to increase as more nitrate is supplied with the medium. One possibility is that they do form an anaerobic methane oxidizing consortium with another bacterium, a conventional denitrifier. This consortium would oxidize methane anaerobically in parallel with *M. oxyfera*. Another possibility is that with excess nitrate, *M. oxyfera* incompletely oxidizes methane to methanol which is then used by the archaea. In that case the archaea would not be methanotrophic but methylotrophic methanogens.

Application in wastewater treatment

Denitrifying methanotrophs are a very recent discovery and the feasibility of their application in wastewater treatment still awaits exploration. Currently, treated wastewater is sometimes 'polished' to remove remaining nitrate. This is usually done by passing the water through sandfilters supplied with methanol to sustain denitrification. It would be much more cost-effective to use methane instead of methanol as the electron donor. Methane is usually available at wastewater treatment plants because it is produced during sludge digestion. The difficulty of using methanotrophs rather than methylotrophs may be the dilution and emission of the substrate methane with the produced dinitrogen.

Role in global nitrogen cycling

If it is possible to make oxygen out of nitric oxide, we should seriously consider the possibility that this is a commonly done by environmental microbes. Such microbes could even outcompete regular denitrifiers for 'normal' heterotrophic substrates. In conventional denitrification, six protons are translocated for each electron pair during the reduction of nitrate to dinitrogen. Via nitric oxide dismutation more than 7.5 protons could be translocated in the same overall reaction (Fig. 6.3). Thus, the new pathway confers a bioenergetic advantage. It may in addition yield a regulatory advantage, because responses to oxic/anoxic transitions in theory require less adaptation of the respiratory chain.

There is some evidence in the literature that methane is oxidized with nitrate in the environment. This activity has for example been observed in an aquifer on Cape Cod (Smith *et al.*, 1991). Also molecular sequences of the 16S gene of *M. oxyfera* or close relatives have been found in various habitats, for example in a pristine coastal aquifer in Doñana National Park, Spain (Lopez-Archilla *et al.*, 2007).

Acknowledgements

Marc Strous is supported by an ERC starting grant entitled 'MASEM'. I also gratefully acknowledge the input and support of many colleagues over the years.

References

Abma, W.R., Schultz, C.E., Mulder, J.W., van der Star, W.R., Strous, M., Tokutomi, T., and van Loosdrecht, M.C. (2007). Full-scale granular sludge anammox process. Water Sci. Technol. 55, 27–33.

Alperin, M.J.J., and Reeburgh, W.S.S. (1985). Inhibition experiments on anaerobic methane oxidation. Appl. Environ. Microbiol. *50*, 940–945.

Boumann, H. A., Longo, M. L., Stroeve, P., Poolman, B., Hopmans, E. C., Stuart, M. C. A., Sinninghe Damsté, J. S., and Schouten, S. (2009a). Biophysical properties of membrane lipids of anammox bacteria: I. ladderane phospholipids form highly organized fluid membranes. Biochim. Biophys. Acta – Biomembranes *1788*, 1444–1451.

Boumann, H. A., Stroeve, P., Longo, M. L., Poolman, B., Kuiper, J. M., Hopmans, E. C., Jetten, M. S., Sinninghe Damsté, J. S., and Schouten, S. (2009b). Biophysical properties of membrane lipids of anammox bacteria: Ii. impact of temperature and bacteriohopanoids. Biochim. Biophys. Acta *1788*,1452–1457.

Broda, E. (1977). Two kinds of lithotrophs missing in nature. Z. Allg. Mikrobiol. *17*, 491–493.

Eisentraeger, A., Klag, P., Vansbotter, B., Heymann, E., and Dotta, W. (2001). Denitrification of groundwater with methane as sole hydrogen donor. Water Res. 35(9):2261–2267.

Ettwig, K.F., Shima, S., van de Pas-Schoonen, K.T., Kahnt, J., Medema, M.H., den Camp, H.J.M., Jetten, M.S.M., and Strous, M. (2008). Denitrifying bacteria anaerobically oxidize methane in the absence of archaea. Environ. Microbiol. *10*, 3164–3173.

Ettwig, K.F., van Alen, T., van de Pas-Schoonen, K.T., Jetten, M.S., and Strous, M. (2009). Enrichment and molecular detection of denitrifying methanotrophic bacteria of the nc10 phylum. Appl. Environ. Microbiol. *75*, 3656–3662.

Hu, S., Zeng, R. J., Burrow, L. C., Lant, P., Keller, J., and Yuan, Z. (2009). Enrichment of denitrifying anaerobic methane oxidizing microorganisms. Environ. Microbiol. Rep. *1*, 377–384.

Jetten M., Op den Camp H.J.M., Kuenen J.G., Strous M. (2010). Order II. 'Candidatus Brocadiales' ord. Nov. In: Krieg NR, Staley JT, Brown DR, Hedlund B, Paster BJ, Ward N, Ludwig W, Whitman WB (eds.) Bergey's Manual of Systematic Bacteriology, 2nd Edition, Volume 4. Springer, New York.

Jørgensen B. B., and Boetius, A. (2007). Feast and famine – microbial life in the deep-sea bed. Nature Rev. Microbiol. 5, 770–781.

Klotz, M.G., Schmid, M.C., Strous, M., den Camp, H.J.M., Jetten, M.S.M., and Hooper, A.B. (2008). Evolution of an octahaem cytochrome c protein family that is key to aerobic and anaerobic ammonia oxidation by bacteria. Environ. Microbiol. *10*, 3150–3163.

Kuypers, M.M., Sliekers, A.O., Lavik, G., Schmid, M., Jørgensen, B.B., Kuenen, J.G., Sinninghe Damsté, J.S., Strous, M., and Jetten, M.S. (2003). Anaerobic ammonium oxidation by anammox bacteria in the black sea. Nature *422*, 608–611.

Hao, X.D., and van Loosdrecht, M. C. (2003). A proposed sustainable bnr plant with the emphasis on recovery of cod and phosphate. Water Sci .Technol. *48*, 77–85.

Medema, M., Zhou, M., van Hijum, S., Gloerich, J., Wessels, H., Siezen, R., and Strous, M. (2010). A predicted physicochemically distinct sub-proteome associated with the intracellular organelle of the anammox bacterium kuenenia stuttgartiensis. BMC Genomics *11*, 299+.

Lam, P., Lavik, G., Jensen, M.M., van de Vossenberg, J., Schmid, M., Woebken, D., Gutierrez, D., Amann, R., Jetten, M.S.M., and Kuypers, M.M.M. (2009). Revising the nitrogen cycle in the Peruvian oxygen minimum zone. Proc. Natl. Acad. Sci. U.S.A. *106*, 4752–4757.

Lopez-Archilla, A. I., Moreira, D., Velasco, S., and Lopez-Garcia, P. (2007). Archaeal and bacterial community composition of a pristine coastal aquifer in donana national park, spain. Aquatic Microbial Ecol. *47*, 123–139.

Mulder, A. (1989). Anoxic ammonia oxidation of wastewater. Patent EP 327184.

Raghoebarsing, A.A., Pol, A., van de Pas-Schoonen, K.T., Smolders, A.J.P., Ettwig, K.F., Irene, Schouten, S., Jaap, Huub, Jetten, M.S.M., and Strous, M. (2006). A microbial consortium couples anaerobic methane oxidation to denitrification. Nature *440*, 918–921.

Rattray, J., van de Vossenberg, J., Hopmans, E., Kartal, B., van Niftrik, L., Rijpstra, W., Strous, M., Jetten, M., Schouten, S., and Damsté, J. (2008). Ladderane lipid distribution in four genera of anammox bacteria. Arch. Microbiol. *190*, 51–66.

Rattray, J., Strous, M., den Camp, H. O., Schouten, S., Jetten, M., and Damste, J. (2009a). A comparative genomics study of genetic products potentially encoding ladderane lipid biosynthesis. Biology Direct *4*, 8

Rattray, J.E., Geenevasen, J.A.J., van Niftrik, L., Irene, Hopmans, E.C., Strous, M., Schouten, S., Jetten, M.S.M., and Sinninghe Damsté, J.S. (2009b). Carbon isotope-labelling experiments indicate that ladderane lipids of anammox bacteria are synthesized by a previously undescribed, novel pathway. FEMS Microbiol. Lett. *292*, 115–122.

Richards, F. A. Anoxic basins and Fjords. In Chemical Oceanography; Ripley, J. P.; Skirrow, G., Eds.; Academic Press: London and New York, 1965.

Sinninghe Damsté, J.S., Strous, M., Rijpstra, W.I., Hopmans, E.C., Geenevasen, J.A., van Duin, A.C., van Niftrik, L.A., and Jetten, M.S. (2002). Linearly concatenated cyclobutane lipids form a dense bacterial membrane. Nature *419*, 708–712.

Smith, R. L., Howes, B. L., and Garabedian, S. P. (1991). In situ measurement of methane oxidation in groundwater by using natural-gradient tracer tests. Appl. Environ. Microbiol. 57, 1997–2004.

Strous, M., Heijnen, J. J., Kuenen, J. G., and Jetten, M. S. M. (1998). The sequencing batch reactor as a powerful tool for the study of slowly growing anaerobic ammonium-oxidizing microorganisms. Appl. Microbiol. Biotechnol. *50*, 589–596.

Strous, M., Kuenen, J. G., Fuerst, J. A., Wagner, M., and Jetten, M. S. (2002). The anammox case-a new experimental manifesto for microbiological eco-physiology. Antonie Van Leeuwenhoek *81*, 693–702.

Strous, M., Fuerst, J.A., Kramer, E.H., Logemann, S., Muyzer, G., van de Pas-Schoonen, K.T., Webb, R., Kuenen, J.G., and Jetten, M.S. (1999). Missing lithotroph identified as new planctomycete. Nature *400*, 446–449.

Strous, M., Pelletier, E., Mangenot, S., Rattei, T., Lehner, A., Taylor, M. W., Horn, M., Daims, H., Bartol-Mavel, D., Wincker, P., *et al.* (2006). Deciphering the evolution and metabolism of an anammox bacterium from a community genome. Nature *440*, 790–794.

Van de Vossenberg, J., Rattray, J.E., Geerts, W., Kartal, B., van Niftrik, L., van Donselaar, E.G. Sinninghe Damsté, J.S., Strous, M., and Jetten, M.S.M. (2008). Enrichment and characterization of marine anammox bacteria associated with global nitrogen gas production. Environ. Microbiol. *10*, 3120–3129.

Van der Star, W.R.L.R., Abma, W.R.R., Blommers, D., Mulder, J.-W.W., Tokutomi, T., Strous, M., Picioreanu,

C., and van Loosdrecht, M.C.M.C. (2007). Startup of reactors for anoxic ammonium oxidation: Experiences from the first full-scale anammox reactor in rotterdam. Water Res. *41*, 4149–4163.

Van der Star, W., Dijkema, C., de Waard, P., Picioreanu, C., Strous, M., and van Loosdrecht, M. (2010). An intracellular ph gradient in the anammox bacterium kuenenia stuttgartiensis as evaluated by 31P NMR. Appl. Microbiol. Biotechnol. *86*, 311–317.

Van Niftrik, L.A., Fuerst, J.A., Sinninghe Damsté, J.S., Kuenen, J.G., Jetten, M.S., and Strous, M. (2004). The anammoxosome: an intracytoplasmic compartment in anammox bacteria. FEMS Microbiol. Lett. *233*, 7–13.

van Niftrik, L., Geerts, W.J., van Donselaar, E.G. Humbel, B.M., Yakushevska, A., Verkleij, A.J., Jetten, M.S., and Strous, M. (2008a). Combined structural and chemical analysis of the anammoxosome: a membrane-bounded intracytoplasmic compartment in anammox bacteria. J. Struct. Biol. *161*, 401–410.

Van Niftrik, L., Geerts, W.J.C., van Donselaar, E.G. Humbel, B.M., Webb, R.I., Fuerst, J.A., Verkleij, A.J., Jetten, M.S.M., and Strous, M. (2008b). Linking ultrastructure and function in four genera of anaerobic ammonium-oxidizing bacteria: Cell plan, glycogen storage, and localization of cytochrome c proteins. J. Bacteriol. *190*, 708–717.

Van Niftrik, L., Geerts, W.J., van Donselaar, E.G. Humbel, B.M., Webb, R.I., Harhangi, H.R., den, H.O., Fuerst, J.A., Verkleij, A.J., Jetten, M.S., and Strous, M. (2009). Cell division ring, a new cell division protein and vertical inheritance of a bacterial organelle in anammox planctomycetes. Mol. Microbiol. 73, 1009–1019.

Van Niftrik, L., Van Helden, M., Kirchen, S., Van Donselaar, E. G., Harhangi, H. R., Webb, R. I., Fuerst, J. A., Op den Camp, H. J. M., Jetten, M. S. M., and Strous, M. (2010). Intracellular localization of membrane-bound atpases in the compartmentalized anammox bacterium 'candidatus kuenenia stuttgartiensis'. Mol. Microbiol. *77*, 701–715.

Zedelius, J., Rabus, R., Grundmann, O., Werner, I., Brodkorb, D., Schreiber, F., Ehrenreich, P., Behrends, A., Wilkes, H., Kube, M., Reinhardt, R., and Widdel, F. (2010). Alkane degradation by a new type of denitrifying bacterium with possible involvement of the electron acceptor in substrate activation. Environ. Microbiol. Rep. *3*, 125–135.

Structure, Function, Regulation and Evolution of the Nitrite and Nitrous Oxide Reductases: Denitrification Enzymes with a Beta-propeller Fold

7

Rob J.M. van Spanning

Abstract

Specialized denitrifiers recruit two β-propeller enzymes in their anaerobic nitrate respiratory electron transfer network, one of which is an iron containing cd_1-type nitrite reductase, termed NirS, and the other is a copper containing nitrous oxide reductase, NosZ. Together they complement a full denitrification pathway along with nitrate and nitric oxide reductases for the sequential reduction of nitrate to dinitrogen gas. These enzymes are tightly controlled on the level of expression and activity not only according to an energetic hierarchy but also to ensure a balanced conversion of the N-oxides and to prevent the accumulation of the toxic intermediates nitrite and nitric oxide. The adaptive response during the switch from aerobic respiration to denitrification is orchestrated by a dedicated signal transduction network that integrates environmental and intracellular signals and passes these on to the DNA. Amongst these signals are oxygen, denitrification intermediates (nitrate, nitrite, nitric oxide), the redox state of the respiratory components, and metal availability (copper, iron). The coordinate acquisition of these metals during the oxic-anoxic shift is a challenge since their bioavailability requires different reduction states and oxygen tensions.

Introduction

Denitrification is an anaerobic mode of cellular respiration during which nitrate is sequentially reduced to dinitrogen gas via the intermediate compounds nitrite, nitric oxide and nitrous oxide. The process occurs in many types of prokaryotes and in some fungi and is crucial to the global geochemical nitrogen cycle that ensures a balanced conversion of nitrogen containing compounds. The four steps within this metabolic pathway are catalysed by the nitrate, nitrite, nitric oxide and nitrous oxide reductases, which are complex multi-subunit metalloenzymes with unique metal-based electron transfer pathways and catalytic sites (Berks *et al.*, 1995). With the exception of nitric oxide reductase, their unique structural and dynamic features have come available from analyses of high-resolution crystal structures (Einsle and Kroneck, 2004). It is noteworthy that different enzymes in different organisms may carry out similar N-oxide reduction steps. In general, the proteins required for denitrification are only produced under (close to) anaerobic conditions, and if anaerobically grown cells are exposed to oxygen then the activities of the proteins are inhibited and the regulated synthesis of them from the genome is blocked. Thus, for many denitrifying organisms respiration with oxygen usually occurs in preference to the use of N-oxides or oxyanions. This mutual exclusion is apparently evolved according to an energetic hierarchy since respiration with oxygen is more efficient in terms of free energy transduction than electron transport during denitrification. At the same time, the switch from aerobic respiration to denitrification confronts the cells with a number of challenges amongst which are (i) management of the toxic intermediates nitrite and nitric oxide, (ii) redox balancing for proper distribution of electrons to the four oxidoreductases and (iii) recruitment of metals, in particular molybdenum, iron and copper, in sufficient amounts to assemble these four oxidoreductases in their active state. In order to achieve this integrated action, complex cross-communicating regulatory networks have evolved in the different species that are ultimately directed to ensure a balanced expression of the four oxidoreductases along with the protein based machineries to cope with the challenges related to denitrification. This review will focus on the structure, metal recruitment, function and possible evolution of two fascinating enzymes of denitrification, the nitrite and nitrous oxide reductases along with the regulatory proteins that control their expression in different types of denitrifying cells. Denitrification has been most studied in *Paracoccus denitrificans, Rhodobacter* and *Pseudomonas* species and therefore their principles will be related to those in other denitrifiers. Since copper and iron containing nitrite reductases are mutually exclusive found in different types of denitrifier, a brief introduction to the chemistry of these metals in the context of biological recruitment in cellular metabolism is presented.

Copper and iron, prevalence in different denitrifiers

Iron and copper are metals that play an important role in cellular metabolism. It has been argued that the early chemistry of life used water soluble ferrous iron while copper was in the water-insoluble cuprous Cu(I) state as highly insoluble sulfides. Most of the ancient iron-enzymes involved in anaerobic metabolism were designed to operate at low redox potentials. The photosynthetic emission of oxygen into the atmosphere changed this situation dramatically since it resulted in the oxidation of iron into insoluble Fe(III) and concomitant loss of its bioavailability. Cellular life responded with the development of chelators of Fe(III), which rendered iron once again accessible, and iron storage proteins (ferritin) along with regulators that control their expression. The bacterial Fur regulatory system apparently evolved to cope with iron scarcity caused by the low solubility of Fe(III) oxides in an oxidizing environment and the toxicity posed by intracellular Fe(II) and free oxygen (Kadner, 2005). However, these inventions made the use of iron more energetically expensive. In addition, the rise of atmospheric oxygen resulted in the oxidation of insoluble Cu(I) to soluble and bioavailable Cu(II) paving the way for the evolution of copper enzymes. The use of copper in biological systems coincides with the advent of an oxygen atmosphere about 1.7 billion years ago. These copper enzymes are ideally suited for redox reactions with oxygen as oxidant since a redox active copper ion can attain redox potentials in the 0 – 800 mV range. As a result, proteins evolved that equipped either iron or copper or even both (Crichton and Pierre, 2001), with the copper containing ones playing a key role in reactions with oxygen. The reactive centres of copper enzymes are mononuclear (type 2), dinuclear (type 3), or trinuclear (type 2 and type 3). Copper also plays a key role in numerous electron transfer proteins. Mononuclear type 1 (blue copper) centres are found in proteins such as plastocyanin, pseudoazurin and azurin (Dennison, 2008). The classification of copper centres into types is derived from optical and electron paramagnetic resonance (EPR) spectroscopic properties, and there are some notable exceptions, including the cysteine-bridged dinuclear Cu_A electron transfer site in cytochrome *c* oxidase (Beinert, 1995) and nitrous oxide reductase, and the tetranuclear catalytic Cu_Z centre in nitrous oxide reductase (Brown *et al.*, 2000a). The coordination chemistry of copper in metalloenzymes is often distinct, with Cu(I) preferring sulfur donor ligands such as cysteine or methionine, whereas Cu(II) prefers nitrogen donors such as histidine or oxygen donors such as glutamate or aspartate (Kim *et al.*, 2008). The evolution of copper enzymes paralleled the evolution of protein-based mechanisms equipped for copper homeostasis. Copper ions are not only essential for the functionality of many respiratory enzymes, but are also potentially reactive and toxic. Accumulation of the free ionic form is expected to be quite detrimental to the cell since they can react with hydrogen peroxide or superoxide to yield hydroxyl radicals that can damage proteins, lipids, and nucleic acids (Halliwell and Gutteridge, 1984). This seemingly paradoxical property of copper ions demands a concerted regulation of its cellular level. Dedicated copper chaperones ensure that copper ions are provided to essential enzymes for their activity even at low concentrations of bioavailable copper since it is a trace element. Other proteins of the copper homeostasis machinery are to detoxify excess copper (Boal and Rosenzweig, 2009). As such, all organisms have devised cellular mechanisms for metal ion sequestration and detoxification that keep free ionic copper at extraordinarily low concentrations. It has for example been shown that in spite of the micromolar quantities of copper that accumulate in a yeast cell, there is not a single copper ion that is freely available in the cytoplasm (Field *et al.*, 2002). Copper chaperones can acquire the metal under conditions where the metalloenzymes cannot, then function to deliver and directly transfer copper to specific cellular targets. In essence, these molecules act to escort copper ions and protect them from copper scavenging detoxification mechanisms. One of the copper trafficking pathways that require the action of copper chaperones involves SenC (SenC is termed PrrC in *Rhodobacter capsulatus*), homologues of which have a role in Cu_A assemblage in subunit II of cytochrome *c* oxidase (Leary *et al.*, 2009) and hence may have a role in copper assemblage in nitrous oxide reductase as well.

Nitrite reductases, copper versus iron

The key step in denitrification is carried out by a respiratory NO-forming nitrite reductase. There are two types of enzyme that can catalyse this step: one is a copper containing nitrite reductase referred to as NirK, which is a homo-trimeric enzyme with a copper-containing reaction centre. The other type of enzyme is an evolutionarily unrelated cytochrome cd_1-type nitrite reductase, which is a dimer with each polypeptide chain forming a predominantly α-helical domain around the *c*-type haem and an eight-bladed β-propeller accommodating the d_1-type haem in its central tunnel ligated to a conserved histidine residue (His200) (Baker *et al.*, 1997). The haem *c* domain is used for electron uptake and the haem d_1 in the catalytic centre for reduction of nitrite to nitric oxide. As yet, there has been no organism found to have both types of reductase, so apparently the presence of either type of reductase excludes the option of gaining the other type. At first sight there does not seem to be a clear-cut correlation between the distributions of either of the two enzymes on the one hand and the taxonomic positions of the species in which they are found on the other (see also Chapter 1). For example, copper containing NirK enzymes are found in many species of the α-, β- and γ-proteobacteria, while many other taxonomically related species have the cytochrome cd_1-type NirS. By implication this indicates that during evolution of species gene clusters encoding either or both types of nitrite reductase are alternatingly lost and gained within each taxonomic clade through horizontal gene transfer (Jones *et al.*, 2008). Since the latter study also indicates that, with some exceptions, the overall taxonomy of either of the enzymes to a great extent correlates with that of the species itself, this would imply that a successful gain of either type might well be passed on vertically to evolutionarily related descendants. If this evolutionarily scenario is correct, then the question becomes apparent where the genetic and functional origin is for making up the NirK or NirS type. Here it must be noted that expression of NirK requires only a single gene, sometimes accompanied with a second one expressing a protein called NirV. The latter enzyme is related to desulfurates and may well be required for proper insertion of the copper reaction centre. Expression of NirS on the other hand requires a suite of genes that are required for maturation and assemblage of the unique haem d_1 in the reaction centre. To understand this in the proper context, the assumption here is that the evolution of the *nirS* locus would be optimized and specialized in the α-proteobacteria as they belong to the most recently developed clades. Here, a full locus comprises at least 10 genes and commonly organized according to the sequence *nirSECFDLGHJN*. The *nirS* gene encodes the functional subunits of the dimeric NirS-type nitrite reductase. All other genes are required for proper synthesis and assemblage of the d_1 haem and related functions. The *nirC* gene encodes a specific cytochrome *c*, *nirF* a truncated homologue of NirS with the d_1 containing WD-40 domain but lacking the N-terminal haem *c*-domain, and *nirN* encoding a full homologue (i.e. including a haem *c* domain) of NirS. Notably, the histidine residue for haem

d_1 ligation in NirS (H200) is conserved in NirF but not in NirN of *Pa. denitrificans*. The *nirJ* gene encodes a cytoplasmic iron–sulfur protein of the radical SAM superfamily. Enzymes of this family generate radicals by combining a 4Fe–4S cluster, which is coordinated by a conserved CxxxCxxC motif, and S-adenosylmethionine (SAM) in close proximity. This allows the transfer of a single electron from the iron–sulfur cluster to SAM, which leads to its reductive cleavage to methionine and a 5'-deoxyadenosyl radical, which, in turn, abstracts a hydrogen atom from the substrate to initiate a radical mechanism (Frey *et al.*, 2008). The *nirE* gene encodes a cytoplasmic uroporphyrin III methylase, member of the tetrapyrrole (corrin/porphyrin) methylase family, which uses SAM in the methylation of diverse substrates. Hence, NirE and NirJ are involved in maturation of the d_1-haem in the cytoplasm, while the NirN and NirF proteins are periplasmic and might act as stager enzymes prior to insertion of the d_1-haem into nitrite reductase. The *nirDLGH* genes encode homologues of the so-called AscN/GntR-type activators. These transcription regulators may initiate a cascade of transcription regulation events to make the cells ready for the challenging mode of nitrite respiration. Possible targets therefore are genes encoding the haem biosynthesis machinery and nitrite and NO-scavenging enzymes. Some specialized denitrifiers have adopted a *nirIX* gene set adjacent to their *nirSECFDLGHJN* gene cluster. The NirIX proteins are flavoproteins with an unknown function in regulation of both expression and activity of the cd_1-type nitrite reductase (see below). The core of such a modern *nirS* locus is found not only in the α-proteobacteria but also with some variations on the theme in the β, γ, and ε-proteobacteria as well as in the Aquificae.

By inference this might implicate that the evolution of NirK would have been much more simple than that of NirS. Moreover, also the possibility of horizontal gene transfer of the *nirK* gene in between taxonomic clades would have been much more likely then that of the much larger *nirS* locus. Indeed, all genomic data indicate that NirK is found in all taxonomically distinct clades including the Archaea and single membrane species of the Actinobacteria and Firmicutes as well as in many species of the different Gram-negative taxa. However, there are two exceptions with regard to the distribution of NirK in that it is not found in species of the ε-proteobacteria and Aquificae. Species from these clades seem to exclusively engage a cd_1-type of nitrite reductase. What makes these two taxa so interesting is that there is an ongoing debate about the relatedness of these clades. In one scenario, the Aquificae are thought to be related to the Thermatoga as an old branch of cellular life as based on comparison of 16S RNA sequences. The other scenario places the Aquificae close to the ε-proteobacteria as judged by the shared presence of specific indents and gene signatures in key genes (Cavalier-Smith, 2002; Griffiths and Gupta, 2004). The exclusive presence of complete *nirS* gene clusters with comparable organizations in both ε-proteobacteria and Aquificae would indeed favour the evolutionarily relatedness of these species. Taxonomic linkage and accompanying vertical gene transfer of their *nirS* locus is then much more likely than the possible occurrence of horizontal gene transfer of such a specific locus to all species within both the ε-proteobacteria and Aquificae. Moreover, they share a specific type of nirN gene encoding a NirS homologue with a unique haem *c* containing domain at the C-terminus. Thus, these proteins are equipped with both N- and C-terminal haem *c* domains, which may well be important for proper turnover of the enzymes from both taxonomic groups.

Evolution of modern NirS from an Archaeal ancestor

The question then remains how the *nirS* locus has evolved and integrated in the metabolism of the ε-proteobacteria and Aquificae and further on to the α, β and γ-proteobacteria. The key to this answer may well be in the observation that ancient homologues of the *nirS* gene are solely and exclusively found in the Archaeal genus of *Pyrobaculum* and not in the Actinobacteria and Firmicutes (Bacilli and Clostridia), which are the oldest clades of bacteria having only a single membrane and thought to be ancestors of modern day proteobacteria. Inspection of the *Pyrobaculum* genomes shows that they contain primitive *nirS* genes along with some accessory genes. Species in that genus are notable both for respiratory versatility, exemplified by *Pyrobaculum aerophilum*, and for diversity in respiratory capabilities between closely related species (Amo *et al.*, 2002; Huber *et al.*, 2000). This is in contrast to species of many other Archaeal genera, which are usually either obligate aerobes (e.g. *Sulfolobus* spp.) or obligate anaerobes (e.g. the Thermococcales), with only limited respiratory flexibility. The *Pyrobaculum nirS* gene phylogenetically clusters with the central homologous domains of the *nirN* gene found in ε-proteobacteria. However, where NirN in these species is equipped with both N- and C-terminal haem *c* domains, the *Pyrobaculum* proteins lack these cytochrome *c*-domains. Moreover they contain a twin-arginine motif in the signal sequence, suggestive of cofactor incorporation in the cytoplasm followed by transport over the membrane via the TAT-translocation machinery. The matured proteins have three histidines that are positionally conserved in the NirS family of proteins. In *Pyrobaculum calidifontis* these are H158, H325 and H368. H158 corresponds to H200 of the *Pa. denitrificans* protein, which is the ligand to the d_1 haem (Baker *et al.*, 1997). The other two histidines are more proximal from the haem and may be involved in proton donation. The *Py. calidifontis nirS* (or perhaps it should be designated *nirN*) gene is present in a respiratory island of about 53 genes including genes encoding a full complement of denitrification genes along with NADH dehydrogenase, cytochrome *c* oxidases, a primitive cytochrome bc_1-complex and *ccm* genes involved in haem *c* biosynthesis and insertion. This genetic potential apparently allows it to denitrify from nitrate to dinitrogen (Cozen *et al.*, 2009). The *nirS* gene in this locus is next to but divergently transcribed from a gene encoding a cytochrome *c*. Another separate locus contains homologues of the *nirJ* and *nirE* genes for haem d_1 biosynthesis, along with a gene encoding sirohaem synthase. A locus with a *nirS* gene accompanied with genes encoding cytochrome *c*, UbiA and NirE is also found in *Pyrobaculum arsenaticum* and *Py. aerophilum*, but not in other Archaeal species. It has been shown that *Py. aerophilum* has the potential to reduce nitrite to NO using menaquinol as electron donor (de Vries and Schroder, 2002). Hence, the minimum requirement for nitrite to nitric oxide catalysis would then be such a primitive supposedly membrane anchored NirS-like protein with a d_1-like haem whose maturation requires NirE and NirJ. It then seems plausible that such an ancient locus has been transferred to Aquificae and/or ε-proteobacteria via horizontal gene transfer. It is notable that the *Pyrobaculum nirS* sequence is most homologous to the central domain of the *nirN* gene in the Aquificae, suggestive for horizontal gene transfer between ancestors of them and those of *Pyrobaculum* species. The Aquificae phylum is a diverse collection of bacteria that live in harsh environmental settings. They have been found in hot springs, sulfur pools, and thermal ocean vents. Members of the genus *Aquifex*, for example, are productive in water between 85 and 95°C. They are the dominant members of most terrestrial neutral to alkaline hot springs above 60°C, autotrophs, and the primary carbon fixers in these environments. As such, these bacteria share extreme environments with the Archaea opening the possibility of genetic exchange between these different clades of bacteria. Then, the Archaeal *nirS* gene may have been duplicated in the Aquificae, one of which is fused to the gene encoding a dedicated type I cytochrome *c*, giving rise to an ancestor of the NirN protein that could use cytochrome *c* as electron donor rather than menaquinol. Indeed, some of the Aquificae and ε-proteobacteria have copies of a *nirF* gene (encoding only a d_1 domain) and *nirN* gene (encoding both haem *c* and d_1 domains), which allows them to denitrify in the absence of a *nirS* gene, indicating that a NirN-type nitrite reductase with *c* and d_1 haems catalyses nitrite reduction in these species. Here it should be noted that this NirN-type has the conserved His residue for haem d_1 ligation whereas it is absent from its counterpart in *Pa. denitrificans* that uses NirS for nitrite reduction. Evolution of the *nirS* gene may then have occurred after a second gene duplication step of the *nirN* gene followed by functionally distinct mutations in each copy, ultimately leading to a *nirS* gene encoding a modern cd_1-type nitrite reductase in for example *Nitratiruptor* sp. SB155-2, and further on to the orthologues found in the α-proteobacteria.

Nitrous oxide reductase

Nitrous oxide reductase (NosZ) is a terminal oxidoreductase of the denitrification electron transfer network that reduces relatively inert nitrous oxide to dinitrogen. The enzyme is dimeric with each monomer composed of two domains: a C-terminal cupredoxin domain where electrons enter the enzyme via the mixed-valent binuclear Cu_A site, and an N-terminal seven-bladed β-propeller domain which hosts the active-site centre with 4 copper ions that make up the μ4-sulfide-bridged Cu_Z reaction centre. The reduction potentials for Cu_A and Cu_Z are estimated at +260 and +60 mV, respectively (Rasmussen *et al.*, 2002). As such, electron flow from the entry site to the reaction centre is thermodynamic unfavourable and suggests an intramolecular electron-gating mechanism. Structures are available from *Pseudomonas nautica* (Brown *et al.*, 2000b), *Pa. denitrificans* (Haltia *et al.*, 2003) and *Achromobacter cycloclastes* (Paraskevopoulos *et al.*, 2006). The biogenesis of a catalytically active NosZ requires auxiliary functions for copper uptake, copper centre assembly and copper centre activity (Zumft, 2005; Zumft and Kroneck, 2007).

Copper management, a role for dedicated copper chaperones

Copper uptake for assemblage of nitrous oxide reductases involves the periplasmic copper binding protein NosL, which has its gene found in virtually all species that perform nitrous oxide reduction. NosL is a 20-kDa periplasmic copper binding lipoprotein, although its function has yet to be determined. NosL from *Achromobacter cycloclastes* contains one Cu(I) ion per monomer, which is lost upon oxidation with concomitant dimerization of NosL via intermolecular disulfide bond formation (McGuirl *et al.*, 2001). The structure of apo-NosL consists of two relatively independent homologous domains that adopt an unusual topology, which displays structural homology to only one other protein, MerB, an organomercury lyase involved in bacterial mercury resistance (Taubner *et al.*, 2006). Alignment of NosL sequences in the NCBI databank does not reveal a discernible copper-binding motif. Of the few conserved residues within this relatively divergent group, there is only one more or less conserved motif, (G/A)XMG surrounding a central Met129 in all NosL sequences relative to the unprocessed one of *Ac. cycloclastes*. However, a more detailed inspection of NosL sequences reveals that they split up in at least 3 distinct groups, each of which with characteristic patterns of conserved methionines, histidines and cysteines, which may ligate copper ions. The *Ac. cycloclastes* NosL protein clusters with those of many α- and β proteobacteria, all of which share an N-terminal domain with a HxCxMxxxH-N7-H amino acid sequence around position 40 along with conserved methionines at positions 94 and 129. Two other groups of NosL are exemplified by the two paralogues of NosL in *Cupriavidus metallidurans*, both of which are lipoproteins, but one of them is equipped with a signal sequence for transport via the TAT-translocon. Apparently, not all lipoproteins are exported as unfolded substrates through the general secretion pathway. This feature has recently been noticed for some lipoproteins in Gram-positive bacteria (Hutchings *et al.*, 2009), but it now seems to occur in Gram-negative species as well. Other species that have a similar set of *nosL* paralogues next to each other in their *nos* gene clusters are *Ralstonia pickettii, Rhodoferax ferrireducens, Acidovorax ebreus, Thiobacillus denitrificans,* and *Thauera sp. MZ1T.* The NosL paralogues lacking the RR signature in their signal sequence share Cys and Met residues around positions 50 (CxxC) and 130 (M). At their C-terminus they also contain a cysteine, histidine or methionine residue or combinations of them at variable protein specific positions. The CxxC motif suggests that these NosL paralogues are thiol-disulfide isomerases. The NosL proteins with the twin-arginine signal sequence have completely different patterns conserved around positions 40 (CxxxxM), 75 (C), 130 (HxxM), 165 (M) and 175 (HxxM). The RR motif suggests that the NosL proteins of this group are translocated to the periplasm in their folded state, perhaps with a metal cofactor already bound to it. Since their allocated genes are both clustered in their respective *nos* operons, it may well be that the two NosL paralogues cooperate during NosZ assemblage, one of which releases the locked copper ion of the other via thiol exchange reactions prior to delivery to its target protein. The *nosL* gene is not always clustered in *nos* operons of some species, and in others, such as *Haloarcula marismortui*, *Geobacillus thermodenitrificans*, and *Opitutaceae bacterium TAV2*, they are even absent. This suggests that the role of NosL may be taken over by other types of protein in these species. A likely candidate is a SenC family protein, the gene of which is clustered with *nos* operons of *Op. bacterium TAV2* and other members of the *Verrucomicrobia*.

Copper and sulfur insertion, a role for dedicated transporters and copper chaperones

Formation of the Cu_Z cluster for maturation of NosZ requires the function of the NosFYD proteins that make up an ATP-binding cassette transmembrane transport system (ABC transporter) to couple ATP/GTP hydrolysis to transport of a yet unknown compound. Without exception, their genes are clustered with all known *nos* gene clusters. NosF is the ATP binding cytoplasmic motor of the translocator, NosY the transmembrane permease, and NosD is a periplasmic protein likely associated with the NosFY translocator (Zumft, 2005). Despite the fact that nitrous oxide reductase in some species is translocated via the TAT translocon for folded proteins, Cu_A and Cu_Z assemblage seems to occur in the periplasm (Zumft, 2005). In addition, NosZ proteins from all the ε-proteobacteria and Aquificae lack the signal sequence for TAT transport and are therefore presumably translocated via the more common Sec pathways for unfolded proteins followed by copper centre assemblage in the periplasm. It has therefore been suggested that the NosFYD translocator is not involved in copper import but in the export of a sulfur compound from cytoplasm to periplasm prior to Cu_Z assemblage (Zumft and Kroneck, 2007). Moreover, NosFY translocators do not resemble copper import pumps like CusA or copper export pumps like CopA (Boal and Rosenzweig, 2009).

Whether or not an exporting role for NosFY is likely depends on the architecture of the translocon. ABC transporters can be roughly divided into three groups, which are (i) exporters, (ii) type I importers and (iii) type II importers (Locher, 2009). ABC importers as well as exporters are dimeric and make use of the same mechanism to expose binding and extrusion sites in the permeases to opposite sides of the membrane, thus moving substrates across the bilayer unidirectionally. Binding of ATP closes the gap between the coupling helices at the cytoplasmic site of the permease and opens one at the periplasmic site. As a consequence, the permeases flip from facing inward to facing outward where a product is released in the case of an ABC exporter or taken up in the case of ABC importers. ATP hydrolysis then triggers the opposite conformation change where the permease opens up again in the cytoplasm to take up (ABC exporter) or release (ABC importer) a product. Importantly, importers recruit a special periplasmic substrate binding protein, whereas exporters for obvious reasons do not. Proteins from the first group, the ABC exporters, all share a common core architecture that consists of 12 transmembrane helices, which extend approximately 25 Å in the cytoplasmic boundary away of the lipid bilayer. The transmembrane permeases of exporters are invariably fused to their cytoplasmic ATP binding partners as illustrated by the high-resolution crystal structure of an ABC exporter, the Sav1866 protein from *Staphylococcus aureus* (Dawson and Locher, 2007).

Many ABC exporters are involved in multidrug extrusion of toxic substances. Proteins from the second group, type I importers, generally contain 12 TM helices, and the fold of this subfamily has been established by the crystal structure of the molybdate/tungstate transporter ModBC from *Archaeoglobus fulgidus* (Hollenstein *et al.*, 2007). ATP binding proteins and their cognate permeases are usually not fused, and they mediate the uptake of ions, sugars, amino acids and other substrates, which are all captured by specific binding proteins that deliver them to the transporters. The structure of the maltose transporter MalFGK has also been determined, revealing that it was similar in conformation to Sav1866 in that it was outward facing when ATP was bound (Oldham *et al.*, 2007).

Proteins from the third group, type II importers, facilitate the uptake of metal chelates that are generally larger than the substrates of type I ABC importers. Examples of substrates are cobalamin (vitamin B_{12}) or haem. Type II importers have a distinct permease architecture with 10 helices each as shown in the crystal structure of the *Escherichia coli* vitamin B_{12} transporter BtuCD (Locher *et al.*, 2002). Structures of other type II translocators have become available (Locher, 2009), but in none of these is there a substrate bound in the permease, nor do they have binding pockets for B_{12} or haem, indicating that the specificity of the transport reaction appears to depend exclusively on the cognate periplasmic substrate-binding protein.

A comparison of the NosF and NosY proteins with those exemplified for the three types of transporter above revealed that NosF and NosY do not conform to a type I exporter since on the one hand there is little sequence similarity with proteins from this class, and on the other hand the NosF and NosY proteins are not fused to a single protein being characteristic for this type of transporter. However, the NosFY proteins do not show abundant similarity with their counterparts from type I and type II importers either. For example, BtuC from *E. coli*, the type II vitamin B12 importer, aligns well with the haem uptake permease (Pden_4204) and with the siderophore uptake system (Pden_4828) from *Pa. denitrificans*, but not with NosY. And the type I maltose transporter MalF from *E. coli* is quite similar to protein Pden_4189, likely involved in uptake of glycerol-3-phosphate, and with protein Pden_2471, likely a transporter for phosphosugars, but also this protein fails to show a comparable degree of similarity with NosY. The same holds for their cognate ATP-binding proteins, which do not share high enough similarity with NosF to explicitly group the NosFY proteins with those of the type I or type II importers. In addition, the ATP binding proteins of type I and type II importers have a size of around 330 and 250 amino acids, respectively, while NosF is made up from around 300 amino acids with a core of 250 amino acids and an N-terminal extension of about 50. Inspection of the *Pa. denitrificans* protein list revealed at least two proteins highly similar to NosF and both with similar sizes (Pden_5056 and Pden_4686). Interestingly, both their cognate permeases have a high degree of similarity with the NosY protein and do not conform to the classical ABC exporters. The gene clusters encoding the ABC transporters at locations 5056 and 4686 do not contain genetic information to synthesize a conventional periplasmic substrate-binding proteins, which would be characteristic for type I and II importers. Instead they share a gene encoding a paralogue of MFP (membrane fusion protein), a component of the RND (resistance, nodulation, and cell division) family of efflux transporters. Note, however, that these MFP paralogues are different from the HlyD secretion protein. Attributed functions of the RND family of efflux transporters imply outward transport including nodulation, acriflavin resistance, heavy metal efflux, and multidrug resistance proteins. Most members of this family are found in Gram-negative bacteria. The proposed function of MFP proteins is to bring the inner and outer membranes together and enable transport to the outside of the outer membrane. These observations all sum up to a hypothesis that the NosFY-like transporters group together in an exporter class different from the classical exporters and type I and II importers. This suggestion would be in accordance with NosFY being implicated in export, may be a sulfur compound, rather than in import.

Maturation of Cu_Z, a role for the β-solenoid protein NosD

Although the NosFY translocator groups together with the ABC transporters encoded by *Pa. denitrificans* genes 5056 and 4686, it is different in that its allocated gene cluster does not contain a gene encoding a MFP paralogue. Instead, *nos* gene clusters consistently integrate a gene encoding the periplasmic NosD protein, which is essential for Cu_Z assemblage and nitrous oxide metabolism (Zumft and Kroneck, 2007). NosD is a 45-kDa hydrophilic periplasmic protein with 2 CASH (carbohydrate-binding and sugar hydrolase) domains. The CASH domain is characterized by internal repetitions of glycines and hydrophobic residues that correspond to the repetitive units of a predicted or observed right-handed β-helix structure of the pectate lyase superfamily (Ciccarelli *et al.*, 2002). Many carbohydrate-binding proteins and sugar hydrolases share the CASH domain. The basic structural unit of this family consists of three β-strands that form a single turn of the β-helix. Each turn contains about 20 amino acids, and is normally repeated between 7 and 11 times to form the elongated helix structure. The repeats show a low degree of sequence identity when compared with each other. The region of homology with the CASH domain corresponds to the core region of the β-helix, covering from the second to the sixth repeat. The tertiary structures of pectate lyases and rhamnogalacturonase A show a stack of parallel β-strands that are coiled into a large helical β-roll where each coil of the helix represents a structural repeat. The elongated shape of β-helical proteins provides enough space to accommodate multiple binding sites for binding a polymeric substrate like polysaccharides and proteins in many members of the β-helical protein family. Proteins containing these repeats most often are enzymes with pectin-like substrates and polysaccharide like alginate. Alginate forms a protective capsule for *Pseudomonas aeruginosa* during chronic pulmonary infections. The structure of alginate, a linear polymer of β1-4-linked O-acetylated D-mannuronate (M) and L-guluronate (G), is important for its activity as a virulence factor. Its formation and protection requires AlgG, a periplasmic C-5 mannuronan epimerase, which contains a right-handed β-helix fold just as the CASH domain (Douthit *et al.*, 2005). AlgG has a long shallow groove that may accommodate alginate, similar to protein/polysaccharide interactions of other CASH domain proteins. A wide variety of other proteins contain these β-helical domains, like autotransporters, haemoglobin protease, glutamate synthase, SufD, which is a stabilizer of the iron transporter, filamentous haemagglutinin and antifreeze protein (Kajava and Steven, 2006). In *Sphaerobacter thermophilus* and *Thermomicrobium roseum*, the *nosD* gene is N-terminally fused to the *nosF* gene resulting in a fusion of periplasmic NosD and the cytoplasmic motor protein. This architecture is suggestive for an unexpected physical and functional interaction of them. The two domains in these fusions are likely connected via an inward moving α-helix just at the C-terminal end of the NosD domain. Since ABC transporters are dimeric, this implies that NosD is also twinned in the translocator. In Archaea, such as *Haloferax volcanii DS2*, *Halogeometricum borinquense* and *Ha. marismortui*, the *nosD* gene is fused to the *nosL* gene, which may then suggest synthesis of a fusion protein in which the NosL domain delivers the copper ions to the NosD domain. Taken together, this may bring us to the hypothesis that NosD gathers copper ions from the NosL protein, and an unknown cargo, perhaps a sulfur compound, via the NosFY proteins (Fig. 7.1). The β-helical structure of NosD could well accommodate the polypeptide chain of nitrous oxide reductase in a chaperone like fashion and facilitate the insertion of the four copper ions along with a proposed sulfur to construct the Cu_Z centre. To do so, it may require active site residues, which would then be conserved throughout the NosD family. Likely candidates are two stretches of more or less conserved sequence located in the centre of these proteins at positions around 200 (HYMY) and 230 (LMN). Though a wealth of experimental and genomic data would support such a scenario for Cu_Z assemblage, the fundamental mechanism for copper centre assembly in nitrous oxide reductases remains unclear and requires perhaps several yet to identify ancillary proteins that participate in its biosynthesis.

CuA assembly, a possible role for SenC

The Cu_A electron entry site of NosZ includes two copper ions bridged by two thiol side chains from two cysteine residues. Maturation of the Cu_A site may well be mediated via SenC like proteins, which are homologues of the mammalian Sco proteins. Members of the Sco/SenC family belong to the thioredoxin-like proteins and they play a role in copper binding and delivery as well as in redox reactions via the reversible oxidation of an active centre disulfide bond. They are usually made up of about 200 amino acids, contain one transmembrane domain with the N-terminus located in the cytoplasm and a periplasmic Cu(I)-binding domain with a central conserved CxxxCP domain and a conserved axial histidyl ligand located C-terminally in a conserved DHS triad (Eraso and Kaplan, 2000; Nittis *et al.*, 2001; Saenkham *et al.*, 2009). Their function, however, also requires the ability to bind Cu(II) (Banci *et al.*, 2007). Sco1 and Sco2 proteins in eukaryotic organisms are required for delivery of copper for assemblage of the binuclear, mixed valence Cu_A centre of cytochrome oxidase in mitochondria.

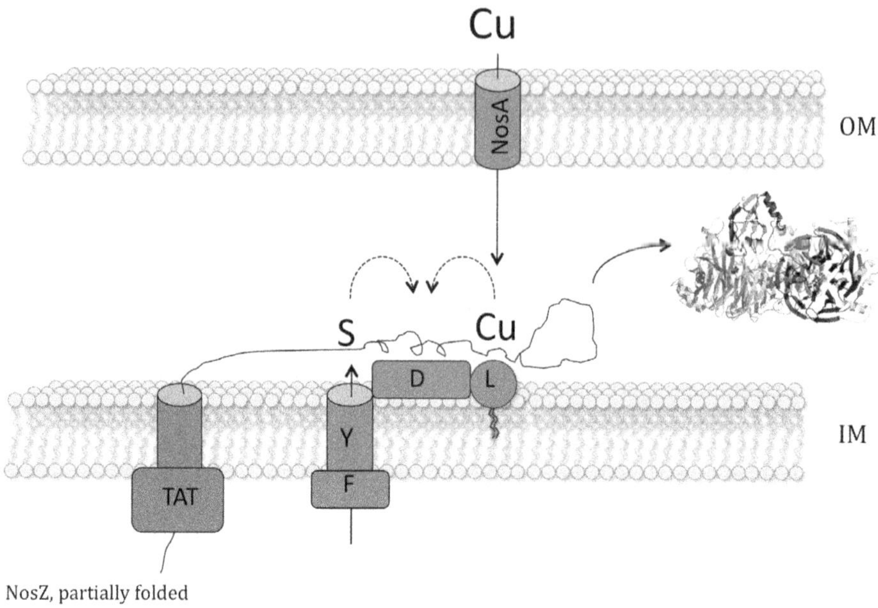

Figure 7.1 Model for the transport and maturation of and assemblage of the Cu_Z reaction centre in nitrous oxide reductase. TAT, twin arginine translocator; Cu, copper ions; S, unknown sulfur compound; NosA, copper uptake protein; FYLD, proteins encoded by *nosFYLD*; OM, outer membrane; IM, inner membrane.

The activity of the Sco paralogues requires the copper chaperone Cox17, originally identified in the yeast *Saccharomyces cerevisiae* as required for cytochrome oxidase assembly (Glerum *et al.*, 1996). It is a small cysteine-rich protein that has been reported to bind between one and four Cu(I) ions (Arnesano *et al.*, 2005). Cox17 operates in the mitochondrial intermembrane space (IMS) where it exists in a partially oxidized form with two S–S bonds and two reduced cysteines, and in this state it is involved in the transfer of Cu(I) and two electrons to Sco1. The copper loaded Sco1 protein then fills the Cu_A site on the Cox2 subunit (Abajian and Rosenzweig, 2006; Horng *et al.*, 2004). The same reaction of copper–electron-coupled transfer from Cox17 does not occur with the Sco2 homologue of Sco1 (Banci *et al.*, 2008). Instead, it was reported that Sco2 acts as a thiol-disulfide oxidoreductase to oxidize the copper-coordinating cysteines in Sco1 during the synthesis of the Cu_A site, making it also indispensable for synthesis of cytochrome *c* oxidase subunit II. The Sco1 and Sco2 metallochaperones therefore seem to have distinct, stage-specific but interrelated functions in copper delivery and the biogenesis of the Cu_A site of cytochrome *c* oxidase subunit II (Leary *et al.*, 2004, 2009). The role of the Sco1 and Sco2 paralogues in prokaryotes is still incomplete. Cu(I) binding in prokaryotic Sco1 of *Bacillus subtilis* also requires the cysteine thiols in the CxxxC motif as well as a conserved axial histidyl ligand. The oxidized form does not seem to bind metal, but can be reduced *in vitro* to bind both Cu(I) and Cu(II) with high affinity at a 1:1 stoichiometry (Andruzzi *et al.*, 2005; Imriskova-Sosova *et al.*, 2005). PrrC is a Sco homologue in *Rhodobacter sphaeroides*, which is monomeric, binds Cu(II) in its reduced form and reduces it to Cu(I), while it also possesses disulfide reductase activity. This is in contrast to the eukaryotic Sco homologues, which are dimeric and bind Cu(I). These results indicate that some bacterial Sco proteins may have biochemical properties that are distinct from those of mitochondrial Sco proteins (Badrick *et al.*, 2007). In many other prokaryotes, the Sco homologue SenC has been shown to be important for copper delivery to cytochrome *c* oxidase, for example in *Rh. capsulatus* and *Agrobacterium tumefaciens* (Krummeck and Rodel, 1990; Nittis *et al.*, 2001; Swem *et al.*, 2005) and in *Pseudomonas aeruginosa*. The SenC mutant phenotype of the latter species could be restored to wild-type levels by the addition of external copper (Frangipani and Haas, 2009), suggesting that SenC is an essential copper chaperone during copper deprivation.

Since overwhelming evidence about the role of Sco/SenC in Cu_A assemblage into cytochrome *c* oxidase is emerging, it might well be that nitrous oxide reductase requires SenC for maturation of its Cu_A site as well. If so, one would expect SenC to be active both under aerobic and anaerobic conditions. Moreover, copper delivery would then rely on a completely different redox environment along with the complications associated with the bioavailability of the copper ions at each different environmental condition. At that switch one would then indeed expect signal transduction mechanisms to operate for redox homeostasis. Williams and co-workers have postulated such a signalling role for mitochondrial Sco proteins (Williams *et al.*, 2005), although the route of signal transduction has not been defined. It has also been suggested that the features reported for Sco of *Ba. subtilis* would allow these proteins to respond to a cell's metal content and redox state, and thus may suit Sco for a role as a redox sensor (Imriskova-Sosova *et al.*, 2005). Indeed it has been shown that the prokaryotic Sco1 homologue SenC communicates with the RegAB signal transduction couple. In *Rh. sphaeroides* 2.4.1 PrrC (SenC) is functionally associated with the PrrBA two-component signal transduction system that induces photosynthesis gene expression in response to a decrease in oxygen tension since its mutation led to low expression of photosynthetic genes under aerobic conditions (Eraso and Kaplan, 2000). Additional evidence revealed that the redox-dependent signalling via SenC/RegAB is also, in part, involved in the regulation of anaerobic respiration in *Rh. sphaeroides* 2.4.1.

Copper centre redox cycling, a role for NosR and NosX proteins

Copper centre maintenance of NosZ is a topic not yet well understood, but it deserves attention. One may regard the enzyme as expensive in terms of its synthesis since it requires quite a few proteins to gather insoluble copper ions under anaerobic conditions and to build in 6 of them in the unique Cu_A and Cu_Z centres of each enzyme. Once assembled, one would then expect the protein to be preserved and catalytically active by mechanisms that maintain the proper state of the reaction centre even when the enzyme is challenged by changes in the cellular environment. For example, when oxygen transiently enters a denitrifying cell it may react with the Cu_Z reaction centre, rendering it inactive. This has been shown for enzymes that were isolated under aerobic conditions, and it turned out that the Cu_Z centre in these proteins had turned into a redox inactive $Cu_Z{}^*$ state (Rasmussen *et al.*, 2002). A redox inert state of Cu_Z has also been observed during studies on the reaction cycle of the *Pseudomonas putida* enzyme (Fig. 7.2) (Dell'acqua *et al.*, 2010). It has been suggested that this inactive state resembles the $Cu_Z{}^*$ state of oxygen challenged NosZ, so that terminology will be used to describe the redox inert form of Cu_Z. The *Ps. putida* study revealed that all 4 copper ions in Cu_Z are in the reduced Cu(I) state prior to binding of nitrous oxide to two of them. Following a two-electron transfer from Cu_Z, nitrous oxide is subsequently reduced to dinitrogen and water. The re-reduction of the two oxidized copper ions in Cu_Z then occurs via single electron transfer steps from the Cu_A centre, which in turn receives electrons from the natural electron donor, either a cytochrome *c* or a blue copper protein such as azurin or pseudoazurin. Here it should be noted that the inflow of the first electron is kinetically fast and turns the centre into a more or less critical $Cu_Z{}^{o}$ state (1Cu(II), 3Cu(I)). It turned out that reduction of the second copper ion in $Cu_Z{}^{o}$ is relatively slow, while in the absence of electron flow, this mixed valence state even turned slowly into the $Cu_Z{}^*$ state by a so far unknown mechanism preventing full reduction of nitrous oxide reductase by its natural electron donor. Only when excess artificial reductant was used could the enzyme return to its active state. Taking these observations into account it suggests that the reaction centre of nitrous oxide reductase in living cells may turn into an inactive $Cu_Z{}^*$

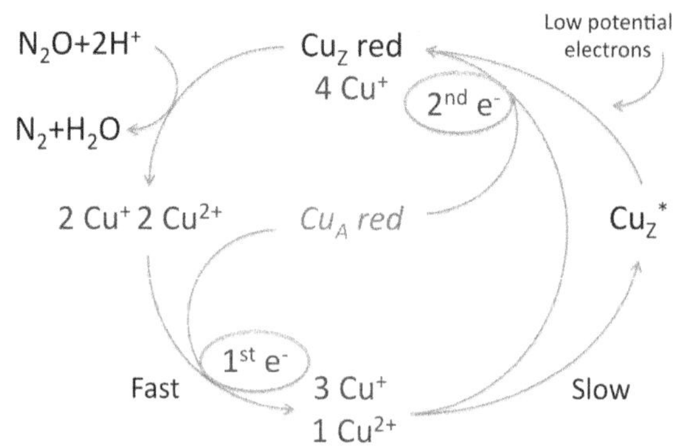

Figure 7.2 Reaction cycle of nitrous oxide reductase (adapted from Dell'acqua *et al.*, 2010). For details, see text. Cu_Z*, resting state of Cu_Z.

state when it encounters an oxygen molecule, or when there is an insufficient supply of the natural electron donors such that the redox machinery becomes incompetent to rapidly and fully reduce the Cu_Z centre. Many denitrifying bacteria may encounter such conditions at the interface of aerobic and anaerobic layers and get occasionally confronted with traces of oxygen or are subjected by deprivation of suitable free energy sources. An essential fast reduction of nitrous oxide reductase may in addition also be prevented by lack of suitable electron donor due to competition with the other enzymes of denitrification. In any case, in order to rescue an already assembled nitrous oxide reductase enzyme it would make sense to mobilize an electron transfer machinery that is able to maintain Cu_Z or to reactivate the Cu_Z* reaction centre via full reduction with low potential electron donors in case the natural electron donors like cytochrome *c* or pseudoazurin are unable to do so. The NosR and NosX proteins may well be proper candidates to make up such an electron donating mechanism. The genes encoding NosR and NosX are usually clustered together in the *nos* operons of many proteobacteria and sometimes even co-transcribed with the *nosZ* gene, suggestive for cooperation and functional importance for NosZ activity. Indeed it has been shown that NosZ function in *Pa. denitrificans* cells depends on both of them and that cells that lack the potential to express either of them express a catalytically inactive nitrous oxide reductase that contains a locked Cu_Z* centre (Zumft, 2005).

NosX

NosX is a small periplasmic protein, homologues of which are widespread in the *nos* gene clusters of the α, β and γ proteobacteria, but not in species of other taxonomic groups. NosX is expressed with a Tat signature in its signal peptide, which strongly suggests that NosX is exported with a redox cofactor (Berks *et al.*, 2000). Loss of the NosX protein from *Pa. denitrificans* resulted in loss of whole-cell N_2O-reducing activity, while the purified reductase was catalytically active *in vitro* and exhibited an increase of the redox-inert, paramagnetic state Cu_z* proportion in NosZ (Wunsch *et al.*, 2005).

NosX is homologous to ApbE, a 36-kDa periplasmic lipoprotein anchored to the inner membrane via a lipid anchor. Although the precise biochemical function of ApbE is not known, mutagenesis studies have indicated that ApbE is important for Fe–S cluster metabolism and its absence in *Salmonella typhimurium* resulted in deficiencies in Fe–S cluster assembly and/or Fe–S cluster repair, which could be suppressed by addition of ferric chloride. More recent studies indicate that ApbE is involved in thiamine biosynthesis in *Sa. typhimurium* (Beck and Downs, 1998, 1999; Han *et al.*, 2006; Skovran and Downs, 2003). Thiamine is an essential cofactor, and ApbE seems to catalyse a crucial step in its synthesis, namely the conversion of aminoimidazole ribotide (AIR), a purine intermediate, to the 4-amino-5-hydroxymethyl-2-methyl pyrimidine (HMP) moiety of thiamine, but the precise role of ApbE is unknown. The crystal structure of ApbE from

Thermotoga maritima has recently come available (Han *et al.*, 2006). The structure includes an unknown ligand that suggests a putative active site location. However, mass spectrometry failed to reveal the precise chemical identity of the ligand, and the ligand electron density was not sufficiently resolved to model any known small molecules with confidence. The structure suggests that the cofactor may be ligated by some amino acids, which are particularly well conserved in homologues of ApbE. Amongst these is a threonine as part of a sequence VATS, which makes part of an FAD binding site in many flavoproteins. Indeed it has been reported that NosX also contains an FAD moiety (Wunsch *et al.*, 2005), suggestive for a role of ApbE and NosX in electron transport.

NosR

NosR is a membrane-bound Fe–S flavoprotein with a large N-terminal periplasmic flavin domain and a hydrophobic domain of five transmembrane helices. The flavin cofactor is presumably bound covalently to an invariant threonine residue as part of a GAT amino acid triad located in the periplasmic domain, which is held in position by the first two helices. The C-terminal region contains four distinct and conserved cysteine patterns in the cytoplasmic regions, two of which are in the loops between the transmembrane helices TM4 and TM6 with a conserved CX3CP sequence that may be in close vicinity in the three dimensional structure of NosR and two of which are located distal and which provide the ligands for 2 [4Fe–4S] clusters. As such, the protein has redox centres positioned at opposite sides of the cytoplasmic membrane. Similarly to the NosX mutations certain mutations in NosR transform the Cu_Z catalytic centre in N_2OR into the Cu_z* state (Wunsch and Zumft, 2005), which strongly indicates interrelated roles for NosX and NosR. Recombinant forms of NosR from *Pseudomonas stutzeri* have been constructed to address the effects of various protein domains on NosZ function and properties. All recombinant NosR proteins allowed the synthesis of NosZ, but a loss of *in vivo* N_2O respiration was observed upon deletion of most of the periplasmic domain or of the C-terminal parts beyond TM2 or upon modification of the cysteine residues in a highly conserved motif, CGWLCP, following TM4. *In vivo* inactive NosZ purified from these recombinant NosR backgrounds contained Cu_Z* and only exhibited *in vitro* catalytic activity.

The periplasmic N-terminal domain of NosR that holds the FMN binding domain resembles to a certain extent the RnfG flavoproteins that make part of a novel, presumably energy-transducing ferredoxin:NAD(+)–oxidoreductase complex composed of at least six different subunits (Biegel *et al.*, 2009; Muller *et al.*, 2008). The Rnf complex is an electron transfer complex with similarity to the Na^+-translocating NADH:quinone-oxidoreductase (Nqr) of, for example, *Vibrio cholerae* (Barquera *et al.*, 2002), and therefore, assumed to be a general coupling site in prokaryotes that connects the pyridine nucleotide pool to ferredoxin (Muller *et al.*, 2008). RnfG is a small hydrophilic protein with a hydrophobic stretch of 30 amino acids at the N-terminus indicating that it may be anchored in the membrane. Covalent binding of FMN to a conserved threonine in RnfG of *Vi. cholerae* has also been demonstrated (Backiel *et al.*, 2008).

The membrane-bound C-terminal region of NosR most likely originates from an ancestor of NapH proteins since it resembles the proteins of the so called NapH/MauM family to a large extent, all of which contain the four distinct and conserved cysteine patterns in the cytoplasmic regions. NapH, together with a periplasmic Fe–S protein designated NapG, has been shown to form a quinol dehydrogenase complex that mediates electron transfer from ubiquinol to the periplasmic nitrate reductase NapAB in *E. coli* (Brondijk *et al.*, 2004). NapG and NapH have similar roles in *Wolinella succinogenes*, but in this bacterium the primary electron donor to the NapGH complex is menaquinol (Simon *et al.*, 2003).

Multidomain proteins are the evolutionary result of gene fusions encoding proteins that physically interact and share functionality. NosR may therefore well be a relatively modern protein, its gene being the likely result of a fusion between ancient *rnfG* and *napH* genes thereby combining the specific functions of separately expressed proteins into the unique properties of a single one. A missing link in the evolution of NosR is the CprC protein (also designated PceC or VcrC), which shares sequence and topological similarities to NosR, although the CprC-like proteins have a smaller periplasmic domain and lack the poly-ferredoxin signature. CprC is co-expressed with the enzymes for halorespiration of *ortho*-chlorophenol in *Desulfitobacterium hafniense* and related species and may well have an electron donating function during catalysis of substrate reduction (Futagami *et al.*, 2006).

All these data add up to the suggestion made by Walter Zumft that the NosR and NosX flavoproteins make up a protein-based transmembrane electron wire (Zumft, 2005). It now seems more than likely that this pathway is recruited for transfer of low potential electrons from quinol to nitrous oxide reductase in order to keep its Cu_Z centre in an active state. The electrons from quinol may be transferred via the NapH domain and the periplasmic flavin domain in NosR to NosX from where the electrons are delivered to nitrous oxide reductase. The possibility that NosX is indeed a redox partner of NosR is further corroborated by the observation that *Verrucomicrobia bacterium Ellin514* has a gene fusion that is composed of NosR and NosX homologues, which is indicative for physical contact and suggests involvement in the same metabolic pathway. Remarkably, the NosX domain of this fusion protein is highly similar to the NosX protein of *Pa. denitrificans*. Perhaps those species that express NosR but do not have a *nosX* gene recruit another periplasmic electron carrier that takes over the role of NosX. A likely candidate is the NosX homologue ApbE, which is widely distributed in bacterial species.

The implications of the proposed scenario are that two electron donating pathways to nitrous oxide reductase operate for optimal performance of the enzyme, one of which is coupled to free energy transduction via the cytochrome bc_1 complex and either a cytochrome *c* or a blue copper protein, the other of which engages the quinol oxidizing NosRX pathway to transfer low potential electrons for Cu_Z maintenance (Fig. 7.3). A balanced but simultaneous distribution of electrons over the quinol and cytochrome *c* oxidizing pathways is essential since a block in either of them prevents nitrous oxide reduction. More specifically, nitrous oxide reductase activity is absent in cells challenged either with inhibitors of the cytochrome bc_1 complex, blocking the cytochrome *c* pathway (unpublished data), or in

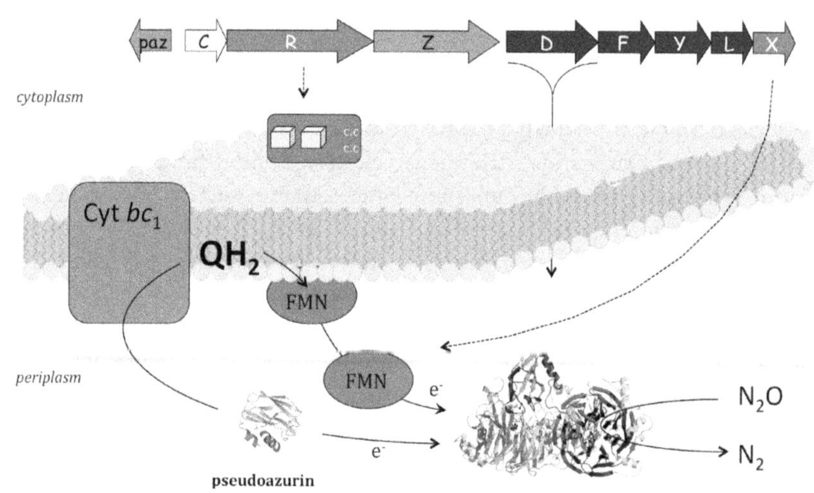

Figure 7.3 The *nosCRZDFYLX* gene cluster from *Paracoccus denitrificans*, *nos* gene products and proposed parallel operating electron transfer pathways from quinol (QH_2) to nitrous oxide reductase, one coupled to free energy transduction via the cytochrome bc_1-complex (Cyt bc_1) and pseudoazurin (encoded by *paz*), the other providing low potential electrons to Cu_Z via NosR and NosX (*nosR* and *nosX* encoded FMN-proteins). *nosZ*, encoding monomers of nitrous oxide reductase; *nosDFYL*, encoding proteins for Cu_Z assemblage in nitrous oxide reductase.

NosRX mutant strains, preventing the operation of the quinol oxidizing pathway.

If the assumption is correct that optimal function of a nitrous oxide reductase requires input of low potential electrons via NosRX bypassing the cytochrome bc_1 complex then one would expect parallel operating electron transfer pathways as suggested for *Pa. denitrificans* and taxonomic neighbours to be present in all species that have the genetic potential for a nitrous oxide reductase. Inspection of available genome sequences revealed that this is not the case. It is striking that NosR proteins are almost exclusively found in the *nos* gene clusters of the α, β and γ proteobacteria only. More specifically, almost all known *nos* gene clusters of these species integrate a *nosR* gene apparently as an essential feature to modern denitrification in specialized denitrifiers, making NosR a relatively recently evolved multifunctional protein. Only *Magnetospirillum magneticum* (but see below) and *Rh. sphaeroides* lack functional *nosRX* genes, but the latter species has a transposon integrated in its *nos* gene cluster and therefore does not require NosR. It would then seem obvious that bacteria other than those belonging to the α, β and γ groups make use of proteins that ensemble an alternative quinol oxidizing electron transfer pathways for fast electron transfer to their corresponding nitrous oxide reductases. Indeed it seems that even 2 types of quinol-oxidizing pathway to nitrous oxide reductase other than the NosRX pathway may have evolved, each of which will be discussed briefly.

Electron transfer pathways from quinol to nitrous oxide reductase

One alternative quinol oxidizing electron transfer pathway to NosZ in ε-proteobacteria may be composed of NosG and NosH proteins, which are paralogues of the NapG and NapH proteins involved in electron transfer from quinol to periplasmic nitrate reductase NapA. A first hint for the involvement of NosG and NosH proteins in electron transfer to nitrous oxide reductase in some species came from the inspection of the genome of the ε-proteobacterium *Wo. succinogenes*. Its *nos* gene cluster does not contain *nosRX* genes, but instead it has homologues of the *napGH* genes along with additional genes encoding small cytochromes *c*. It was suggested that *Wolinella* is able to couple the oxidation of hydrogen to the reduction of nitrous oxide via the quinol oxidizing NosGH couple and bypassing the cytochrome bc_1 complex (Simon *et al.*, 2004). This hypothesis has not been experimentally verified, so the question remains if a parallel operating electron pathway via its cytochrome bc_1 complex has physiological significance as well. A similar make up of the *Wolinella nos* gene cluster is found in all denitrifying ε-proteobacteria. Remarkably, nitrous oxide reductase from all these species is extended with a C-terminal cytochrome *c* domain, which may indicate that indeed a parallel electron transfer pathway to Cu_A and Cu_Z, which engages the cytochrome bc_1 complex and this haem *c* domain in nitrous oxide reductase, operates just as in the other proteobacteria. Apart from the ε-proteobacteria, *nos* gene clusters with *nosGH* genes are found in the α-proteobacterium *Ma. magneticum* AMB-1, the β-proteobacteria *Dechloromonas aromatica* and *Candidatus Accumulibacter phosphatis*, and in one species of the Deferribacteres, *Denitrovibrio acetiphilus*. NosZ proteins in these species cluster phylogenetically with those of the ε-proteobacteria and all have a C-terminal haem *c* domain, which suggests that these species have received their *nos* gene clusters from species of the ε-proteobacteria via relatively recent horizontal gene transfer events.

The *nos* gene clusters of bacteria from taxonomic clades other than the proteobacteria lack *nosRX* genes as well as *nosGH* genes, prompting for a quest to identify a third type of quinol-oxidizing pathway present in these bacteria. Perhaps these bacteria recruit endogenous NapGH proteins for NosZ maintenance, but that seems unlikely since not all of them have the genetic potential to express such proteins. Inspection of their genomes revealed that their *nos* gene clusters consistently integrate a gene, tentatively designated *nosB*, with a so far unknown function but located adjacent to their corresponding *nosZ* gene. The *nos* gene clusters of *Persephonella marina* and *Hydrogenobacter thermophilus*, which belong to the Aquificae, integrate another two genes encoding cytochromes *c* in addition to *nosB*. Copies of the *nosB* gene are also found in *Hydrogenivirga* sp. and *Sulfurihydrogenibium* sp., but their *nos* gene clusters are truncated and do not contain full sized *nosZ* genes. In addition to the Aquificae, the *nosB* gene is found in the *nos* gene clusters of δ- and ε-proteobacteria, and in members of virtually all remaining taxonomic clades including the Archaea and Firmicutes (note: *nos* genes are absent in the Actinobacteria), but not in the α-, β- and γ-proteobacteria with the exceptions mentioned above. NosB is a hydrophobic protein of about 200 amino acids. It has been notated in GenBank as a membrane protein, in accordance with some limited similarity of the N-terminus with helices 10, 11, and 12 of cytochrome c oxidases, although the conserved histidines for ligation of the haems are absent. The C-terminus may then be periplasmic as suggested by the observation that in some species (*Opitutaceae bacterium, Opitutis terrae*) NosB is equipped with an RR containing signal sequence. It would even suggest periplasmic transport of that domain in the folded state via the TAT translocon implicating that a cofactor has been inserted into it. Alignment of NosB homologues revealed 2 regions with highly conserved stretches of amino acid. They are located at positions 60 (LNHYIGM), and 125 (YGHN) relative to the *Hy. thermophilus* sequence. The conservation of histidines and the methionine suggests a role in metal or cofactor ligation. In yet other species NosB is fused to a haem *c* domain (*Sulfurimonas denitrificans, Nitratiruptor, Wo. succinogenes*), which is indicative for a role of NosB in electron transport. Taken these data together, it might well be that the NosB protein in these species mediates electron transfer from a postulated quinol binding site in its membrane domain via a dedicated cytochrome *c* to NosZ for proper Cu_Z turnover. Remarkably, NosB in *Dyadobacter fermentans* is fused to the copper binding protein NosL, suggesting that NosB has a combined function not only in electron transport to NosZ but also in copper management.

In summary, and with the assumption that NosZ function depends in part on a quinol dependent electron transfer pathway to prevent the Cu_Z centre from entering a locked-state, it is now proposed that three different systems have independently evolved to realize this activity in different clades of Bacteria and Archaea. One is an old and widely distributed pathway that engages a postulated NosB protein along with a dedicated cytochrome *c*. The second is specifically evolved in the ε-proteobacteria and made up from NosGH proteins and dedicated cytochromes *c*. The third, and most modern Cu_Z maintenance system is exclusively present in the α-, β- and γ-proteobacteria and makes use of the NosRX pathway for transport of low potential electrons to the reaction centre of NosZ. In this scenario it must then be that the ε-proteobacteria bridge the evolution of *nos* gene clusters between the ancient (Archaea and Firmicutes) and modern (α-, β-, and γ-proteobacteria) prokaryotes that respire nitrous oxide, since they contain both the NosB and NosGH proteins. Ancestors of them may then have lost the *nosB* gene, and further developed ancestral NosH to present day NosR.

Just as in *Pa. denitrificans* may a combined and compulsory action of two parallel operating pathways, one engaging the cytochrome bc_1 complex for steady state electron transport coupled to free energy transduction and the other engaging a quinol dependent pathway for Cu_Z maintenance, be a common design in all species that perform nitrous oxide reduction. The *nos* gene clusters of *Anaeromyxobacter* species (δ-proteobacteria) and of *Desulfitobacterium hafniense* (Clostridia), species from two taxonomically distinct clades, for example, contain a *nosB* gene and they are all clustered with genes encoding a cytochrome bc_1 complex, suggestive for co-expression and operation of both electron transfer pathways to NosZ in these species. In yet other species, such as the Archaea, *nos* gene clusters are associated with a gene encoding a blue copper protein that is likely to mediate steady state electron transfer from the cytochrome bc_1 complex to nitrous oxide reductase in addition to the parallel quinol oxidizing pathway via NosB and a cytochrome *c*.

NosR for NosZ maintenance, NirI for NirS maintenance

Pa. denitrificans and some related species of the α-proteobacteria also express paralogues of the NosR and NosX proteins, which are designated NirI and NirX, respectively, and found clustered with the *nir* genes encoding the cd_1-type nitrite reductase (Saunders *et al.*, 1999). Perhaps these proteins are specifically required to make up a quinol depending electron transfer pathway in order to prevent their respective nitrite reductases to end up in a locked state just as has been observed for nitrous oxide reductase. It has been shown, that haem-ligand switches occur during the catalytic cycle of NirS that affect the redox potential of the haems. This observation suggests a mechanism by which internal electron transfer is regulated in order to stream the redox energy into conformational changes (Williams *et al.*, 1997), perhaps to recycle the enzyme from resting state to active state. The phenomenon of resting states in respiratory enzymes is not new. It is noteworthy that the reaction cycle of cytochrome *c* oxidase also involves a main fast catalytic cycle with active state intermediates of the reaction centre and a slow parallel cycle involving a so-called locked, pulsed or resting state (Wikstrom and Verkhovsky, 2002). If the oxidized enzyme is allowed to relax in the absence of sufficient electron donor, the binuclear haem–copper centre attains a state that lies outside the main catalytic cycle. Reduction of this form of the enzyme is not linked to proton translocation, but is necessary for a return to the main cycle.

An alternative nitrous oxide reductase

Py. aerophilum has the potential to reduce nitrite to NO using menaquinol as electron donor (de Vries and Schroder, 2002) and even displays a low and variable reduction of nitrite all the way to gaseous NO, N_2O, and N_2, indicating that it has a full operating denitrification pathway (Cozen *et al.*, 2009; de Vries and Schroder, 2002). The *Py. aerophilum* genome contains genes for a NarG-type nitrate reductase, a primitive cd_1-type nitrite reductase, and a quinol oxidizing nitric oxide reductase. However, the *Py. aerophilum* genome lacks a recognizable homologue of nosZ, which encodes nitrous oxide reductase in bacteria. Expression studies during growth on nitrate, however, revealed specific upregulation of a multicopper oxidase with a predicted N-terminal twin-arginine translocation signal, a Cu-2-type centre similar to those in copper-containing nitrite reductases, and a mononuclear C-terminal Cu-1-type centre (Cozen *et al.*, 2009). On the basis of these findings, the authors suggested that this multicopper oxidase is a likely candidate for nitrous oxide reduction in Py. aerophilum. A BLAST search did not reveal orthologues of its allocated gene in the other *Pyrobaculum* or *Thermoproteales* genomes, while the closest orthologues encode bilirubin- and multicopper oxidases in diverse types of bacteria including *Nitrosomonas* and *Nitrosospira*. The authors also recognized similarity between the gene encoding the multicopper oxidase in *Py. aerophilum* and genes in *Nitrosomonas* species, which also reduce nitrous oxide and lack *nosZ*-type nitrous oxide reductase genes, which would support the hypothesis that these genes encode alternative nitrous oxide reductases.

Regulation of transcription of the denitrification genes

Denitrification is a form of respiration during which electron transfer to the N-oxides is coupled to the generation of a proton motive force across the cytoplasmic membrane of the cell. But many if not all denitrifying bacteria are facultative aerobic chemoorganotrophic organisms and their respiratory networks are designed such that electrons flow to oxygen under aerobic growth conditions, as it is the most efficient route with respect to free energy transduction. Due to the organization and structural features of the denitrification enzymes the maximum efficiency of free energy transduction during denitrification is only 60% of that during aerobic respiration where oxygen is the terminal electron acceptor as will be briefly explained. An important parameter is the number of net positive charges that are moved outwards from the cytoplasm to the periplasm for each pair of electrons moving from an ubiquinol to a terminal e^- acceptor (Richardson *et al.*, 2005; Simon *et al.*, 2008). In the case of oxygen, the effective stoichiometry is 10 protons transferred per two electrons transferred from NADH to oxygen (four by NADH dehydrogenase, two by the bc_1-complex, and another 4 by the aa_3-type cytochrome *c* oxidase). In the case of nitrate respiration, a two electron transfer from NADH to the nitrate reductase not only bypasses the bc_1-complex, but also the net proton translocation by membrane-bound

NAR is only two making up a stoichiometry of six protons moved outside the cell per two electrons transferred. Further, a two-electron transfer from NADH to any of the other three reductases engages NADH dehydrogenase (four protons per two electrons) and the bc_1 complex (two protons per two electrons), and since all three enzymes themselves are non-electrogenic these pathways have also an overall stoichiometry of six proton movements per two electrons transferred. This means that the ATP yield per $2e^-$ flowing to a terminal acceptor from any given reductant available to the cells will be 40% lower for denitrification as compared to aerobic respiration, and hence the regulation of expression of either type of respiration occurs according to an energetic hierarchy explaining why oxygen respiration is preferred over denitrification. Current knowledge about well-studied denitrifiers such as *Pa. denitrificans, Rhodobacter, Bradyrhizobium* and *Pseudomonas* species are paradigms for a fundamental understanding of transcription regulation during the switch from aerobic respiration to denitrification. In all species it seems that the onset of denitrification is triggered by oxygen depletion and nitrate availability (Baker *et al.*, 1998; Zumft, 1997). Also on the metabolic level there is rapid cessation of denitrification when denitrifying cells are confronted with free oxygen. For example, when a suspension of cells of *Pa. denitrificans* growing under anaerobic, denitrifying conditions is exposed to O_2 there is usually an immediate inhibition of denitrification and the electrons flowing through the ubiquinone/ubiquinol couple is then diverted to standby oxidase activities. There is some evidence that the presence of oxygen blocks the transport of nitrate across the cytoplasmic membrane and thus restricts the activity of the membrane-bound nitrate reductase and subsequent steps of denitrification (Ferguson, 1994).

Oxygen and nitrate are not the only signals for cells to decide whether or not to proceed with denitrification. The fact that the reduction steps during denitrification are sequential implies that the product of the one enzyme is a substrate for the next one. Fine-tuned regulation of the concentration and activity of the denitrification enzymes is therefore required in order to keep the free concentrations of nitrite and nitric oxide (NO) below cytotoxic levels. In physiological growth experiments with *Pa. denitrificans* cells, it was noted that the nitrate and nitrous oxide reductases are the first enzymes that are induced in response to oxygen depletion and nitrate availability, while the nitrite and NO- reductases were recruited significantly later (Baumann *et al.*, 1996). Apparently, the nitrite and NO reductases are co-ordinately expressed and their concentrations and activities tuned such that the free nitrite concentration drops to the micromolar range while NO is kept in the nanomolar range. It also indicates that specific regulators other than possible shared ones control the sequential temporal expression of the denitrifying enzymes. The molecular basis of the regulatory networks in denitrifying species that control the oxic–anoxic shift as well as the fine-tuning of transcriptional activation is beginning to emerge. Not surprisingly, some of the key molecules that act as signals to these regulation pathways are nitrite and NO in addition to oxygen and nitrate. It has now become evident that more than one type of regulatory protein is involved in sensing each of these key molecules. Analogous systems for detecting similar signals in denitrification are used in different combinations in denitrifying species. In addition to the regulatory proteins that can monitor these key molecules, regulation of the on-set and fine-tuning of denitrification in some denitrifiers involves copper responsive regulators, redox sensing mechanisms and the NirI and NosR proteins.

Although there does not seem to be a universal mode of regulation of denitrification in different denitrifiers, these analogous networks apparently operate sufficiently enough such that homeostasis of nitrite and NO concentrations is realized in all these different types of cell.

Oxygen sensors

The two most important types of O_2 sensor involved in regulation of denitrification are FixL and FNR.

FixL is a membrane-bound O_2 sensor found in *Bradyrhizobium japonicum* and related Rhizobia species. Together with its cognate response regulator FixJ, these proteins belong to the group of two-component regulatory systems. The N-terminal domain of FixL contains a PAS domain with an O_2 responsive haem group. This is an independent domain found in the PAS (Per-ARNT-Sim) superfamily of proteins of about 100 amino acids, suitable for transmitting a signal from the receptive site to other domains of the same protein or to partner proteins through proper conformational changes. Despite the low sequence similarity of PAS domains, a high conservation in the fold and topology of the known structures suggests a strong evolutionary conservation of some functional features. The C-terminal domain of FixL has histidine kinase activity for autophosphorylation and subsequent P-group transfer to FixJ, which is the response regulator that activates transcription of its target genes. Binding of oxygen to the haem in FixL switches the iron from high to low spin, resulting in a conformational change, which inactivates the kinase activity and terminates the cascade.

FNR is an oxygen responsive transcription regulator that belongs to an expanding family of Crp/FNR like transcriptional activators (Spiro, 1994). All these regulators contain a signal sensing domain, a dimerization domain, a helix–turn–helix DNA-binding domain, and up to three sites that are involved in contacting RNA polymerase. FNR of *E. coli* contains a sensory domain with four conserved cysteine residues that ligate an oxygen sensitive [4Fe–4S] cluster. This cluster disintegrates by direct interaction with oxygen rendering FNR inactive. Once the oxygen concentration drops below threshold levels, the cluster is assembled. This event triggers a conformational change, resulting in dimerization of FNR. The dimer binds with its DNA binding helix–turn–helix motif to a conserved target sequence TTGAT-N4-ATCAA (FNR-box). The DNA binding motif contains conserved glutamate and serine residues, which make contact with the thymidine and guanine residues of the FNR-box half-site, 5′-TTGAT-3′, at position 1 and 3, respectively. The FNR-box is usually located directly upstream of the RNA-polymerase binding site to facilitate contact between FNR and RNA polymerase. The physical contact between the dimer and RNA-polymerase on target promoters initiates a change in the geometry of the ternary transcription initiation complex, which results in transcription of the downstream gene. Homologues of FNR are widespread in Nature and have been encountered in a variety of prokaryotes that have to cope with changes in oxygen availability, including pathogens. Only some of these protein members have cysteine signatures diagnostic for binding an oxygen sensitive 4Fe–4S cluster. Most of them lack the cluster, indicating that these FNR-homologues have different sensing domains and respond to signals other than oxygen.

Nitrate and nitrite sensors

Three types of nitrate/nitrite sensing systems have been characterized in denitrifying species, NarXL, NarQP and NarR. NarXL and NarQP are members of two component regulatory systems (Unden *et al.*, 1995). The NarX and NarQ proteins are the signal sensors, which both respond to nitrate and nitrite although with different affinities. NarX is more specific for nitrate and NarQ for nitrite. The NarL and NarP proteins are their cognate response regulators, respectively. In *E. coli*, they bind DNA to control induction of the genes encoding membrane-bound and periplasmic nitrate reductases and repression of genes encoding alternative anaerobic respiratory enzymes. The NarX and NarQ proteins communicate with both the NarP and NarL proteins in *E. coli*. Denitrifiers from the β- and γ-proteobacteria have genes encoding orthologues of NarXL. Their products regulate a nitrate-induced expression of Nar. Expression of the *Neisseria* nitrite- and NO-reductase genes are subjected to a nitrite response mediated by a NarQP couple.

NarR is a member of the FNR family of transcriptional activators, but it lacks the cysteines to incorporate a [4Fe–4S] cluster. NarR of *Paracoccus pantotrophus* (Wood *et al.*, 2001) and *Pa. denitrificans* (van Spanning, unpublished) is specifically required for transcription of the *narKGHJI* genes most likely in response to nitrite. The mechanism of the response is not clear, but it is notable that NarR can also be activated by azide, which normally binds to metal centres, raising the possibility that NarR is a metalloprotein. Genes encoding NarR are exclusively found in the α-proteobacteria usually adjacent and divergently transcribed from their *narGHJI* gene clusters such as noted in *Brucella suis*, *Brucella melitensis*, *Pa. denitrificans*, and *Pa. pantotrophus*. There are no indications that these organisms have counterparts of *narXL* and therefore NarR might substitute for the NarXL system in the α-proteobacteria. The implication of regulation by a nitrite sensor is that the onset of denitrification in these bacteria is based on a product induction mechanism rather than on a substrate induction mechanism as mediated by the NarXL nitrate sensors. Hence, every molecule of nitrite that is formed by low but existing amounts of Nar or Nap will then induce the NarR mediated synthesis of more nitrate reductase molecules resulting in an enhanced catalysis of nitrate reduction and nitrite formation and progressive increase of Nar expression. A product induction mode of regulation is decision driven and apparently characteristic for the α-proteobacteria, which are specialized in denitrification. The rationale behind this response is apparently to switch on the denitrification apparatus rapidly but only then when oxygen has been fully consumed as to swiftly proceed with respiration on the one hand, yet to prevent the simultaneous operation of nitrate and oxygen respiration on the other hand. The latter situation might be detrimental to these cells since oxygen radicals may react with nitric oxide to peroxynitrite, which is extremely reactive towards biomolecules.

Nitric oxide sensors

As yet, three different types of NO-responsive transcriptional regulators are characterized in denitrifying species and species that are confronted with nitric oxide production in their natural environment. These are NNR (sometimes referred to as NnrR, DNR, or DnrD), NsrR and NorR, all of which have a more or less dedicated task in the coordinate transcription of *nir* and/or *nor* gene clusters in response to NO. Their structural and functional features along with their target operons and prevalence in different species is described in Chapter 9. Here it is notable to refer to an earlier observation by Zumft, who noticed that NO responsive DNR proteins that direct transcription of genes encoding the NirK-type nitrite reductase in the proteobacteria are homologous to but clearly distinct from those that control expression of the *nirS* locus. This observation matches well with the assumption of horizontal gene transfer of either the *nirS* locus or the *nirK* gene in between taxonomically distinct species. For understanding of this it is notable that the genes encoding either type of DNR protein are usually found in the close proximity of their target genes, perhaps for efficient and/or specific gene regulation but also allowing efficient horizontal gene transfer of both the structural gene(s) along with their regulator. In that likely scenario it is then obvious that the evolution of either type branches in two separate groups but still retaining the NO-responsive property.

Redox sensors

The RegAB proteins control denitrification processes both in *Rh. sphaeroides* and in *Ag. tumefaciens* where they regulate the activity of the *nirK* promoter, which expresses a copper containing nitrite reductase. They do that in both species in a concerted action with the nitric oxide sensor NnrR (Baek *et al.*, 2008; Laratta *et al.*, 2002).

The RegBA proteins of *Rh. capsulatus* and *Rh. sphaeroides* (also referred to as PrrBA, RegSR, ActSR or RoxSR) make up a two-component regulatory system that responds to the cellular redox state and controls the expression of many target genes involved in free energy transduction such as photosynthesis, carbon fixation, nitrogen fixation, hydrogen utilization, aerobic and anaerobic respiration, denitrification, electron transport and aerotaxis (Wu and Bauer, 2008). In addition to controlling expression of respiratory enzymes, RegBA also controls the genes encoding their cofactors like haem *c* (Smart *et al.*, 2004). RegAB proteins in *Ps. putida* coordinate expression of proteins involved in sugar and amino acid metabolism and the sulfur starvation response and elements of the respiratory chain (a cbb_3-type cytochrome oxidase, Fe–S clusters, and cytochrome c-related proteins) and genes participating in the maintenance of the redox balance in response to redox signalling. In addition they are required for the cell density-dependent gene *ddcA*, involved in bacterial colonization of plant surfaces (Fernandez-Pinar *et al.*, 2008).

The redox signal that is detected by the membrane-bound sensor kinase, RegB, originates from the ubiquinone pool in the membrane which is therefore in turn strongly affected by the electron flux through the respiratory network coupled to the generation of a proton motive force. A highly conserved quinone binding site, a heptapeptide sequence of GGXXNPF, is located in the membrane-spanning domain of RegB (Swem *et al.*, 2006). Binding of quinone to this site renders RegB inactive perhaps by allosteric changes resulting in the inability of autophosphorylation. Thus any change in electron in- and/or output rate that would lower the quinone concentration would make RegB increasingly active for RegA mediated activation or repression of target genes. Such a situation where high levels of phosphorylated RegA are to be expected may occur for example when oxygen levels are low or when the free energy source is highly reduced, or when possible redox balancing mechanisms are not properly expressed. This is consistent with the observation that mutations in the *cbb*$_3$-type oxidase result in enhanced expression of RegAB controlled genes in *Rh. capsulatus*, even under aerobic growth conditions (Eraso and Kaplan, 1994). The cytosolic domain of RegB contains the autophosphorylation site (H-box) along with the ATP-binding domain (Chen *et al.*, 2000).

Regulation of RegB autophosphorylation also involves a conserved redox box with a cysteine that is present in the cytosolic region of RegB (Elsen *et al.*, 2004). This redox-active cysteine is prone to formation of an intermolecular disulfide bridge between two RegB dimers in response to the redox state, resulting in the formation of inactive tetramers (Swem *et al.*, 2003). The mechanism for disulfide bridge formation or breakage is not known, but functions independent of redox regulation via quinone binding to RegB and perhaps integrates two types of redox signal. As a result, the (integrated) signal as sensed by RegB is passed on via its kinase activity to RegA, which activates or represses a plethora of target genes. It does that by binding to a weakly conserved RegA-box, which conforms to the sequence GCG-Xn-CGC, where Xn represents a variable number of nucleotides, usually 9 (Laguri *et al.*, 2003).

Under anaerobic conditions, RegA acts as a repressor of the *dor* operon encoding DMSO reductase in *Rh. capsulatus* but only when the cells grow on malate. However, repression is abolished when the cells switch to pyruvate as carbon and free energy source (Kadenbach, 1996), which might well be a consequence of a change in redox state of the Q-pool due to the type of free energy source.

Highly conserved homologues of RegB and RegA have been found in a wide number of photosynthetic and non-photosynthetic bacteria with evidence suggesting that RegB/RegA have a fundamental role in the transcription of redox-regulated genes in many bacterial species (Wu and Bauer, 2008). Apparently, the RegAB signal transduction network has been universally developed to monitor a cellular redox state on the level of the quinone/quinol pool in addition to those involving oxygen sensors. To a certain extent, the redox state of the Q-pool reflects changes in environmental oxygen tension, but other factors contribute to that as well since the steady state concentrations of quinone and quinol are determined by electron in- and output rates, which in turn are affected by the concentrations and activities of electron donating dehydrogenases and electron consuming oxido-reductases. It may well be essential for all living cells to continuously monitor the redox state of the Q-pool since it reflects the reduction state of all respiratory components and hence provides the means by regulated expression via a global RegAB regulatory circuit to adapt to situations that interfere with cellular redox homeostasis. These situations could well be detrimental to the cell as a result of unwanted chemical reactions of the participating metalloenzymes with oxygen or other oxidants that could lead to reactive oxygen species. In that perspective it is likely that the RegAB proteins communicate with other regulatory circuits that control expression of metabolic enzymes whose activities affect the cellular redox state. Indeed it has been shown that RegA of *Rh. capsulatus* interacts with the response regulator NtrX of the NtrY/NtrX two-component system, which is believed to be involved in regulation of nitrogen fixation genes encoding nitrogenase (Gregor *et al.*, 2007). The activity of nitrogenase demands a subtle balance between low oxygen on the one hand and ATP requirement on the other hand, and apparently this type of physical interaction connects the regulatory systems for redox control and nitrogen control to meet these demands.

It has also been demonstrated that RegB binds a single copper ion but the way in which this is connected to the redox activity of this cysteine residue is not clear (Swem *et al.*, 2003). Perhaps this property relates the function of the RegAB couple to that of the universal copper chaperone SenC. In *Rh. sphaeroides* and in some other proteobacteria (for example *Pa. denitrificans*, *Roseobacter denitrificans*, *Silicibacter lacuscaerulensis*, *Dinoroseobacter shibae*), the *senC* gene is sandwiched in between the *regA* and *regB* genes, suggestive for co-expression and co-functionality of SenC with the RegAB couple.

CopY, a copper-responsive transcription regulator

CopY is a repressor of the *cop* operon in many organisms amongst which is *Enterococcus hirae* (Solioz and Stoyanov, 2003). Intracellular copper routing in this species is accomplished by the CopZ copper chaperone. Under copper stress, CopZ donates Cu(I) to the CopY repressor, thereby releasing its bound zinc and abolishing repressor–DNA interaction. This in turn induces the expression of the *cop* operon, which encodes CopY and CopZ, in addition to two copper ATPases, CopA and CopB. Since CopZ physically interacts with CopA, it has been suggested that CopZ scavenges copper from the cytoplasm and passes it on to the copper ATPase, CopA. CopZ also interacts with a protein designated Gls24, a member of a family of stress response regulators, which finding links copper homeostasis to stress adaptation (Stoyanov *et al.*, 2010). A *copY* gene precedes the *nos* operons in the Archaeal bacteria *Ha. borinquense*, *Ha. marismortui* and *Halorubrum lacusprofundi*, suggesting that CopY unlocks the *nos* promoters in these species in response to copper availability.

NirI and NosR, regulators of expression of nitrite and nitrous oxide reductases

NosR seems to have a dual functionality. Apart from having a role in maintenance of the Cu_Z centre in N_2O reductase, the protein is also essential for its expression as judged by the fact that *nosZ* transcripts are missing in NosR mutants of *Ps. stutzeri*, *Sinorhizobium meliloti* and *Pa. denitrificans* or halved in the NosR mutant of *Ps. aeruginosa*. In these species, NosR expression in turn is controlled by an NO-sensing NnrR orthologue, and in *Pa. denitrificans* both by FnrP and NnrR (Fig. 7.4). NirI, a paralogue of NosR in *Pa. denitrificans* has a specific role in transcription of the *nir* gene cluster encoding cd_1-type nitrite reductase in response to oxygen limitation and the presence of N-oxides (Saunders *et al.*, 1999). The *nirI* gene is in a two-gene operon together with *nirX* (encoding a NosX homologue) just upstream the *nirS* gene cluster. Transcription of the *nirIX* gene cluster is controlled by NnrR. NirI mutants were unable to express NirS. Remarkably, attempts to complement a NirI mutation *in trans* failed and were only successful when the *nirI* gene was reintegrated into the chromosome of the NirI-deficient mutant via homologous recombination in such a way that the wild-type *nirI* gene was present directly upstream of the *nir* operon. This suggests a transcriptional and translational coupling of NirI and its target promoter. The background of this functionality is far from clear since on the one hand NosR and NirI do not have signatures diagnostic for binding to DNA, and on the other hand they are membrane-bound proteins, which makes them unlikely candidates to interfere directly with the DNA (although DNA binding of membrane-bound proteins has been shown in exceptional cases). Just like as has been shown for NirI and its target promoter in *Pa. denitrificans*, a site specific position of its *nosR* gene relative to the *nosZ* promoter is also essential (R. van Spanning, unpublished data), indicating that both regulatory pathways share common features of unknown origin in transcription regulation of their respective targets.

Sigma factors?

Three FNR-type transcriptional regulators FnrP (an oxygen sensor), NarR (a nitrite sensor) and NNR (a NO sensor) orchestrate the transcriptional changes during the shift from aerobic respiration to denitrification in *Pa. denitrificans*. Although the three FNR orthologues are highly homologous to one another, they display a remarkable degree of specificity towards their target promoters. In spite of these specificities, their binding sites on the DNA resemble each

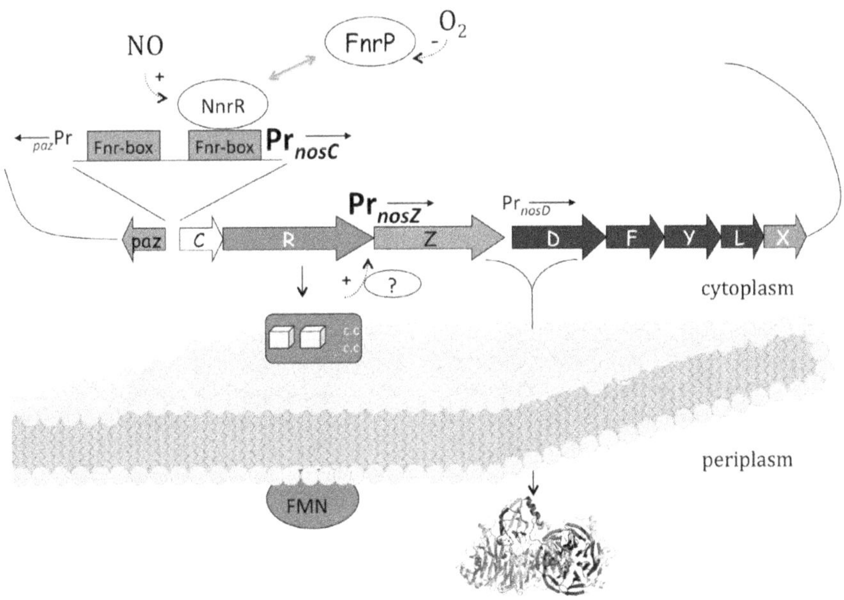

Figure 7.4 Model for the stepwise regulation of the *nos* gene cluster of *Paracoccus denitrificans*. Pr, promoter with the direction of the arrows pointing to the direction of transcription; +, stimulation; –, inhibition; NO, nitric oxide;?, unknown mechanism. For further details, see legend to Fig. 7.3.

other to a large extent and are sometimes even identical (Van Spanning *et al.*, 1997). These sites are correctly positioned relative to the RNA-polymerase binding site as judged from transcript start analyses. Perhaps subtle differences in the binding strength between the regulator and the DNA underlie the observed specificity. Alternatively, the FNR family members interact with specific yet unknown transcription factors, perhaps a dedicated sigma factor (Veldman *et al.*, 2006).

Concluding remarks

Specialized denitrifiers in the class of α-proteobacteria such as *Pa. denitrificans, Ro. denitrificans, Si. lacuscaerulensis, Di. shibae*, the denitrifying *Pseudomonades* from the γ-proteobacteria and many of the ε-proteobacteria and Aquificae contain gene clusters with a total of over 50 genes, which encode the proteins that make up a full denitrification pathway. These proteins include the structural enzymes and cognate electron donors, transporters, dedicated transcription regulators as well as many accessory genes required for assembly, cofactor synthesis and insertion into the enzymes. In contrast, opportunistic denitrifiers such as pathogens and many members of the *Flavobacteriacea* respire more simply with a copper-type nitrite reductase and quinol dependent nitric oxide reductase, the expression and activity of which requires only a few genes. Recent years have seen the emergence of overwhelming structural and genome information that allows a much better insight into the functional aspects, regulation, phylogenetic distribution and evolution of the denitrification enzymes. Of these, the cd_1-type nitrite reductases and nitrous oxide reductases are special in that they share a β-propeller fold. They evolved independently, the first with a complex ligated iron in the catalytic centre, the second with an even more complex reaction centre with four bridged copper ions. In the specialized denitrifiers, they meet again.

References

Abajian, C., and Rosenzweig, A.C. (2006). Crystal structure of yeast Sco1. J. Biol. Inorg. Chem. *11*, 459–466.

Amo, T., Pajé, M.L., Inagaki, A., Ezaki, S., Atomi, H., and Imanaka, T. (2002). *Pyrobaculum calidifontis* sp. nov., a novel hyperthermophilic archaeon that grows in atmospheric air. Archaea *1*, 113–121.

Andruzzi, L., Nakano, M., Nilges, M.J., and Blackburn, N.J. (2005). Spectroscopic studies of metal binding and metal selectivity in *Bacillus subtilis* BSco, a homologue of the yeast mitochondrial protein Sco1p. J. Am. Chem. Soc. *127*, 16548–16558.

Arnesano, F., Balatri, E., Banci, L., Bertini, I., and Winge, D.R. (2005). Folding studies of Cox17 reveal an important interplay of cysteine oxidation and copper binding. Structure *13*, 713–722.

Backiel, J., Juarez, O., Zagorevski, D.V., Wang, Z., Nilges, M.J., and Barquera, B. (2008). Covalent binding of flavins to RnfG and RnfD in the Rnf complex from *Vibrio cholerae*. Biochemistry *47*, 11273–11284.

Badrick, A.C., Hamilton, A.J., Bernhardt, P.V., Jones, C.E., Kappler, U., Jennings, M.P., and McEwan, A.G. (2007). PrrC, a Sco homologue from *Rhodobacter sphaeroides*, possesses thiol-disulfide oxidoreductase activity. FEBS Lett. *581*, 4663–4667.

Baek, S.H., Hartsock, A., and Shapleigh, J.P. (2008). *Agrobacterium tumefaciens C58* uses ActR and FnrN to control *nirK* and *nor* expression. J. Bacteriol. *190*, 78–86.

Baker, S.C., Ferguson, S.J., Ludwig, B., Page, M.D., Richter, O.M., and van Spanning, R.J.M. (1998). Molecular genetics of the genus *Paracoccus*: metabolically versatile bacteria with bioenergetic flexibility. Microbiol Mol. Biol. Rev. *62*, 1046–1078.

Baker, S.C., Saunders, N.F., Willis, A.C., Ferguson, S.J., Hajdu, J., and Fulop, V. (1997). Cytochrome *cd*1 structure: unusual haem environments in a nitrite reductase and analysis of factors contributing to beta-propeller folds. J. Mol. Biol. *269*, 440–455.

Banci, L., Bertini, I., Ciofi-Baffoni, S., Gerothanassis, I.P., Leontari, I., Martinelli, M., and Wang, S. (2007). A structural characterization of human SCO2. Structure *15*, 1132–1140.

Banci, L., Bertini, I., Ciofi-Baffoni, S., Hadjiloi, T., Martinelli, M., and Palumaa, P. (2008). Mitochondrial copper(I) transfer from Cox17 to Sco1 is coupled to electron transfer. Proc. Natl. Acad. Sci. U.S.A. *105*, 6803–6808.

Barquera, B., Hellwig, P., Zhou, W., Morgan, J.E., Hase, C.C., Gosink, K.K., Nilges, M., Bruesehoff, P.J., Roth, A., Lancaster, C.R., *et al.* (2002). Purification and characterization of the recombinant Na(+)-translocating NADH:quinone oxidoreductase from *Vibrio cholerae*. Biochemistry *41*, 3781–3789.

Baumann, B., Snozzi, M., Zehnder, A.J.B., and Vandermeer, J.R. (1996). Dynamics of denitrification activity of *Paracoccus denitrificans* in continuous culture during aerobic-anaerobic changes. J. Bacteriol. *178*, 4367–4374.

Beck, B.J., and Downs, D.M. (1998). The *apbE* gene encodes a lipoprotein involved in thiamine synthesis in *Salmonella typhimurium*. J. Bacteriol. *180*, 885–891.

Beck, B.J., and Downs, D.M. (1999). A periplasmic location is essential for the role of the ApbE lipoprotein in thiamine synthesis in *Salmonella typhimurium*. J. Bacteriol. *181*, 7285–7290.

Beinert, H. (1995). Crystals and structures of cytochrome *c* oxidases – the end of an arduous road. Chem. Biol. *2*, 781–785.

Berks, B.C., Ferguson, S.J., Moir, J.W.B., and Richardson, D.J. (1995). Enzymes and associated electron transport systems that catalyse the respiratory reduction of nitrogen oxides and oxyanions. Bba-Bioenergetics *1232*, 97–173.

Berks, B.C., Sargent, F., and Palmer, T. (2000). The Tat protein export pathway. Mol. Microbiol. *35*, 260–274.

Biegel, E., Schmidt, S., and Muller, V. (2009). Genetic, immunological and biochemical evidence for a Rnf complex in the acetogen *Acetobacterium woodii*. Environ. Microbiol. *11*, 1438–1443.

Boal, A.K., and Rosenzweig, A.C. (2009). Structural biology of copper trafficking. Chem. Rev. *109*, 4760–4779.

Brondijk, T.H., Nilavongse, A., Filenko, N., Richardson, D.J., and Cole, J.A. (2004). NapGH components of the periplasmic nitrate reductase of *Escherichia coli K-12*: location, topology and physiological roles in quinol oxidation and redox balancing. Biochem. J. *379*, 47–55.

Brown, K., Djinovic-Carugo, K., Haltia, T., Cabrito, I., Saraste, M., Moura, J.J., Moura, I., Tegoni, M., and Cambillau, C. (2000a). Revisiting the catalytic Cu_Z cluster of nitrous oxide (N_2O) reductase. Evidence of a bridging inorganic sulfur. J. Biol. Chem. *275*, 41133–41136.

Brown, K., Tegoni, M., Prudencio, M., Pereira, A.S., Besson, S., Moura, J.J., Moura, I., and Cambillau, C. (2000b). A novel type of catalytic copper cluster in nitrous oxide reductase. Nat. Struct. Biol. *7*, 191–195.

Cavalier-Smith, T. (2002). The neomuran origin of archaebacteria, the negibacterial root of the universal tree and bacterial megaclassification. Int. J. Syst. Evol. Microbiol. *52*, 7–76.

Chen, W., Jager, A., and Klug, G. (2000). Correction of the DNA sequence of the *regB* gene of *Rhodobacter capsulatus* with implications for the membrane topology of the sensor kinase RegB. J. Bacteriol. *182*, 818–820.

Ciccarelli, F.D., Copley, R.R., Doerks, T., Russell, R.B., and Bork, P. (2002). CASH – a beta-helix domain widespread among carbohydrate-binding proteins. Trends Biochem. Sci. *27*, 59–62.

Cozen, A.E., Weirauch, M.T., Pollard, K.S., Bernick, D.L., Stuart, J.M., and Lowe, T.M. (2009). Transcriptional map of respiratory versatility in the hyperthermophilic crenarchaeon *Pyrobaculum aerophilum*. J. Bacteriol. *191*, 782–794.

Crichton, R.R., and Pierre, J.L. (2001). Old iron, young copper: from Mars to Venus. Biometals *14*, 99–112.

Dawson, R.J., and Locher, K.P. (2007). Structure of the multidrug ABC transporter Sav1866 from *Staphylococcus aureus* in complex with AMP-PNP. FEBS Lett. *581*, 935–938.

de Vries, S., and Schroder, I. (2002). Comparison between the nitric oxide reductase family and its aerobic relatives, the cytochrome oxidases. Biochem. Soc. Trans. *30*, 662–667.

Dell'acqua, S., Pauleta, S.R., Paes de Sousa, P.M., Monzani, E., Casella, L., Moura, J.J., and Moura, I. (2010). A new Cu_Z active form in the catalytic reduction of N_2O by nitrous oxide reductase from *Pseudomonas nautica*. J. Biol. Inorg. Chem. *15*, 967–976.

Dennison, C. (2008). The role of ligand-containing loops at copper sites in proteins. Nat Prod Rep *25*, 15–24.

Douthit, S.A., Dlakic, M., Ohman, D.E., and Franklin, M.J. (2005). Epimerase active domain of *Pseudomonas aeruginosa* AlgG, a protein that contains a right-handed beta-helix. J. Bacteriol. *187*, 4573–4583.

Einsle, O., and Kroneck, P.M. (2004). Structural basis of denitrification. Biol. Chem. *385*, 875–883.

Elsen, S., Swem, L.R., Swem, D.L., and Bauer, C.E. (2004). RegB/RegA, a highly conserved redox-responding global two-component regulatory system. Microbiol. Mol. Biol. Rev. *68*, 263–279.

Eraso, J.M., and Kaplan, S. (1994). PrrA, a putative response regulator involved in oxygen regulation of photosynthesis gene expression in *Rhodobacter sphaeroides*. J. Bacteriol. *176*, 32–43.

Eraso, J.M., and Kaplan, S. (2000). From redox flow to gene regulation: role of the PrrC protein of *Rhodobacter sphaeroides 2.4.1*. Biochemistry *39*, 2052–2062.

Ferguson, S.J. (1994). Denitrification and its control. Anton Leeuwenhoek Int. J. Gen. M *66*, 89–110.

Fernandez-Pinar, R., Ramos, J.L., Rodriguez-Herva, J.J., and Espinosa-Urgel, M. (2008). A two-component regulatory system integrates redox state and population density sensing in *Pseudomonas putida*. J. Bacteriol. *190*, 7666–7674.

Field, L.S., Luk, E., and Culotta, V.C. (2002). Copper chaperones: personal escorts for metal ions. J. Bioenerg. Biomembr. *34*, 373–379.

Frangipani, E., and Haas, D. (2009). Copper acquisition by the SenC protein regulates aerobic respiration in *Pseudomonas aeruginosa PAO1*. FEMS Microbiol. Lett. *298*, 234–240.

Frey, P.A., Hegeman, A.D., and Ruzicka, F.J. (2008). The Radical SAM Superfamily. Crit. Rev. Biochem. Mol. Biol. *43*, 63–88.

Futagami, T., Yamaguchi, T., Nakayama, S., Goto, M., and Furukawa, K. (2006). Effects of chloromethanes on growth of and deletion of the *pce* gene cluster in dehalorespiring *Desulfitobacterium hafniense* strain Y51. Appl. Environ. Microbiol. *72*, 5998–6003.

Glerum, D.M., Shtanko, A., and Tzagoloff, A. (1996). Characterization of COX17, a yeast gene involved in copper metabolism and assembly of cytochrome oxidase. J. Biol. Chem. *271*, 14504–14509.

Gregor, J., Zeller, T., Balzer, A., Haberzettl, K., and Klug, G. (2007). Bacterial regulatory networks include direct contact of response regulator proteins: interaction of RegA and NtrX in *Rhodobacter capsulatus*. J Mol. Microbiol. Biotechnol *13*, 126–139.

Griffiths, E., and Gupta, R.S. (2004). Signature sequences in diverse proteins provide evidence for the late divergence of the Order Aquificales. Int Microbiol *7*, 41–52.

Halliwell, B., and Gutteridge, J.M. (1984). Oxygen toxicity, oxygen radicals, transition metals and disease. Biochem. J. *219*, 1–14.

Haltia, T., Brown, K., Tegoni, M., Cambillau, C., Saraste, M., Mattila, K., and Djinovic-Carugo, K. (2003). Crystal structure of nitrous oxide reductase from *Paracoccus denitrificans* at 1.6 A resolution. Biochem. J. *369*, 77–88.

Han, G.W., Sri Krishna, S., Schwarzenbacher, R., McMullan, D., Ginalski, K., Elsliger, M.A., Brittain, S.M., Abdubek, P., Agarwalla, S., Ambing, E., *et al.* (2006). Crystal structure of the ApbE protein (TM1553) from *Thermotoga maritima* at 1.58 A resolution. Proteins *64*, 1083–1090.

Hollenstein, K., Frei, D.C., and Locher, K.P. (2007). Structure of an ABC transporter in complex with its binding protein. Nature *446*, 213–216.

Horng, Y.C., Cobine, P.A., Maxfield, A.B., Carr, H.S., and Winge, D.R. (2004). Specific copper transfer from the Cox17 metallochaperone to both Sco1 and Cox11 in the assembly of yeast cytochrome *c* oxidase. J. Biol. Chem. *279*, 35334–35340.

Huber, R., Sacher, M., Vollmann, A., Huber, H., and Rose, D. (2000). Respiration of arsenate and selenate by hyperthermophilic archaea. Syst. Appl. Microbiol. *23*, 305–314.

Hutchings, M.I., Palmer, T., Harrington, D.J., and Sutcliffe, I.C. (2009). Lipoprotein biogenesis in Gram-positive bacteria: knowing when to hold 'em, knowing when to fold 'em. Trends Microbiol *17*, 13–21.

Imriskova-Sosova, I., Andrews, D., Yam, K., Davidson, D., Yachnin, B., and Hill, B.C. (2005). Characterization of the redox and metal binding activity of BsSco, a protein implicated in the assembly of cytochrome *c* oxidase. Biochemistry *44*, 16949–16956.

Jones, C.M., Stres, B., Rosenquist, M., and Hallin, S. (2008). Phylogenetic analysis of nitrite, nitric oxide, and nitrous oxide respiratory enzymes reveal a complex evolutionary history for denitrification. Mol. Biol. Evol. *25*, 1955–1966.

Kadenbach, B. (1996). X-ray crystal structures of cytochrome *c* oxidases from *Paracoccus denitrificans* and bovine heart and their implications for the molecular mechanism of cell respiration. Angew Chem. Int. Ed. *34*, 2635–2637.

Kadner, R.J. (2005). Regulation by iron: RNA rules the rust. J. Bacteriol. *187*, 6870–6873.

Kajava, A.V., and Steven, A.C. (2006). Beta-rolls, beta-helices, and other beta-solenoid proteins. Adv. Protein Chem. *73*, 55–96.

Kim, B.E., Nevitt, T., and Thiele, D.J. (2008). Mechanisms for copper acquisition, distribution and regulation. Nat. Chem. Biol *4*, 176–185.

Krummeck, G., and Rodel, G. (1990). Yeast SCO1 protein is required for a post-translational step in the accumulation of mitochondrial cytochrome *c* oxidase subunits I and II. Curr Genet *18*, 13–15.

Laguri, C., Phillips-Jones, M.K., and Williamson, M.P. (2003). Solution structure and DNA binding of the effector domain from the global regulator PrrA (RegA) from *Rhodobacter sphaeroides*: insights into DNA binding specificity. Nucleic Acids Res. *31*, 6778–6787.

Laratta, W.P., Choi, P.S., Tosques, I.E., and Shapleigh, J.P. (2002). Involvement of the PrrB/PrrA two-component system in nitrite respiration in *Rhodobacter sphaeroides* 2.4.3: evidence for transcriptional regulation. J. Bacteriol. *184*, 3521–3529.

Leary, S.C., Kaufman, B.A., Pellecchia, G., Guercin, G.H., Mattman, A., Jaksch, M., and Shoubridge, E.A. (2004). Human SCO1 and SCO2 have independent, cooperative functions in copper delivery to cytochrome *c* oxidase. Hum. Mol. Genet. *13*, 1839–1848.

Leary, S.C., Sasarman, F., Nishimura, T., and Shoubridge, E.A. (2009). Human SCO2 is required for the synthesis of CO II and as a thiol-disulphide oxidoreductase for SCO1. Hum. Mol. Genet. *18*, 2230–2240.

Locher, K.P. (2009). Review. Structure and mechanism of ATP-binding cassette transporters. Phil. Trans. R. Soc. Lond. B Biol. Sci. *364*, 239–245.

Locher, K.P., Lee, A.T., and Rees, D.C. (2002). The *E. coli* BtuCD structure: a framework for ABC transporter architecture and mechanism. Science *296*, 1091–1098.

McGuirl, M.A., Bollinger, J.A., Cosper, N., Scott, R.A., and Dooley, D.M. (2001). Expression, purification, and characterization of NosL, a novel Cu(I) protein of the nitrous oxide reductase (*nos*) gene cluster. J. Biol. Inorg. Chem. *6*, 189–195.

Muller, V., Imkamp, F., Biegel, E., Schmidt, S., and Dilling, S. (2008). Discovery of a ferredoxin:NAD+-oxidoreductase (Rnf) in *Acetobacterium woodii*: a novel potential coupling site in acetogens. Ann. N. Y. Acad. Sci. *1125*, 137–146.

Nittis, T., George, G.N., and Winge, D.R. (2001). Yeast Sco1, a protein essential for cytochrome *c* oxidase function is a Cu(I)-binding protein. J. Biol. Chem. *276*, 42520–42526.

Oldham, M.L., Khare, D., Quiocho, F.A., Davidson, A.L., and Chen, J. (2007). Crystal structure of a catalytic intermediate of the maltose transporter. Nature *450*, 515–521.

Paraskevopoulos, K., Antonyuk, S.V., Sawers, R.G., Eady, R.R., and Hasnain, S.S. (2006). Insight into catalysis of nitrous oxide reductase from high-resolution structures of resting and inhibitor-bound enzyme from *Achromobacter cycloclastes*. J. Mol. Biol. *362*, 55–65.

Rasmussen, T., Berks, B.C., Butt, J.N., and Thomson, A.J. (2002). Multiple forms of the catalytic centre, Cu_Z, in the enzyme nitrous oxide reductase from *Paracoccus pantotrophus*. Biochem. J. *364*, 807–815.

Richardson, D.J., Sawers, G., and van Spanning, R.J.M. (2005). The bioenergetics of periplasmic electron transport systems in bacteria, Vol 3 (Elsevier Science).

Saenkham, P., Vattanaviboon, P., and Mongkolsuk, S. (2009). Mutation in *sco* affects cytochrome *c* assembly and alters oxidative stress resistance in *Agrobacterium tumefaciens*. FEMS Microbiol. Lett. 293, 122–129.

Saunders, N.F., Houben, E.N., Koefoed, S., de Weert, S., Reijnders, W.N., Westerhoff, H.V., De Boer, A.P., and Van Spanning, R.J. (1999). Transcription regulation of the *nir* gene cluster encoding nitrite reductase of *Paracoccus denitrificans* involves NNR and NirI, a novel type of membrane protein. Mol. Microbiol. *34*, 24–36.

Simon, J., Einsle, O., Kroneck, P.M., and Zumft, W.G. (2004). The unprecedented *nos* gene cluster of *Wolinella succinogenes* encodes a novel respiratory electron transfer pathway to cytochrome *c* nitrous oxide reductase. FEBS Lett. *569*, 7–12.

Simon, J., Sanger, M., Schuster, S.C., and Gross, R. (2003). Electron transport to periplasmic nitrate reductase (NapA) of *Wolinella succinogenes* is independent of a NapC protein. Mol. Microbiol. *49*, 69–79.

Simon, J., van Spanning, R.J., and Richardson, D.J. (2008). The organisation of proton motive and non-proton motive redox loops in prokaryotic respiratory systems. Biochim. Biophys. Acta *1777*, 1480–1490.

Skovran, E., and Downs, D.M. (2003). Lack of the ApbC or ApbE protein results in a defect in Fe–S cluster metabolism in *Salmonella enterica serovar Typhimurium*. J. Bacteriol. *185*, 98–106.

Smart, J.L., Willett, J.W., and Bauer, C.E. (2004). Regulation of *hem* gene expression in *Rhodobacter capsulatus* by redox and photosystem regulators RegA, CrtJ, FnrL, and AerR. J. Mol. Biol. *342*, 1171–1186.

Solioz, M., and Stoyanov, J.V. (2003). Copper homeostasis in *Enterococcus hirae*. FEMS Microbiol. Rev. *27*, 183–195.

Spiro, S. (1994). The FNR family of transcriptional regulators. Anton Leeuwenhoek Int. J. Gen. M *66*, 23–36.

Stoyanov, J.V., Mancini, S., Lu, Z.H., Mourlane, F., Poulsen, K.R., Wimmer, R., and Solioz, M. (2010). The stress response protein Gls24 is induced by copper and interacts with the CopZ copper chaperone of *Enterococcus hirae*. FEMS Microbiol. Lett. *302*, 69–75.

Swem, D.L., Swem, L.R., Setterdahl, A., and Bauer, C.E. (2005). Involvement of SenC in assembly of cytochrome *c* oxidase in *Rhodobacter capsulatus*. J. Bacteriol. *187*, 8081–8087.

Swem, L.R., Gong, X., Yu, C.A., and Bauer, C.E. (2006). Identification of a ubiquinone-binding site that affects autophosphorylation of the sensor kinase RegB. J. Biol. Chem. *281*, 6768–6775.

Swem, L.R., Kraft, B.J., Swem, D.L., Setterdahl, A.T., Masuda, S., Knaff, D.B., Zaleski, J.M., and Bauer, C.E. (2003). Signal transduction by the global regulator RegB is mediated by a redox-active cysteine. EMBO J. *22*, 4699–4708.

Taubner, L.M., McGuirl, M.A., Dooley, D.M., and Copie, V. (2006). Structural studies of Apo NosL, an accessory protein of the nitrous oxide reductase system: insights from structural homology with MerB, a mercury resistance protein. Biochemistry *45*, 12240–12252.

Unden, G., Becker, S., Bongaerts, J., Holighaus, G., Schirawski, J., and Six, S. (1995). O_2-sensing and O_2-dependent gene regulation in facultatively anaerobic bacteria. Arch Microbiol *164*, 81–90.

Van Spanning, R.J.M., De Boer, A.P., Reijnders, W.N., Westerhoff, H.V., Stouthamer, A.H., and Van Der Oost, J. (1997). FnrP and NNR of *Paracoccus denitrificans* are both members of the FNR family of transcriptional activators but have distinct roles in respiratory adaptation in response to oxygen limitation. Mol. Microbiol. 23, 893–907.

Veldman, R., Reijnders, W.N., and van Spanning, R.J.M. (2006). Specificity of FNR-type regulators in *Paracoccus denitrificans*. Biochem. Soc. Trans. 34, 94–96.

Wikstrom, M., and Verkhovsky, M.I. (2002). Proton translocation by cytochrome *c* oxidase in different phases of the catalytic cycle. Biochim. Biophys. Acta *1555*, 128–132.

Williams, J.C., Sue, C., Banting, G.S., Yang, H., Glerum, D.M., Hendrickson, W.A., and Schon, E.A. (2005). Crystal structure of human SCO1: implications for redox signaling by a mitochondrial cytochrome *c* oxidase 'assembly' protein. J. Biol. Chem. *280*, 15202–15211.

Williams, P.A., Fulop, V., Garman, E.F., Saunders, N.F., Ferguson, S.J., and Hajdu, J. (1997). Haem-ligand switching during catalysis in crystals of a nitrogen-cycle enzyme. Nature *389*, 406–412.

Wood, N.J., Alizadeh, T., Bennett, S., Pearce, J., Ferguson, S.J., Richardson, D.J., and Moir, J.W. (2001). Maximal expression of membrane-bound nitrate reductase in Paracoccus is induced by nitrate via a third FNR-like regulator named NarR. J. Bacteriol. *183*, 3606–3613.

Wu, J., and Bauer, C.E. (2008). RegB/RegA, a global redox-responding two-component system. Adv Exp Med Biol *631*, 131–148.

Wunsch, P., Korner, H., Neese, F., van Spanning, R.J., Kroneck, P.M., and Zumft, W.G. (2005). NosX function connects to nitrous oxide (N2O) reduction by affecting the Cu(Z) center of NosZ and its activity *in vivo*. FEBS Lett *579*, 4605–4609.

Wunsch, P., and Zumft, W.G. (2005). Functional domains of NosR, a novel transmembrane iron–sulfur flavoprotein necessary for nitrous oxide respiration. J. Bacteriol. *187*, 1992–2001.

Zumft, W.G. (1997). Cell biology and molecular basis of denitrification. Microbiol. Mol. Biol. Rev. *61*, 533–616.

Zumft, W.G. (2005). Biogenesis of the bacterial respiratory CuA, Cu-S enzyme nitrous oxide reductase. J. Mol. Microbiol. Biotechnol *10*, 154–166.

Zumft, W.G., and Kroneck, P.M. (2007). Respiratory transformation of nitrous oxide (N_2O) to dinitrogen by Bacteria and Archaea. Adv. Microb. Physiol. *52*, 107–227.

8 Assembly of Respiratory Proteins of the Nitrogen Cycle

Shilpa Bali and Stuart J. Ferguson

Abstract

The respiratory reactions of the nitrogen cycle are those of denitrification, the successive reductions of nitrate, nitrite, nitric oxide and nitrous oxide to nitrogen gas, and those of nitrification, oxidation of ammonium first to nitrite and then to nitrate. These reactions are catalysed by enzymes containing one or more of the cofactors, haem (non-covalent as in *b*-type haems, or covalent as in *c*-type haems), iron–sulfur, molybdenum and copper centres. With the exception of molybdenum, these redox active cofactors are also integral to the operation of the respiratory chain systems that deliver electrons to and from the individual enzymes. This chapter gives an overview of current knowledge about how each of these cofactor types is attached to their respective apo-proteins, in the context that many of the proteins are located on the periplasmic side of the membrane, and delivery to that compartment is either as an unfolded protein, and thus mediated by the sec system, or as a folded protein and thus facilitated by the tat system.

Introduction

The simplest form of the biological nitrogen cycle comprises the fixation of nitrogen to give ammonium, followed by aerobic oxidation of the latter to first nitrite and then nitrate; under anaerobic conditions the latter is reduced via nitrite, nitric oxide, and nitrous oxide back to nitrogen gas in the process known as denitrification (Bothe *et al.*, 2007). Apart from nitrogen fixation, all these reactions are examples of respiratory processes. The nitrogen cycle can be extended to include additional processes, again involving respiratory processes. These include the direct reduction of nitrite to ammonium, which effectively short circuits the N-cycle, and the anammox process through which nitrite combines with ammonium under anaerobic conditions to give nitrogen gas (Bothe *et al.*, 2007).

All these respiratory processes require underlying electron transport systems to transfer electrons to or from specific enzymes for individual reactions. Essentially all these proteins fall into a small number of classes. Thus the electron transport systems comprise iron–sulfur proteins, haem proteins (cytochromes) and copper proteins, whilst the specific enzymes contain similar cofactors with a limited number of other redox centres such as molybdopterins. This chapter is concerned with aspects of the assembly of these proteins and thus is structured in terms of these groups of proteins rather than in terms of the processes catalysed.

Many of the enzymes and associated electron transport proteins are located in the periplasm or have their important functional globular domains in the periplasm, with at least one transmembrane helix to provide anchoring to the cytoplasmic membrane. Some of these proteins are delivered in an unfolded conformation to their final location by the ubiquitous sec system and the cofactor added either spontaneously or through the action of post-translational activities. However, there are some cofactors which are unstable and are thus inserted into their proteins before delivery to the periplasmic side of the membrane and these require a different transport process as described later. Other general targeting systems may be required. YidC is a membrane insertase that is known to insert relatively hydrophobic, negatively charged proteins into the membranes of bacteria. In cooperation with the Sec-YEG translocase, it can also mediate the membrane integration of some proteins (Pop *et al.*, 2009). YidC is thus likely to be involved in the assembly of some of the integral membrane subunits of the respiratory proteins of the nitrogen cycle.

Assembly of haem proteins

b-Type cytochromes

Non-covalent attachment of haem (Fe-protoporphyrin-IX) to a polypeptide can generate a *b*-type cytochrome. Pathways for haem biosynthesis are well established and not discussed here. The most general *b*-type cytochrome functioning in the nitrogen cycle is the cytochrome bc_1 complex, usually involved in every type of step, including the anammox process, except nitrogen fixation. There is a transmembrane subunit to which two haem groups are non-covalently bound by the axial ligation of the haem iron with conserved histidines on the helices; this is the *b*-type cytochrome component. The haem is thought to insert spontaneously. Although this has not been studied specifically for the bc_1 complex, there is evidence from work on a related protein, cytochrome b_6f from photosynthetic membranes, that haem insertion is spontaneous (Dreher *et al.*, 2008; Volkmer *et al.*, 2006).

Another important example is in the Nar-type of nitrate reductase in which the transmembrane subunit has two *b*-type haems sandwiched between helices. There is some old evidence that the incorporation of haem into this subunit is ATP-dependent (Reid *et al.*, 1981) but no gene products responsible for this insertion have ever been identified. Another important example of the occurrence of *b*-haem is in the NO reductase. There are two types of *b*-haem containing NO reductases, cNor (cytochrome *c*/pseudoazurin dependent) and qNor (quinol dependent), They both also have EPR silent non haem iron in the transmembrane region. The structural similarities that are observed in cNor and qNor are mainly in their large, integral membranous subunit that harbours the haems involved in electron transfer and in their binuclear catalytic centres (haem iron/non-haem iron), where the NO reduction occurs. However, the main difference between them is that cNOR consists of two subunits whereas qNOR is single-subunit enzyme.

NO reductases are members of the haem–copper oxidase family for which it is known that biogenesis factors are required. The same is likely to be true for NO reductase; indeed, expression of the *Paracoccus denitrificans* NorCB protein in *Escherichia coli* is dependent on expression of the adjacent *norQDEF* genes suggestive of a role in assembly for at least some of these proteins (Butland *et al.*, 2001). To what extent these proteins are needed for haem and non-haem iron insertion remains to be investigated.

c-Type cytochromes

c-Type cytochromes underpin many respiratory reactions of the nitrogen cycle through two types of function. The first relates to their roles as electron transfer proteins. The second is the part that they play in specific enzymes, for example hydroxylamine oxidoreductase and cytochrome cd_1 nitrite reductase. The definition of a *c*-type cytochrome is that one or more haem groups are attached to their polypeptide chain *via* two thioether bonds, formed between the original vinyl groups of haem and the sulfur atoms of two cysteine residues found in a CXXCH motif. The simplest *c*-type cytochromes have one haem group and act to transfer electrons from one protein to another, analogously to the best known *c*-type cytochrome, mitochondrial cytochrome *c*. However, there are more complex *c*-type cytochromes, for example the hydroxylamine oxidoreductase that functions in nitrifying bacteria to catalyse oxidation of hydroxylamine to nitrite. This enzyme has eight haems (7 *c*-type and 1 novel P_{460}-type haem) per polypeptide chain (Igarashi *et al.*, 1997). Many of the *c*-type cytochromes that participate in reactions of the nitrogen cycle are water soluble proteins that are located in the periplasm. Others are anchored to the cytoplasmic membrane, but the haem groups are always located in a globular domain on the periplasmic side of the membrane. A prominent example of such a *c*-type cytochrome is the cytochrome c_1 of the cytochrome bc_1 complex that is very common in bacterial species that catalyse the reactions of the nitrogen cycle. Another important example is the NapC type of protein which delivers electrons to and from periplasmic reductases or dehydrogenases.

The same biosynthetic apparatus for *c*-type cytochromes functions irrespective of whether the cytochrome is water-soluble or membrane bound. In either case it is known that the protein is translocated to the periplasm as the apo-form *via* the sec system for protein translocation. This means that the protein is moved across the membrane in an unfolded conformation; the haem is covalently attached on the periplasmic side of the membrane. Two types of biogenesis system for making *c*-type cytochromes are found in eubacteria; these are termed Systems I and II. The System I is probably more common amongst organisms that play roles in the nitrogen cycle. The current understanding of the two systems will be discussed in turn.

System I is also known as the Ccm system and comprises the proteins Ccm A, B, C, D, E, F,G, H and I (but sometimes H and I are fused into one protein such that the C terminal domain of the fusion corresponds to I, as for example in CcmH of *Escherichia coli*; Fig. 8.1). Although much progress has been made in understanding this system, there are still many uncertainties. The interested reader is referred to detailed reviews (Ferguson *et al.*, 2008; Kranz *et al.*, 2009). Here, we provide a brief account of current knowledge. CcmE appears to play a central role and binds haem covalently *via* a novel bond between a histidine side chain and a vinyl group of haem. This bond then breaks as the haem is transferred to the thiol groups of the apo-cytochrome. The detailed chemistry of this covalent bond formation and subsequent breakage presents many puzzles. CcmA and CcmB constitute a protein of the ABC class but the

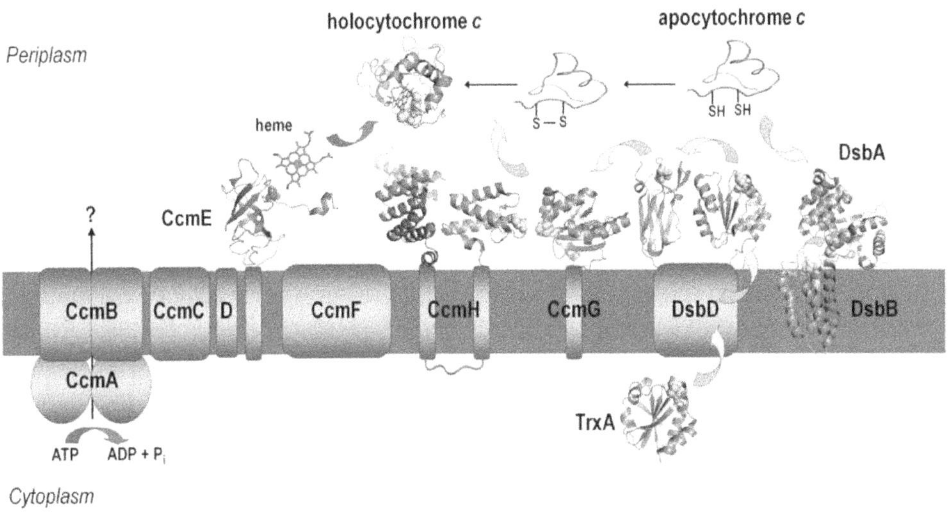

Figure 8.1 The cytochrome *c* biogenesis system I of *E. coli* showing the eight Ccm proteins and the disulfide bond (Dsb) proteins DsbA, B and D. The cysteine residues in the thiol oxidoreductase are indicated by yellow spheres; there is a pair of cysteines in the transmembrane domain of DsbD that is involved in the disulfide cascade. The flow of electrons is indicated with pale grey arrows. The transported substrate, if any, of CcmAB is unknown. The dark grey arrow schematically represents transfer of haem from ComE to apocytochrome *c*. The haem-binding histidine of CcmE is shown as spheres. The structures shown, which are from *E. coli* unless otherwise indicated, are from the following PDB entries: CcmE (1SR3), CcmH (N-terminus from *Pseudomonas aeruginosa* (2HL7), C-terminus is *E. coli* NrfG (2E2E) equivalent to CcmI – see text), CcmG (2BLK), nDsbD (1JPE), cDsbD (2FWH), thioredoxin (TrxA) (2TRX), DsbAB (2HI7). The cytochrome *c* shown is cytochrome c_{550} from *Paracoccus denitrificans* (155C). The structures were rendered in Pymol (DeLano, W.L. The Pymol Molecular Graphics System (2002) http://www.pymol.org). (This figure was kindly provided by Dr J.M. Stevens.)

abolition of the ATPase activity does not prevent haem from being attached to CcmE (Christensen *et al.*, 2007; Feissner *et al.*, 2006; Kranz *et al.*, 2009). This and other lines of evidence indicate that haem is not transported across the membrane by this ABC protein. CcmC is a transmembrane protein with six predicted transmembrane helices and which is needed for haem attachment to CcmE; it is possible that haem crosses the membrane via CcmC although haem can still reach the periplasm in the absence of any Ccm protein (Ahuja and Thony-Meyer, 2005; Ren and Thony-Meyer, 2001). CcmC may, with assistance from the single transmembrane helical protein CcmD, deliver haem from the cytoplasm to CcmE.

Whereas the ATPase activity of CcmAB is not needed for haem attachment to CcmE, this activity is needed for transfer of haem onwards to an apo-*c*-type cytochrome (Christensen *et al.*, 2007). Thus it appears that CcmAB might be one of the examples of an ABC protein that is not a transporter (Feissner *et al.*, 2006; Kranz *et al.*, 2009). The CcmF protein has fifteen predicted transmembrane helices and is generally regarded as the key component in catalysing haem transfer to the apo-cytochrome. Recently it has been found that CcmF binds quinone and that an electron transfer activity is associated with the protein, with a possible role being ensuring that the haem bound to CcmE remains in the ferrous state which should promote transfer to the apo-protein (Richard-Fogal *et al.*, 2009). Much remains to be learned about this process. The CcmH protein has a CXXC motif which is reminiscent of a thioredoxin protein but whether an oxidation/reduction activity, typical of a thioredoxin, is actually the role remains uncertain. On the other hand, CcmG is clearly a periplasmic protein of the thioredoxin type; it is generally considered that CcmG acts either alone or in concert with other proteins to reverse the tendency of the two cysteines in the CXXC motif to form a disulfide bond, a process that is catalysed in the periplasm by the DsbAB system. Any such reduction of a disulfide means that CcmG must in turn be supplied with reducing equivalents. This supply is delivered by the transmembrane protein DsbD which is in turn supplied by cytoplasmic thioredoxin (Fig. 8.1). Finally, CcmI is a protein of the TRP (tetratricopeptide repeat) family. This type of protein is thought to have a chaperone-like function in various biosynthetic assembly processes but the exact role in *c*-type cytochrome assembly is not understood. Thus System I is a complex apparatus; it appears, based on several lines of evidence, to recognize little more than the CXXCH motif which is presumably presented to it as part of an unfolded polypeptide. The only other motifs which appear to be handled are CXXXCH and CXXXXCH. A notable variation on the System I apparatus, is the set of proteins that is needed to attach haem to the unusual CXXCK motif that is found in the NrfA type of nitrite reductase that catalyses reduction of nitrite to ammonium in what is effectively a short cut in the nitrogen cycle. This attachment needs a variant of CcmF, known as NrfE and also a variant of CcmI known as NrfG (Eaves *et al.*, 1998). The key features of NrfE and NrfG that distinguish them from their Ccm counterparts have not yet been recognized. We should also mention that a variant of system I, known as system I*, in which the essential histidine in CcmE of System I is replaced by a cysteine and CcmH is absent (Goddard *et al.*, 2010a). However, at this stage it is not known that this variant system functions in any N-cycle organisms.

Compared with system I, system II (Fig. 8.2) presents a much simpler picture (Kranz *et al.*, 2009). A protein known as ResA is a thioredoxin-like protein and, like CcmG in System I, is supplied with reductant by DsbD or a related protein called CcdA. ResA is not required if the disulfide bond forming system is absent, thus strongly indicating that the role of ResA is to reverse any formation of a disulfide in the CXXCH motif. The core of System II is two proteins, sometimes fused as one, called ResB (or Ccs1/CcsB) and ResC (CcsA). Recent evidence indicates that ResB transports haem across the membrane in a step that might be substantially separate from the attachment of haem to the apo-cytochrome (Goddard *et al.*, 2010b; Richard-Fogal *et al.*, 2009). It is striking that System II is simpler than System I; thus there is no requirement for ATP hydrolysis, no analogue of CcmE with its covalently bound haem [although there is evidence that haem can be covalently attached to ResB (Ahuja *et al.*, 2009)], and no equivalent to the CcmI or CcmH proteins.

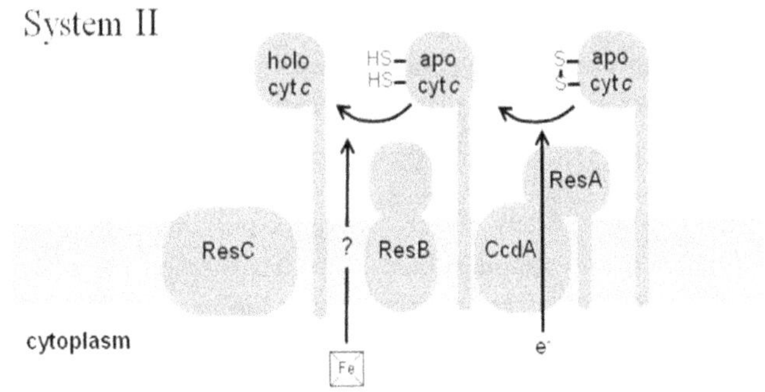

Figure 8.2 The cytochrome *c* biogenesis system II showing the cartoon representation of four Ccm proteins of system II. ResB and ResC are involved in haem handling and haem ligation, whereas ResA and CcdA have a role in the disulfide cascade. CcdA is reduced by cytoplasmic thioredoxin. It is important to note the system II is much simplified relative to system I, which has at least eight dedicated proteins (see Fig. 8.1).

In common with system I, there is a variant System II that is required for attaching haem to the CXXCK motif of the NrfA-type nitrite reductase involving NrfI which is an orthologue of ResC. Exactly how the CXXCK is recognized instead of CXXCH is not clear since a variety of experiments have failed to detect attachment of haem to CXXCK, except when the latter motif is part of NrfA (Kern *et al.*, 2010). In contrast to System I, there is less experimental evidence to indicate that System II simply recognizes CXXCH and that folding of the protein follows attachment of haem. Nevertheless, this is the likely sequence of events.

Assembly of the cytochrome bc_1 complex

The cytochrome bc_1 is a dimeric protein complex, which includes two copies of three proteins that contain redox prosthetic groups; cytochrome *b*, cytochrome c_1 and the Rieske Fe–S protein. This protein complex is located in the bacterial cytoplasmic membrane, where it participates in cell respiration, for example in transferring electrons from the ubiquinol pool on towards reductases for nitrite, nitric oxide and nitrous oxide in denitrifiers, from ubiquinol towards oxygen in ammonia-oxidizing bacteria, and in reverse from cytochrome c to ubiquinone in organisms that oxidize nitrite to nitrate. The Rieske protein is anchored in one monomer of bc_1 by a transmembrane helix, whereas the peripheral domain, on the periplasmic side of the membrane, that includes the 2Fe–2S cluster is incorporated into the other monomer. In the case of the mitochondrial cytochrome bc_1 complex it is known that assembly is dependent upon a number of proteins, for example Bcs1p in yeast and humans, as discussed by Zara *et al.* (2007). At present there does not appear to be any clear evidence that prokaryotic cytochrome bc_1 complexes require assembly factors, other than those required generically for *c*-type cytochromes and FeS centres (see later). The latter is inserted into the Rieske protein on the cytoplasmic side of the membrane such that the folded protein has to be moved to the periplasmic side. Such export cannot be facilitated by the ubiquitous sec system but rather relies upon a specialized system known as Tat (twin arginine translocator, the name relating to the RR sequence that is found in the targeting sequences of proteins using this system), which is able to move folded proteins across the membrane. For details of current knowledge of the Tat proteins, and how the transport might be driven by the proton motive force, the reader is referred to (Berks *et al.*, 2005; Mueller and Kloesgen, 2005).

The d_1 haem of cytochrome cd_1 nitrite reductase

This enzyme catalyses the one electron reduction of nitrite to nitric oxide in the denitrification part of the nitrogen cycle. As the name suggests, it contains a *c*-type haem centre which is assembled by the *c*-type cytochrome biogenesis apparatus, a process that can occur without incorporation of the non-covalent d_1 haem (Ferguson, 2009) which is, therefore, presumably inserted second. Cytochromes cd_1 are targeted to the periplasm by a sec signal sequence consistent with addition, first of haem to the CXXCH motif of the unfolded protein, and then partial folding to allow capture of the d_1 haem. The d_1 haem is only found in this particular enzyme and is not truly a haem but rather a dioxo-isobacteriochlorin, distinguished by presence of two electron withdrawing carbonyl groups at C3 and C8 and two reduced pyrrole rings involved in making the tetrapyrrole skeleton with an acrylate side chain at position C-17, features which appear to make the molecule well tuned to bind nitrite and release nitric oxide in its ferrous state (Allen *et al.*, 2005; Cutruzzola *et al.*, 2001; Jafferji *et al.*, 2000; Rinaldo *et al.*, 2011). Synthesis of d_1 haem requires a dedicated set of proteins that occur contiguous to the gene, *nirS*, for the enzyme itself.

It is clear that the first step in d_1 haem synthesis is catalysed by NirE which catalyses formation of pre-corrin2 from uroporphyrinogen III, a reaction which also occurs in other tetrapyrrole biosyntheses, for example corrins (Storbeck *et al.*, 2009; Zajicek *et al.*, 2009). The route from pre-corrin2 to the d_1 haem is still a matter of speculation. One of the required proteins, NirJ, is a radical SAM protein, with an FeS centre, but the reaction catalysed is not known (Brindley *et al.*, 2010). Other required proteins include NirD, NirG NirH and NirL. There are very few clues as to what these proteins do and it has been suggested that they are transcription factors with presumably no catalytic role (Xiong *et al.*, 2007). Sequence analyses of NirF and NirN suggest that both might be related to the C-terminal domain of cytochrome cd_1, which is the site of d_1 haem binding. Consistent with this prediction NirN has been shown to bind d_1 haem (Zajicek *et al.*, 2009). This periplasmic protein also has a *c*-type haem centre but its role in d_1 synthesis is perplexing since although it is ever present in sets of *nir* genes that flank *nirS*, it can be inactivated through targeted mutation without any consequence for the production of d_1 haem and nitrite reductase activity, as judged by reduction of nitrite and formation of gas bubbles in cultures of cells with lack NirN (Zajicek *et al.*, 2009). NirC, which codes for a periplasmic *c*-type cytochrome is also not essential for d_1 haem synthesis. On the other hand NirF is essential for d_1 haem production and it has been shown to bind d_1 haem (Bali *et al.*, 2010). The binding of d_1 haem by NirF is lost if a histidine corresponding to the proximal histidine ligand in cytochrome cd_1 itself is altered to several other residues, including methionine and cysteine, which can in principle act as axial ligands to haem iron; these observations, together with the finding that these variants do not allow d_1 haem synthesis *in vivo*, support the idea that binding of d_1 haem is physiologically significant and that d_1 haem might be the product of any catalytic activity of NirF. For example, NirF might insert Fe given the analogy that the final step in haem synthesis is insertion of Fe into protoporphyrin-IX. A curious feature of NirF is that sequence analysis suggests that it is a periplasmic protein in many, but not all, organisms. It has recently been shown that NirF is indeed located in the periplasm of *Paracoccus pantotrophus* (Bali *et al.*, 2010). This location means that its substrate must be transported to the periplasm and NirF must presumably catalyse the last step of d_1 haem synthesis, given the non-essential nature of the two other known periplasmic proteins, NirC and NirN. It remains the case that for the frequently studied *Pseudomonas aeruginosa* there is not a very strongly indicated N-terminal sec targeting sequence. Further work is needed to determine the location of *P. aeruginosa* NirF. It would be very strange if it is cytoplasmic since that might imply that the final step of d_1 synthesis can be catalysed in either compartment depending upon the organism. The *P. pantotrophus* NirF protein is inactive, and thus d_1 haem synthesis abolished, when expressed in the cytoplasm of a *nirF* deletion strain from a gene without the targeting sequence. Interestingly (S. Bali and S.J. Ferguson, unpublished work), the full-length *nixF* gene from *P. aeruginosa* will complement a *nirF* deletion in *P. pantotrophus;* this implies that targeting of the *P. aeruginosa* protein to the periplasm occurred since, as mentioned above, cytoplasmic *P. pantotrophus* NirF does not restore d_1 haem synthesis to the same deletion strain.

In either case, cytoplasmic or periplasmic, it is puzzling that sets of *nir* genes do not contain a member that codes for a transmembrane protein that could be a transporter; intuitively non-catalysed diffusion of the d_1 haem, or an immediate precursor, across the membrane seems unlikely given the biosynthetic expense involved in its manufacture. Recall that as the apo-cytochrome cd_1 is transported by the sec system, which carries unfolded proteins, then the d_1 haem cannot be taken across the membrane with the polypeptide.

Expression of cytochrome cd_1 nitrite reductase, including production of the d_1 haem is dependent on the activated form of a transcription factor that belongs to the Fnr family (Zajicek *et al.*, 2006). In many organisms this transcription factor is known as Nnr and it is distinct from Fnr in that it lacks an iron sulfur centre. Nitric oxide is needed for the expression of the *nir* genes and it is currently speculated that Nnr is a haem protein that is activated through the binding of nitric oxide. In other organisms, for example *P. aeruginosa*, a protein known as Dnr is an important transcription factor. In this case the presence of a haem group has been established (Castiglione *et al.*, 2009).

Assembly of molybdenum proteins

Nitrate reductases contain a complex organic cofactor that provides some of the ligands to the active site molybdenum. The cofactor is not stable outside of a binding site in a protein and so it has to be delivered from its biogenesis system to enzymes that use it. The molybdenum cofactor itself is a pterin (hence molybdopterin) that is made using a complex four step process involving ModABCD and additional proteins in *E. coli* and other bacteria. Interestingly, the sulfide required for incorporation of the cofactor into proteins is provided by the same IscS type of protein that participates in the assembly of FeS centres (see later). For more details the reader is referred to recent publications (Mendel and Bittner, 2006; Schwarz *et al.*, 2009; Wuebbens and Rajagopalan, 2003; Zhang *et al.*, 2010).

There are three types of dissimilatory nitrate reductase found in denitrifying bacteria. One, called Nar, is a membrane-bound heterotrimeric enzyme composed of the subunits NarG, NarH and NarI, with its active site in a globular domain facing the cytoplasm. The Mo-*bis*-molybdopterin guanine dinucleotide (Mo-*bis*-MGD) cofactor is delivered to this protein by a chaperone protein known as NarJ that participates in NarGH assembly before its attachment to NarI in the membrane (Blasco *et al.*, 1998; Lanciano *et al.*, 2007). The most recent work has shown that NarJ interacts with the helical arrangement adopted by residues 1–15 of the *E. coli* molybdopterin protein. This interaction between a chaperone with either the N-terminus or the N-terminal tat signal sequence is a common feature for this class of molybdoprotein (Zakian *et al.*, 2010). Interestingly, in *E. coli*, sequence analysis of cytoplasmic NarG revealed that it has an N-terminal region, which bears some resemblance to twin-arginine signal peptides, leading to the hypothesis that this protein might be exported to periplasm by Tat machinery in a common ancestor. Indeed, mutations of six residues out of 36 residues in the N-terminal region of NarG led to formation of active Tat export signals (Ize *et al.*, 2009). However, these changes had bigger implications on the activity of the translocated protein, the mutated NarG could no longer bind to its chaperone, NarJ, and it also failed to make a functional dimer with NarH owing to its rapid degradation, suggesting that these signals were retained in these cytoplasmic proteins for chaperone binding.

The second type of enzyme is the periplasmic nitrate reductase (Nap), which has to be exported to the periplasm in a folded conformation with its molybdenum and FeS cofactors bound. This export is catalysed by the Tat system described earlier. Coordination of protein export and cofactor insertion by this system requires a proof-reading process, in which chaperones bind twin-arginine motif of the signal peptide in order to prevent premature export. The *napA* gene encodes catalytic subunit NapA that contains the Mo-*bis*-MGD active site. Thus it is important that NapA is not translocated prematurely before the molybdopterin cofactor has been inserted. The NapD protein plays a role here since it can bind tightly and specifically to the signal peptide on NapA and prevent export until the NapA is fully assembled (Maillard *et al.*, 2007). In contrast to *nar* operon, the *nap* operons present heterogeneity in gene composition and gene order. The minimal operon organization consists of *napDGAH* and some operons have been identified to contain maximum of nine different genes, some of which may code for proteins involved in the biosynthesis of the complex.

The third type of nitrate reductase is a more recently recognized variant of Nar in which the topology is inverted such that the globular domain is on the external surface of the protein. This type of nitrate reductase is presumably assembled by the same pathway as for Nar, including an NarJ analogue, but needs to be exported carrying the cofactor and thus by the Tat system (Martinez-Espinosa *et al.*, 2007).

Nitrite oxidases, as found in nitrifying bacteria, show significant similarities to nitrate reductases with some, as in *Nitrobacter* (Bothe *et al.*, 2007), having the globular catalytic site domain facing the cytoplasm, but with others, such as in *Nitrospira*, where catalytic domain faces the periplasmic space (Lucker *et al.*, 2010); in each case the same type of molybdenum cofactor is present as in nitrate reductases. The tat system is required for the export of the outward facing enzyme.

The expression of respiratory nitrate reductases is under the control of several factors. In general the Nar type reductase is controlled by a transcription of Fnr type which ensures that expression only occurs under anaerobic conditions. The availability of nitrate also needs to be signalled and this is done in different ways in different organisms. In some cases this is dependent upon a two component regulatory system in which a transmembrane protein, typically NarL transfers phosphate to NarX, which then directly activates transcription. In other cases, such as in alpha group of bacteria, a member of CRP family, known as NarR, is involved as the nitrate sensor (Wood *et al.*, 2001).

Assembly of FeS centres

Fe–S centres are important for the nitrogen cycle. Not only are they found in some enzymes, for example nitrate reductases, but they also occur in some proteins involved in the assembly of various enzymes important for the nitrogen cycle. They are found, for example, in the proteins involved in assembly of Nap, and the d_1 haem of the cd_1 nitrite reductase.

The reader is referred elsewhere for a detailed discussion of FeS assembly in bacteria (Ayala-Castro *et al.*, 2008). The basic principle is that a sulfur atom is first transferred from cysteine to a cysteine sidechain of a protein typified by IcsS, with the resulting formation of a persulfide. Iron sulfur centres then form on a protein called IcsU, using Fe delivered by a still uncertain pathway, followed by sulfide liberation from the persulfide on IscS. Once the cluster is assembled it is transferred to an apo protein, assisted in at least some cases by an ATPase activity, for example provided by HscAB proteins.

Respiratory enzymes of the nitrogen cycle include NapA which is an example of a protein that has two redox centres. The molybdopterin of the active site has already been mentioned but there is also an FeS centre. This will be inserted in the cytoplasm before the NapA also containing the Mo cofactors is exported to the periplasm *via* the Tat system. NapF, itself an FeS protein may play a role in the insertion of the FeS centre into NapA.

In the Nar-type nitrate reductase (and the related outward facing enzyme), the NarH subunit also contains four Fe–S centres; one 3Fe–4S and three 4Fe–4S. NarH is associated with membrane-bound NarI subunit, which carries two haem *b*, located at the opposite sides of membranes. Electrons are transferred from cytoplasmic oriented haem *b* of NarI to the 3Fe–4S cluster of NarH. From there they flow *via* the Fe–S cluster in NarH to 4Fe–4S cluster of NarG, which is the direct electron donor to Mo-bis-MGD cofactor, the site where two electron reduction of nitrate occurs.

NirJ is a monomeric cytoplasmic Fe–S cluster containing protein known to be essential in haem d_1 biogenesis for cytochrome cd_1 nitrite reductase. Multiple sequence alignments reveal that there are two cysteine-based motifs conserved in NirJ; the N-terminal motif CX_3CX_2C corresponds exactly to the signature motif of the Radical SAM superfamily. The second C-terminal motif CX_2CX_5C is suspected to bind a second iron–sulfur cluster (Zumft, 1997). Electron paramagnetic resonance (EPR) spectroscopy confirmed that the N-terminal motif of NirJ binds a Radical SAM centre

(Brindley *et al.*, 2010). The axial lineshape, *g* value, and temperature sensitivity of the spectra presented are diagnostic of the presence of a $[4Fe–4S]^+$ cluster at the N-terminal motif; evidence has been provided for SAM binding to this motif. It is generally hypothesized that the Radical SAM centre could enable the elimination of propionate at C-3 and C-8 by activating the alkyl leaving groups in a radical mechanism.

Assembly of copper proteins

No specific apparatus is needed for the acquisition of copper by some periplasmic proteins. This can be deduced from the observations that azurin, pseudoazurin and copper nitrite reductase are readily expressed from their respective genes in *E. coli* (Glockner and Zumft, 1996; Karlsson *et al.*, 1989; Leung *et al.*, 1997). As the latter does not have any of these proteins, it seems reasonable to infer that copper insertion is spontaneous. This leaves open the question of whether the copper needs to be reduced to the cuprous state before insertion. There is also evidence that the binuclear Cu_A centre of nitrous oxide reductase is assembled spontaneously, in contrast to the more complex Cu_Z centre (see below).

Usually, the copper nitrite reductase, NirK is exported to the periplasm in an unfolded conformation and is routed *via* the sec system. This is entirely consistent with the copper being inserted spontaneously as the protein folds in the periplasm. However, curiously, in some bacterial species (e.g. *B. japonicum*), the nitrite reductase is exported *via* the Tat system; this suggestion is based on the presence of RR motif in the signal sequence and is not experimentally verified yet (Bedmar *et al.*, 2005). In some bacteria such as *Rhodobacter* and *Neisseria*, NirV is thought to be involved in the biosynthesis and assembly of NirK. However, this protein is thought to be a sulfatase modifying factor-2 precursor and its actual function is, so far, unknown.

Assembly of nitrous oxide reductase

Nitrous oxide reductase is a very distinct copper protein, coded for by the *nosZ* gene and usually delivered to the periplasm via the Tat system. This enzyme contains two types of copper centre. One known as Cu_A is a binuclear electron transfer centre which is also found in the cytochrome aa_3 oxidases of mitochondrial and bacterial membranes. The catalytic site is provided by the Cu_Z centre, specific for this type of enzyme, which comprises a cluster of four copper ions with an inorganic sulfide bridging ligand. On the basis of the standard paradigm of Tat-dependent enzyme delivery pathways, one might assume that one or both coppers is inserted in the cytoplasm before export of the folded protein. However, matters are not so simple and several lines of evidence indicate that the insertion of the metal centres occurs in the periplasm (Zumft and Kronek, 2007). These include (1) export of NosZ proceeds undisturbed in accessory gene mutants, (2) assembly of nitrous oxide reductase requires periplasmic components such as NosL and NosD, (3) mutagenesis of Tat protein transporter/signal sequence retains the NosZ, with low copper content, in cytoplasm and finally (4) copper deficiency does not affect the location of NosZ, but lowers the enzyme activity. Activity of the periplasmic enzyme thus formed could be fully restored by exogenous copper. Also, most notably, the NosZ of *W. succinogenes* is exported via the sec system. Thus we have a strange situation in which the apo-form of NosZ in many organisms is sufficiently folded to be a Tat substrate and yet in *W. succinogenes* must be sufficiently unfolded to be carried by the sec system!

In organisms such as *P. denitrificans* and *P. stutzeri* there are several genes whose products are required for enzyme assembly. The best evidence for the minimal gene set comes from the work of Wunsch and coworkers who showed that the ability to reduce nitrous oxide was conferred on *Pseudomonas putida* by supplying the *nosD, F, Y* and *R* genes in addition to the structural gene *nosZ* (Wunsch *et al.*, 2003).

NosF is the component of an ABC-type cytoplasmic protein that contains the ATPase activity and NosY is a predicted accompanying six-helix integral membrane protein. NosD is a member of the carbohydrate-binding proteins and sugar hydrolases (CASH) protein family. The genetic organization and the common occurrence of periplasmic binding proteins with ABC transporters, suggests that it functions in tandem with NosF and NosD. Periplasmic binding proteins are normally associated with ABC transporters with an import function and so a tentative view of the role of the NosDFY proteins is that they function to import something. At this point it is apposite to note that not all ABC transporters are involved in importing anything and it is possible that despite the periplasmic protein component this system might be an exporter. It is interesting to recall that another ABC protein, CcmAB of the cytochrome *c* assembly pathway, also appears to have a role in a process other than transport (see earlier).

What might be the function of NosDFY? There would seem to be two possibilities, although neither are mutually exclusive. Some role in copper provision to the nitrous oxide reductase can be contemplated but is seemingly ruled out because the Cu_A site certainly assembles in the absence of the putative ABC system and there is evidence that exogenous copper can supply the Cu_Z site. For instance, as long as NosDFY are present apo-nitrous oxide reductase in the periplasm can acquire activity if copper is added exogenously. Zumft has argued that the system is transporting sulfur or sulfide (or a donor of sulfide) (Zumft and Kroneck, 2007). This is speculative.

Beyond NosDFY several other proteins are often implicated in the assembly of nitrous oxide reductase but apart from factors possibly needed for recruitment of copper to periplasmic proteins in general (see later) an obligatory role for these other proteins has not been demonstrated. NosL is one such protein. Sequence analysis suggests that it is a periplasmic lipoprotein anchored to the outer membrane. Based on the properties of NosL expressed in *E. coli*, NosL binds Cu(I) tightly but can release Cu(II). Thus it is regarded as a copper chaperone; perhaps the reason that it is dispensable is that there are other periplasmic proteins involved in copper delivery, for example ScoP, which presumably might be able to supply the Cu_A sites of both cytochrome aa_3 oxidase and nitrous oxide reductase.

NosA has the predicted properties of a beta barrel outer membrane protein. At one time it was thought to be required for production of a functional NosZ protein but that view has now been rescinded (Zumft and Kronek, 2007). Nevertheless it may well, possibly in association with the TonB transport system, be involved with copper import through the outer membrane.

NosR poses several dilemmas (Zumft and Kronek, 2007). It is nearly always found upstream of *nosZ* and was originally thought, based on work with *P. stutzeri*, to be involved in the transcription of genes for nitrous oxide reductase, including NosZ the structural gene. However, there is now evidence that in strains carrying mutated versions of NosR the NosZ protein still assembles both its copper sites but is in an inactive form in cells. However, *in vitro* the form of NosZ obtained from such cells is categorized as NosZ* 'redox inactive state', which is catalytically active for turning over the N_2O to N_2 but does not show the same kind of midpoint redox potential as seen for the wild type NosZ (Rasmussen *et al.*, 2002). The organization of NosR is very intriguing. It has complex topology with different cofactors at either side of the membrane-bound protein. The cytoplasmically located C-terminus contains CX_3CP motif and other cysteine-containing motifs, which are thought to bind iron–sulfur clusters with poly-ferredoxin type structure. There are five transmembrane helices predicted and the N-terminal periplasmic domain is a flavoprotein with a flavin covalently attached at threonine T163 (in *P. stutzeri*). A current view is that activity of NosR is somehow required to maintain nitrous oxide reductase in an active state in the cell; an electron donation event is postulated. There are other puzzles concerning NosR; it has a sec signal sequence and yet apparently has a flavin bound in the periplasmic domain. The presence of a flavin cofactor would normally suggest transport via the tat system. If the flavin is redox active what is its redox potential and what serves as its reductant *in vivo*? One possibility is that the NosR is required to act as a low potential reductant to maintain the active state of NosZ (see also Chapter 7). The steady-state electron flow to the enzyme would be supplied by the higher potential cytochromes *c* or cupredoxins. So NosR would be an essential component for the assembly of the Cu_Z catalytic site.

NosX is even more puzzling than NosR as it is not present in all denitrifiers by any means. It is present in *P. denitrificans* and its absence seems to result in much the same phenotype as does the absence of *nosR*; that is the NosZ* form of nitrous oxide reductase is formed. NosX is also deduced to have a flavin in the periplasmic domain and this is congruent with the presence of a Tat target sequence. Broadly, NosX seems to have postulated roles similar to those for NosR mentioned above. Notably in *R. capsulatus* the gene for *nosX* to fused to that of *nosR* which suggests that the two proteins interact. Interestingly, there is a related protein, NirX, in *Paracoccus* species which is associated with the nitrite reductase genes. The role of this protein is also not known.

What signals are needed for the transcription of genes coding for nitrous oxide reductase and its biogenesis genes? This is a complex and unresolved issue. There is evidence that NO might stimulate transcription through the action of Nnr or Dnr but that the general anaerobic transcription factor Fnr might also play a role. Certainly for many species of bacteria growth is possible with nitrous oxide as sole electron acceptor; thus NO is not an obligatory factor for at least some extent of transcription of the genes. It is thought that N_2O is a weak signal for *nosZ* and *nosD* promoters. Additional signals, in some bacteria might be provided by the redox status or oxygen tension that are processed by an Fnr type Fe–S proteins. In *B. japonicum* dependence of *nos* gene on the haem-based oxygen sensor FixLZ is suggested in cooperation with FixKZ (Velasco *et al.*, 2004).

Ammonia mono-oxygenase

The ammonia mono-oxygenase enzyme is essential for nitirifiers that oxidize ammonium to nitrite. This membrane-bound enzyme is generally believed to be analogous to a methane mono-oxygenase which has a dicopper active site in a soluble domain of the protein (Balasubramanian *et al.*, 2010). Whether or not insertion of copper into this class of protein is spontaneous or not is not yet known.

References

Ahuja, U., and Thony-Meyer, L. (2005). CcmD is involved in complex formation between CcmC and the heme chaperone CcmE during cytochrome *c* maturation. J. Biol. Chem. *280*, 236–243.

Ahuja, U., Kjelgaard, P., Schulz, B.L., Thony-Meyer, L., and Hederstedt, L. (2009). Haem delivery proteins in cytochrome *c* maturation system II. Mol. Microbiol. *73*, 1058–1071.

Allen, J.W.A., Barker, P.D., Daltrop, O., Stevens, J.M., Tomlinson, E.J., Sinha, N., Sambongi, Y., and Ferguson, S.J. (2005). Why isn't 'standard' heme good enough for *c*-type and d_1-type cytochromes? Dalton Trans. 3410–3418.

Ayala-Castro, C., Saini, A., and Outten, F.W. (2008). Fe–S cluster assembly pathways in bacteria. Microbiol. Mol. Biol. Rev. *72*, 110–125.

Balasubramanian, R., Smith, S.M., Rawat, S., Yatsunyk, L.A., Stemmler, T.L., and Rosenzweig, A.C. (2010). Oxidation of methane by a biological dicopper centre. Nature *465*, 115–131.

Bali, S., Warren, M.J., and Ferguson, S.J. (2010). NirF is a periplasmic protein that binds d_1 heme as part of its essential role in d_1 heme biogenesis. FEBS J. *277*, 4944–4955.

Bedmar, E.J., Robles, E.F., and Delgado, M.J. (2005). The complete denitrification pathway of the symbiotic, nitrogen-fixing bacterium *Bradyrhizobium japonicum*. Biochem. Soc. Trans. *33*, 141–144.

Berks, B.C., Palmer, T., and Sargent, F. (2005). Protein targeting by the bacterial twin-arginine translocation (Tat) pathway. Curr. Opin. Microbiol. *8*, 174–181.

Blasco, F., Dos Santos, J.P., Magalon, A., Frixon, C., Guigliarelli, B., Santini, C.L., and Giordano, G. (1998). NarJ is a specific chaperone required for molybdenum cofactor assembly in nitrate reductase A of *Escherichia coli*. Mol. Microbiol. *28*, 435–447.

Bothe, H., Ferguson, S.J., and W.N., N. (2007). Biology of the Nitrogen Cycle. Amsterdam: Elsevier, pp. 37–56.

Brindley, A.A., Zajicek, R., Warren, M.J., Ferguson, S.J., and Rigby, S.E.J. (2010). NirJ, a radical SAM family member of the d_1 heme biogenesis cluster. FEBS Lett. *584*, 2461–2466.

Butland, G., Spiro, S., Watmough, N.J., and Richardson, D.J. (2001). Two conserved glutamates in the bacterial nitric oxide reductase are essential for activity but not assembly of the enzyme. J. Bacteriol. *183*, 189–199.

Castiglione, N., Rinaldo, S., Giardina, G., and Cutruzzola, F. (2009). The transcription factor DNR from Pseudomonas aeruginosa specifically requires nitric oxide and haem for the activation of a target promoter in *Escherichia coli*. Microbiology-Sgm *155*, 2838–2844.

Christensen, O., Harvat, E.M., Thorny-Meyer, L., Ferguson, S.J., and Stevens, J.M. (2007). Loss of ATP hydrolysis activity by CcmAB results in loss of *c*-type cytochrome synthesis and incomplete processing of CcmE. FEBS J. *274*, 2322–2332.

Cutruzzola, F., Brown, K., Wilson, E.K., Bellelli, A., Arese, M., Tegoni, M., Cambillau, C., and Brunori, M. (2001). The nitrite reductase from *Pseudomonas aeruginosa*: Essential role of two active-site histidines in the catalytic and structural properties. Proc. Natl. Acad. Sci. U.S.A. *98*, 2232–2237.

Dreher, C., Prodohl, A., Hielscher, R., Hellwig, P., and Schneider, D. (2008). Multiple step assembly of the transmembrane cytochrome b_6. J. Mol. Biol. *382*, 1057–1065.

Eaves, D.J., Groves, J., Standerman, W., James, P., Poole, R.K., White, S.A., Griffiths, I., and Cole, J.A. (1998). Involvement of the products of the nrfEFG genes in the covalent attachment of haem o to a novel cysteine–lysine motif in cytochrome *c* (552) nitrate reductase from *Escherichia coli*. Mol. Microbiol. *28*, 205–216.

Feissner, R.E., Richard-Fogal, C.C, Frowley, E.R., and Kranz, R.G. (2006). ABC transporter-medicated release of a haem chaperone allows cytochrome *c* biogenesis. Mol. Microbiol. *61*, 219–231.

Ferguson, S.J. (2009). The role of heme d_1 in denitrification. In Tetrapyrroles: birth life and death, M.J. Warren and A.J. Smith, eds. (Landes Bioscience Springer Science), pp.390–396.

Ferguson, S.J., Stevens, J.M., Allen, J.W.A., and Robertson, I.B. (2008). Cytochrome *c* assembly: A tale of ever increasing variation and mystery? Biochim. Biophys. Acta –Bioenergetics *1777*, 980–984.

Glockner, A.B., and Zumft, W.G. (1996). Sequence analysis of an internal 9.72-kb segment from the 30-kb denitrification gene cluster of *Pseudomonas stutzeri*. Biochim. Biophys. Acta –Bioenergetics *1277*, 6–12.

Goddard, A.D., Stevens, J.M., Rao, F., Mavridou, D.A., Chan, W., Richardson, D.J., Allen, J.W., and Ferguson, S.J. (2010a) *c*-Type cytochrome biogenesis can occur via a natural Ccm system lacking CcmH, CcmG, and the heme-binding histidine of CcmE. J. Biol. Chem. *285*, 22882–22889.

Goddard, A.D., Stevens, J.M., Rondelet, A., Nomerotskaia, E., Allen, J.W.A., and Ferguson, S.J. (2010b). Comparing the substrate specificities of cytochrome *c* biogenesis Systems I and II Bioenergetics. FEBS J. *277*, 726–737.

Igarashi, N., Moriyama, H., Fujiwara, T., Fukumori, Y., and Tanaka, N. (1997). The 2.8 angstrom structure of hydroxylamine oxidoreductase from a nitrifying chemoautotrophic bacterium, *Nitrosomonas europaea*. Nat. Struct. Biol. *4*, 276–284.

Ize, B., Coulthurst, S.J., Hatzixanthis, K., Caldelari, I., Buchanan, G., Barclay, E.C., Richardson, D.J., Palmer, T., and Sargent, F. (2009). Remnant signal peptides on non-exported enzymes: implications for the evolution of prokaryotic respiratory chains. Microbiology-Sgm *155*, 3992–4004.

Jafferji, A., Allen, J.W.A., Ferguson, S.J., and Fulop, V. (2000). X-ray crystallographic study of cyanide binding provides insights into the structure–function relationship for cytochrome cd_1 nitrite reductase from *Paracoccus pantotrophus*. J. Biol. Chem. *275*, 25089–25094.

Karlsson, B.G., Pascher, T., Nordling, M., Arvidsson, R.H.A., and Lundberg, L.G. (1989). Expression of the Blue copper protein Azurin from *Pseudomonas-aeruginosa* in *Escherichia-coli*. FEBS Lett.*246*, 211–217.

Kern, M., Eisel, F., Scheithauer, J., Kranz, R.G., and Simon, J. (2010). Substrate specificity of three cytochrome *c* haem lyase isoenzymes from Wolinella succinogenes: unconventional haem *c* binding motifs are not sufficient for haem c attachment by Nrfl and CcsA1. Mol. Microbiol. *75*, 122–137.

Kranz, R.G., Richard-Fogal, C., Taylor, J.S., and Frawley, E.R. (2009). Cytochrome *c* biogenesis: mechanisms for covalent modifications and trafficking of heme and for heme-iron redox control. Microbiol. Mol. Biol. Rev. *73*, 510–528.

Lanciano, P., Vergnes, A., Grimaldi, S., Guigliarelli, B., and Magalon, A. (2007). Biogenesis of a respiratory complex is orchestrated by a single accessory protein. J. Biol. Chem. *282*, 17468–17474.

Leung, Y.C., Chan, C., Reader, J.S., Willis, A.C., vanSpanning, R.J.M., Ferguson, S.J., and Radford, S.E. (1997). The pseudoazurin gene from *Thiosphaera pantotropha*: Analysis of upstream putative regulatory sequences and overexpression in *Escherichia coli*. Biochemical Journal *321*, 699–705.

Lücker, S., Wagner, M., Maixner, F., Pelletier, E., Koch, H., Vacherie, B., Rattei, T., Damsté, J.S., Spieck, E., Le Paslier, D., and Daims, H. (2010). A Nitrospira metagenome illuminates the physiology and evolution of globally important nitrite-oxidizing bacteria. Proc. Natl. Acad. Sci. U.S.A. *107*, 13479–13484.

Maillard, J., Spronk, C., Buchanan, G., Lyall, V., Richardson, D.J., Palmer, T., Vuister, G.W., and Sargent, F. (2007). Structural diversity in twin-arginine signal peptide-binding proteins. Proc. Natl. Acad. Sci. U.S.A. *104*, 15641–15646.

Martinez-Espinosa, R.M., Dridge, E.J., Bonete, M.J., Butt, J.N., Butler, C.S., Sargent, F., and Richardson, D.J. (2007). Look on the positive side! The orientation, identification and bioenergetics of 'Archaeal' membrane-bound nitrate reductases. FEMS Microbiol. Lett. *276*, 129–139.

Mendel, R.R., and Bittner, F. (2006). Cell biology of molybdenum. Biochim. Biophys. Acta Mol. Cell Res. *1763*, 621–635.

Mueller, M., and Kloesgen, R.B. (2005). The Tat pathway in bacteria and chloroplasts (Review). Mol. Membr. Biol. *22*, 113–121.

Pop, O.I., Soprova, Z., Koningstein, G., Scheffers, D.J., van Ulsen, P., Wickstrom, D., de Gier, J.W., and Luirink, J. (2009). YidC is required for the assembly of the MscL homopentameric pore. FEBS J. *276*, 4891–4899.

Rasmussen, T., Berks, B.C., Butt, J.N., and Thomson, A.J. (2002). Multiple forms of the catalytic centre, Cu-z, in the enzyme nitrous oxide reductase from P*aracoccus pantotrophus*. Biochem. J. *364*, 807–815.

Reid, G.A., Haddock, B.A., and Ingledew, W.J. (1981). Assembly of functional b-type cytochromes in membranes from a 5-aminolevulinic acid requiring mutant of *Escherichia coli*. FEBS Lett. *131*, 346–350.

Ren, Q., and Thony-Meyer, L. (2001). Physical interaction of CcmC with heme and the heme chaperone CcmE during cytochrome c maturation. J. Biol.Chem. *276*, 32591–32596.

Richard-Fogal, C.L., Frawley, E.R., Bonner, E.R., Zhu, H.F., Francisco, B.S., and Kranz, R.G. (2009). A conserved haem redox and trafficking pathway for cofactor attachment. EMBO J. *28*, 2349–2359.

Rinaldo, S., Sam, K.A., Stelitano, V., Arcovito, A., Brunori, M., Allen, J.W.A., Ferguson, S.J., and Cutruzzola, F. (2011). Observation of fast release of Na from ferrous *d*, haem allows formulation of a unified reaction mechanism for cytochrome *cd*, nitrite reductase. Biochem. J. *435*, 217–225.

Schwarz, G., Mendel, R.R., and Ribbe, M.W. (2009). Molybdenum cofactors, enzymes and pathways. Nature *460*, 839–847.

Storbeck, S., Walther, J., Muller, J., Parmar, V., Schiebel, H.M., Kemken, D., Dulcks, T., Warren, M.J., and Layer, G. (2009). The *Pseudomonas aeruginosa* nirE gene encodes the S-adenosyl-L-methionine-dependent uroporphyrinogen III methyltransferase required for heme d_1 biosynthesis. FEBS J. *276*, 5973–5982.

Velasco, L., Mesa, S., Xu, C.A., Delgado, M.J., and Bedmar, E.J. (2004). Molecular characterization of nosRZDFYLX genes coding for denitrifying nitrous oxide reductase of *Bradyrhizobium japonicum*. Antonie Van Leeuwenhoek Int. J. Gen. Mol. Microbiol.*85*, 229–235.

Volkmer, T., Becker, C., Prodohl, A., Finger, C., and Schneider, D. (2006). Assembly of a transmembrane b-Type cytochrome is mainly driven by transmembrane helix interactions. Biochim. Biophys. Acta – Biomembr. *1758*, 1815–1822.

Wood, N.J., Alizadeh, T., Bennett, S., Pearce, J., Ferguson, S.J., Richardson, D.J., and Moir, J.W. (2001). Maximal expression of membrane-bound nitrate reductase in Paracoccus is induced by nitrate via a third FNR-like regulator named NarR. J. Bacteriol. *183*, 3606–3613.

Wuebbens, M.M., and Rajagopalan, K.V. (2003). Mechanistic and mutational studies of *Escherichia coli* molybdopterin synthase clarify the final step of molybdopterin biosynthesis. J. Biol. Chem. *278*, 14523–14532.

Wunsch, P., Herb, M., Wieland, H., Schiek, U.M., and Zumft, W.G. (2003). Requirements for Cu-A and Cu-S center assembly of nitrous oxide reductase deduced from complete periplasmic enzyme maturation in the nondenitrifier *Pseudomonas putida*. J. Bacteriol. *185*, 887–896.

Xiong, J., Bauer, C.E., and Pancholy, A. (2007). Insight into the haem d_1 biosynthesis pathway in heliobacteria through bioinformatics analysis. Microbiology *153*, 3548–3562.

Zajicek, R.S., Bali, S., Arnold, S., Brindley, A.A., Warren, M.J., and Ferguson, S.J. (2009). d_1 haem biogenesis – assessing the roles of three nir gene products. FEBS J. *276*, 6399–6411.

Zajicek, R.S., Cartron, M.L., and Ferguson, S.J. (2006). Probing the unusual oxidation/reduction behavior of *Paracoccus pantotrophus* cytochrome cd_1 nitrite reductase by replacing a switchable methionine heme iron ligand with histidine. Biochemistry *45*, 11208–11216.

Zakian, S., Lafitte, D., Vergnes, A., Pimentel, C., Sebban-Kreuzer, C., Toci, R., Claude, J.B., Guerlesquin, F., and Magalon, A. (2010). Basis of recognition between the NarJ chaperone and the N-terminus of the NarG subunit from Escherichia coli nitrate reductase. FEBS J. 277, 1886–1895.

Zara, V., Conte, L., and Trumpower, B.L. (2007). Identification and characterization of cytochrome bc_1 subcomplexes in mitochondria from yeast with single and double deletions of genes encoding cytochrome bc(1) subunits. FEBS J. *274*, 4526–4539.

Zhang, W.J., Urban, A., Mihara, H., Leimkuhler, S., Kurihara, T., and Esaki, N. (2010). IscS Functions as a primary sulfur-donating enzyme by interacting specifically with MoeB and MoaD in the biosynthesis of molybdopterin in *Escherichia coli*. J. Biol.Chem. *285*, 2302–2308.

Zumft, W.G. (1997). Cell biology and molecular basis of denitrification. Microbiol. Mol. Biol.Rev. *61*, 533–616.

Zumft, W.G., and Kroneck, P.M.H. (2007). Respiratory transformation of nitrous oxide (N_2O) to dinitrogen by Bacteria and Archaea. Adv. Microb. Physiol. 52, 107–227.

Nitric Oxide Metabolism: Physiology and Regulatory Mechanisms

9

Stephen Spiro

Abstract

Nitric oxide (NO) is synthesized in bacteria as a product of the reduction of nitrite or the oxidation of arginine. NO is growth inhibitory, due to its ability to inhibit respiratory oxidases and [Fe–S] cluster containing dehydratases. NO also reacts with oxygen and biologically relevant oxygen radicals (such as superoxide) to generate a number of other toxic reactive nitrogen species. NO detoxification is typically accomplished by oxidation to nitrate, or reduction to nitrous oxide or ammonia, and these activities have been associated with a variety of enzymes. In many cases, expression of the genes encoding NO detoxification activities is controlled by NO-sensitive regulatory proteins. This chapter describes the different pathways for NO synthesis and consumption in bacteria, and the mechanisms and roles of the associated regulatory proteins. The chapter also reviews the growing body of evidence that implicates NO in regulating other physiological processes, including [Fe–S] cluster biogenesis, metabolism, motility and biofilm development.

Introduction

The first postulated and then demonstrated biological role for nitric oxide (here abbreviated as NO) was an intermediate in the respiratory pathway denitrification. In denitrification, NO functions as a terminal electron acceptor, and is synthesized and consumed by electron transport linked respiratory enzymes (Zumft, 1997; Watmough *et al.*, 1999). This role for NO, and the biochemistry, physiology and molecular biology involved, is described in other chapters in this book. Bacteria that do not denitrify may nevertheless be exposed to NO from a number of different sources. Firstly, since NO is freely diffusible and crosses biological membranes readily, non-denitrifiers in mixed communities may be exposed to NO made by their neighbours. Secondly, some non-denitrifiers synthesize NO using a homologue of the eukaryotic NO synthase that oxidizes arginine to citrulline and NO (Sudhamsu and Crane, 2009). Thirdly, bacteria that use nitrite as a terminal electron acceptor, reducing it to ammonia, make some NO as an adventitious by-product of this metabolism. Finally, pathogenic bacteria are exposed to NO made by the NO synthases of host cells (Fang, 2004). NO is a reactive free radical that is quite soluble in aqueous systems. It reacts with multiple targets in biological systems, having a particular avidity for metal centres in proteins. NO also reacts with oxygen and biologically relevant oxygen radicals (such as superoxide) to generate a plethora of reactive and toxic downstream products (sometimes collectively referred to as NO congeners, or reactive nitrogen species). NO congeners react with a wide range of cellular targets, including lipids, protein tyrosines and thiols, and DNA (Fang, 2004). Thus, organisms that are exposed to NO, whatever the source, probably require activities that detoxify NO, and/or activities that repair the damage mediated by NO and its congeners.

The goal of this chapter is to provide a review of the pathways for NO synthesis and consumption, and of the regulatory proteins that control the expression of NO detoxification activities in response to NO. Other activities that are regulated by NO will also be described briefly. The focus will be on the non-denitrifying bacteria, since NO metabolism and its regulation in denitrifying bacteria and in pathogenic strains is described in detail in other Chapters in this book.

Pathways for NO synthesis

NO synthesis by the enzymes of nitrate and nitrite respiration

In the denitrifying bacteria, the one-electron reduction of nitrite to NO is catalysed by a cytochrome cd_1 or copper-containing nitrite reductase (Watmough *et al.*, 1999). The NO generated in this reaction is subsequently reduced to N_2O by one of several types of respiratory NO reductase, depending on the species (Zumft, 2005). The biochemistry of these enzymes is discussed in Chapter 2 and will not be considered further here. Rather, the focus will be pathways for NO synthesis and consumption in non-denitrifying bacteria, such as members of the Enterobacteriaceae, especially *Escherichia coli* and *Salmonella enterica* serovar Typhimurium (*S.* Typhimurium). In the enteric bacteria, nitrate and nitrite can also be used as electron acceptors for anaerobic respiration, but are reduced to ammonia by the sequential action of nitrate reductase (either the periplasmic Nap, or membrane-associated Nar) and nitrite reductase. *E. coli* expresses a multi-subunit electron transport linked nitrite reductase, Nrf (for nitrite reduction by formate), and a cytoplasmic NADH-linked sirohaem containing nitrite reductase called NirB. In both cases the reaction product is ammonia, the physiological role of NirB is probably to remove excess nitrite (making it available for assimilation in the form of ammonia), while Nrf allows nitrite reduction to be coupled to electron transport pathways.

Reports in the 1980s first suggested that the reduction of nitrate and nitrite to ammonia in *E. coli* is accompanied by the formation of some NO. Maximum concentrations of NO measured in growing cultures were in the hundreds of micromolar range, and reduction of NO to N_2O was also detected (Ji and Hollocher, 1988a,b). In these early experiments, NO formation was suggested to arise from the reduction of nitrite, as a side-reaction of the membrane-associated respiratory nitrate reductase, Nar (Ji and Hollocher, 1989). Measurements of nitrosation of model compounds (which depends upon reactive nitrogen species derived from NO) also implicated Nar as the major source of NO (Metheringham and Cole, 1997). More recently, NO formation from nitrite has instead been proposed to be an activity of Nrf and NirB (Corker and Poole, 2003); indirect evidence also supports a role for NirB in NO formation (Weiss, 2006). This issue has been further clouded by a report that NO formation from nitrite in *S.* Typhimurium does not involve NirB or Nrf, but instead is an activity of the membrane-associated nitrate reductase (Gilberthorpe and Poole, 2008), as was first suggested for *E. coli*. Given the similarities between the enzymes in these two organisms, it seems unlikely (though not impossible) that the mechanisms of NO formation are different. Indeed, there may be multiple pathways for NO formation in the enteric bacteria, which may account for these apparently unresolved discrepancies in the literature. Nevertheless, it is clear that some NO is made in cultures respiring nitrate and/or nitrite, one consequence of which is that regulatory systems that are believed to be specific for NO are activated in cultures grown under these conditions (Hutchings *et al.*, 2000, 2002; Bodenmiller and Spiro, 2006; Constantinidou *et al.*, 2006). The endogenous production of NO as a by-product of nitrate and nitrite respiration is not confined to *E. coli* and *S.* Typhimurium, since there is evidence of the same phenomenon in, for example, *Campylobacter jejuni* (Pittman *et al.*, 2007) and *Shewanella oneidensis* (Price *et al.*, 2007).

NO synthesis by NO synthase

Some bacteria express an orthologue of the mammalian NO synthase (NOS), enzymes which oxidize arginine to form NO (Sudhamsu *et al.*, 2009; Crane *et al.*, 2010). Recent investigations have begun to shed some light on the diverse physiological roles of these enzymes. In *Streptomyces scabies*, the NO produced by NOS is required to nitrate the tryptophan moiety of the dipeptide phytotoxin thaxtomin (Kers *et al.*, 2004). The *Deinococcus radiodurans* NOS co-purifies with a tryptophanyl tRNA synthetase (one of two isoenzymes made by the organism), and the complex charges tRNA with 4-nitro-tryptophan (Buddha *et al.*, 2004b). The role of this unusual amino acid in *D. radiodurans* is not known; in *S. scabies* 4-nitro-tryptophan is a precursor for thaxtomin synthesis (Johnson *et al.*, 2009). The *D. radiodurans nos* gene is up-regulated by UV irradiation, and NOS-derived NO is required for UV-resistance. NO stimulates recovery when supplied hours after the UV exposure, and it is suggested that NO acts as a signal that activates the expression of genes required for growth recovery (Patel *et al.*, 2009). Some Gram-positive bacteria such as *Bacillus* species and *Staphylococcus aureus* also express a NOS. In *B. subtilis*, NOS-derived NO protects against the oxidative stress mediated by H_2O_2 or organic peroxides. This effect is rapid and does not require new protein synthesis (Gusarov and Nudler, 2005). Two mechanisms were suggested to account for the protective effect of NO. Firstly, NO (or, more likely, its congeners) activates a catalase, apparently by *S*-nitrosating the free cysteine which acts as a catalase inhibitor. Secondly, NO inhibits the Fenton reaction, in which H_2O_2 reacts with ferrous iron to form hydroxyl radicals. NO does this by directly inhibiting a thioredoxin reductase, which is required for the generation of reduced thiols, which are in turn required to reduce Fe^{3+} for sustained Fenton chemistry (Gusarov *et al.*, 2005). In the related pathogenic organism *Bacillus anthracis*, NOS-derived NO protects against the oxidative stress exerted by macrophages, and is thus a virulence determinant (Shatalin *et al.*, 2008). Deletion of the *nos* gene sensitizes *B. subtilis* and *S. aureus* to several cytotoxic antibiotics (Gusarov *et al.*, 2009). For some antibiotics, direct modification of the compound by NO (or its congeners) contributes to this effect. Secondly, by the mechanisms described above, NO is suggested to alleviate the oxidative stress that is believed to account (at least partly) for the killing effect of antibiotics (Kohanski *et al.*, 2007; Gusarov *et al.*, 2009). Interestingly, there is evidence to suggest that NO also counteracts oxidative stress in plants and animal cells. The mechanism of protection may involve direct reaction between NO and oxygen radicals, and, in the case of plants, an NO-stimulated increase in catalase activity (Wink *et al.*, 1996; Beligni and Lamattina, 2002).

NO detoxification

NO, at least when supplied exogenously, is somewhat growth inhibitory to *E. coli* and *S.* Typhimurium. Numerous cellular targets for NO have been described in the literature (Fang, 2004), though it is not always easy to distinguish the effects of NO from those of its congeners that arise from the reaction of NO with other species, especially oxygen. For example, it is clear that exposure of bacteria to NO causes damage to their DNA, though it is the NO congeners rather than NO itself that are the active species in this case (Weiss, 2006; Richardson *et al.*, 2009). A detailed consideration of the damage mediated by reactive nitrogen species other than NO is beyond the scope of this article. In terms of NO *per se*, it now seems likely that the most sensitive targets for inhibition are the respiratory oxidases, and the dehydratases that contain labile [4Fe–4S] clusters of the type found in aconitase (Hyduke *et al.*, 2007). Since no physiological role has been ascribed to the NO that is made endogenously in enteric bacteria, it can be regarded as being analogous to superoxide, in the sense that both are potentially harmful free radical by-products of normal respiratory processes. The endogenous production of superoxide in aerobic organisms requires the expression of detoxification activities (superoxide dismutases, catalases and peroxidases) that remove superoxide and its downstream products (Imlay, 2008). Similarly, it might be imagined that the endogenous production of NO requires the expression of enzymes that protect the cell against this NO, by converting it to a less harmful product. In *E. coli*, at least three enzymes have been shown to detoxify NO by oxidizing it (to nitrate) or reducing it (to ammonia or N_2O). These enzymes are the flavohaemoglobin, Hmp, the flavorubredoxin, FlRd and the respiratory nitrite reductase, Nrf, each of which will be briefly considered in turn in the following sections.

Flavohaemoglobin and related proteins

The flavohaemoglobin (Hmp) is the best characterized of the NO detoxification activities in non-denitrifying bacteria. Having said that, the first evidence for a role for Hmp in nitrogen metabolism came from an organism capable of denitrification, *Ralstonia eutropha*. In this case, an *hmp* mutant reduced nitrite to dinitrogen normally during anaerobic growth, but failed to show a transient accumulation of N_2O that was observed in the wild type (Cramm *et al.*, 1994). The authors speculated that Hmp might catalyse the reduction of NO to N_2O, but were unable to detect this activity *in vitro*. The expression of *hmp* in *E. coli* was shown to be up-regulated by exposure of cultures to NO, suggesting a role for Hmp in NO metabolism (Poole *et al.*, 1996). *E. coli* mutants selected for resistance to NO (which was present in the gas phase) over-expressed *hmp*, and Hmp was associated with an NO-inducible oxygen-dependent activity that oxidized NO to nitrate (Gardner *et al.*, 1998). Hmp has a C-terminal FAD and NAD(P)H binding domain, fused to an N-terminal globin-like domain. In the so-called NO dioxygenase mechanism, oxy-Hmp is proposed to react with NO to form nitrate (Gardner *et al.*, 1998; Gardner, 2005). On the other hand, it has been argued that NO binds to Hmp preferentially at physiological oxygen and NO concentrations, and that the haem-bound nitroxyl (NO^-) equivalent is then oxidized to nitrate. This is referred to as the O_2 nitroxylase or denitrosylase mechanism (Hausladen *et al.*, 2001). Under anaerobic conditions, the reaction product is N_2O, though this reaction is much slower and its physiological significance has been a matter of debate. A detailed description of Hmp biochemistry is beyond the scope of this article and interested readers may consult a number of reviews (Poole and Hughes, 2000; Wu *et al.*, 2003; Gardner, 2005; Poole, 2005).

Besides the flavohaemoglobins, the bacterial single-domain globins and the truncated globins have also been implicated in NO metabolism. The truncated haemoglobins are distantly related to vertebrate haemoglobin, though shorter by 20–40 residues (Vuletich and Lecomte, 2006). *C. jejuni* expresses a single domain globin (Cgb) and a truncated globin (Ctb), both of which are up-regulated by exposure to *S*-nitrosoglutathione (GSNO) and NO (Elvers *et al.*, 2005; Wainwright *et al.*, 2005; Monk *et al.*, 2008). Cgb is believed to catalyse the NO dioxygenase reaction (Lu *et al.*, 2007), and *cgb* mutants are sensitive to NO and nitrosative stress (Elvers *et al.*, 2005; Pittman *et al.*, 2007). The role of Ctb is less clear; mutants are not sensitive to NO, and Ctb may have a role in oxygen metabolism (Wainwright *et al.*, 2005). *Mycobacterium* species express two truncated haemoglobins, designated HbN (or trHbN) and HbO (trHbO). The HbN, when expressed in an *hmp* mutant of *E. coli*, confers an NO consumption activity and protects against NO (Pathania *et al.*, 2002). The HbN of *M. bovis* converts NO to nitrate and is required for NO metabolism *in vivo* (Ouellet *et al.*, 2002). The role of the HbO is not known, though the HbO gene (unlike HbN) fails to complement an *hmp* mutant of *S.* Typhimurium, suggesting that HbO is not involved in NO detoxification (Pawaria *et al.*, 2007).

Flavorubredoxin

The *norV* gene of *E. coli* encodes a flavorubredoxin (FlRd), a flavo di-iron protein fused to a C-terminal rubredoxin-like domain (Wasserfallen *et al.*, 1998). FlRd is similar to the two separate flavo di-iron and rubredoxin subunits of a soluble oxygen–reducing complex from *Desulfovibrio gigas* (Gomes *et al.*, 1997). The *norV* gene is co-transcribed with *norW*, which encodes a rubredoxin reductase that mediates electron transfer from NADH to FlRd (Gomes *et al.*, 2000). The FlRd/FlRd–reductase complex couples NADH oxidation to oxygen reduction *in vitro*, suggesting it might be an oxidase responsible for scavenging low concentrations of oxygen, as is the case for the similar *D. gigas* enzyme (Gomes *et al.*, 2000). The FlRd/FlRd–reductase complex also acts as an NO reductase *in vitro*, with a turnover number comparable to those of the respiratory NO reductases (Gomes *et al.*, 2002). Mutation of *norV* or *norW* eliminated or impaired, respectively, an NO inducible NO detoxification activity, and growth of the mutants was more sensitive to exogenous NO (Gardner *et al.*, 2002). Thus, the *norVW* gene products are a physiological NO reductase, which operates under anoxic or micro-oxic conditions, perhaps because the activity is oxygen sensitive (Gardner *et al.*, 2002). The *D. gigas* flavo di-iron protein/rubredoxin complex also has NO reductase activity, and complements the NO-sensitive phenotype of an *E. coli norV* mutant (Rodrigues *et al.*, 2006). Expression of *norVW* in *E. coli* is activated in the presence of NO by the product of the *norR* regulatory gene, which is divergently transcribed from *norVW* (Hutchings *et al.*, 2002; Gardner *et al.*, 2002, 2003). The properties of the NorR protein are described below.

The *norVW* genes are taxonomically widely distributed, though less so than *hmp* (Rodionov *et al.*, 2005). In pathogenic strains of *E. coli*, it is quite common for *norV* to be a pseudogene that is inactivated by one or more mutations. In the uropathogenic strain CFT073, *norV* is inactivated by a frameshift mutation, which along with several missense mutations creates a novel open reading frame on the opposite strand. In the closely related uropathogen UTI89, *norV* is intact. In 100 isolates of *E. coli* O157:H7, 58% have a 204 bp in-frame deletion in the *norV* gene that was presumed (though not proven) to inactivate FlRd. Occurrence of the *norV* deletion correlated strongly with the presence of the Shiga toxin stx_1 gene, leading to speculation that there is a selective pressure for *norV* inactivation in strains that have stx_1 (Kulasekara *et al.*, 2009). In this context, it is interesting that NO inhibits expression of the stx_2 gene in *E. coli* O157:H7 strain EDL933, an isolate which has the 204 bp deletion in *norV* (Vareille *et al.*, 2007; Kulasekara *et al.*, 2009).

Nitrite reductase

The NrfA catalytic subunit of the respiratory nitrite reductase is a periplasmically located homodimer of two penta-haem cytochromes, which interacts with membrane-associated subunits of the Nrf complex. The six electron reduction of nitrite to ammonia proceeds through enzyme-bound intermediates, NO and hydroxylamine (Einsle *et al.*, 2002). This mechanism is consistent with observations that purified NrfA can reduce both NO and hydroxylamine to ammonia (Costa *et al.*, 1990; Rudolf *et al.*, 2002; van Wonderen *et al.*, 2008). Indeed, the estimated turnover number for the reduction of NO by NrfA is much higher than those previously reported for respiratory NO reductases, Hmp and FlRd (van Wonderen *et al.*, 2008). The Nrf complex functions as a physiological NO reductase in *E. coli in vivo*, if cultures are grown under anaerobic conditions that favour its expression, and Nrf protects against exogenous sources of NO (Poock *et al.*, 2002). Transcription of the *nrf* operon is regulated by the NO-sensitive repressor NsrR (Filenko *et al.*, 2007; Browning *et al.*, 2010). However, mechanisms that regulate *nrf* expression according to the function of Nrf as a nitrite reductase seem to dominate, so that although Nrf is an NO reductase, its expression is not strongly up-regulated by exposure to NO. In other words, the role of NsrR appears to be restricted to fine-tuning of *nrf* expression (Browning *et al.*, 2010). In *S.* Typhimurium, Nrf also protects against NO in anaerobic cultures, and there is evidence that Nrf synthesis is up-regulated by exposure to NO (Mills *et al.*, 2008). Thus, in this case, Nrf may be a *bona fide* NO-inducible NO detoxification system. However, up-regulation of *nrf* expression by NO in *S.* Typhimurium is probably not mediated by NsrR, since there is a substitution in the NsrR binding site in the *nrf* promoter, which eliminates regulation by NsrR (Browning *et al.*, 2010). In *C. jejuni*, Nrf is also a physiological NO reductase that protects against exogenous sources of NO, and NO produced endogenously from nitrite (Pittman *et al.*, 2007).

As mentioned above, Nrf has also been implicated as a source of NO in *E. coli* (Corker *et al.*, 2003). Further biochemical characterization of the enzyme may be required to provide an understanding of how its two reported activities (as an NO-generating nitrite reductase and an NO reductase that produces ammonia) can be reconciled.

NO responsive regulatory networks

At least some of the NO detoxification systems described above are synthesized in response to the exposure of cells to a source of NO. This effect can be seen in very simple experiments. For example, a culture exposed to NO and allowed to recover from the resulting transient growth inhibition becomes resistant to a second dose of NO (Hyduke *et al.*, 2007). Similarly, oxygen uptake by washed cell suspensions becomes largely resistant to NO if cultures are pre-treated with a source of

NO prior to harvest (Stevanin *et al.*, 2000). These and other observations point to the existence of regulatory proteins that, in the presence of NO, switch on the expression of genes encoding enzymes that detoxify NO. In the following sections, the various regulatory proteins that mediate transcriptional responses to NO are described. First, the known regulators of detoxification activities are described (FNR/CRP family members, NorR and NsrR). Secondly, NO-responsive regulators of other activities, or of unknown activities are described (H-NOX proteins, and DosS and DosT). Finally, some regulators are mentioned briefly that react with NO, but that mediate responses to other signals as their principal role. Some relevant information derived from studies on denitrifying bacteria is included, but for a detailed description of the mechanisms that regulate the expression of denitrification genes, the reader should consult Chapter 7.

FNR/CRP family members

The FNR/CRP family is a very large group of structurally related transcriptional regulators that have diverse functions, in terms of the signals that they recognize and the processes they regulate (Korner *et al.*, 2003). A phylogenetic analysis has organized the FNR/CRP family into 21 branches (Korner *et al.*, 2003), members of at least 4 of these branches have been shown to be responsive to NO. In other words, proteins that sense NO are somewhat scattered in the very large FNR/CRP family, as is described below.

In a denitrifying strain of the photosynthetic organism *Rhodobacter sphaeroides*, an FNR/CRP family member designated NnrR is required for the expression of the genes encoding the respiratory nitrite and NO reductases. Strong circumstantial evidence implicated NO as the co-activator for NnrR (Tosques *et al.*, 1996; Kwiatkowski and Shapleigh, 1996; Kwiatkowski *et al.*, 1997). Other NO-responsive FNR/CRP family members were subsequently described in other denitrifying bacteria, including *Paracoccus denitrificans, Pseudomonas stutzeri, Ps. aeruginosa, Bradyrhizobium japonicum* and *Sinorhizobium meliloti* (Arai *et al.*, 1997; Hasegawa *et al.*, 1998; Van Spanning *et al.*, 1999; Vollack and Zumft, 2001; Mesa *et al.*, 2003; de Bruijn *et al.*, 2006). These proteins have been variously designated DNR, DnrD, NNR, and NnrR, a confusing nomenclature which turns out not to reflect phylogenetic position (Korner *et al.*, 2003). In fact, these proteins fall into two distinct branches of the family named Dnr (which includes the proteins from *Pa. denitrificans, Ps. stutzeri, Ps. aeruginosa*) and NnrR (*R. sphaeroides, B. japonicum* and *S. meliloti*). Members of the two branches are not distributed according to species phylogeny, for example representatives of both groups are found in the alpha proteobacteria (Dufour *et al.*, 2010). Rather, the distribution of Dnr and NnrR like proteins seems to correlate with the expression of copper or cytochrome cd_1 nitrite reductases (Mesa *et al.*, 2003).

A mechanism for NO sensing by Dnr/NnrR proteins has not yet been described. DnrD of *Ps. stutzeri* was reported to purify with a non-covalently bound haem (Vollack *et al.*, 2001). NO signalling mediated by NNR (*Pa. denitrificans*) and DNR (*Ps. aeruginosa*) is defective in an *E. coli* strain that cannot synthesize haem (Lee *et al.*, 2006; Castiglione *et al.*, 2009). Purified DNR binds haem, and NO binds to the haem-containing of DNR. Haem is apparently required for the DNA binding activity of DNR *in vitro*, though it is not yet known whether or how NO regulates that activity (Giardina *et al.*, 2008).

FNR from *E. coli*, a founding member of the FNR/CRP family, is a sensor of molecular oxygen and contains an oxygen labile [4Fe–4S] cluster (Kiley and Beinert, 2003). Besides O_2, FNR can also be inactivated by NO, by the formation of dinitrosyl iron complexes that result from reaction of NO with the [Fe–S] cluster (Cruz-Ramos *et al.*, 2002). Since FNR is a repressor of the *hmp* gene (Poole *et al.*, 1996), inactivation of FNR by NO provides a mechanism for NO-mediated de-repression of a defence activity. Inactivation of FNR by NO may also lead to up-regulation of the *cydAB* genes encoding the cytochrome *bd* oxidase, which endows a degree of NO resistance on respiration (Pullan *et al.*, 2007; Mason *et al.*, 2009). An alternative possibility is that *cydAB* upregulation results from an increased activity of the ArcBA sensor–regulator pair when respiration is inhibited by NO (Hyduke *et al.*, 2007). The FNR orthologue of *Azotobacter vinelandii* is also sensitive to NO, in this case inactivation of

CydR by NO would also lead to de-repression of a cytochrome *bd* oxidase, in cultures growing at low oxygen tensions (Wu *et al.*, 2000).

The NssR protein of *C. jejuni* is another member of the FNR/CRP family that is NO responsive (Elvers *et al.*, 2005). Genes in the NssR regulon include those encoding both Cgb and Ctb (see above) and are up-regulated by exposure to GSNO and NO (Monk *et al.*, 2008). Thus it is likely that NssR is a direct sensor of NO and its congeners, though NssR belongs in a different branch of the FNR/CRP family to those proteins described above (Elvers *et al.*, 2005).

NorR

NorR was discovered as an activator of the respiratory NO reductase gene in *R. eutropha* (Pohlmann *et al.*, 2000). NorR activity required the endogenous production of NO from nitrite, or exogenous provision of sodium nitroprusside, suggesting that NorR is an NO sensor (Pohlmann *et al.*, 2000). In *E. coli*, *norR* is divergently transcribed from the *norVW* genes encoding FlRd and its redox partner, and NorR activates transcription of *norVW* in response to sources of NO or nitrosative stress (Hutchings *et al.*, 2002; Gardner *et al.*, 2002, 2003; Mukhopadhyay *et al.*, 2004). There are three NorR binding sites upstream of *norVW*, and the *norAB* genes of *R. eutropha* (Tucker *et al.*, 2004; Büsch *et al.*, 2004). All three binding sites are required for transcription activation (Büsch *et al.*, 2004; Tucker *et al.*, 2010a). In *Ps. aeruginosa*, the product of the *norR* gene activates expression of the divergently transcribed *fhp* (*hmp*) gene encoding flavohaemoglobin, in response to NO (Arai *et al.*, 2005). Thus, in different organisms, NorR regulates the expression of three different enzymes that act on NO (NO reductase, FlRd, and Hmp) and it is common for *norR* to be linked to and divergently transcribed from the predicted target for NorR regulation (Rodionov *et al.*, 2005).

NorR is a σ^{54}-dependent activator with a three domain structure typical for that family of proteins: a C-terminal DNA binding domain, a central domain from the AAA$^+$ family that interacts with RNA polymerase to drive the ATP dependent transition from closed to open complex, and an N-terminal signalling domain. In NorR, the N-terminal GAF domain contains a mono-nuclear non-haem iron that is the binding site for a single molecule of NO. NO binding relieves the intramolecular inhibition exerted by the GAF domain on the central domain, which then becomes active for ATP hydrolysis and makes a productive interaction with RNA polymerase (D'Autréaux *et al.*, 2005). Site-directed mutagenesis, assays of iron binding, spectroscopy and structural modelling of the NorR proteins from *E. coli* and *R. eutropha* have suggested that the iron is coordinated by the side chains of three aspartate residues, an arginine and a cysteine (Klink *et al.*, 2007; Tucker *et al.*, 2008a).

NsrR

NsrR is a transcriptional repressor from the Rrf2 family; it was discovered in the nitrifying organism *Nitrosomonas europaea*, where it behaves as a nitrite-sensitive repressor of the gene encoding a copper nitrite reductase that reduces nitrite to NO (Beaumont *et al.*, 2004). In *Rhodobacter capsulatus*, *nsrR* is linked to genes involved in the assimilatory metabolism of nitrate, nitrite and hydroxylamine, including the *hcp* gene encoding a putative hydroxylamine reductase (Cabello *et al.*, 2004). NsrR was later shown to be an NO-sensitive repressor of the *E. coli hmp* gene, as was also predicted by comparative genomics (Rodionov *et al.*, 2005; Bodenmiller and Spiro, 2006). The balance of evidence suggests that the effector for NsrR is NO, in the case of *N. europaea* it is probably the NO generated by nitrite reduction, rather than nitrite itself (Spiro, 2007). NsrR has since been shown to regulate the expression of *hmp* in *S.* Typhimurium and *B. subtilis* (Nakano *et al.*, 2006; Bang *et al.*, 2006; Rogstam *et al.*, 2007; Gilberthorpe *et al.*, 2007) and the *norB* gene encoding a respiratory NO reductase in *Neisseria* species and *Moraxella catarrhalis* (Overton *et al.*, 2006; Rock *et al.*, 2007; Wang *et al.*, 2008). Predicted targets for NsrR regulation in other organisms include *norVW* (in *Photobacterium profundum*); thus, like NorR, NsrR regulates expression of the three dedicated NO-metabolizing enzymes (NO reductase, FlRd and Hmp) in different species (Rodionov *et al.*, 2005). As far as is known, NorR only activates a single target, while NsrR controls a regulon of multiple genes in cases that have been examined experimentally and computationally.

Confirmed additional targets for NsrR regulation include the *hcp-hcr* genes encoding the hybrid cluster protein and its redox partner, homologous genes variously designated *ytfE*, *dnrN* and *norA*, genes encoding a nitrite reductase, and a gene of unknown function, *ygbA* (Bodenmiller and Spiro, 2006; Filenko *et al.*, 2007; Heurlier *et al.*, 2008). In *Vibrio fischeri*, NsrR regulates the expression of *aox*, which encodes an NO-resistant alternative oxidase (Dunn *et al.*, 2010). The functions of some of these NsrR-regulated genes in the response to NO are discussed below. In *E. coli*, chromatin immunoprecipitation and microarray analysis has identified ~60 NsrR binding sites in intergenic regions in the chromosome, implicating the adjacent genes as possible NsrR targets. Confirmed NsrR targets from this dataset include genes involved in degradation of aromatic amines and genes involved in motility (Rankin *et al.*, 2008; Partridge *et al.*, 2009).

NsrR is similar in sequence to the *E. coli* IscR protein, a repressor of the *isc* operon, which encodes proteins involved in iron–sulfur cluster biogenesis. IscR contains a [2Fe–2S] cluster, and is proposed to regulate *isc* expression according to the cellular iron sulfur status (Schwartz *et al.*, 2001). IscR also regulates other genes, and functions as a DNA binding protein in both the [2Fe–2S] and cluster-free forms, interestingly the two forms of the protein bind to different DNA sequences (Giel *et al.*, 2006; Nesbit *et al.*, 2009). IscR and NsrR share three conserved cysteine residues, in both cases the only three cysteines in the protein. Thus it is very likely that the cysteines act as ligands to the cluster in IscR, in which case NsrR is very likely also to be an [Fe–S] protein, which immediately suggests a possible mechanism of reaction with NO. Indeed, NsrR proteins from three organisms expressed and purified from *E. coli* contain [Fe–S] clusters (see below). In the case of IscR, there is a prediction that the fourth ligand to the [Fe–S] cluster is a conserved glutamate residue in the DNA binding domain (Zeng *et al.*, 2008). However, there is no direct evidence to support this assignment, which is based only on the observation that some substitutions of the glutamate residue prevent incorporation of the cluster. It has been predicted that the fourth ligand to the cluster in NsrR is a conserved histidine in the DNA binding domain (Tucker *et al.*, 2010b), the residue at position 101 in the sequence logo shown in Fig. 9.1A. There is currently no evidence to support the assignment of this residue as a cluster ligand.

The NsrR proteins from *Neisseria gonorrhoeae* and *Streptomyces coelicolor* purified from recombinant *E. coli* in the presence of oxygen contained a [2Fe–2S] cluster (Tucker *et al.*, 2008b; Isabella *et al.*, 2009). The *Bacillus subtilis* protein, modified with a hexa-histidine tag and purified in the absence of oxygen contained a [4Fe–4S] cluster (Yukl *et al.*, 2008). The proteins from *S. coelicolor* and *B. subtilis* react with NO to form dinitrosyl iron complexes, and reaction of the *S. coelicolor* and *N. gonorrhoeae* proteins with NO lowers their affinity for DNA (Tucker *et al.*, 2008b; Yukl *et al.*, 2008; Isabella *et al.*, 2009). Exposure of the *B. subtilis* NsrR to oxygen led to conversion of the [4Fe–4S] cluster to a [2Fe–2S] cluster, which was itself unstable to oxygen (Yukl *et al.*, 2008). In contrast, the [2Fe–2S] proteins from *S. coelicolor* and *N. gonorrhoeae* appear to be air stable (Tucker *et al.*, 2008b; Isabella *et al.*, 2009). These observations suggest either that there is heterogeneity in NsrR proteins from different organisms in the nature of the [Fe–S] cluster, or that the [2Fe–2S] proteins are degradation products that result from isolating the protein in the presence of oxygen. In the latter case, it is noteworthy that the [2Fe–2S] proteins retain the ability to bind to DNA in an NO-sensitive fashion (Tucker *et al.*, 2008b; Isabella *et al.*, 2009).

NsrR is a widespread protein, with >250 proteins being annotated as NsrR in sequenced genomes (although there is some degree of uncertainty associated with sequence-based annotations, in other words these proteins are not necessarily NsrR orthologues). Fig. 9.1 shows a sequence logo generated from an alignment of 273 proteins annotated as NsrR. This analysis reveals conservation of residues in the DNA recognition helix of the predicted DNA binding domain; these residues are presumably involved in DNA binding. There is considerable flexibility in the region of the protein containing the conserved cysteine residues, where the spacing between cysteines is somewhat variable. Near the C-terminus of these proteins there are several conserved or partly conserved hydrophobic residues with a spacing of

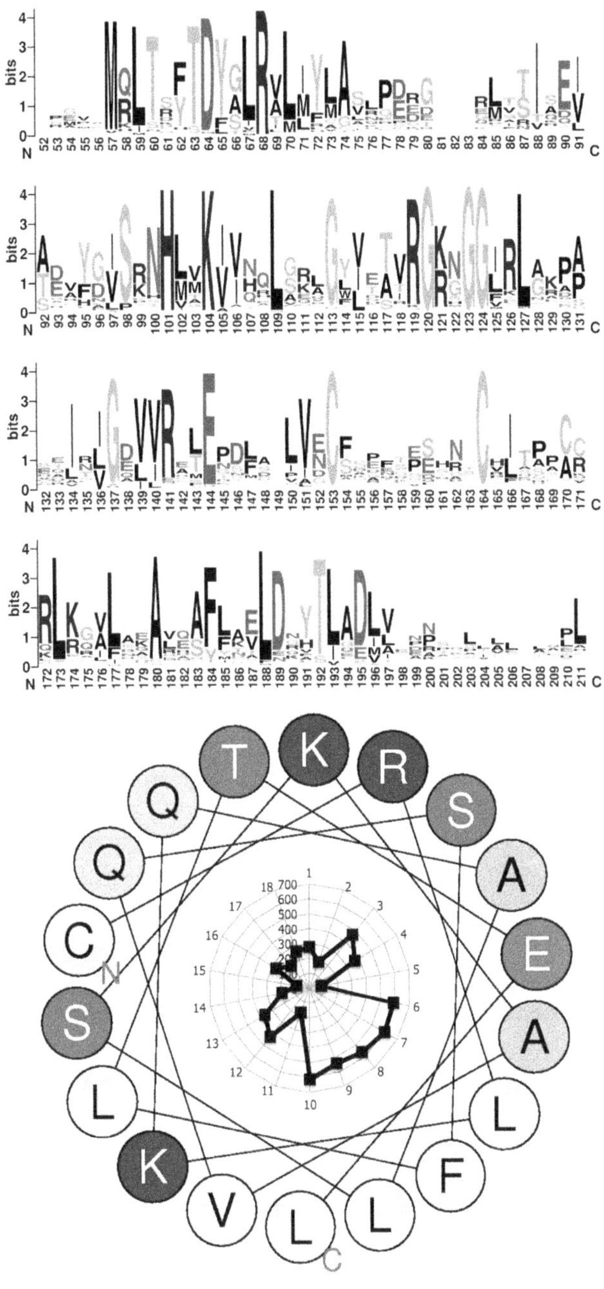

Figure 9.1 Sequence analysis of NsrR proteins. (A) Sequence logo (weblogo.berkeley.edu) of 277 sequences annotated as NsrR in the National Microbial Pathogen Data Resource (www.nmpdr.org). Four sequences were excluded because they introduce long insertions into the alignment; the remaining 273 sequences were aligned with ClustalW2 (www.ebi.ec.uk). The *E. coli* NsrR sequence initiates with the methionine at position 57 in this diagram, and ends with a glutamate at position 210. Residues 158–163 are absent from the *E. coli* protein, which therefore has a spacing of four amino acids between the first two conserved cysteines. The predicted helix–turn–helix motif of the DNA binding domain extends approximately from positions 87 to 107. (B) Helical wheel representation of residues Cys-102 to Leu-119 of the *E. coli* NsrR (positions 171–188 in Fig. 9.1A). The helical wheel was generated with HeliQuest (heliquest.ipmc.cnrs.fr/). Inside the helical wheel is shown a plot of the 'quality' of the alignment (the quality score calculated by ClustalW) for each of the 18 positions of the helical wheel, illustrating the high degree of conservation of hydrophobic residues at 5 positions of the putative hydrophobic surface. The numbers around the circumference of this plot refer to positions around the flat representation of the helix, not to positions in the linear sequence.

six or seven other amino acids. This arrangement is suggestive of a hydrophobic interaction surface, indeed this region of the protein might form an amphipathic helix, as suggested by a helical wheel projection (Fig. 9.1B). It is suggested that this helix represents a dimerization domain, NsrR has been reported to be dimeric in solution and when bound to DNA (Tucker *et al.*, 2008b).

H-NOX domain proteins

The major receptor for NO in eukaryotes is the haem-containing regulatory domain of the soluble guanylate cyclase (sGC). The haem binding domain of sGC has homologues in bacterial genomes, which, when originally detected, were named HNOB, for haem NO binding (Iyer *et al.*, 2003). Bacterial HNOB domains are found fused to the methyl accepting domains of chemoreceptors, or as single domain proteins. In the latter case, the proteins are typically encoded in operons together with genes for two component regulatory proteins, or for GGDEF proteins with roles in signalling via cyclic di-GMP (Iyer *et al.*, 2003). An HNOB domain from *Clostridium botulinum* (that is fused to a methyl accepting chemotaxis domain in the native protein) contains haem and binds NO with extremely high affinity (Nioche *et al.*, 2004). The prediction that this protein mediates taxis towards or away from NO has not been tested. A 'standalone' HNOB protein from *Vibrio cholerae* is encoded in an operon with a histidine kinase, and also binds NO (Karow *et al.*, 2004). On the other hand, the homologous domain from a predicted chemoreceptor in *Thermoanaerobacter tengcongensis*, in addition to binding NO, also forms a stable ferrous-oxy complex (Karow *et al.*, 2004). Thus, the domain was renamed H-NOX, for haem-nitric oxide/oxygen (Karow *et al.*, 2004). Despite quite extensive biophysical characterization (Boon and Marletta, 2005) the physiological roles of bacterial H-NOX domain proteins remain poorly understood. As mentioned above, genome context information suggests roles in chemotaxis, and signal transduction pathways involving sensory histidine kinases and GGDEF proteins. But the regulatory outputs of these pathways are not well characterized in any organism. In *Shewanella oneidensis*, an H-NOX protein, when bound by NO, inhibits the activity of a sensory histidine kinase (the two proteins are encoded by adjacent genes that are likely co-transcribed). The clear implication is that NO regulates the activity of a response regulator via the H-NOX protein/sensor kinase complex, but the response regulator and its target genes have not yet been identified (Price *et al.*, 2007). An H-NOX protein from the squid symbiont *Vibrio fischeri* was recently characterized (Wang *et al.*, 2010). This protein is also encoded by a gene linked to a gene encoding a histidine kinase, and the protein binds NO but not oxygen. Transcriptomic analysis identified 40 genes differentially regulated by NO in an H-NOX dependent manner. Further analysis revealed that Fur-repressed genes in *V. fischeri* are down-regulated by NO, and that this down-regulation is abolished in a mutant lacking the H-NOX protein. The authors suggest that the H-NOX protein, together with its cognate histidine kinase and a still to be identified response regulator, somehow interact with Fur (either directly or indirectly) to cause enhanced repression of Fur regulated genes in the presence of NO (Wang *et al.*, 2010).

Recent work has also provided experimental evidence for a role for H-NOX proteins in the regulation of biofilm formation, via the second messenger cyclic di-GMP. *Legionella pneumophila*, unusually, encodes two H-NOX proteins. One, designated H-NOX1, is encoded by a gene linked to the gene encoding a GGDEF-EAL protein implicated in cyclic di-GMP metabolism. The latter protein has a diguanylate cyclase activity (albeit rather low), which is inhibited by H-NOX1 only when it is in the ferrous-nitrosyl state. An *hnox1* mutant has a hyper-biofilm phenotype, leading to a model in which NO inhibits biofilm formation via H-NOX1 mediated inhibition of cyclic di-GMP synthesis (Carlson *et al.*, 2010).

DosS/DosT

Although regarded as an obligate aerobe, *Mycobacterium tuberculosis* can survive long periods of oxygen starvation (hypoxia). Persistence under hypoxic conditions is thought to be important in the host, since *M. tuberculosis* can survive in a latent form in granulomas, where the oxygen tension is low (Rustad *et al.*, 2009). The DosR (also called DevR) protein of *M. tuberculosis* is a response regulator that controls a regulon of genes expressed in response to hypoxia, NO and carbon monoxide (CO), and a variety of other signals (Rustad *et al.*, 2009). The role that the DosR regulon plays in persistence under hypoxic conditions and long-term survival in the host is controversial (Rustad *et al.*, 2009). DosR is phosphorylated by two similar histidine kinases, DosS (also called DevS) and DosT. Both proteins have two N-terminal GAF domains, the first binds haem and is the signal receptor, the function of the second is not known. Two slightly different views have emerged of the signal-sensing mechanisms of DosS and DosT. In one study, DosT and DosT were isolated in the ferrous form and DosT was extremely resistant to oxidation, DosS somewhat less so. In both cases, auto-phosphorylation of the histidine kinase was inhibited by oxygen. While both NO and CO bound avidly, the CO and NO bound forms were catalytically active. According to these results, both DosS and DosT are sensors of molecular oxygen, and the ability of NO (and presumably CO) to activate the DosR regulon results from the fact that NO and CO bind to the proteins with high affinity, preventing them from being inhibited by O_2 (Sousa *et al.*, 2007). In another study, oxygen-exposed DosS was found to be readily oxidized to the ferric (met) form, in which it was inactive for auto-phosphorylation. The ferrous-NO form of DosS was as active as the deoxy form. Thus, DosS was suggested to be a redox (as opposed to oxygen) sensor. This model requires a reductant to convert the inactive (met) form of DosS to the active (deoxy) form. In this study, DosT was again found to be resistant to oxidation, and the NO bound form had a similar auto-phosphorylation activity as the deoxy form (Kumar *et al.*, 2007). A subsequent study has also reported that DosS is prone to oxidation in air, and that it can be reduced by $FMNH_2$ but not by NADH (Cho *et al.*, 2009). In the context of NO, one conclusion that can be drawn is that NO is not a true activity switching ligand for DosS/DosT. Rather, NO activates the DosR regulon because it binds avidly to these proteins and locks them in an active state in which they are resistant to inhibition by molecular oxygen or by oxidation. In *Mycobacterium smegmatis* only a single histidine kinase is known to phosphorylate DosR. That protein is now called DosT, since like the *M. tuberculosis* protein, it is resistant to oxidation and is regulated by molecular oxygen. Interestingly, the ferrous-NO form of this DosT seems to be significantly more active for auto-phosphorylation than the deoxy form, suggesting that it may be a *bona fide* NO sensor (Lee *et al.*, 2008).

Other regulators that react with NO

NO is a reactive species which readily interacts with (for example) metal centres in proteins, especially haem, non-haem iron, [Fe–S] clusters and zinc. A consequence is that regulatory proteins containing these metals may have activities that are sensitive to NO, even if the proteins do not necessarily mediate a physiological response to NO exposure. Deciding whether a particular protein falls into this category or is a true NO sensor is not necessarily straightforward. A case in point is SoxR, which is activated by nitrosylation of its [2Fe–2S] cluster (Ding and Demple, 2000). Genes activated by SoxR (via SoxS) were reported to be required for resistance to macrophage-derived NO (Nunoshiba *et al.*, 1993). However, SoxRS regulated genes are not consistently up-regulated by NO in microarray data (Spiro, 2007), casting some doubt on the physiological relevance of the reaction of SoxR with NO.

IscR is another [2Fe–2S] protein that reacts with NO (at least as judged by *in vivo* data), a reaction that that might be physiologically significant. IscR-regulated genes are de-repressed by exposure of cultures to a source of NO (Pullan *et al.*, 2007; Hyduke *et al.*, 2007). Since the IscR regulon includes genes involved in the biogenesis of [Fe–S] clusters (Schwartz *et al.*, 2001), a possible physiological rationale for this response is that IscR-regulated genes play a role in the repair of clusters damaged or modified by NO (Hyduke *et al.*, 2007).

Iron-containing regulatory proteins that have been reported to be sensitive to NO include the ferric uptake regulator, Fur, and its homologue from *Bacillus subtilis*, the peroxide sensor, PerR. These and other cases are discussed in more detail in Spiro (2007) where citations to the original literature can also be found. NO also reacts with haem-containing regulatory proteins that sense

CO and control genes involved in CO metabolism (Clark *et al.*, 2006; Marvin *et al.*, 2008), though this reaction may have no physiological significance. Many investigators interested in responses to NO have used compounds such as *S*-nitrosoglutathione (GSNO) as an NO mimic or source. In fact, while GSNO does release NO slowly by homolytic decomposition, it may also nitrosate cysteine thiols directly, an activity not normally associated with NO *per se*. Thus, GSNO and related compounds (such as *S*-nitrosocysteine) impose a 'nitrosative stress', this being originally defined as a stress caused by a treatment that results in nitrosation of protein thiols (Hausladen *et al.*, 1996). Thus NO itself does not cause nitrosative stress, *sensu stricto* (Hausladen *et al.*, 1996). Indeed, comparison of array data for cultures treated with NO (Pullan *et al.*, 2007; Hyduke *et al.*, 2007) and GSNO (Flatley *et al.*, 2005; Jarboe *et al.*, 2008) reveals overlapping but non-identical effects. Importantly, some regulatory proteins may be sensitive to nitrosative stress (because they contain reactive thiols) but not to NO *per se*. One such example may be OxyR (Hausladen *et al.*, 1996; Kim *et al.*, 2002), a case which has been extensively discussed on the literature (Poole *et al.*, 2004). While the major role of OxyR is to mediate the response to H_2O_2, recent evidence implicates SsrB of *S*. Typhimurium as a specific sensor of reactive nitrogen species that uses a thiol-based mechanism (Husain *et al.*, 2010). Other regulators that might be sensitive to nitrosative stress include MetR (Flatley *et al.*, 2005), the zinc protein NmlR (Seib *et al.*, 2006), and zinc proteins in general (Schapiro *et al.*, 2003).

Other processes regulated by NO

So far, the discussion of physiological processes that are regulated by NO has been largely confined to the NO detoxification activities. But, it is becoming increasingly clear that the regulatory effects of NO extend beyond the NO detoxification pathways, to include repair processes, and other activities that ostensibly have little or nothing to do with NO. Some of these NO-regulated processes will be reviewed briefly in this section.

[Fe–S] cluster repair

As has been mentioned, [Fe–S] clusters are sensitive targets for NO, and the ability to return nitrosylated clusters to their native state may be an important component of the recovery from NO exposure. NO reacts *in vitro* with purified [Fe–S] proteins to form protein bound dinitrosyl iron complexes (DNIC) with a characteristic signal in electron paramagnetic resonance (EPR) spectra. The formation of this signal has been observed *in vivo* in cells over-expressing dihydroxyacid dehydratase and exposed to a source of NO, and in SoxR (Ding *et al.*, 2000; Ren *et al.*, 2008). Other dehydratases including fumarase, aconitase and 6-phosphogluconate dehydratase are sensitive to NO *in vivo*, presumably because of nitrosylation of their [Fe–S] clusters (Gardner *et al.*, 1997; Woodmansee and Imlay, 2003; Justino *et al.*, 2007). Enzymes inactivated by DNIC formation can regain activity *in vivo*, in the absence of new protein synthesis, with the repair being considerably accelerated or only observed in the presence of oxygen (Gardner *et al.*, 1997; Ren *et al.*, 2008). Two rather different (though not necessarily mutually exclusive) views have emerged in the recent literature concerning the mechanism of repair of nitrosylated clusters. On the one hand, the di-iron rubrethryin-like protein YtfE has been suggested (based on evidence from *in vivo* and *in vitro* experiments) to be required for the repair of clusters damaged by H_2O_2 or NO (Justino *et al.*, 2007). Expression of the *E. coli ytfE* gene is regulated by NsrR (Bodenmiller and Spiro, 2006), which is consistent with a role for YtfE in the repair of clusters damaged by NO, though the biochemical details of the suggested repair process have not been described. YtfE activity may be required to return NsrR to its resting state in cells recovering from NO exposure (Overton *et al.*, 2008). On the other hand, the repair of nitrosylated clusters has been reported to occur in aerobically growing cells in the presence of a protein synthesis inhibitor, with repair requiring only L-cysteine and no protein factors (Ren *et al.*, 2008). Also, as mentioned above, the IscR-regulated genes encoding cluster biogenesis proteins are also up-regulated by exposure to NO. In *R. eutropha*, a YtfE homologue is encoded by a gene designated *norA* that is co-transcribed with the NO reductase gene *norB* under the control of NorR (Pohlmann *et al.*, 2000). In this organism, NorA has been suggested to act to lower the concentration of free NO in the cytoplasm (Strube *et al.*, 2007).

Motility and biofilm formation

Several reports indicate that NO influences motility and the formation and/or dispersal of bacterial biofilms. The motile and sessile lifestyles are typically subject to reciprocal regulation, such that signals that stimulate one also inhibit the other. In *Pseudomonas aeruginosa*, mutants lacking the quorum sensing regulator RhlR are defective for biofilm formation under anaerobic conditions, because of dysregulation of denitrification genes, which leads to an accumulation of NO (Yoon *et al.*, 2002). *P. aeruginosa* mutants lacking the respiratory nitrite reductase (which reduces nitrite to NO) showed a reduced ability to disperse from biofilm; whereas mutants lacking the NO reductase exhibited an enhanced dispersal. These experiments point to a role for NO as a signal that stimulates biofilm dispersal, and stimulates swimming motility (Barraud *et al.*, 2006). Another study has suggested that endogenously generated NO stimulates swarming but not swimming motility in *P. aeruginosa* (Van Alst *et al.*, 2007). A mechanistic explanation for the effects of NO on biofilm development has recently begun to emerge. NO is proposed to stimulate dispersal by lowering the intracellular concentration of the second messenger cyclic di-GMP. NO effects on cyclic di-GMP levels require a protein called BdlA, which has two PAS domains fused to a methyl accepting chemotaxis domain. The PAS domains are candidate receptors for NO, but BdlA must regulate cyclic d-GMP levels indirectly, since it does not have domains associated with cyclic di-GMP synthesis or degradation (Morgan *et al.*, 2006; Barraud *et al.*, 2009). Thus, some components of this NO-stimulated signal transduction pathway remain to be identified. As mentioned above, NO may inhibit biofilm formation in *L. pneumophila* by indirectly inhibiting the synthesis of cyclic di-GMP (Carlson *et al.*, 2010).

Preliminary evidence suggests that NO also regulates motility and/or biofilm development in other organisms. In *Staphylococcus aureus*, biofilm formation was impaired by the respiratory reduction of nitrate or nitrite, or by an exogenous source of NO. Nitrite (perhaps *via* the endogenous generation of NO) caused a down-regulation of the *icaABCD* operon, which is required for the production of polysaccharide intercellular adhesin, which is required for cell accumulation during the maturation of biofilm (Schlag *et al.*, 2007). In the ammonia oxidizer *N. europaea*, NO stimulated biofilm formation, an effect that was correlated with a reduced abundance of some flagellar proteins (Schmidt *et al.*, 2004). Finally, in *E. coli*, positive and negative effects of NO on motility and biofilm formation, respectively, may at least in part be explained by NsrR regulation of the *fliAZ* genes (Partridge *et al.*, 2009). The *fliA* gene product is an alternative sigma factor, σ^{28}, which is required for the transcription of genes encoding structural components of the flagellum, and chemotaxis proteins. The *fliZ* gene product is, indirectly, a negative regulator of sessile growth (Pesavento *et al.*, 2008).

Reprogramming metabolism

The inhibition of growth by NO exposure can be explained, at least partly, by the inhibition of metabolic processes by NO, especially aerobic respiration and the reactions catalysed by the [Fe–S] containing dehydratases. Thus it, is perhaps not surprising that NO can stimulate a reprogramming of metabolism, by changing patterns of gene expression. One example has already been described: the up-regulation of the *cydAB* genes encoding the cytochrome *bd* oxidase that endows a degree of NO resistance on respiration (Hyduke *et al.*, 2007; Mason *et al.*, 2009). In *V. fischeri*, the gene encoding an NO-resistant alternative oxidase is a member of the NsrR regulon (Rodionov *et al.*, 2005; Dunn *et al.*, 2010). Paradoxically, NO mediated inhibition of respiration may protect against oxidative stress, by causing an accumulation of NADH that is then available for detoxification activities (Husain *et al.*, 2008).

Besides respiration, NO may exert effects on the expression of genes encoding enzymes involved in carbon and nitrogen metabolism. In *S. aureus*, an NO-inducible lactate dehydrogenase allows for redox-balanced metabolism in cells exposed to NO (Richardson *et al.*, 2008). The NsrR regulon of *E. coli* and other organisms includes the *hcp-hcr* genes, which encode the hybrid cluster protein (Hcp, also called the prismane protein) and its redox partner, and *hcp* is regulated by other NO sensors in different genomes (Rodionov *et al.*, 2005; Filenko *et al.*, 2007). The physiological role of Hcp is not well understood, it has been suggested to act as a hydroxylamine reductase (Wolfe *et al.*, 2002; Cabello *et al.*, 2004), though it is not clear why this activity would need to be NO-inducible. On the other hand, Hcp has been suggested to act as a peroxidase (Almeida *et al.*, 2006), though again a rationale for NO-inducibility is not obvious. Since NO is known to regulate genes of unknown function, such as *ygbA* (Bodenmiller and Spiro, 2006), it seems possible that other effects of NO on metabolism remain to be discovered.

NO and virulence

In response to infection, phagocytic cells of the innate immune system synthesize superoxide and NO, the enzymes involved being the phagocyte oxidase and the inducible NO synthase (Fang, 2004). These activities are partly responsible for the efficiency with which phagocytes kill invading pathogens. Conversely, some pathogens are resistant to phagocyte killing, and there is some evidence that the ability to detoxify host generated O and N radicals is an important component of this resistance. The synthesis of NO by host cells and its metabolism by pathogens is described in detail in Chapter 12, and so will not be further discussed here. Suffice to say, that at least some of the NO detoxification systems described in this Chapter are virulence determinants in pathogens. For example, the *Salmonella enterica* Hmp is required for virulence in mouse and for survival in macrophages (Bang *et al.*, 2006; Gilberthorpe *et al.*, 2007). Thus, an attractive hypothesis is that the enzymes and their regulators that evolved to protect against the NO that is endogenously generated have been recruited by pathogenic strains to protect against the NO made by host cells.

Conclusions

Bacteria have evolved multiple pathways for metabolizing NO, and a variety of regulatory mechanisms for controlling the expression of those pathways. However, NO does not only regulate NO detoxification pathways, and the list of other physiological processes that are influenced by NO continues to grow. The diversity of regulatory proteins that mediate genetic and behavioural responses NO is similarly impressive and is, presumably, a testament to the importance of NO in the biology of many bacteria. Doubtless, much remains to be learned about the interaction of NO with biological systems, at both the molecular and physiological levels.

Acknowledgements

Work in the author's laboratory is supported by a grant from the National Science Foundation (MCB-0702858). I am grateful to James Moir for his comments on the manuscript.

Note added in proof

We recently resequenced the norV gene of *Escherichia coli* strain CFT073 and found that the frameshift mutation in sequence databases is an error. Our data indicate that the norV gene is intact in this organism and is not a pseudogene. We are able to detect FlRd activity in CFT073 extracts, confirming that the norVW genes express functional products.

WhiB-like (Wbl) proteins from Actinomycetes (such as *Streptomyces coelicolor* and *Mycobacterium tuberculosis*) are DNA-binding regulatory proteins that contain a [4Fe-4S] cluster. WhiD of *S. coelicolor* and WhiB1 of *M. tuberculosis* react with NO extremely rapidly. The stoichiometry of this reaction is eight NO per cluster, the product being an octa-nitrosylated cluster in which the inorganic sulfurs have been expelled and replaced by bridging cysteine thiolates (Crack *et al.*, 2011). Reaction of NO with the [4Fe-4S] cluster of WhiB1 increases the protein's affinity for DNA (Smith *et al.*, 2010).

References

Almeida, C.C., Romao, C.V., Lindley, P.F., Teixeira, M., and Saraiva, L.M. (2006). The role of the hybrid cluster protein in oxidative stress defense. J. Biol. Chem. *281*, 32445–32450.

Arai, H., Hayashi, M., Kuroi, A., Ishii, M., and Igarashi, Y. (2005). Transcriptional regulation of the flavohemoglobin gene for aerobic nitric oxide detoxification by the second nitric oxide-responsive regulator of *Pseudomonas aeruginosa*. J. Bacteriol. *187*, 3960–3968.

Arai, H., Kodama, T., and Igarashi, Y. (1997). Cascade regulation of the two CRP/FNR-related transcriptional regulators (ANR and DNR) and the denitrification enzymes in *Pseudomonas aeruginosa*. Mol. Microbiol. *25*, 1141–8.

Bang, I.-S., Liu, L., Vazquez-Torres, A., Crouch, M.-L., Stamler, J.S., and Fang, F.C. (2006). Maintenance of nitric oxide and redox homeostasis by the *Salmonella* flavohemoglobin Hmp. J. Biol. Chem. *281*, 28039–47.

Barraud, N., Hassett, D.J., Hwang, S.-H., Rice, S.A., Kjelleberg, S., and Webb, J.S. (2006). Involvement of nitric oxide in biofilm dispersal of *Pseudomonas aeruginosa*. J. Bacteriol. *188*, 7344–7353.

Barraud, N., Schleheck, D., Klebensberger, J., Webb, J.S., Hassett, D.J., Rice, S.A., and Kjelleberg, S. (2009). Nitric oxide signaling in *Pseudomonas aeruginosa* biofilms mediates phosphodiesterase activity, decreased cyclic di-GMP levels, and enhanced dispersal. J. Bacteriol. *191*, 7333–7342.

Beaumont, H.J.E., Lens, S.I., Reijnders, W.N.M., Westerhoff, H.V., and van Spanning, R.J.M. (2004). Expression of nitrite reductase in *Nitrosomonas europaea* involves NsrR, a novel nitrite-sensitive transcription repressor. Mol. Microbiol. *54*, 148–158.

Beligni, M.V., and Lamattina, L. (2002). Nitric oxide interferes with plant photo-oxidative stress by detoxifying reactive oxygen species. Plant Cell Environ. *25*, 737–748.

Bodenmiller, D.M., and Spiro, S. (2006). The *yjeB* (*nsrR*) gene of *Escherichia coli* encodes a nitric oxide sensitive transcriptional regulator. J. Bacteriol. *188*, 874–881.

Boon, E.M., and Marletta, M.A. (2005). Ligand discrimination in soluble guanylate cyclase and the H-NOX family of heme sensor proteins. Curr. Opin. Chem. Biol. *9*, 441–446.

Browning, D.F., Lee, D.J., Spiro, S., and Busby, S.J.W. (2010). Down-regulation of the *Escherichia coli* K-12 *nrf* promoter by binding of the NsrR nitric oxide-sensing transcription repressor to an upstream site. J. Bacteriol. *192*, 3824–3828.

de Bruijn, F.J., Rossbach, S., Bruand, C., and Parrish, J.R. (2006). A highly conserved *Sinorhizobium meliloti* operon is induced microaerobically via the FixLJ system and by nitric oxide (NO) via NnrR. Env. Microbiol. *8*, 1371–1381.

Buddha, M.R., Keery, K.M., and Crane, B.R. (2004a). An unusual tryptophanyl tRNA synthetase interacts with nitric oxide synthase in *Deinococcus radiodurans*. Proc. Natl. Acad. Sci. U.S.A. *101*, 15881–15886.

Buddha, M.R., Tao, T., Parry, R.J., and Crane, B.R. (2004b). Regioselective nitration of tryptophan by a complex between bacterial nitric-oxide synthase and tryptophanyl-tRNA synthetase. J. Biol. Chem *279*, 49567–49570.

Büsch, A., Pohlmann, A., Friedrich, B., and Cramm, R. (2004). A DNA region recognized by the nitric oxide-responsive transcriptional activator NorR is conserved in ß- and γ-Proteobacteria. J. Bacteriol. *186*, 7980–7987.

Cabello, P., Pino, C., Olmo-Mira, M.F., Castillo, F., Roldan, M.D., and Moreno-Vivian, C. (2004). Hydroxylamine assimilation by *Rhodobacter capsulatus* E1F1: requirement of the *hcp* gene (hybrid cluster protein) located in the nitrate assimilation *nas* gene region for hydroxylamine reduction. J. Biol. Chem. *279*, 45485–45494.

Carlson, H.K., Vance, R.E., and Marletta, M.A. (2010). H-NOX regulation of c-di-GMP metabolism and biofilm formation in *Legionella pneumophila*. Mol. Microbiol. *77*, 930–942.

Castiglione, N., Rinaldo, S., Giardina, G., and Cutruzzolà, F. (2009). The transcription factor DNR from *Pseudomonas aeruginosa* specifically requires nitric oxide and haem for the activation of a target promoter in *Escherichia coli*. Microbiology *155*, 2838–2844.

Cho, H.Y., Cho, H.J., Kim, Y.M., Oh, J.I., and Kang, B.S. (2009). Structural insight into the heme-based redox sensing by DosS from *Mycobacterium tuberculosis*. J. Biol. Chem *284*, 13057–13067.

Clark, R.W., Lanz, N.D., Lee, A.J., Kerby, R.L., Roberts, G.P., and Burstyn, J.N. (2006). Unexpected NO-dependent DNA binding by the CooA homolog from *Carboxydothermus hydrogenoformans*. Proc. Natl. Acad. Sci. U.S.A. *103*, 891–896.

Constantinidou, C., Hobman, J.L., Griffiths, L., Patel, M.D., Penn, C.W., Cole, J.A., and Overton, T.W. (2006). A reassessment of the FNR regulon and transcriptomic analysis of the effects of nitrate, nitrite, NarXL, and NarQP as *Escherichia coli* K12 adapts from aerobic to anaerobic growth. J. Biol. Chem. *281*, 4802–4815.

Corker, H., and Poole, R.K. (2003). Nitric oxide formation by *Escherichia coli*: dependence on nitrite reductase, the NO-sensing regulator Fnr, and flavohemoglobin Hmp. J. Biol. Chem. *278*, 31584–31592.

Costa, C., Moura, J.J., Moura, I., Liu, M.Y., Peck, H.D., LeGall, J., Wang, Y.N., and Huynh, B.H. (1990). Hexaheme nitrite reductase from *Desulfovibrio desulfuricans*. Mössbauer and EPR characterization of the heme groups. J. Biol. Chem. *265*, 14382–14388.

Cramm, R., Siddiqui, R.A., and Friedrich, B. (1994). Primary sequence and evidence for a physiological function of the flavohemoprotein of *Alcaligenes eutrophus*. J. Biol. Chem. *269*, 7349–54.

Crane, B.R., Sudhamsu, J., and Patel, B.A. (2010). Bacterial nitric oxide synthases. Annu. Rev. Biochem *79*, 445–470.

Cruz-Ramos, H., Crack, J., Wu, G., Hughes, M.N., Scott, C., Thomson, A.J., Green, J., and Poole, R.K. (2002). NO sensing by FNR: regulation of the *Escherichia coli* NO-detoxifying flavohaemoglobin, Hmp. EMBO J. *21*, 3235–44.

D'Autréaux, B., Tucker, N.P., Dixon, R., and Spiro, S. (2005). A non-haem iron centre in the transcription factor NorR senses nitric oxide. Nature *437*, 769–72.

Ding, H., and Demple, B. (2000). Direct nitric oxide signal transduction via nitrosylation of iron–sulfur centers in the SoxR transcription activator. Proc. Natl. Acad. Sci. U.S.A. *97*, 5146–50.

Dufour, Y.S., Kiley, P.J., and Donohue, T.J. (2010). Reconstruction of the core and extended regulons of global transcription factors. PLoS Genet. *6*, e1001027.

Dunn, A.K., Karr, E.A., Wang, Y., Batton, A.R., Ruby, E.G., and Stabb, E.V. (2010). The alternative oxidase (AOX) gene in *Vibrio fischeri* is controlled by NsrR and upregulated in response to nitric oxide. Mol. Microbiol. *77*, 44–55.

Einsle, O., Messerschmidt, A., Huber, R., Kroneck, P.M.H., and Neese, F. (2002). Mechanism of the six electron reduction of nitrite to ammonia by cytochrome *c* nitrite reductase. J. Am. Chem. Soc. *124*, 11737–11745.

Elvers, K.T., Turner, S.M., Wainwright, L.M., Marsden, G., Hinds, J., Cole, J.A., Poole, R.K., Penn, C.W., and Park, S.F. (2005). NssR, a member of the Crp-Fnr superfamily from *Campylobacter jejuni*, regulates a nitrosative stress-responsive regulon that includes both a single-domain and a truncated haemoglobin. Mol. Microbiol. *57*, 735–750.

Fang, F.C. (2004). Antimicrobial reactive oxygen and nitrogen species: concepts and controversies. Nat. Rev. Microbiol. *2*, 820–32.

Filenko, N., Spiro, S., Browning, D.F., Squire, D., Overton, T.W., Cole, J., and Constantinidou, C. (2007). The NsrR regulon of *Escherichia coli* K-12 includes genes encoding the hybrid cluster protein and the periplasmic, respiratory nitrite reductase. J. Bacteriol. *189*, 4410–7.

Flatley, J., Barrett, J., Pullan, S.T., Hughes, M.N., Green, J., and Poole, R.K. (2005). Transcriptional responses of *Escherichia coli* to *S*-Nitrosoglutathione under defined chemostat conditions reveal major changes in methionine biosynthesis. J. Biol. Chem. *280*, 10065–10072.

Gardner, A.M., Gessner, C.R., and Gardner, P.R. (2003). Regulation of the nitric oxide reduction operon (*norRVW*) in *Escherichia coli*. Role of NorR and σ^{54} in the nitric oxide stress response. J. Biol. Chem. *278*, 10081–10086.

Gardner, A.M., Helmick, R.A., and Gardner, P.R. (2002). Flavorubredoxin, an inducible catalyst for nitric oxide reduction and detoxification in *Escherichia coli*. J. Biol. Chem. *277*, 8172–8177.

Gardner, P.R. (2005). Nitric oxide dioxygenase function and mechanism of flavohemoglobin, hemoglobin, myoglobin and their associated reductases. J. Inorg. Biochem. *99*, 247–266.

Gardner, P.R., Costantino, G., Szabo, C., and Salzman, A.L. (1997). Nitric oxide sensitivity of the aconitases. J. Biol. Chem. *272*, 25071–25076.

Gardner, P.R., Gardner, A.M., Martin, L.A., and Salzman, A.L. (1998). Nitric oxide dioxygenase: an enzymic function for flavohemoglobin. Proc. Natl. Acad. Sci. U.S.A. *95*, 10378–10383.

Giardina, G., Rinaldo, S., Johnson, K.A., Di Matteo, A., Brunori, M., and Cutruzzolà, F. (2008). NO sensing in *Pseudomonas aeruginosa*: structure of the transcriptional regulator DNR. J. Mol. Biol. *378*, 1002–1015.

Giel, J.L., Rodionov, D., Liu, M., Blattner, F.R., and Kiley, P.J. (2006). IscR-dependent gene expression links iron–sulphur cluster assembly to the control of O_2-regulated genes in *Escherichia coli*. Mol. Microbiol. *60*, 1058–1075.

Gilberthorpe, N.J., Lee, M.E., Stevanin, T.M., Read, R.C., and Poole, R.K. (2007). NsrR: a key regulator circumventing *Salmonella enterica* serovar Typhimurium oxidative and nitrosative stress *in vitro* and in IFN-γ-stimulated J774.2 macrophages. Microbiology *153*, 1756–1771.

Gilberthorpe, N.J., and Poole, R.K. (2008). Nitric oxide homeostasis in *Salmonella typhimurium*: roles of respiratory nitrate reductase and flavohemoglobin. J. Biol. Chem *283*, 11146–11154.

Gomes, C.M., Silva, G., Oliveira, S., LeGall, J., Liu, M.Y., Xavier, A.V., Rodrigues-Pousada, C., and Teixeira, M. (1997). Studies on the redox centers of the terminal oxidase from *Desulfovibrio gigas* and evidence for its interaction with rubredoxin. J. Biol. Chem. *272*, 22502–22508.

Gomes, C.M., Vicente, J.B., Wasserfallen, A., and Teixeira, M. (2000). Spectroscopic studies and characterization of a novel electron-transfer chain from *Escherichia coli* involving a flavorubredoxin and its flavoprotein reductase partner. Biochemistry *39*, 16230–16237.

Gomes, C.M., Giuffre, A., Forte, E., Vicente, J.B., Saraiva, L.M., Brunori, M., and Teixeira, M. (2002). A novel type of nitric-oxide reductase. *Escherichia coli* flavorubredoxin. J. Biol. Chem. *277*, 25273–25276.

Gusarov, I., and Nudler, E. (2005). NO-mediated cytoprotection: Instant adaptation to oxidative stress in bacteria. Proc. Natl. Acad. Sci. U.S.A. *102*, 13855–13860.

Gusarov, I., Shatalin, K., Starodubtseva, M., and Nudler, E. (2009). Endogenous nitric oxide protects bacteria against a wide spectrum of antibiotics. Science *325*, 1380–1384.

Hasegawa, N., Arai, H., and Igarashi, Y. (1998). Activation of a consensus FNR-dependent promoter by DNR of *Pseudomonas aeruginosa* in response to nitrite. FEMS Microbiol. Lett. *166*, 213–217.

Hausladen, A., Gow, A., and Stamler, J.S. (2001). Flavohemoglobin denitrosylase catalyzes the reaction of a nitroxyl equivalent with molecular oxygen. Proc. Natl. Acad. Sci. U.S.A. *98*, 10108–10112.

Hausladen, A., Privalle, C.T., Keng, T., DeAngelo, J., and Stamler, J.S. (1996). Nitrosative stress: activation of the transcription factor OxyR. Cell *86*, 719–729.

Heurlier, K., Thomson, M.J., Aziz, N., and Moir, J.W.B. (2008). The nitric oxide (NO)-sensing repressor NsrR of *Neisseria meningitidis* has a compact regulon of genes involved in NO synthesis and detoxification. J. Bacteriol. *190*, 2488–2495.

Husain, M., Bourret, T.J., McCollister, B.D., Jones-Carson, J., Laughlin, J., and Vázquez-Torres, A. (2008). Nitric oxide evokes an adaptive response to oxidative stress by arresting respiration. J. Biol. Chem. *283*, 7682–7689.

Husain, M., Jones-Carson, J., Song, M., McCollister, B.D., Bourret, T.J., and Vázquez-Torres, A. (2010). Redox sensor SsrB Cys203 enhances *Salmonella* fitness against nitric oxide generated in the host immune response to oral infection. Proc. Natl. Acad. Sci. U.S.A. *107*, 14396–14401.

Hutchings, M.I., Shearer, N., Wastell, S., van Spanning, R.J., and Spiro, S. (2000). Heterologous NNR-mediated nitric oxide signaling in *Escherichia coli*. J. Bacteriol. *182*, 6434–6439.

Hutchings, M.I., Mandhana, N., and Spiro, S. (2002). The NorR protein of *Escherichia coli* activates expression of the flavorubredoxin gene *norV* in response to reactive nitrogen species. J. Bacteriol. *184*, 4640–4643.

Hyduke, D.R., Jarboe, L.R., Tran, L.M., Chou, K.J.Y., and Liao, J.C. (2007). Integrated network analysis identifies nitric oxide response networks and dihydroxyacid dehydratase as a crucial target in *Escherichia coli*. Proc. Natl. Acad. Sci. U.S.A. *104*, 8484–8489.

Imlay, J.A. (2008). Cellular defenses against superoxide and hydrogen peroxide. Annu. Rev. Biochem. *77*, 755–776.

Isabella, V.M., Lapek Jr., J.D., Kennedy, E.M., and Clark, V.L. (2009). Functional analysis of NsrR, a nitric oxide-sensing Rrf2 repressor in *Neisseria gonorrhoeae*. Mol. Microbiol. *71*, 227–239.

Iyer, L.M., Anantharaman, V., and Aravind, L. (2003). Ancient conserved domains shared by animal soluble guanylyl cyclases and bacterial signaling proteins. BMC Genomics *4*, 5.

Jarboe, L.R., Hyduke, D.R., Tran, L.M., Chou, K.J.Y., and Liao, J.C. (2008). Determination of the *Escherichia coli* S-Nitrosoglutathione response network using integrated biochemical and systems analysis. J. Biol. Chem. *283*, 5148–5157.

Ji, X.B., and Hollocher, T.C. (1988a). Mechanism for nitrosation of 2,3-diaminonaphthalene by *Escherichia coli*: enzymatic production of NO followed by O_2-dependent chemical nitrosation. Appl. Environ. Microbiol. *54*, 1791–1794.

Ji, X.B., and Hollocher, T.C. (1988b). Reduction of nitrite to nitric oxide by enteric bacteria. Biochem. Biophys. Res. Comm. *157*, 106–108.

Ji, X.B., and Hollocher, T.C. (1989). Nitrate reductase of *Escherichia coli* as a NO-producing nitrite reductase. Biochem. Arch. *5*, 61–66.

Johnson, E.G. Krasnoff, S.B., Bignell, D.R.D., Chung, W.-C., Tao, T., Parry, R.J., Loria, R., and Gibson, D.M. (2009). 4-Nitrotryptophan is a substrate for the non-ribosomal peptide synthetase TxtB in the thaxtomin A biosynthetic pathway. Mol. Microbiol. 73, 409–418.

Justino, M.C., Almeida, C.C., Teixeira, M., and Saraiva, L.M. (2007). *Escherichia coli* di-iron YtfE protein is necessary for the repair of stress-damaged iron–sulfur clusters. J. Biol. Chem. *282*, 10352–10359.

Karow, D.S., Pan, D., Tran, R., Pellicena, P., Presley, A., Mathies, R.A., and Marletta, M.A. (2004). Spectroscopic characterization of the soluble guanylate cyclase-like heme domains from *Vibrio cholerae* and *Thermoanaerobacter tengcongensis*. Biochemistry *43*, 10203–10211.

Kers, J.A., Wach, M.J., Krasnoff, S.B., Widom, J., Cameron, K.D., Bukhalid, R.A., Gibson, D.M., Crane, B.R., and Loria, R. (2004). Nitration of a peptide phytotoxin by bacterial nitric oxide synthase. Nature *429*, 79–82.

Kiley, P.J., and Beinert, H. (2003). The role of Fe–S proteins in sensing and regulation in bacteria. Curr. Opin. Microbiol. *6*, 181–185.

Kim, S.O., Merchant, K., Nudelman, R., Beyer, W.F., Keng, T., DeAngelo, J., Hausladen, A., and Stamler, J.S. (2002). OxyR: a molecular code for redox-related signaling. Cell *109*, 383–396.

Klink, A., Elsner, B., Strube, K., and Cramm, R. (2007). Characterization of the signaling domain of the NO-responsive regulator NorR from *Ralstonia eutropha* H16 by site-directed mutagenesis. J. Bacteriol. *189*, 2743–2749.

Kohanski, M.A., Dwyer, D.J., Hayete, B., Lawrence, C.A., and Collins, J.J. (2007). A common mechanism of cellular death induced by bactericidal antibiotics. Cell *130*, 797–810.

Korner, H., Sofia, H.J., and Zumft, W.G. (2003). Phylogeny of the bacterial superfamily of Crp-Fnr transcription regulators: exploiting the metabolic spectrum by controlling alternative gene programs. FEMS Microbiol. Rev. *27*, 559–592.

Kulasekara, B.R., Jacobs, M., Zhou, Y., Wu, Z., Sims, E., Saenphimmachak, C., Rohmer, L., Ritchie, J. M., Radey, M., McKevitt, M., *et al.* (2009). Analysis of the genome of the *Escherichia coli* O157:H7 2006 spinach-associated outbreak isolate indicates candidate genes that may enhance virulence. Infect. Immun. *77*, 3713–3721.

Kumar, A., Toledo, J.C., Patel, R.P., Lancaster, J.R., and Steyn, A.J.C. (2007). *Mycobacterium tuberculosis* DosS is a redox sensor and DosT is a hypoxia sensor. Proc. Natl. Acad. Sci. U.S.A. *104*, 11568–11573.

Kwiatkowski, A.V., Laratta, W.P., Toffanin, A., and Shapleigh, J.P. (1997). Analysis of the role of the *nnrR* gene product in the response of *Rhodobacter sphaeroides* 2.4.1 to exogenous nitric oxide. J. Bacteriol. *179*, 5618–5620.

Kwiatkowski, A.V., and Shapleigh, J.P. (1996). Requirement of nitric oxide for induction of genes whose products are involved in nitric oxide metabolism in *Rhodobacter sphaeroides* 2.4.3. J. Biol. Chem. *271*, 24382–24388.

Lee, J., Cho, H.Y., Cho, H.J., Ko, I., Park, S. W., Baik, H., Oh, J., Eom, C., Kim, Y.M., Kang, B.S., *et al.* (2008). O_2- and NO-sensing mechanism through the DevSR two-component system in *Mycobacterium smegmatis*. J. Bacteriol. *190*, 6795–6804.

Lee, Y.-Y.., Shearer, N., and Spiro, S. (2006). Transcription factor NNR from *Paracoccus denitrificans* is a sensor of both nitric oxide and oxygen: isolation of *nnr** alleles encoding effector-independent proteins and evidence for a haem-based sensing mechanism. Microbiology *152*, 1461–1470.

Lu, C., Mukai, M., Lin, Y., Wu, G., Poole, R.K., and Yeh, S. (2007). Structural and functional properties of a single domain hemoglobin from the food-borne pathogen *Campylobactor jejuni*. J. Biol. Chem. *282*, 25917–25928.

Marvin, K.A., Kerby, R.L., Youn, H., Roberts, G.P., and Burstyn, J.N. (2008). The transcription regulator RcoM-2 from *Burkholderia xenovorans* is a cysteine-ligated hemoprotein that undergoes a redox-mediated ligand switch. Biochemistry *47*, 9016–9028.

Mason, M.G., Shepherd, M., Nicholls, P., Dobbin, P.S., Dodsworth, K.S., Poole, R.K., and Cooper, C.E. (2009). Cytochrome *bd* confers nitric oxide resistance to *Escherichia coli*. Nat. Chem. Biol. *5*, 94–96.

Mesa, S., Bedmar, E.J., Chanfon, A., Hennecke, H., and Fischer, H.-M. (2003). *Bradyrhizobium japonicum* NnrR, a denitrification regulator, expands the FixLJ-FixK2 regulatory cascade. J. Bacteriol. *185*, 3978–3982.

Metheringham, R., and Cole, J.A. (1997). A reassessment of the genetic determinants, the effect of growth conditions and the availability of an electron donor on the nitrosating activity of *Escherichia coli* K-12. Microbiology *143*, 2647–2656.

Mills, P.C., Rowley, G., Spiro, S., Hinton, J.C.D., and Richardson, D.J. (2008). A combination of cytochrome *c* nitrite reductase (NrfA) and flavorubredoxin (NorV) protects *Salmonella enterica* serovar Typhimurium against killing by NO in anoxic environments. Microbiology *154*, 1218–1228.

Monk, C.E., Pearson, B.M., Mulholland, F., Smith, H.K., and Poole, R.K. (2008). Oxygen- and NssR-dependent globin expression and enhanced iron acquisition in the response of *Campylobacter* to nitrosative stress. J. Biol. Chem. *283*, 28413–28425.

Morgan, R., Kohn, S., Hwang, S., Hassett, D.J., and Sauer, K. (2006). BdlA, a chemotaxis regulator essential for biofilm dispersion in *Pseudomonas aeruginosa*. J. Bacteriol. *188*, 7335–7343.

Mukhopadhyay, P., Zheng, M., Bedzyk, L.A., LaRossa, R.A., and Storz, G. (2004). Prominent roles of the NorR and Fur regulators in the *Escherichia coli* transcriptional response to reactive nitrogen species. Proc. Natl. Acad. Sci. U.S.A. *101*, 745–750.

Nakano, M.M., Geng, H., Nakano, S., and Kobayashi, K. (2006). The nitric oxide-responsive regulator NsrR controls ResDE-dependent gene expression. J. Bacteriol. *188*, 5878–5887.

Nesbit, A.D., Giel, J.L., Rose, J.C., and Kiley, P.J. (2009). Sequence-specific binding to a subset of IscR-regulated promoters does not require IscR Fe–S cluster ligation. J. Mol. Biol. *387*, 28–41.

Nioche, P., Berka, V., Vipond, J., Minton, N., Tsai, A.-L., and Raman, C.S. (2004). Femtomolar sensitivity of a NO sensor from *Clostridium botulinum*. Science *306*, 1550–1553.

Nunoshiba, T., DeRojas-Walker, T., Wishnok, J.S., Tannenbaum, S.R., and Demple, B. (1993). Activation by nitric oxide of an oxidative-stress response that defends *Escherichia coli* against activated macrophages. Proc. Natl. Acad. Sci. U.S.A. *90*, 9993–9997.

Ouellet, H., Ouellet, Y., Richard, C., Labarre, M., Wittenberg, B., Wittenberg, J., and Guertin, M. (2002). Truncated hemoglobin HbN protects *Mycobacterium bovis* from nitric oxide. Proc. Natl. Acad. Sci. U.S.A. *99*, 5902–5907.

Overton, T.W., Whitehead, R., Li, Y., Snyder, L.A.S., Saunders, N.J., Smith, H., and Cole, J.A. (2006). Coordinated regulation of the *Neisseria gonorrhoeae* truncated denitrification pathway by the nitric oxide-sensitive repressor, NsrR, and nitrite-insensitive NarQ-NarP. J. Biol. Chem. *281*, 33115–33126.

Overton, T.W., Justino, M.C., Li, Y., Baptista, J.M., Melo, A.M.P., Cole, J.A., and Saraiva, L.M. (2008). Widespread distribution in pathogenic bacteria of di-iron proteins that repair oxidative and nitrosative damage to iron–sulfur centers. J Bacteriol. *190*, 2004–2013.

Partridge, J.D., Bodenmiller, D.M., Humphrys, M.S., and Spiro, S. (2009). NsrR targets in the *Escherichia coli* genome: new insights into DNA sequence requirements for binding and a role for NsrR in the regulation of motility. Mol. Microbiol. 73, 680–694.

Patel, B.A., Moreau, M., Widom, J., Chen, H., Yin, L., Hua, Y., and Crane, B.R. (2009). Endogenous nitric oxide regulates the recovery of the radiation-resistant bacterium *Deinococcus radiodurans* from exposure to UV light. Proc. Natl. Acad. Sci. U.S.A *106*, 18183–18188.

Pathania, R., Navani, N.K., Gardner, A.M., Gardner, P.R., and Dikshit, K.L. (2002). Nitric oxide scavenging and detoxification by the *Mycobacterium tuberculosis* haemoglobin, HbN in *Escherichia coli*. Mol. Microbiol. *45*, 1303–1314.

Pawaria, S., Rajamohan, G., Gambhir, V., Lama, A., Varshney, G.C., and Dikshit, K.L. (2007). Intracellular growth and survival of *Salmonella enterica* serovar Typhimurium carrying truncated hemoglobins of *Mycobacterium tuberculosis*. Microb. Pathog. *42*, 119–128.

Pesavento, C., Becker, G., Sommerfeldt, N., Possling, A., Tschowri, N., Mehlis, A., and Hengge, R. (2008). Inverse regulatory coordination of motility and curli-mediated adhesion in *Escherichia coli*. Genes Dev. *22*, 2434–2446.

Pittman, M.S., Elvers, K.T., Lee, L., Jones, M.A., Poole, R.K., Park, S.F., and Kelly, D.J. (2007). Growth of *Campylobacter jejuni* on nitrate and nitrite: electron transport to NapA and NrfA via NrfH and distinct roles for NrfA and the globin Cgb in protection against nitrosative stress. Mol. Microbiol. *63*, 575–590.

Pohlmann, A., Cramm, R., Schmelz, K., and Friedrich, B. (2000). A novel NO-responding regulator controls the reduction of nitric oxide in *Ralstonia eutropha*. Mol. Microbiol. *38*, 626–638.

Poock, S.R., Leach, E.R., Moir, J.W., Cole, J.A., and Richardson, D.J. (2002). Respiratory detoxification of nitric oxide by the cytochrome *c* nitrite reductase of *Escherichia coli*. J. Biol. Chem. *277*, 23664–23669.

Poole, L.B., Karplus, P.A., and Claiborne, A. (2004). Protein sulfenic acids in redox signaling. Ann. Rev. Pharmacol. Toxicol. *44*, 325–347.

Poole, R.K. (2005). Nitric oxide and nitrosative stress tolerance in bacteria. Biochem. Soc. Trans. *33*, 176–180.

Poole, R., Anjum, M.F., Membrillo-Hernandez, J., Kim, S.O., Hughes, M.N., and Stewart, V. (1996). Nitric oxide, nitrite, and Fnr regulation of *hmp* (flavohemoglobin) gene expression in *Escherichia coli* K-12. J. Bacteriol. *178*, 5487–5492.

Poole, R.K., and Hughes, M.N. (2000). New functions for the ancient globin family: bacterial responses to nitric oxide and nitrosative stress. Mol. Microbiol. *36*, 775–83.

Price, M.S., Chao, L.Y., and Marletta, M.A. (2007). *Shewanella oneidensis* MR-1 H-NOX regulation of a histidine kinase by nitric oxide. Biochemistry *46*, 13677–13683.

Pullan, S.T., Gidley, M.T., Jones, R.A., Barrett, J., Stevanin, T.M., Read, R.C., Green, J., and Poole, R.K. (2007). Nitric oxide in chemostat-cultured *Escherichia coli* is sensed by Fnr and other global regulators: unaltered methionine biosynthesis indicates lack of *S* nitrosation. J. Bacteriol. *189*, 1845–1855.

Rankin, L.D., Bodenmiller, D.M., Partridge, J.D., Nishino, S.F., Spain, J.C., and Spiro, S. (2008). *Escherichia coli* NsrR regulates a pathway for the oxidation of 3-nitrotyramine to 4-hydroxy-3-nitrophenylacetate. J. Bacteriol. *190*, 6170–6177.

Ren, B., Zhang, N., Yang, J., and Ding, H. (2008). Nitric oxide-induced bacteriostasis and modification of iron–sulphur proteins in *Escherichia coli*. Mol. Microbiol. *70*, 953–964.

Richardson, A.R., Libby, S.J., and Fang, F.C. (2008). A nitric oxide-inducible lactate dehydrogenase enables *Staphylococcus aureus* to resist innate immunity. Science *319*, 1672–1676.

Richardson, A.R., Soliven, K.C., Castor, M.E., Barnes, P.D., Libby, S.J., and Fang, F.C. (2009). The base excision repair system of *Salmonella enterica* serovar Typhimurium counteracts DNA damage by host nitric oxide. PLoS Pathog. *5*, e1000451.

Rock, J.D., Thomson, M.J., Read, R.C., and Moir, J.W.B. (2007). Regulation of denitrification genes in *Neisseria meningitidis* by nitric oxide and the repressor NsrR. J. Bacteriol. *189*, 1138–1144.

Rodionov, D.A., Dubchak, I.L., Arkin, A.P., Alm, E.J., and Gelfand, M.S. (2005). Dissimilatory metabolim of nitrogen oxides in Bacteria: comparative reconstruction of transcriptional networks. PLOS Comp. Biol. *1*, e55.

Rodrigues, R., Vicente, J.B., Félix, R., Oliveira, S., Teixeira, M., and Rodrigues-Pousada, C. (2006). *Desulfovibrio gigas* flavodiiron protein affords protection against nitrosative stress *in vivo*. J. Bacteriol. *188*, 2745–2751.

Rogstam, A., Larsson, J.T., Kjelgaard, P., and von Wachenfeldt, C. (2007). Mechanisms of adaptation to nitrosative stress in *Bacillus subtilis*. J. Bacteriol. *189*, 3063–3071.

Rudolf, M., Einsle, O., Neese, F., and Kroneck, P.M.H. (2002). Pentahaem cytochrome *c* nitrite reductase: reaction with hydroxylamine, a potential reaction intermediate and substrate. Biochem. Soc. Trans. *30*, 649–653.

Rustad, T.R., Sherrid, A.M., Minch, K.J., and Sherman, D.R. (2009). Hypoxia: a window into *Mycobacterium tuberculosis* latency. Cell. Microbiol. *11*, 1151–1159.

Schapiro, J.M., Libby, S.J., and Fang, F.C. (2003). Inhibition of bacterial DNA replication by zinc mobilization during nitrosative stress. Proc. Natl. Acad. Sci. U.S.A. *100*, 8496–8501.

Schlag, S., Nerz, C., Birkenstock, T.A., Altenberend, F., and Gotz, F. (2007). Inhibition of Staphylococcal biofilm formation by nitrite. J. Bacteriol. *189*, 7911–7919.

Schmidt, I., Steenbakkers, P.J.M., op den Camp, H.J.M., Schmidt, K., and Jetten, M.S.M. (2004). Physiologic and proteomic evidence for a role of nitric oxide in biofilm formation by *Nitrosomonas europaea* and other ammonia oxidizers. J. Bacteriol. *186*, 2781–2788.

Schwartz, C.J., Giel, J.L., Patschkowski, T., Luther, C., Ruzicka, F.J., Beinert, H., and Kiley, P.J. (2001). IscR, an Fe–S cluster-containing transcription factor, represses expression of *Escherichia coli* genes encoding Fe–S cluster assembly proteins. Proc. Natl. Acad. Sci. U.S.A. *98*, 14895–900.

Seib, K.L., Wu, H.-J., Kidd, S.P., Apicella, M.A., Jennings, M.P., and McEwan, A.G. (2006). Defenses against oxidative stress in *Neisseria gonorrhoeae*: a system tailored for a challenging environment. Microbiol. Mol. Biol. Rev. *70*, 344–361.

Shatalin, K., Gusarov, I., Avetissova, E., Shatalina, Y., McQuade, L.E., Lippard, S.J., and Nudler, E. (2008). *Bacillus anthracis*-derived nitric oxide is essential for pathogen virulence and survival in macrophages. Proc. Natl. Acad. Sci. U.S.A. *105*, 1009–1013.

Sousa, E.H.S., Tuckerman, J.R., Gonzalez, G., and Gilles-Gonzalez, M. (2007). DosT and DevS are oxygen-switched kinases in *Mycobacterium tuberculosis*. Protein Sci. *16*, 1708–1719.

Spiro, S. (2007). Regulators of bacterial responses to nitric oxide. FEMS Microbiol. Rev. *31*, 193–211.

Stevanin, T.M., Ioannidis, N., Mills, C.E., Kim, S.O., Hughes, M.N., and Poole, R.K. (2000). Flavohemoglobin Hmp affords inducible protection for *Escherichia coli* respiration, catalyzed by cytochromes *bo'* or *bd*, from nitric oxide. J. Biol. Chem. *275*, 35868–35875.

Strube, K., de Vries, S., and Cramm, R. (2007). Formation of a dinitrosyl iron complex by NorA, a nitric oxide-binding di-iron protein from *Ralstonia eutropha* H16. J. Biol. Chem. *282*, 20292–20300.

Sudhamsu, J., and Crane, B.R. (2009). Bacterial nitric oxide synthases: what are they good for? Trends Microbiol. *17*, 212–218.

Tosques, I.E., Shi, J., and Shapleigh, J.P. (1996). Cloning and characterization of *nnrR*, whose product is required for the expression of proteins involved in nitric oxide metabolism in *Rhodobacter sphaeroides* 2.4.3. J. Bacteriol. *178*, 4958–64.

Tucker, N.P., D'Autréaux, B., Studholme, D.J., Spiro, S., and Dixon, R. (2004). DNA binding activity of the *Escherichia coli* nitric oxide sensor NorR suggests a conserved target sequence in diverse proteobacteria. J. Bacteriol. *186*, 6656–6660.

Tucker, N.P., D'Autreaux, B., Yousafzai, F.K., Fairhurst, S.A., Spiro, S., and Dixon, R. (2008a). Analysis of the nitric oxide-sensing non-heme iron center in the NorR regulatory protein. J. Biol. Chem. *283*, 908–918.

Tucker, N.P., Hicks, M.G., Clarke, T.A., Crack, J.C., Chandra, G., Le Brun, N.E., Dixon, R., and Hutchings, M.I. (2008b). The transcriptional repressor protein NsrR senses nitric oxide directly via a [2Fe–2S] cluster. PLoS ONE *3*, e3623.

Tucker, N.P., Ghosh, T., Bush, M., Zhang, X., and Dixon, R. (2010a). Essential roles of three enhancer sites in σ^{54}-dependent transcription by the nitric oxide sensing regulatory protein NorR. Nucl. Acids Res. *38*, 1182–1194.

Tucker, N.P., Le Brun, N.E., Dixon, R., and Hutchings, M.I. (2010b) There's NO stopping NsrR, a global regulator of the bacterial NO stress response. Trends Microbiol. *18*, 149–156.

Van Alst, N.E., Picardo, K.F., Iglewski, B.H., and Haidaris, C.J. (2007). Nitrate sensing and metabolism modulate motility, biofilm formation, and virulence in *Pseudomonas aeruginosa*. Infect. Immun. *75*, 3780–3790.

Van Spanning, R.J., Houben, E., Reijnders, W.M., Spiro, S., Westerhoff, H.V., and Saunders, N. (1999). Nitric oxide is a signal for NNR-mediated transcription activation in *Paracoccus denitrificans*. J. Bacteriol. *181*, 4129–4132.

Vareille, M., de Sablet, T., Hindré, T., Martin, C., and Gobert, A.P. (2007). Nitric oxide inhibits Shiga-toxin synthesis by enterohemorrhagic *Escherichia coli*. Proc. Natl. Acad. Sci. U.S.A. *104*, 10199–10204.

Vollack, K.U., and Zumft, W.G. (2001). Nitric oxide signaling and transcriptional control of denitrification genes in *Pseudomonas stutzeri*. J. Bacteriol. *183*, 2516–2526.

Vuletich, D.A., and Lecomte, J.T.J. (2006). A phylogenetic and structural analysis of truncated hemoglobins. J. Mol. Evol. *62*, 196–210.

Wainwright, L.M., Elvers, K.T., Park, S.F., and Poole, R.K. (2005). A truncated haemoglobin implicated in oxygen metabolism by the microaerophilic food-borne pathogen *Campylobacter jejuni*. Microbiology *151*, 4079–4091.

Wang, W., Richardson, A.R., Martens-Habbena, W., Stahl, D.A., Fang, F.C., and Hansen, E.J. (2008). Identification of a repressor of a truncated denitrification pathway in *Moraxella catarrhalis*. J. Bacteriol. *190*, 7762–7772.

Wang, Y., Dufour, Y.S., Carlson, H.K., Donohue, T.J., Marletta, M.A., and Ruby, E.G. (2010). H-NOX-mediated nitric oxide sensing modulates symbiotic colonization by *Vibrio fischeri*. Proc. Natl. Acad. Sci. U.S.A. *107*, 8375–8380.

Wasserfallen, A., Ragettli, S., Jouanneau, Y., and Leisinger, T. (1998). A family of flavoproteins in the domains Archaea and Bacteria. Eur. J. Biochem. *254*, 325–332.

Watmough, N.J., Butland, G., Cheesman, M.R., Moir, J.W., Richardson, D.J., and Spiro, S. (1999). Nitric oxide in bacteria: synthesis and consumption. Biochim. Biophys. Acta. *1411*, 456–474.

Weiss, B. (2006). Evidence for mutagenesis by nitric oxide during nitrate metabolism in *Escherichia coli*. J. Bacteriol. *188*, 829–833.

Wink, D.A., Cook, J.A., Pacelli, R., DeGraff, W., Gamson, J., Liebmann, J., Krishna, M. C., and Mitchell, J.B. (1996). The effect of various nitric oxide-donor agents on hydrogen peroxide-mediated toxicity: a direct correlation between nitric oxide formation and protection. Arch. Biochem. Biophys. *331*, 241–248.

Wolfe, M.T., Heo, J., Garavelli, J.S., and Ludden, P.W. (2002). Hydroxylamine reductase activity of the hybrid cluster protein from *Escherichia coli*. J. Bacteriol. *184*, 5898–5902.

van Wonderen, J.H., Burlat, B., Richardson, D.J., Cheesman, M.R., and Butt, J.N. (2008). The nitric oxide reductase activity of cytochrome *c* nitrite reductase from *Escherichia coli*. J. Biol. Chem. *283*, 9587–9594.

Woodmansee, A.N., and Imlay, J.A. (2003). A mechanism by which nitric oxide accelerates the rate of oxidative DNA damage in *Escherichia coli*. Mol. Microbiol. *49*, 11–22.

Wu, G., Cruz-Ramos, H., Hill, S., Green, J., Sawers, G., and Poole, R.K. (2000). Regulation of cytochrome *bd* expression in the obligate aerobe *Azotobacter vinelandii* by CydR (Fnr). Sensitivity to oxygen, reactive oxygen species, and nitric oxide. J. Biol. Chem. *275*, 4679–4686.

Wu, G., Wainwright, L.M., and Poole, R.K. (2003). Microbial globins. Adv. Micro. Physiol. *47*, 255–310.

Yoon, S.S., Hennigan, R.F., Hilliard, G.M., Ochsner, U.A., Parvatiyar, K., Kamani, M.C., Allen, H.L., DeKievit, T.R., Gardner, P.R., Schwab, U., *et al.* (2002). *Pseudomonas aeruginosa* anaerobic respiration in biofilms: relationships to cystic fibrosis pathogenesis. Dev. Cell 3, 593–603.

Yukl, E.T., Elbaz, M.A., Nakano, M.M., and Moënne-Loccoz, P. (2008). Transcription factor NsrR from *Bacillus subtilis* senses nitric oxide with a 4Fe-4S cluster. Biochemistry *47*, 13084–13092.

Zeng, J., Zhang, X., Wang, Y., Ai, C., Liu, Q., and Qiu, G. (2008). Glu43 is an essential residue for coordinating the [Fe_2S_2] cluster of IscR from *Acidithiobacillus ferrooxidans*. FEBS Lett. *582*, 3889–3892.

Zumft, W.G. (1997). Cell biology and molecular basis of denitrification. Microbiol. Mol. Biol. Rev. *61*, 533–616.

Zumft, W.G. (2005). Nitric oxide reductases of prokaryotes with emphasis on the respiratory, heme-copper oxidase type. J. Inorg. Biochem. *99*, 194–215.

Denitrification in Legume-associated Endosymbiotic Bacteria

10

Cristina Sánchez, Eulogio J. Bedmar and María J. Delgado

Abstract

Rhizobia are soil, Gram-negative bacteria with the unique ability to establish a dinitrogen-fixing symbiosis on legume roots and on the stems of some aquatic legumes. During this interaction bacteroids, as rhizobia are called in the symbiotic state, are contained in intracellular compartments within a specialized organ, the nodule, where they fix dinitrogen. When faced with a shortage of oxygen some rhizobia species are able to switch from oxygen respiration to using nitrates to support respiration in a process known as denitrification. The complete denitrification pathway comprises the sequential reduction of nitrate or nitrite to dinitrogen, via the gaseous intermediates nitric oxide and nitrous oxide. The enzymes involved in denitrification are nitrate-, nitrite-, nitric oxide- and nitrous oxide reductase, encoded by *nar/nap*, *nir*, *nor* and *nos* genes, respectively. In recent years it has emerged that many rhizobia species have genes for enzymes of some or all of the four reductase reactions for denitrification. In fact, denitrification can be readily observed in many rhizobia species, in their free-living form, in legume root nodules, or in isolated bacteroids. This chapter will focus on update progress on denitrification by rhizobia under free-living and symbiotic conditions.

Introduction

Biological nitrogen fixation by symbiotic, associative and free-living bacteria, is of tremendous importance to the environment and to world agriculture. Nitrogen fixation is an important part of the nitrogen cycle as it replenishes the overall nitrogen content of the biosphere and compensates for the losses that are incurred owing to denitrification.

The growth of most plants is limited by the availability of nitrogen. Many legumes can bypass this limitation by entering into a symbiosis with nitrogen-fixing bacteria that can reduce atmospheric nitrogen (N_2) to ammonia (NH_4^+), a form of nitrogen that is usable by the plant. The genera *Allorhizobium, Azorhizobium, Bradyrhizobium, Mesorhizobium, Rhizobium,* and *Sinorhizobium* (*Ensifer*), collectively referred to as rhizobia, are members, among others, of the bacterial order *Rhizobiales* of the α-Proteobacteria (Sawada *et al.*, 2003; Willems, 2006; Zakhia and de Lajudie, 2006). Rhizobia are soil, Gram-negative bacteria with the unique ability to establish a N_2-fixing symbiosis on legume roots and on the stems of some aquatic legumes. During this interaction bacteroids, as rhizobia are called in the symbiotic state, are contained in intracellular compartments within a specialized organ, the nodule, where they fix N_2.

When faced with a shortage of oxygen (O_2) many bacterial species are able to switch from O_2-respiration to using nitrate or its derived-nitrogen oxides to support respiration in a process known as denitrification. During this process the water-soluble nitrate (NO_3^-) is converted into gaseous nitrogen-containing gases. These are the: (i) cytotoxic and ozone-depleting nitric oxide (NO), (ii) potent and long-lived greenhouse gas nitrous oxide (N_2O), and (iii) the relatively inert dinitrogen gas (N_2). The enzymes involved in denitrification are nitrate-, nitrite-, nitric oxide- and nitrous oxide reductase, encoded by *nar*/*nap*, *nir*, *nor* and *nos* genes, respectively (van Spanning *et al.*, 2007).

In recent years it has emerged that many rhizobia species have genes for enzymes of some or all of the four reductase reactions for denitrification (Delgado *et al.*, 2007). In fact, denitrification can be readily observed in many rhizobia species, in their free-living form, in legume root nodules, or in isolated bacteroids (Delgado *et al.*, 2007). The soybean symbiont *Bradyrhizobium japonicum* is the only rhizobia species able to denitrify under both free-living and symbiotic conditions where denitrification has been characterized. This chapter will focus on updating progress on denitrification by rhizobia, especially *B. japonicum*, under free-living and symbiotic conditions.

The rhizobia–legume nitrogen-fixing symbiosis

Nodule formation

To initiate the symbiosis, rhizobia usually attach to the root-hair cells of legumes and induce root-hair deformation. In a few cases, the root hairs deform in such a way as to entrap bacteria within a curl. Infection is initiated from these curled root hairs, with infection threads growing as tunnel-like invaginations of the host cell from the centre of the curl. The infection thread is reminiscent of a growing cylinder of plant cell wall in which rhizobia replicate to remain at the growing tip. In parallel with this infection process, cell division is initiated in the cortical cells of the root, and this leads to the formation of a nodule. The infection thread continues its growth towards the cortex to reach the newly dividing cells, where it ramifies into the newly forming nodule primordium. Release of the bacteria from the infection thread resembles endocytosis by resulting in the formation of a membrane-bound compartment in which bacteria exist as intracellular symbionts. This membrane-bound compartment has been termed the symbiosome and is the unit of biological N_2 fixation because its membrane (across which there is metabolite exchange, including uptake of dicarboxylic acids, export of ammonia and cycling of amino acids) mediates interaction with the host cell, whereas the bacteria induce all the machinery for N_2 fixation. Establishing this symbiosis requires an exchange of molecular signals between the plant and the bacteria. Plant perceives a critical bacterial signal (the Nod factor) and transduces the signal for the activation of downstream responses, leading to infection and nodule morphogenesis. Several reviews summarize the findings on the involvement of signals during nodule development (Jones *et al.*, 2007; Stacey, 2007; Oldroy and Downie, 2008). The nodule is a highly specialized organ and exhibits cellular specialization. Two morphological types of nodules are known and they are determined by the plant host. The first type is called indeterminate such as those of pea (*Pisum sativum*), alfalfa (*Medicago sativa*), and broad bean (*Vicia faba*). They have a persistent meristem and, in longitudinal section, can be divided into four zones: (i) the meristem at the nodule tip, (ii) the invasion zone immediately behind the meristem with cells invaded by rhizobial-containing infection threads, (iii) the N_2-fixing zone, where the cells contain fully differentiated bacteroids, and (iv) the senescent zone. The second type, called determinate nodules, which form on plants such as soybean (*Glycine max*) and common bean (*Phaseolus vulgaris*) have a globular structure. Typical determinate legume nodules possess: (i) a central zone formed by bacteroids containing infected cells and uninfected cells, (ii) an inner cortex containing small cells with large intercellular spaces and a boundary layer of closely packed cells, (iii) a mid-cortex containing large cells with thickened but unlignified cell walls and, (iv) an outer cortex with large loosely packed cells with large intercellular spaces, sometimes surrounded by a periderm. Rhizobia within the symbiosomes differentiate and induce a variety of new enzyme systems (i.e. nitrogenase and the high-affinity cytochrome *cbb*$_3$-type oxidase) and often take on a larger, more extended shape. For these reasons, the special term bacteroid is used to define the intracellular symbiont. Nitrogenases are complex metalloenzymes with conserved structural and mechanistic features that catalyse the biological reduction of N_2 to ammonia (for reviews about nitrogenases, see Dixon and Kahn, 2004; Seefeldt *et al.*, 2009). Rhizobia possess a conventional molybdenum-based nitrogenase, which contain two components that are named according to their metal composition. The smaller dimeric component, known as the iron (Fe) protein, functions as an ATP-dependent electron donor to the larger heterotetrameric component, known as the molybdenum–iron (MoFe) protein, which contains the enzyme catalytic site. Both of the nitrogenase component proteins are extremely O_2 sensitive. The O_2 sensitivity of the Fe protein is conferred by a surface-exposed [4Fe–4S] cluster that bridges the two subunits of the dimer. The MoFe protein contains two types of metal centres: the P cluster (an [8Fe-7S] cluster) and the FeMo cofactor ($MoFe_7S_9$·homocitrate), which is the site of substrate reduction. The overall stoichiometry of N_2 reduction under optimal conditions is as follows:

$$N_2 + 8e^- + 8H^+ + 16MgATP \rightarrow 2NH_3 + H_2 + 16MgADP + 16P_i \qquad (10.1)$$

Control of oxygen diffusion in nodules

In the nodule, maintenance of nitrogenase activity is subject to a delicate equilibrium. A high rate of O_2 respiration is necessary to supply the energy demands of the N_2-reduction process but, on the other hand, O_2 also irreversibly inactivates the nitrogenase complex. A delicate balance in O_2 supply and demand solves this apparent conflict in order to keep the steady-state concentration of free O_2 low. The cortex of nodules acts as a diffusion barrier, which greatly limits permeability to O_2 (Hunt and Layzell, 1993). That barrier is a complex of structures involving intercellular space occlusions in the mid-cortex and osmocontractile responses in the inner cortex. There are also several metabolic or structural changes which could occur within the infected region (for a review, see Minchin *et al.*, 2008).

Oxygen is delivered to the symbiosomes by the plant O_2-carrier leghaemoglobin (Lb), which transports O_2 at a low but stable concentration allowing for the simultaneous operation of nitrogenase activity and bacteroid respiration (Downie, 2005 and references therein). The O_2-binding characteristics of Lbs are unusual in that they have an extremely fast O_2 association rate and a relatively slow O_2 dissociation rate, and so can buffer the free O_2 concentration at around 7–11 nM. The presence of millimolar concentrations of Lb within the cytoplasm of nodule cells serves to buffer free O_2 in the nanomolar range while ensuring rapid transport of O_2 to the sites of respiration (Appleby, 1984, 1992; Bergersen and Turner, 1993). It has been shown that removal of Lb from *Lotus* nodules via RNA interference (RNAi) resulted in an increase in nodule free O_2, destabilization of nitrogenase, and a failure of symbiotic N_2 fixation (Ott *et al.*, 2005). Additionally, a decrease in hydrogen peroxide and alterations of the plant antioxidative machinery, probably due to the lack of Lb autooxidation, were observed (Günther *et al.*, 2007).

High affinity *cbb*$_3$ oxidase

N_2-fixing bacteroids deal with the low levels of free O_2 by inducing a high-affinity cytochrome *cbb*$_3$-type oxidase. Genes encoding the *cbb*$_3$ oxidase complex were isolated initially from rhizobial species and named *fixNOQP* for their role in symbiotic N_2 fixation (Preisig *et al.*, 1993; Mandon *et al.*, 1994). Since then, orthologous genes have been identified in other Gram-negative bacteria and called *ccoNOQP* (for a review, see Pitcher and Watmough, 2004). The *cbb*$_3$-type oxidase is made up of three to four subunits: subunit I is encoded by *ccoN* and is a membrane-integral *b*-type cytochrome with a high-spin haem-Cu_B binuclear centre and a low-spin haem. Subunits II and III are encoded by *ccoO* and *ccoP*, respectively, and are membrane-anchored *c*-type cytochromes. As shown in *Rhodobacter sphaeroides* and *B. japonicum*, the *ccoN*, *ccoO* and *ccoP* gene products are essential for both the activity and the assembly of the *cbb*$_3$ oxidase. In contrast, *ccoQ*, which encodes subunit IV, is sometimes missing. The product of the *ccoQ* gene is not required either for assembly or for activity of the oxidase at low O_2 concentration (Zufferey *et al.*, 1996; Oh and Kaplan, 1999), but may play a protective, stabilizing role under oxic conditions (Oh and Kaplan, 2002). CcoNOQP has been extensively studied in *R. sphaeroides*, where it has multiple roles. It functions not only as a terminal oxidase (García-Horsman *et al.*, 1994) but also as a redox sensor in a signal transduction pathway controlling photosynthesis gene expression (Oh and Kaplan, 2002; Kim *et al.*, 2007).

Like many other anaerobic facultative bacteria, *B. japonicum* adapts to different environmental O_2 concentrations by inducing multiple terminal oxidases with different affinities for O_2 (Delgado *et al.*, 1998). Free living bacteria growing microaerobically and N_2-fixing bacteroids use a cytochrome cbb_3-type oxidase encoded by the *fixNOQP* operon (Preisig *et al.*, 1993). Electron transfer to this high-affinity oxidase is via the cytochrome bc_1 complex (Thöny-Meyer *et al.*, 1989). Cytochrome cbb_3 oxidases have been purified from several organisms including *B. japonicum* (Preisig *et al.*, 1996; Pitcher and Watmough, 2004), where the experimentally determined K_m for O_2 is in the order of 7 nM, a value which is consistent with its function in the bacteroid. Concentration of O_2 is the primary effector of *fixNOQP* expression in *B. japonicum* (see 'Regulation of nitrogen fixation', below). In addition to O_2, recent results showed that expression levels of a *fixP′–′lacZ* translational fusion under denitrifying conditions are largely dependent on the oxidized or reduced nature of the carbon source in the medium, suggesting that a change in the electron flow through the denitrification pathway may affect the cellular redox state, leading to alterations in cbb_3 expression (Bueno *et al.*, 2009).

Regulation of nitrogen fixation

Microaerobiosis is a prerequisite not only for nitrogenase activity but also for induction of N_2 fixation and symbiosis-related genes (i.e. *nif* and *fix*) (Fischer, 1994). Perception and transduction of the 'low-O_2 signal' is mediated by conserved regulatory proteins that are integrated into species-specific networks in different rhizobia (Fischer, 1994; Dixon and Kahn, 2004). In the Proteobacteria, *nif* genes are invariably subject to transcriptional activation by the enhancer-binding protein NifA, together with the RNA polymerase sigma factor σ^{54}. Transcription of *nifA* and *fix* genes in rhizobia is predominantly controlled by the O_2-responsive two component FixL-FixJ system, together with FixK, which is a member of the Fnr (fumarate and nitrate reductase)-Crp (cAMP receptor protein) superfamily, and by the redox-sensing system RegS-RegR (Dixon and Kahn, 2004). In *B. japonicum*, the FixLJ-FixK$_2$ and the RegSR-NifA O_2-responsive regulatory cascade are present (Sciotti *et al.*, 2003). In the FixLJ-FixK$_2$ cascade, the 'low-O_2 signal' is sensed by the haem-based sensory kinase FixL, which auto-phosphorylates and then transfers the phosphoryl group to the FixJ response regulator, which activates transcription of *fixK$_2$*. FixK$_2$ has been shown to activate expression of genes associated with microaerobic metabolism, including those involved in denitrification (Bedmar *et al.*, 2005), the *fixNOQP* genes for biogenesis of the high-affinity cbb_3-type terminal oxidase, and the *rpoN$_1$* gene encoding sigma factor (σ^{54}) (Nellen-Anthamatten *et al.*, 1998; Mesa *et al.*, 2008), among others. Induction of the second pathway is mediated by the RegSR two-component regulatory system. Activation proceeds by a phosphorylation process and induces expression of the *fixR-nifA* operon, which encodes the NifA regulatory protein. NifA activates gene expression in concert with the RNA polymerase containing σ^{54}, which in *B. japonicum*, is encoded by the two highly similar and functionally equivalent genes *rpoN$_1$* and *rpoN$_2$* (Kullik *et al.*, 1991). Thus, RpoN$_{1/2}$ represents the link between the two regulatory cascades (Sciotti *et al.*, 2003). Targets of NifA include the *nif* and *fix* genes, which are directly involved in N_2 fixation, and also genes that are indirectly related to N_2 fixation (i.e. *groESL$_3$*) or have unknown function in this process (i.e. *nrgA*, *nrgBC*) (Fischer, 1994; Nienaber *et al.*, 2000). Microarray based experiments have led to a substantial expansion of the NifA regulon, revealing a total of 65 genes for N_2 fixation and other diverse processes (Hauser *et al.*, 2007). In *B. japonicum* free-living cultures, the NifA protein is only active at O_2 concentration below 2% in the gas phase, whereas induction of FixLJ-controlled targets progressively increased when O_2 concentration was lowered from 20% to 0.5%. Thus, expression of FixL-FixJ targets allows the bacteria to adapt their respiratory metabolism to the microoxic environment of the nodule, whereas the O_2 sensitivity of NifA is compatible with the very low-O_2 conditions that are required for nitrogenase activity in particular zones of the nodule (Sciotti *et al.*, 2003). This disparate oxygen-responsive expression pattern reflects a response to a gradient of increasing oxygen deprivation as bacteria invade their host during root nodule development.

Denitrification in rhizobia

In addition to fixing N_2 in symbiosis with legumes, some rhizobia are able to grow and reduce nitrate to N_2 through denitrification, when cultured under O_2-limiting conditions with nitrate as the terminal electron acceptor. The complete denitrification pathway comprises the sequential reduction of nitrate or nitrite (NO_2^-) to N_2, via the gaseous intermediates NO and nitrous oxide. Denitrification reactions are catalysed by periplasmic (Nap) or membrane-bound (Nar) nitrate reductase, nitrite reductases (CuNir/cd_1Nir), nitric oxide reductases (cNor, qNor, or qCuANor) and nitrous oxide reductases (Nos) encoded by *nap/nar, nirK/nirS, nor* and *nos* genes, respectively. Reviews covering the physiology, biochemistry and molecular genetics of denitrification have been published elsewhere (Zumft *et al.*, 1997; van Spanning *et al.*, 2005, 2007).

Denitrification among rhizobia is rare, and most species do not contain the whole set of denitrification genes. *Pseudomonas* sp. G-179 (actually *Rhizobium galegae*) (Bedzyk *et al.*, 1999) has been shown to contain Nap, Nor and CuNir. *R. sullae* (formerly *R. hedysari*) only expresses CuNir (Toffanin *et al.*, 1996). The genetic determinants for expression of CuNir and cNor are present in *R. etli* (Bueno *et al.*, 2005). *S. meliloti* (Galibert *et al.*, 2001; Holloway *et al.*, 1996), and *B. japonicum* (Kaneko *et al.*, 2002; Bedmar *et al.*, 2005) contain *nap, nirK, nor,* and *nos* genes (see http://www.kazusa.or.jp/rhizobase). Denitrification genes are not present either in the complete genome sequence of *M. loti* strain MAFF303099 or in the symbiotic island of *M. loti* strain R7A (see http://www.kazusa.or.jp/rhizobase). However, *M. loti* fast-growing strains isolated from *Lotus* sp. showed a hybridization band with the *B. japonicum nirK* (Monza *et al.*, 2006).

R. sullae strain HCNT1, which induces nodule formation on *Hedysarum coronarium,* has been shown to contain only a CuNir encoded by *nirK.* Expression of *nirK* is atypical in that it does not require the presence of a nitrogen oxide (NOx), but only a decrease in O_2 concentration (Casella *et al.*, 1986). *R. sullae* strain HCNT1 has been shown to enter a viable-but-not-culturable status when O_2 is limiting, and NirK produces NO from nitrite present in the growth medium (Toffanin *et al.*, 2000). A physiological role for the truncated denitrification chain in *R. sullae* is not obvious. It is possible that Nir activity allows the bacteria to convert into a viable-but-not-culturable form, which, in turn, would survive for long periods under stress conditions without loss of the ability to recover the vegetative state (Casella *et al.*, 2006).

Sequence analysis of the type strain *R. etli* CFN42, which forms N_2-fixing symbiosis with *Phaseolus vulgaris,* permitted the identification of *norCBQD* genes coding for the cytochrome *c*-type Nor and the *nirK* gene encoding a CuNir. Since nitrite reductase activity and nitrite-uptake were very low in a *R. etli nirK* mutant, it is possible that the Cu-containing Nir may have both a respiratory and a nitrite detoxifying role in *R. etli* (Bueno *et al.*, 2005).

In *S. meliloti* strain 1021, *nap, nir, nor,* and *nos* denitrification genes are present in the symbiotic plasmid pSymA (http://www.kazusa.or.jp/rhizobase). Intriguingly, despite possessing the complete set of denitrification genes, *S. meliloti* 1021 does not grow under O_2-limiting conditions with either nitrate or nitrite as terminal electron acceptors.

Although the ability to denitrify may enhance bacterial survival and growth capability in anaerobic soils, only *B. japonicum, Pseudomonas* sp. G-179, and *A. caulinodans* have been shown to grow under O_2-limiting conditions with nitrate through denitrification pathway.

Bradyrhizobium japonicum as a model of free-living rhizobial denitrification

In *B. japonicum* USDA110, denitrification depends on the *napEDABC* (Delgado *et al.*, 2003), *nirK* (Velasco *et al.*, 2001), *norCBQD* (Mesa *et al.*, 2002), and *nosRZDFYLX* (Velasco *et al.*, 2004) genes encoding nitrate-, nitrite-, nitric oxide-, and nitrous oxide-reductase, respectively

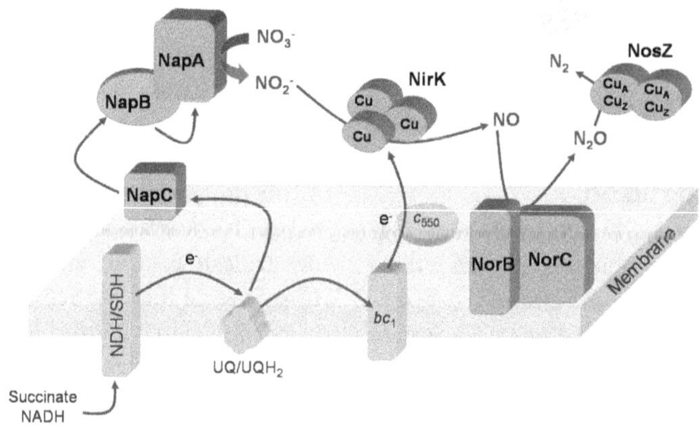

Figure 10.1 Scheme of a full denitrification process in *Bradyrhizobium japonicum*. NDH, NADH dehydrogenase; SDH, succinate dehydrogenase; UQ, ubiquinone; UQH_2, ubiquinol; bc_1, cytochrome bc_1 complex; c_{550}, cytochrome *c*; Nap, periplasmic nitrate reductase; NirK, copper-containing nitrite reductase; Nor, *c*-type nitric oxide reductase; Nos, nitrous oxide reductase.

(see Fig. 10.1) (Bedmar *et al.*, 2005; Delgado *et al.*, 2007).

Periplasmic nitrate reductase

In a shotgun cloning experiment, a DNA fragment of *B. japonicum* USDA110 was sequenced and found to contain the *napEDABC* genes (Delgado *et al.*, 2003). The *napA* gene encodes the catalytic subunit (90 kDa) containing the molybdopterin guanine-dinucleotide cofactor (MGD) and a [4Fe–4S] cluster, *napB* an electron-transfer subunit, dihaem cytochrome *c* of about 15 kDa, and *napC* a membrane-bound *c*-type tetrahaem cytochrome of about 25 kDa, respectively (see Fig. 10.1) (Delgado *et al.*, 2003). The N-terminal region of NapA contains a twin-arginine motif diagnostic for a large number of metalloproteins that are exported to the periplasm through the Tat (twin arginine translocon) system (Berks *et al.*, 2000). The *napE* gene encodes a transmembrane protein of unknown function, and the *napD* gene product is a soluble protein which is assumed to play a role in the maturation of NapA. Because a *napA* mutant was incapable of growing under nitrate-respiring conditions, lacked nitrate reductase activity, and did not show the NapA, NapB and NapC protein components, the *B. japonicum* Nap system is the primary enzyme respiring nitrate under anoxic conditions.

Copper-containing nitrite reductase

In *B. japonicum* USDA110, a *nirK* gene, responsible for the synthesis of CuNir, was identified by Velasco *et al.* (2001). NirK is soluble and apparently exported to the periplasm via the Tat translocon, as judged by a twin-arginine motif present in its N-terminal sequence. Implication of *B. japonicum nirK* in denitrification occurred when a *nirK* mutant was shown to be incapable of growing when cultured under microaerobic conditions in the presence of either nitrate or nitrite. In those mutants, however, Nap was an active enzyme because cells accumulated nitrite during incubation (Velasco *et al.*, 2001).

Recently, it has been shown that cytochrome c_{550}, encoded by the *cycA* gene, is involved in electron transfer from the cytochrome bc_1 complex to the CuNir of *B. japonicum* USDA110, as revealed by the inability of a *cycA* mutant strain to consume nitrite and, consequently, to grow under denitrifying conditions with nitrite as the electron acceptor (see Fig. 10.1) (Bueno *et al.*, 2008). Mutation of *cycA* had no apparent effect on methylviologen-dependent nitrite reductase activity. However, succinate-dependent nitrite reduction was largely inhibited, suggesting that c_{550} is the *in vivo* electron donor to CuNir (Bueno *et al.*, 2008).

Nitric oxide reductase

In *B. japonicum* USDA110, the *nor* genes are organized in the *norCBQDE* gene cluster, in which *norC* and *norB* encode the 16 kDa membrane-bound cytochrome *c*-containing subunit II and the cytochrome *b*-containing subunit I of cNor respectively (see Fig. 10.1); the *norQ* gene encodes a protein with an ATP/GTP-binding motif, and the predicted *norD* gene product is of unknown function (Mesa *et al.*, 2002). Mutational analysis indicated that the two structural *norC* and *norB* genes are required for microaerobic growth under nitrate-respiring conditions. Each *norC* and *norB* insertional mutant accumulated NO during incubation under microaerobic conditions, and the *norC* mutant lacked the 16 kDa *c*-type cytochrome found in membranes from wild-type cells (Mesa *et al.*, 2002). Inspection of the complete genome sequence of *B. japonicum* USDA110 shows the existence of an open reading frame, *blr3212*, whose sequence has more than 60% identity with *norE* genes from various denitrifiers. The NorE protein also has similarity to CoxIII, the third subunit of the aa_3-type cytochrome *c* oxidases.

Nitrous oxide reductase

The *B. japonicum* USDA110 *nos* genes were identified using a major internal portion of the *Pseudomonas stutzeri nosZ* gene as a probe (Viebrock and Zumft, 1987), and found to be organized in the *nosRZDFYLX* gene cluster (Velasco *et al.*, 2004). The *nosR* gene encodes an integral membrane protein with six transmembrane helices and a C-terminal cytoplasmic domain containing two cysteine clusters similar to the [4Fe–4S] binding motifs of several bacterial ferredoxins. *nosZ* encodes the monomers of NosZ, whose deduced primary sequence has between 50 and 77% identity with the translated sequences of the *nosZ* gene from other denitrifiers. In its N-terminus, a twin-arginine motif indicates that NosZ is exported via the Sec-independent Tat translocon. The NosZ sequence also contains the site signature believed to bind the copper metal ligand. The NosD, -F, -Y and L proteins show significant homology to the bacterial ATP-binding-cassette (ABC) transporter systems, which are typically composed of a cytoplasmic ATP/GTP binding protein (NosF), a transmembrane protein (NosY) and one or two periplasmic components (NosD and NosL). Because NosZ has its own Tat-related periplasmic signal sequence, it is unlikely that the ABC-transporter is used for transport of Nos. *B. japonicum* strains carrying either a *nosZ* or a *nosR* mutation grew well when cultured microaerobically with nitrate as the final electron acceptor. Nevertheless, Nos was not an active enzyme because cells accumulated nitrous oxide during growth (Velasco *et al.*, 2004).

Regulation of denitrification

Maximal expression of denitrification genes requires simultaneously both the absence of oxygen and the presence of nitrate or a derived NOx. Microaerobic induction of transcription from the *nap*, *nir*, and *nor* promoter regions depends on the FixLJ-FixK$_2$-NnrR regulatory cascade (Mesa *et al.*, 2003; Bedmar *et al.*, 2005; Robles *et al.*, 2006). Recent results showed that NifA, which induces gene expression at a very low-O_2 concentration, is required for maximal expression of *napEDABC*, *nirK* and *norCBQD* genes, suggesting a role for RegSR-NifA regulatory cascade in the control of the denitrification process in *B. japonicum* USDA110 (Bueno *et al.*, 2010). In this study disruption of *nifA* caused a growth defect in *B. japonicum* cells when grown under denitrifying conditions along with a decreased activity of the Nap and Nir enzymes. Furthermore, expression of *napE-lacZ*, *nirK-lacZ* and *norC-lacZ* transcriptional fusions, as well as levels of *nirK* transcripts was significantly reduced in the *nifA* mutant after incubation under nitrate-respiring conditions. Haem *c* staining analyses revealed that NifA is required for full synthesis of the NapC and NorC proteins, which are required for denitrification. A *B. japonicum* *rpoN*$_{1/2}$ mutant lacking both copies of the gene encoding the alternative sigma factor σ^{54} was able to grow anaerobically with nitrate as terminal electron acceptor, and showed wild-type levels of nitrate and nitrite reductase activities. That result suggest that the influence of NifA on denitrification genes is independent of σ^{54} (Bueno *et al.*, 2010).

Denitrification in the rhizobia–legume symbiosis

The significance of denitrification in rhizobia–legume symbiosis can be appreciated when O_2 concentration in soils decreases during environmental stress such as flooding of the roots, which causes hypoxia. Under these conditions, denitrifying activity could work as a mechanism to generate ATP for survival of rhizobia in the rhizosphere and also to maintain nodule functioning. An associated role of denitrification in nodules could also be detoxification of the cytotoxic compounds nitrite and NO produced as intermediates during denitrification reactions or emerging from the host plant. In fact, nitrite and NO have been reported as inhibitors of nitrogenase activity (Trinchant and Rigaud, 1980, 1982; Sasakura *et al.*, 2006; Kato *et al.*, 2009; Shimoda *et al.*, 2009).

Expression of *B. japonicum* USDA110 *nirK*, *norC* and *nosZ* denitrification genes in soybean root nodules has been reported by *in situ* histochemical detection of β-galactosidase activity due to transcriptional fusions of the *nirK*, *norC* and *nosZ* promoter regions to the reporter gene *lacZ* (Mesa *et al.*, 2004). Similarly, β-galactosidase analyses in isolated bacteroids also demonstrated that *nirK*, *norC* and *nosZ* genes are expressed in bacteroids (Mesa *et al.*, 2004). Levels of β-galactosidase activity were similar in both bacteroids and nodule sections from plants that were solely N_2-dependent or grown in the presence of 4 mM KNO_3, which suggest that O_2, and not nitrate, is the main factor controlling expression of denitrification genes in soybean nodules (Mesa *et al.*, 2004).

The symbiotic phenotype of *B. japonicum* strains carrying a mutation in any of the *nirK*, *norC*, or *nosZ* structural genes has also been reported (Mesa *et al.*, 2004). In soybean plants not amended with nitrate, *nirK*, *norC* or *nosZ* genes are not essential for symbiotic N_2 fixation. Similarly, nodulation, plant growth, and rates of N_2 fixation in *H. coronarium* were similar after inoculation of the wild-type and a *nirK*-deficient strain (Casella *et al.*, 1986). None of the Tn5 insertions in the *S. meliloti* strain JJ1c10 *nos* region affected N_2-fixing ability in symbiosis with alfalfa, which demonstrated that denitrification is not essential for N_2 fixation (Holloway *et al.*, 1996).

However, in soybean plants grown with nitrate, mutation of either the *nirK* or *norC* genes confers on *B. japonicum* a reduced ability for nodulation (Mesa *et al.*, 2004). Nodules formed by the parental and mutant strains exhibited similar nitrogenase activities suggesting that denitrification enzymes play a role in nodule formation rather than in nodule function.

Nitrogen oxides in root nodules

Denitrification as a source of nitrogen oxides in nodules

NO production in nodules has been reported by several authors. NO may originate from the plant root tissue or from the bacteroids. UV-visible and electron paramagnetic resonance (EPR) spectroscopy studies (Kanayama *et al.*, 1991; Mathieu *et al.*, 1998; Meakin *et al.*, 2007) have demonstrated the presence of NO bound to Lb (nitrosylleghaemoglobin, LbNO) within soybean nodules. In addition, diaminofluoresceins have been used to detect free NO within nodules (Shimoda *et al.*, 2009; Sánchez *et al.*, 2010). These frequently and widely used fluorescence indicators are available in cell impermeable forms (DAF-FM) and as cell permeable diacetates (DAF-2DA and DAF-FM DA) (Kojima *et al.*, 1999), which are hydrolysed inside cells and thereby trapped.

In plant cells, NO can be produced from nitrite via both non-enzymatic (Bethke *et al.*, 2004) and enzymatic pathways (see Fig. 10.2), i.e., those catalysed by nitrate reductase (Meyer *et al.*, 2005) or by the plasma membrane-bound nitrite: NO-reductase (Stöhr *et al.*, 2001). Assimilatory nitrate reductase (NR) normally reduces nitrate to nitrite at the expense of NAD(P)H, but also catalyses a one-electron transfer from NAD(P)H to nitrite resulting in NO formation (for reviews, see Meyer *et al.*, 2005; Planchet and Kaiser, 2006). NO may also be produced in plant root tissue from arginine via an unidentified nitric oxide synthase (NOS)-like enzyme, or via an as yet uncharacterized process that uses polyamines as substrates (for a review, see Besson-Bard *et al.*, 2008). In this context, a NOS-like activity has been identified in nodules of *Lupinus albus* (Cueto *et al.*, 1996) Furthermore, a NOS inhibitor (*N*-methyl-L-arginine)

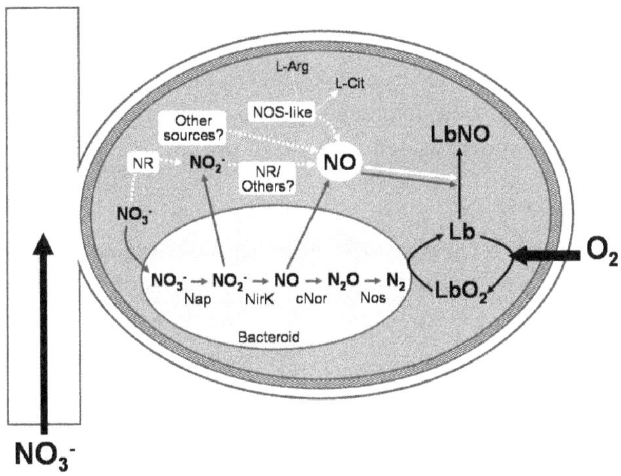

Figure 10.2 Schematic representation of the sources of NO in soybean nodules. The dashed arrows indicate the hypothetical sources. NR, assimilatory nitrate reductase; NOS-like, nitrous oxide synthase-like enzyme; L-Arg, L-arginine; L-Cit, L-citrulline; Lb, leghaemoglobin; LbO_2, oxyleghaemoglobin; LbNO, nitrosylleghaemoglobin; Nap, periplasmic nitrate reductase; NirK, copper-containing nitrite reductase; cNor, *c*-type nitric oxide reductase; Nos, nitrous oxide reductase.

impaired NO detection in alfalfa nodules (Baudouin *et al.*, 2006), suggesting that a NOS-like enzyme may participate to NO production in nodules.

Since nitrite and NO are produced as intermediates during the reduction of nitrate to N_2, bacterial denitrification is another likely source of NO in nodules (see Fig. 10.2). Meakin and associates (2007) have demonstrated that nitrate reduction by Nap in *B. japonicum* USDA110 bacteroids contributes to the formation of LbNO in soybean nodules in response to hypoxia. In agreement with these findings, LbNO and free NO formation in response to nitrate and flooding conditions is significantly reduced in nodules formed by a *napA* deficient mutant (Sánchez *et al.*, 2010). Furthermore, LbNO formation in soybean nodules in response to flooding is caused by nitrite and NO generated from Nap and CuNir denitrification enzymes, respectively (Sánchez *et al.*, 2010) (see Fig. 10.2). In contrast to these observations, the denitrification pathway is not involved in NO production in nodules formed by *S. meliloti* in *Medicago truncatula*. The infection of roots with a *S. meliloti nirK* mutant strain impaired in nitrite reductase activity lead to the formation of nodules with an unaffected NO level, which indicates that the denitrification pathway is not required for NO production in alfalfa nodules (Baudouin *et al.*, 2010). Therefore, it may be possible that bacterial denitrification plays a role in NO production within nodules when roots are subjected to specific physiological conditions such as flooding stress and the presence of nitrate.

Basal levels of LbNO and free NO have been detected in soybean nodules produced by a *B. japonicum napA* mutant in response to flooding and nitrate conditions (Sánchez *et al.*, 2010). This suggests the presence of other denitrification-independent sources of NO, more likely those arising from the action of plant enzymes such as nitrate reductase and/or NOS-like enzyme (see Fig. 10.2).

In addition to NO, several studies have reported the production of nitrous oxide from soybean nodules, and that its reduction to N_2 was mediated by nitrous oxide reductase encoded by the *nosZ* gene (Mesa *et al.*, 2004; Sameshima-Saito *et al.*, 2006).

Role of nitrogen oxides in nodules

Several studies have clearly shown the production of NO during early stages of the symbiosis and in mature nodules. A rapid and transient NO production has been reported to occur in roots of *Lotus japonicus* inoculated with *M. loti* (Shimoda *et al.*, 2005), and modulation of NO levels after inoculation has been demonstrated to be involved in the establishment of the symbiosis (Nagata *et al.*, 2008). NO has also been shown to be involved in the auxin-signalling pathway controlling nodule formation and number in legumes with indeterminate nodules (Pii *et al.*, 2007). Accordingly, a recent transcriptosomal study on NO-responsive genes in *M. truncatula* revealed that NO may regulate nodule development and functioning (Ferrarini *et al.*, 2008).

Production of NO in functioning nodules has been associated with the bacteroid-containing cells of the fixing zone in alfalfa (Baudouin *et al.*, 2006) and soybean nodules (Meakin *et al.*, 2007; Sánchez *et al.*, 2010). A direct inhibition of nitrogenase activity by NO has been demonstrated *in vitro* (Trinchant and Rigaud, 1982), and application of any of the artificial NO donor sodium nitroprusside (SNP) and S-nitro-N-acetylpenicillamine (SNAP) decreased N_2 fixation activity in *L. japonicus* (Kato *et al.*, 2009; Shimoda *et al.*, 2009) and *Alnus firma* root nodules (Sasakura *et al.*, 2006).

The inhibition of nitrogenase activity in legume nodules by nitrate might be due to the toxic effects of nitrite or NO produced as intermediates during denitrification reactions or emerging from the host plant. Since the inhibitory effect of nitrate was suppressed by scavenging NO, it is more likely involved in inhibition of nitrogenase activity by nitrate in *Lotus* root nodules (Kato *et al.*, 2009). Other authors have indeed observed that a decrease in NO production in root nodules results in an increase in N_2 fixation activity (Shimoda *et al.*, 2009; Sánchez *et al.*, 2010; Tominaga *et al.*, 2010), which suggests that adequate concentrations of NO might be necessary for nitrogenase activity. It has also been suggested that plant nitrate reductase in the nodule cytosol is involved in inhibition of nitrogenase activity by nitrate (Streeter *et al.*, 1985; Kaiser *et al.*, 1997; Kanayama *et al.*, 1999). NO formed by *B. japonicum* NirK enzyme in soybean nodules in response to flooding and nitrate has a negative effect on both nitrogenase activity and expression of the *nifH* and *nifD* genes, which encode the Fe protein and the α-chain of the FeMo protein (Sánchez *et al.*, 2010). In that study, nitrogenase activity and gene expression were affected in nodules produced by a *B. japonicum norC* mutant, which accumulates NO produced through denitrification, but not in those nodules produced by a *nirK* mutant, where NO production from denitrification is blocked (Sánchez *et al.*, 2010).

As such, NO could also interfere with N_2 fixation by binding to Lb, which would therefore impair Lb functionality by competing with O_2 for binding sites, thus diminishing the O_2 supply available to bacteroids and thereby reducing N_2 fixation (Kanayama *et al.*, 1990). Other authors, however, have proposed that because only a small proportion if Lb is bound to O_2, and given that affinity of Lb for NO is higher than that for O_2, then Lb could act as NO scavenger modulating NO bioactivity (Herold and Puppo, 2005). Supporting this hypothesis, Sánchez and associated (2010) have recently proposed that Lb has a major role in detoxifying NO and nitrite produced by bacteroidal denitrification in response to flooding conditions. OxyLb is also able to scavenge peroxynitrite ($ONOO^-$), which is likely to be produced from the reaction of NO with superoxide (O_2^-) formed in nodules, precluding any damaging effect of this species (Herold and Puppo, 2005).

NO can also bind to the haem-based sensor FixL (Gilles-González *et al.*, 1994; Winkler *et al.*, 1996; Gong *et al.*, 2000), which is involved in O_2 regulation of N_2 fixation and denitrification genes (Fischer, 1994; Dixon and Kahn, 2004; Gilles-González and González, 2005; Bedmar *et al.*, 2005). In FixL, a haem-binding domain couples to a histidine-protein kinase such that the unliganded (deoxy, ferrous haem) form is the 'on-state' active kinase, and the O_2-bound (K_d ~ 50–140 mM) form is the 'off-state' (Gilles-González and González, 1993; Lois *et al.*, 1993; Mukai *et al.*, 2000). The FixL-catalysed phosphorylation of a receiver domain in FixJ promotes this response regulator to a dimeric form that activates transcription of target genes. Although all ligands (O_2, NO, CO) of ferrous haem bind to *B. japonicum* FixL, O_2 abolishes their kinase activity (> 200-fold inhibition), but CO (carbon monoxide) and NO do not (< 3-fold inhibition) (Gilles-González *et al.*, 1994; Dunham *et al.*, 2003; Sousa *et al.*, 2007).

Alternatively, it is also possible that NO reacts

directly with the N_2 fixation regulatory protein NifA (Fischer, 1994), inactivating the protein in a similar way as it has been demonstrated for the O_2-responsive regulator FNR in *Escherichia coli* (Cruz-Ramos *et al.*, 2002; Pullan *et al.*, 2007). Since NifA is the direct transcriptional activator of N_2 fixation genes, a possible effect of NO on the regulation of the N_2 fixation genes via this activator would result in a widespread effect on the transcription of all targets of NifA. The mechanism involved in the downregulation of nitrogenase genes by NO is the focus of future investigations.

References

Appleby, C.A. (1984). Leghemoglobin and *Rhizobium* respiration. Annu. Rev. Plant Physiol. Plant Mol. Biol. *35*, 443–478.

Appleby, C.A. (1992). The origin and functions of hemoglobin in plants. Sci. Prog. *76*, 365–398.

Baudouin, E., Pleuchot, L., Engler, G., Pauly, N., and Puppo, A. (2006). Nitric oxide is formed in *Medicago truncatula-Sinorhizobium meliloti* functional nodules. Mol. Plant-Microbe Interact. *19*, 970–975.

Bedmar, E.J., Robles, E.F., and Delgado, M.J. (2005). The complete denitrification pathway of the symbiotic, nitrogen-fixing bacterium *Bradyrhizobium japonicum*. Biochem. Soc. Trans. *33*, 141–144.

Berks, B.C., Sargent, F., and Palmer, T. (2000). The Tat protein export pathway. Mol. Microbiol. *35*, 260–274.

Bergersen, F.J., and Turner, G.L. (1993). Effects of concentration of substrates supplied to N_2 fixing soybean bacteroids in flow chamber reaction. Proc. R. Soc. London Ser. B. *251*, 103–109.

Besson-Bard, A., Pugin, A., and Wendehenne, D. (2008). New Insights into Nitric Oxide Signaling in Plants. Annu. Rev. Plant Biol. *59*, 21–39.

Bethke, P.C., Badger, M.R., and Jones, R.L. (2004). Apoplastic synthesis of nitric oxide by plant tissues. Plant Cell. *16*, 332–341.

Bueno, E., Gómez-Hernández, N., Girard, L., Bedmar, E.J., and Delgado, M.J. (2005). Function of the *Rhizobium etli* CFN42 *nirK* gene in nitrite metabolism. Biochem. Soc. Trans. *35*, 166–167.

Bueno, E., Bedmar, E.J., Richardson, D.J., and Delgado, M.J. (2008). Role of *Bradyrhizobium japonicum* cytochrome c_{550} in nitrite and nitrate respiration. FEMS Microbiol. Lett. *279*, 188–194.

Bueno, E., Bedmar, E.J., Richardson, D.J., and Delgado, M.J. (2009). Expression of *Bradyrhizobium japonicum* cbb_3 terminal oxidase under denitrifying conditions is subjected to redox control. FEMS Microbiol. Lett. *298*, 20–28.

Bueno, E., Mesa, S., Sánchez, C., Bedmar, E.J., and Delgado, M.J. (2010). NifA protein is required for maximal expression of denitrification genes in *Bradyrhizobium japonicum*. Environ. Microbiol. *12*, 393–400.

Casella, S., Shapleigh, J.P., and Payne, W.J. (1986). Nitrite reduction in *Rhizobium 'hedysari'* strain HCTN1. Arch. Microbiol. *146*, 233–238.

Casella, S., Shapleigh, J.P., Toffanin, A., and Basaglia, M. (2006). Investigation into the role of the truncated denitrification chain in *Rhizobium sullae* strain HCNT1. Biochem. Soc. Trans. *34*, 130–132.

Cruz-Ramos, H., Crack, J., Wu, G., Hughes, M.N., Scott, C., Thomson1, A.J., Green, J., and Poole, R.K. (2002). NO sensing by FNR: regulation of the *Escherichia coli* NO-detoxifying flavohaemoglobin, Hmp. The EMBO J. *21*, 3235–3244

Cueto, M., Hernández-Perera, O., Martín, R., Bentura, M.L., Rodigo, J., Lamas, S., and Golvano, M.P. (1996). Presence of nitric oxide synthase activity in roots and nodules of *Lupinus albus*. FEBS Lett. *398*, 159–164.

Delgado, M.J., Bedmar, E.J., and Downie. J.A. (1998). Genes involved in the formation and assembly of rhizobial cytochromes and their role in symbiotic nitrogen fixation. Adv. Microb. Physiol. *40*, 191–231.

Delgado, M.J., Bonnard, N., Tresierra-Ayala, A., Bedmar, E.J., and Müller, P. (2003). The *Bradyrhizobium japonicum napEDABC* genes encoding the periplasmic nitrate reductase are essential for nitrate respiration. Microbiology. *149*, 3395–3403.

Delgado, M.J., Casella, S., and Bedmar, E.J. (2007). Denitrification in Rhizobia-Legume Symbiosis. In Biology of the Nitrogen Cycle. H. Bothe, S.J. Ferguson and W.E. Newton, eds. (Elsevier), pp. 83–92.

Dixon, R., and Kahn, D. (2004). Genetic regulation of the biological nitrogen fixation. Nat. Rev. Microbiol. *2*, 621–631.

Downie, J.A. (2005). Legume haemoglobins: symbiotic nitrogen fixation needs bloody nodules. Curr. Biol. *15*, 196–198.

Dunham, C.M., Dioum, E.M., Tuckerman, J.R., Gonzalez, G., Scott, W.G., and Gilles-González, M.A. (2003). A distal arginine in oxygen-sensing heme-PAS domains is essential to ligand binding, signal transduction, and structure. Biochemistry. *42*, 7701–7708.

Ferrarini, A., De Stefano, M., Baudouin, E., Pucciariello, C., Polverari, A., Puppo, A., and Delledonne, M. (2008). Expression of *Medicago truncatula*genes responsive to nitric oxide in pathogenic and symbiotic conditions. Mol. Plant-Microbe Interact. *21*, 781–790.

Fischer, H.M. (1994). Genetic regulation of nitrogen fixation in rhizobia. Microbiol. Rev. *58*, 352–386.

Galibert, F., Finan, T.M., Long, S.R., and Puhler, A. (2001). The composite genome of the legume symbiont *Sinorhizobium meliloti*. Science. *293*, 668–672.

García-Horsman, J.A., Berry, E., Shapleigh, J.P., Alben, J.O., and Gennis, R.B. (1994). A novel cytochrome *c* oxidase from *Rhodobacter sphaeroides* that lacks CuA. Biochemistry. *33*, 3113–3119.

Gilles-González, M.A., and González, G. (1993). Regulation of the kinase activity of heme protein FixL from the two-component system FixL/FixJ of *Rhizobium meliloti*. J. Biol Chem. *268*, 16293–16297.

Gilles-González, M.A., González, G., and Perutz, M.F. (1994). Heme-based sensors, exemplified by the kinase FixL, are a new class of heme protein with distinctive ligand binding and autoxidation. Biochemistry. *33*, 8067–8073.

Gilles-González, M.A., and González, G. (2005). Heme-based sensors: defining characteristics, recent developments, and regulatory hypotheses. J. Inorg. Biochem. *99*, 1–22.

Gong, W., Hao, B., and Chan, M.K. (2000). New mechanistic insights from structural studies of oxygen-sensing domain of *Bradyrhizobium japonicum* FixL, Biochemistry *39*, 3955–3962.

Günther, C., Schlereth, A., Udvardi, M., and Ott, T. (2007). Metabolism of reactive oxygen species is attenuated in leghemoglobin-deficient nodules of *Lotus japonicus* Mol. Plant-Microbe Interact. *20*, 1596–1603.

Hauser, F., Pessi, G., Friberg, M., Weber, C., Rusca, N., Lindemann, A., Fischer, H.-M., and Hennecke, H. (2007). Dissection of the *Bradyrhizobium japonicum* NifA+σ^{54} regulon, and identification of a ferredoxin gene (*fdxN*) for symbiotic nitrogen fixation. Mol. Genet. Genomics. *278*, 255–271.

Herold, S., and Puppo, A. (2005). Oxyleghemoglobin scavenges nitrogen monoxide and peroxynitrite: a possible role in functioning nodules? J. Biol. Inorg. Chem. *10*, 935–945.

Holloway, P., McCormick, W., Watson, R.J., and Chan Y.K. (1996). Identification and analysis of the dissimilatory nitrous oxide reduction genes, *nosRZDFY*, of *Rhizobium meliloti*. J. Bacteriol. *178*, 1505–1514.

Hunt, S., and Layzell, D.B. (1993). Gas exchange of legume nodules and the regulation of nitrogenase activity. Annu. Rev. Plant Physiol. Plant Mol. Biol. *44*, 483–511.

Jones, K.M., Kobayashi, H., Davies, B.W., Taga, M.E., and Walker, G.C. (2007). How rhizobial symbionts invade plants: the *Sinorhizobium-Medicago* model. *Nat. Rev. Microbiol. 5*, 619–633.

Kaiser, B.N., Layzell, D.B., and Shelp, B.J. (1997). Role of oxygen limitation and nitrate metabolism in the nitrate inhibition of nitrogen fixation by pea. Physiol. Plant. *101*, 45–50.

Kanayama, Y., Watanabe, I., and Yamamoto, Y. (1990). Inhibition of nitrogen fixation in soybean plants supplied with nitrate I. Nitrite accumulation and formation of nitrosylleghemoglobin in nodules. Plant Cell Physiol. *31*, 341–346.

Kanayama, Y., Kimura, K., Nakamura, Y., and Ike, T. (1999). Purification and characterization of ntirate reductase from nodule cytosol of soybean plants. Physiol. Plant. *105*, 396–401.

Kaneko, T., Nakamura, Y., Sato, S., Minamisawa, K., Uchiumi, T., Sasamoto, S., Watanabe, A., Idesawa, K., Iriguchi, M., Kawashima, K., Kohara, M., Matsumoto, M., Shimpo, S., Tsuruoka, H., Wada, T., Yamada, M., and Tabata, S. (2002). Complete genomic sequence of nitrogen-fixing symbiotic bacterium *Bradyrhizobium japonicum* USDA110. DNA Res. *9*, 189–197.

Kato, K., Kanahama, K., and Kanayama, Y. (2009). Involvement of nitric oxide in the inhibition of nitrogenase activity by nitrate in *Lotus* root nodules. J. Plant. Physiol. *167*, 38–241.

Kim, Y.J., Ko, I.J., Lee, J.M., Kang, H.Y., Kim, Y.M., Kaplan, S., and Oh, J.I. (2007). Dominant role of the cbb_3 oxidase in regulation of photosynthesis gene expression through the PrrBA system in *Rhodobacter sphaeroides* 2.4.1. J. Bacteriol. *189*, 5617–5625.

Kojima, H., Urano, Y., Kikuchi, K., Higuchi, T., and Nagano, T. (1999). Fluorescent indicators for imaging nitric oxide production. Angew. Chem. Int. Ed. *38*, 3209–3212.

Kullik, I., Fritsche, S., Knobel, H., Sanjuan, J., Hennecke, H., and Fischer, H.M. (1991). *Bradyrhizobium japonicum* has two differentially regulated, functional homologs of the σ^{54} gene (*rpoN*). *J. Bacteriol. 173*, 1125–1138.

Lois, A.F., Weinstein, M., Ditta, G.S., and Helinski, D.R. (1993). Autophosphorylation and phosphatase activities of the oxygen-sensing protein FixL of *Rhizobium meliloti* are coordinately regulated by oxygen. J Biol Chem. *268*, 4370–4375.

Mandon, K., Kaminski, P.A., and Elmerich, C. (1994). Functional analysis of the *fixNOQP* region of *Azorhizobium caulinodans*. J. Bacteriol. *176*, 2560–2568.

Mathieu, C., Moreau, S., Frendo, P., Puppo, A., and Davies, M.J. (1998). Direct detection of radicals in intact soybean nodules: presence of nitric oxide-leghemoglobin complexes. Free. Radic. Biol. Med. *24*, 1242–1249.

Meakin, G.E., Jepson, B., Bedmar, E.J., Richardson, D.J., and Delgado, M.J. (2006). The role of *Bradyrhizobium japonicum* nitric oxide reductase in nitric oxide detoxification in soya bean root nodules. Biochem. Soc. Trans. *34*, 195–196.

Meakin, G.E., Bueno, E., Jepson, B., Bedmar, E.J., Richardson, D.J., and Delgado, M.J. (2007). The contribution of bacteroidal nitrate and nitrite reduction to the formation of nitrosylleghaemoglobin complexes in soybean root nodules. Microbiology *153*, 411–419.

Mesa, S., Velasco, L., Manzanera, M.E., Delgado, M.J., and Bedmar, E.J. (2002). Characterization of the *norCBQD* genes, encoding nitric oxide reductase, in the nitrogen fixing bacterium *Bradyrhizobium japonicum*. Microbiology *148*, 3553–3560.

Mesa, S., Bedmar, E.J., Chanfon, A., Hennecke H., and Fischer H.M. (2003). *Bradyrhizobium japonicum* NnrR, a denitrification regulator, expands the FixLJ-FixK$_2$ regulatory cascade. J. Bacteriol. *185*, 3978–3982.

Mesa, S., de Dios Alché, J., Bedmar, E.J., and Delgado, M.J. (2004). Expression of the *nir*, *nor* and *nos* denitrification genes from *Bradyrhizobium japonicum* in soybean root nodules. Physiol. Plant. *120*, 205–211.

Mesa, S., Hauser, F., Friberg, M., Malaguti, E., Fischer, H.M., and Hennecke, H. (2008). Comprehensive assesment of the regulons controlled by the FixLJ-FixK$_2$-FixK$_1$ cascade in *Bradyrhizobium japonicum*. J. Bacteriol. *190*, 6568–6579.

Meyer, C., Lea, U.S., Provan, F., Kaiser, W.M., and Lillo, C. (2005). Is nitrate reductase a major player in the plant NO (nitric oxide) game? Photosyn. Res. *83*, 181–189.

Minchin, F.R., James, E.K., and Becana, M. (2008). Oxygen diffusion, production of reactive oxygen and nitrogen species, and antioxidants in legume nodules. In Nitrogen-Fixing Leguminous Symbioses. M.J. Dilworth, E.K. James, J.I. Sprent and W.E. Newton, eds. (Springer Science), pp. 321–362.

Monza, J., Irisarri, P., Díaz, P., Delgado, M.J., Mesa, S., and Bedmar, E.J. (2006). Denitrification ability of rhizobial strains isolated from *Lotus* sp. Antonie Van Leeuwenhoek. *89,* 479–84.

Mukai, M., Nakamura, K., Nakamura, H., Iizuka, T., and Shiro, Y. (2000). Roles of Ile209 and Ile210 on the heme pocket structure and regulation of histidine kinase activity of oxygen sensor FixL from *Rhizobium meliloti*. Biochemistry. *39,* 13810–13816.

Nagata, M., Murakami, E., Shimoda, Y., Shimoda-Sasakura, F., Kucho, K., Suzuki, A., Abe, M., Higashi, S., and Uchiumi, T. (2008). Expression of a class 1 hemoglobin gene and production of nitric oxide in response to symbiotic and pathogenic bacteria in *Lotus japonicus*. Mol. Plant-Microbe Interact. *21,* 1175–1183.

Nellen-Anthamatten, D., Rossi, P., Preisig, O., Kullik, I., Babst, M., Fischer, H.M., and Hennecke, H. (1998). *Bradyrhizobium japonicum* $FixK_2$, a crucial distributor in the FixLJ-dependent regulatory cascade for control of genes inducible by low oxygen evels. J. Bacteriol. *180,* 5251–5255.

Nienaber, A., Huber, A., Göttfert, M., Hennecke, H., and Fischer, H.M. (2000). Three new NifA-regulated genes in the *Bradyrhizobium japonicum* symbiotic gene region discovered by competitive DNA–RNA hybridization. J. Bacteriol. *182,*1472–1480.

Oh, J.I., and Kaplan, S. (1999). The cbb_3 terminal oxidase of *Rhodobacter sphaeroides* 2.4.1.: structural and functional implications for the regulation of spectral complex formation. Biochemistry. *38,* 2688–2696.

Oh, J.I., and Kaplan, S. (2002). Oxygen adaptation. The role of the CcoQ subunit of the cbb_3 cytochrome *c* oxidase of *Rhodobacter sphaeroides* 2.4.1. J. Biol. Chem. 277: 16220-16228.

Oldroy, G.E.D., and Downie, J.A. (2008). Coordination nodule morphogenesis with rhizobial infection in legumes. Annu. Rev. Plant Biol. *59,* 519–546.

Ott, T., van Dongen, J.T., Günther, C., Krusell, L., Desbrosses, G., Vigeolas, H., Bock, V., Czechowski, T., Geigenberger, P., and Udvardi, M.K. (2005). Symbiotic leghemoglobins are crucial for nitrogen fixation in legume root nodules but not for general plant growth and development. Curr. Biol. *15,* 531–535

Pii, Y., Crimi, M., Cremonese, G., Spena, A., and Pandolfini, T. (2007). Auxin and nitric oxide control indeterminate nodule formation. BMC Plant Biol. *8, 7–21.*

Pitcher, R.S., and Watmough, N.J. (2004). The bacterial cytochrome cbb_3 oxidases. Biochim. Biophys. Acta. *30,* 653–658.

Planchet, E., and Kaiser, W.M. (2006). Nitric oxide detection by DAF fluorescence and chemiluminescence: A comparison using abiotic and biotic NO sources. J. Exp. Bot. 57, 3043–3055.

Preisig, O., Anthamatten, D., and Hennecke, H. (1993). Genes for a microaerobically induced oxidase complex in *Bradyrhizobium japonicum* are essential for a nitrogen-fixing endosymbiosis. Proc. Natl. Acad. Sci. USA. *90,* 3309–3313.

Preisig, O., Zufferey, R., Thöny-Meyer, L., Appleby, C.A., and Hennecke, H. (1996). A high-affinity cbb_3-type cytochrome oxidase terminates the symbiosis-specific respiratory chain of *Bradyrhizobium japonicum*. J. Bacteriol. *178,* 1532–1538.

Pullan, S.T., Gidley, M.D., Jones, R.A., Barrett, J., Stevanin, T.M., Read, R.C., Green, J., and Poole, R.K. (2007). Nitric oxide in chemostat-cultured *Escherichia coli* is sensed by Fnr and other global regulators: unaltered methionine biosynthesis indicates lack of *S* nitrosation. J. Bacteriol. *189,* 1845–1855.

Robles, E.F., Sánchez, C., Bonnard, N., Delgado, M.J., and Bedmar, E.J. (2006). The *Bradyrhizobium japonicum napEDABC* genes are controlled by the FixLJ-$FixK_2$-NnrR regulatory cascade. Biochem. Soc. Trans. *34,* 108–110.

Sameshima-Saito, R., Chiba, K., Hirayama, J., Itakura, M., Mitsui, H., Eda, S., and Minamisawa, K. (2006). Symbiotic *Bradyrhizobium japonicum* reduces N_2O surrounding the soybean root system via nitrous oxide reductase. Appl. Environm. Microbiol. 72, 2526–2532.

Sánchez, C., Gates, A.J., Meakin, G.E., Uchiumi, T., Girard, L., Richardson, D.J., Bedmar, E.J., and Delgado, M.J. (2010). Production of nitric oxide and nitrosylleghaemoglobin complexes in soybean nodules in response to flooding. Mol. Plant-Microbe Interact. 23, 702–711.

Sasakura, F., Uchiumi, T., Shimoda, Y., Suzuki, A., Takenouchi, K., Higashi, S., and Abe, M. (2006). A class 1 hemoglobin gene from *Alnus firma* functions in symbiotic and nonsymbiotic tissues to detoxify nitric oxide. Mol. Plant-Microbe Interact. *19,* 441–450.

Sawada, H., Kuykendall, L.D., and Young, J.M. (2003). Changing concepts in the systematics of bacterial nitrogen-fixing legume symbionts. J. Gen. Appl. Microbiol. *49,* 155–179.

Sciotti, M.-A., Chanfon, A., Hennecke, H., and Fischer, H.M. 2003. Disparate oxygen responsiveness of two regulatory cascades that control expression of symbiotic genes in *Bradyrhizobium japonicum*. J. Bacteriol. *185,* 5639–5642.

Seefeldt, L.C., Hoffman, B.M., and Dean, D.R. (2009). Mechanism of Mo-Dependent Nitrogenase. Annu. Rev. Biochem. *78,* 701–722.

Shimoda, Y., Nagata, M., Suzuki, A., Abe, M., Sato, S., Kato, T., Tabata, S., Higashi, S., and Uchiumi, T. (2005). Symbiotic *Rhizobium* and nitric oxide induce gene expression of non-symbiotic hemoglobin in *Lotus japonicus*. Plant Cell Physiol. *46,* 99–107.

Shimoda, Y., Shimoda-Sasakura, F., Kucho, K., Kanamori, N., Nagata, M., Suzuki, A., Abe, M., Higashi, S., and Uchiumi, T. (2009). Overexpression of class 1 plant hemoglobin genes enhances symbiotic nitrogen fixation activity between *Mesorhizobium loti* and *Lotus japonicus*. Plant J. 57, 254–263.

Sousa, E.H.S., Tuckerman, J.R., González, G., and Gilles-González, M.A. (2007). DosT and DevS are oxygen-switched kinases in Mycobacterium tuberculosis. Protein Sci. *16*, 1708–1719.

Stacey, G. (2007). The *Rhizobium*-Legume Nitrogen Fixing Symbiosis. In Biology of the nitrogen cycle. H. Bothe, S.J. Ferguson and W.E. Newton, eds. (Elsevier), pp. 147–163.

Stöhr, C., Strube, F., Marx, G., Ullrich, W. R., and Rockel, P. (2001). A plasma membrane-bound enzyme of tobacco roots catalyses the formation of nitric oxide from nitrite. Planta. *212*, 835–841.

Streeter, J.G. (1985). Nitrate inhibition of legume nodule growth and activity I. Long term studies with a continuous supply of nitrate. Plant Physiol. *77*, 321–324.

Thöny-Meyer, L., Stax, D., and Hennecke, H. (1989). An unusual gene cluster for the cytochrome bc_1 complex in *Bradyrhizobium japonicum* and its requirement for effective root nodule symbiosis. Cell. *57*, 683–697.

Toffanin, A., Wu, Q., Maskus, M., Casella, S., Abruna, H.D., and Shapleigh, J.P. (1996). Characterization of the gene encoding nitrite reductase and the physiological consequences of its expression in the nondenitrifying *Rhizobium 'hedysari'* strain HCNT1. Appl. Environ. Microbiol. *62*, 4019–4025.

Toffanin, M., Basaglia, C., Ciardi, P., Vian, S., Povolo, S., and Casella, S. (2000). Energy content decrease and viable-not-culturable status induced by oxygen limitation coupled to the presence of nitrogen oxides in *Rhizobium 'hedysari'*. Biol. Fertil. Soils. *31*, 484–488.

Tominaga, A., Nagata, M., Futsuki, K., Abe, H., Uchiumi, T., Abe, M., Kucho, K., Hashiguchi, M., Akashi, R., Hirsch, A.M., Arima, S., and Suzuki, A. (2010). Enhaced nodulation and nitrogen fixation in the abscisic acid low-sensitive mutant *enhaced nitrogen fixation1* of *Lotus japonicus*. Plant Physiol. *151*, 1965–1976.

Trinchant, J.C., and Rigaud, J. (1980). Nitrite inhibition of nitrogenase from soybean bacteroids. Appl. Environ. Microbiol. *124*, 49–54.

Trinchant, J.C., and Rigaud, J. (1982). Nitrite and nitric oxide as inhibitors of nitrogenase from soybean bacteroids. Appl. Environ. Microbiol. *44*, 1385–1388.

van Spanning, R.J., Delgado, M.J., y Richardson, D.J. (2005). The nitrogen cycle: denitrification and its relationship to N_2 fixation. In Nitrogen fixation in agriculture, forestry, ecology and the environment. D. Werner and W.E. Newton, eds. (Springer Netherlands), pp. 277–342.

van Spanning, R.J., Richardson, D.J., y Ferguson, S.J. (2007). Introduction to the biochemistry and molecular biology of denitrification. In Biology of the nitrogen cycle. H. Bothe, S.J. Ferguson and W.E. Newton, eds. (Elsevier), pp. 3–20.

Velasco, L., Mesa, S., Delgado, M.J., and Bedmar, E.J. (2001). Characterization of the *nirK* gene encoding the respiratory, Cu-containing nitrite reductase of *Bradyrhizobium japonicum*. Biochim. Biophys. Acta. *1521*, 130–134.

Velasco, L., Mesa, S., Xu, C., Delgado, M.J., and Bedmar, E.J. (2004). Molecular characterization of *nosRZDFYLX* genes coding for denitrifying nitrous oxide reductase of *Bradyrhizobium japonicum*. Antonie van Leeuwenhoek. *85*, 229–235.

Velázquez, E., Mateos, P.F., Trujillo, M., Rivas, R., and Martínez-Molina, E. (2006). Diversidad de bacterias rizoendosimbióticas de plantas. Bedmar, E.J., González, J., Lluch, C. and Rodelas, B., eds. (Sociedad Española de Fijación de Nitrógeno), pp. 45–52.

Viebrock, A., and Zumft, W.G. (1987). Physical mapping of transposon TnS insertions defines a gene cluster functional in nitrous oxide respiration by *Pseudomonas stutzeri*. J. Bacteriol. *169*, 4577–4580.

Willems, A. (2006). The taxonomy of rhizobia: an overview. Plant. Soil *287*, 3–14

Winkler, W.C., González, G., Wittenberg J.B., Hille, R., Dakappagari, N., Jacob, A., González, L.A., and Gilles, M.A. (1996). Nonsteric factors dominate binding of nitric oxide, azide, imidazole, cyanide, and fluoride to the Rhizobial heme-based oxygen sensor FixL. Chem. Biol. 3, 841–850.

Zakhia, F., and de Lajudie, P. (2006). Modern bacterial taxonomy: techniques review application to bacteria that nodulate leguminous plants (BNL). Can. J. Microbiol. 52, 169–181.

Zufferey, R., Preisig, O., Hennecke, H., and Thöny-Meyer, L. (1996). Assembly and function of the cytochrome cbb_3 oxidase subunits in *Bradyrhizobium japonicum*. J. Biol. Chem. *271*, 9114–9119.

Zumft, W.G. (1997). Cell biology and molecular basis of denitrification. Microbiol. Mol. Biol. Rev. *61*, 533–616.

Nitrous Oxide Production in the Terrestrial Environment

11

Elizabeth M. Baggs and Laurent Philippot

Abstract

Terrestrial ecosystems are a major source of nitrous oxide (N_2O), with soils accounting for ~70% of the atmospheric loading of this greenhouse gas. Here we provide a synthesis of current understanding of the environmental regulation of N_2O production and reduction through different microbial pathways, presenting examples of where measured emissions have been related to characterizations of the underpinning microbial communities. We explore the direct and indirect influence of plants on rhizosphere N_2O production, reduction and net emission, and the interactions between N_2O production and methane oxidation, as examples of coupling between the C and N cycles that need to be considered when developing appropriate and more targeted strategies for greenhouse gas mitigation.

Nitrous oxide – a potent greenhouse gas

Terrestrial systems are major sources of atmospheric nitrous oxide (N_2O), a greenhouse gas accounting for 6% of the present greenhouse forcing rate ascribed to anthropogenically derived gases (CO_2, perfluorocarbons, hydrofluorocarbons, sulfur hexafluoride, CH_4 and N_2O) (IPCC, 2007). Because of the strong absorption of these gases in the infrared window (8–12 μm) only about 5% of long-wave radiation can escape from the earth's surface into space, with more than 90% of the radiation being radiated back to the earth's surface. Therefore, an increase in the concentration of a greenhouse gas initially reduces the flux of long-wave terrestrial radiation to space, as more becomes trapped in the troposphere. This results in a rise in temperature at the earth's surface, known as the greenhouse effect (Wang *et al.*, 1976). N_2O has a global warming potential (GWP) of 298 per 1 kg of gas (over 100 years), referenced to CO_2, which has a GWP of 1 (IPCC, 2007).

Nitrous oxide is also a major natural regulator of stratospheric O_3, which effectively controls the earth's ultraviolet-B radiation balance. In the stratosphere N_2O is destroyed by direct photolysis by UV radiation:

$$N_2O + hv \rightarrow N_2 + O_2 + O(^1D) \qquad (11.1)$$

and by reaction with excited oxygen atoms formed in ozone photolysis:

$$N_2O + O(^1D) \rightarrow 2NO \ \ (60\%) \qquad (11.2)$$

$$N_2O + O(^1D) \rightarrow N_2 + O_2 \ \ (40\%) \qquad (11.3)$$

The NO formed catalyses the decomposition of stratospheric O_3, and therefore N_2O is considered as the dominant ozone-depleting substance emitted in the twenty-first century (Crutzen, 1970; Ravishankara *et al.*, 2009). The rate of these reactions varies throughout the stratosphere, with an average lifetime of over 100 years.

N_2O emission from soils

Soils account for ~ 70% of the atmospheric loading of N_2O. Emissions from agricultural systems constitute the largest single global N_2O source (6.8 Tg N_2O-N/year) with 4.2 (0.6–14.8) Tg N_2O-N/year from N added to soil and indirect emissions, 2.1 (0.6–3.1) Tg N_2O-N/year from manure management, and 0.5 (0.2–1.0) Tg N_2O-N/year from biomass burning. In most global assessments the high N_2O fluxes from agriculture are assumed to result from fertilizer use, with peak emissions usually measured in the first 3 or 4 weeks after N fertilizer application (Skiba *et al.*, 1996). Typically, increases in N_2O emissions after fertilization are either immediate or lag a few days behind the date of application (Conrad *et al.*, 1983; McTaggart *et al.*, 1994). Peak emissions may last between several days and a few weeks, often being ended with a sharp decrease (Conrad *et al.*, 1983; Van Cleemput *et al.*, 1994). Secondary peaks may occur following rainfall, until there is a depletion of fertilizer in the upper soil layers (Seiler and Conrad, 1981; Conrad *et al.*, 1983). Fertilizer type, the fertilizer's physical state when applied (dry or liquid), the application procedure and timing of application are important in determining the magnitude and timing of N_2O emissions (Mosier and Hutchinson, 1981; Mosier *et al.*, 1982, 1983; Duxbury and McConnaughey, 1986; Bouwman, 1996). In the UK, soil temperature and soil water have been identified as the main controls on N_2O emissions after N input. This explains regional differences in N_2O emissions between the warmer, wetter south-west, and the cooler, drier north-east, as well as inter-annual variation (Skiba *et al.*, 1996). Direct fluxes of N_2O from agricultural sources may also result from leaching and runoff into aquatic systems, and the hydrologic transport of dissolved N_2O through ground water to surface waters (Weymann *et al.*, 2008), although the magnitude of this contribution to atmospheric N_2O is uncertain.

The global emission of N_2O from natural grassland, wetlands and forest soils is low compared to agricultural soils, and there are uncertainties over their contributions to the global atmospheric N_2O budget. The uncertainties are mainly due to paucity in data compared to agricultural soils, and the high spatial and temporal variability associated with these emissions. Average N_2O emissions from EU forest soils are estimated to be about ten times lower than from agricultural soils (Kesik *et al.*, 2005; Schulte-Bisping, 2003). The magnitude of emissions from forests may be dependent not only on environmental variables, but also on the species diversity within the forest (Guckland *et al.*, 2010). Given that fungi can dominate forest soil microbiology (Ananyeva *et al.*, 2010) it is often assumed that fungal denitrification is important in forest systems, some of which may be attributable to ecto-mycorrhizal fungi (Prendergast-Miller *et al.*, 2011). There is also the possibility that heterotrophic nitrification contributes to N_2O fluxes in these systems (Wolf and Brumme, 2002), although this remains to be verified.

N_2O from arable agriculture

Emission factors vary from 0.1% to 7% of nitrogen applied in different agricultural systems (Skiba and Smith, 2000), reflecting differences in vegetation type, crop management and climate. Measured emissions can vary significantly with crop type, for example ranging from 0.2 to 0.7 kg N_2O-N 100/kg N applied for small grain cereals, from 0.3 to 5.8 kg N_2O-N 100/kg N applied from cut grassland (Dobbie *et al.*, 1999), and from 3.9 to 8.7 kg N_2O-N/ha/year from maize fields (Sehy *et al.*, 2003). In legume fields, emissions range from 0.34 to 4.6 kg N_2O-N/ha/year (Eichner, 1990), including natural emissions, those associated with cultivation, and those derived from nitrogen fixed by the legume. These differences may arise from crop specific water demand, rhizodeposition, residue input and root turnover, or may reflect different climatic or management regimes required for different crops.

Nitrous oxide emissions from plant residue amended soil vary with the type of residue, the residue composition, and biomass incorporated (Reinertsen *et al.*, 1984; Paul and Beauchamp, 1989). The chemical composition, or quality, of the residue is important in regulating N release and subsequent N_2O production (Baggs *et al.*, 2000; Harrison *et al.*, 2002; Millar *et al.*, 2004; Gentile *et al.*, 2008; Millar and Baggs, 2004, 2005; Huang *et al.*, 2004; Garcia-Ruiz and Baggs, 2007). Strong negative relationships have been observed between N_2O emission and residue C-to-N ratio in (Millar *et al.*, 2004) and with residue lignin content and (lignin + polyphenol)-to-N ratio after applying tropical agroforestry residues and olive crop weed residues to soil under controlled environment conditions (Millar and Baggs, 2004; Garcia-Ruiz and Baggs, 2007). Addition of plant biomass to soil also enhances N_2O production through creation of sub-oxic microsites resulting from increased heterotrophic respiration, physical placement of residues, and retention of soil moisture (Drury *et al.*, 1991; McKenney *et al.*, 1995; Harper and Lynch, 1981).

N_2O emission from livestock agriculture

Livestock agriculture has been estimated to account for up to 65% of global N_2O emissions (Steinfeld *et al.*, 2006), mainly as a result of fertilizer N application and N_2 fixation inputs, and N excretion as dung or urine, either directly returned to soil, or applied to land as manure or effluent from animal housing. Grazing accounts for 45% of this total livestock emission, because these fertilized pastures are often located in regions of high rainfall where soils are too wet for arable, and soil compaction associated with animal traffic also provides conditions favourable for high N_2O production during denitrification (Skiba *et al.*, 1996; Oenema *et al.*, 2005). Strategies for mitigation from livestock systems include changes in animal feed (Miller *et al.*, 2001; Misselbrook *et al.*, 2005), animal breeding for N conversion efficiency within the rumen (Coffey, 1996) to manipulate N excretion (de Klein and Eckard, 2008), optimizing fertilizer N input and drainage, application of nitrification inhibitors (Di and Cameron, 2002) and grazing management (Schils *et al.*, 2006).

Microbial sources of N_2O in soil

Soil microbial N_2O production occurs via nitrification (ammonia oxidation) and nitrate dissimilation (denitrification and nitrate ammonification) pathways (Fig. 11.1). These processes rarely occur in isolation, with possible competition for substrates under near limiting conditions, and the possibility of transfer of N_2O or intermediary products from one process to another depending on prevailing environmental conditions, microbial community structure and location within the soil matrix. This means that N_2O produced during several processes form one pool that might be reduced to N_2 during denitrification.

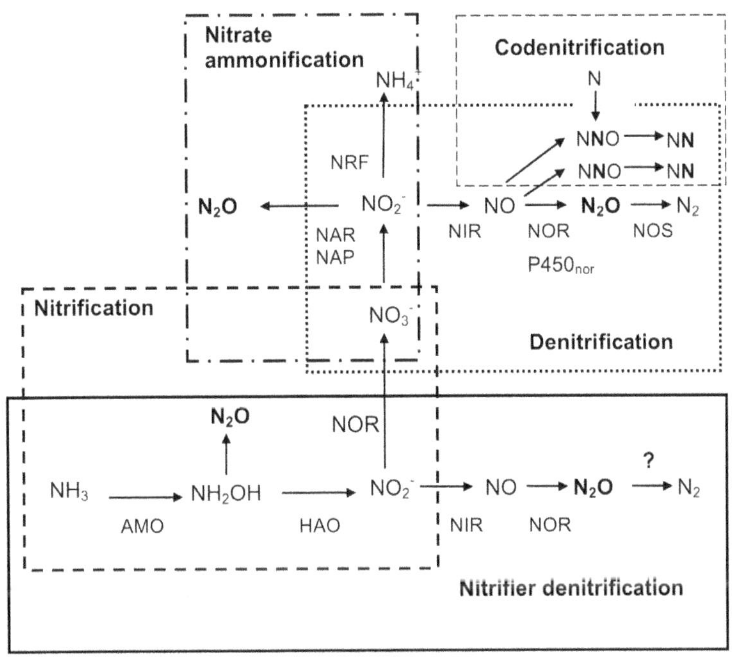

Figure 11.1 Soil microbial pathways to N_2O production (modified from Baggs, 2008).

Denitrification

Denitrification is the reduction of NO_3^- or NO_2^- to the gases NO, N_2O and N_2 under anaerobic conditions (Robertson and Tiedje, 1987; Bremner, 1997). The process is catalysed by the enzymes nitrate reductase, nitrite reductase, nitric oxide reductase and nitrous oxide reductase, and the transport of electrons to NO_3^- or other N oxides is coupled to the synthesis of ATP (adenosine triphosphate) (Hochstein and Tomlinson, 1988). The nature and regulation of the reductase enzymes involved in denitrification have been well characterized (Zumft, 1997). More than 60 genera of denitrifying microorganisms have been identified within bacteria but also archaea, fungi and other eukaryotes (Philippot *et al.,* 2007; Shoun *et al.,* 1992; Risgaard-Petersen *et al.,* 2006; Piña-Ochoa *et al.,* 2010), The density of denitrifiers in soil can be up to 10^9 cells per gram of soil (Babic *et al.,* 2008; Henry *et al.,* 2008; Dandie *et al.,* 2008) and both cultivation-dependent and -independent methods showed that the proportion of denitrifiers represent up to 5% of the total soil microbial community (Henry *et al.,* 2006; Tiedje, 1988).

Since the ability to denitrify is sporadically distributed both within and between different genera, the genes encoding the catalytic subunit of the different denitrification reductases are commonly used as molecular markers in studies investigating the diversity of denitrifiers. Thus membrane-bound and periplasmic nitrate reductase encoding genes (*napA, narG;* Gregory *et al.,* 2000; Philippot *et al.,* 2002), cytochrome cd_1 and Cu-containing nitrite reductase encoding genes (*nirS, nirK;* Hallin and Lindgren, 1999) and N_2O reductase encoding genes (*nosZ;* Stres *et al.,* 2004) have successfully been used in terrestrial environments for fingerprinting analyses to identify the environmental factors driving the denitrifier community composition.

The activity of denitrifying bacteria is controlled by C availability, O_2 concentration (mainly through soil water content), N availability, pH and temperature. The contribution of denitrification to N_2O emissions from soils is traditionally thought to be greatest under sub-oxic conditions, and in soils at a water-filled pore space (WFPS) >70% where soil NO_3^- and available C are non-limiting (Davidson, 1991; Bateman and Baggs, 2005). High O_2 concentrations are known to suppress the activity and synthesis of the denitrification reductases. The N_2O reductase is thought to be the most sensitive to O_2 (Firestone at al, 1980; Otte *et al.,* 1996), reflected in a higher N_2O/N_2 ratio with increasing O_2 availability (Weier *et al.,* 1993). When re-exposed to O_2 after an anaerobic phase, all denitrification enzymes but the N_2O reductase remain active (Morley *et al.,* 2008), and conversely when aerobic soils become anaerobic, such as following heavy rainfall, the NO_3^- and NO_2^- reductases are suggested to be activated sooner than the N_2O reductase, explaining why the denitrifier N_2O/N_2 ratio can be raised up to one or two days after rainfall (Knowles, 1982). Several bacteria isolated from soils and sediment are capable of denitrifying in the presence of O_2 (Lloyd *et al.,* 1987; Patureau *et al.,* 2000). Although N_2 and N_2O production is common during aerobic denitrification by cultured isolates (Patureau *et al.,* 2000; Takaya *et al.,* 2003), their contribution to N_2O emissions from soils has yet to be proven, and in the meantime, denitrification is still modelled as an anaerobic process.

Denitrification rates are influenced by the availability of N electron acceptors, and this is why they are stimulated by addition of N, usually in the form of inorganic fertilizers (Eichner, 1990). N availability also affects the N_2O/N_2 ratio, which can be raised with NO_3^- concentrations >10 μg/g soil, where NO_3^- is preferred over N_2O as an electron acceptor (Blackmer and Bremner, 1978; Baggs *et al.,* 2003). The enzyme kinetics and affinity for substrates by denitrifiers is highly variable, possibly reflecting the wide phylogenetic diversity of this group, with K_m values for N_2O reduction of between 0.1 and 100 μM (Conrad, 1996; Holtan-Hartwig *et al.,* 2000).

As most denitrifiers are heterotrophic, the availability of organic C, which can be used as an electron donor, is also a major factor determining denitrification rates (Knowles, 1982; Bremner, 1997). However, organic C availability is the least well understood factor controlling denitrifier N_2O production and reduction, despite the fact that it has traditionally been thought a reliable index for predicting a soil's denitrification capacity (Burford and Bremner, 1975). The effects of C are not only mediated directly, but also indirectly through the creation of anaerobiosis generated as a result of microbial respiration (Azam *et al.*, 2002). With regards to direct effects, there is likely to be strong coupling between plant exudate-C and denitrification, through the provision of the reductant for denitrification, and changing the size, structure and activity of the denitrifier community. Indeed, Smith and Tiedje (1979) found a positive effect of roots on denitrification when soil NO_3^- concentrations were high, which was thought to be due to release of organic C from the roots, with denitrifying activity decreasing rapidly in the first few millimetres away from the roots. Denitrification has been shown to increase following addition of easily oxidizable organic matter to soils, often very rapidly, and significant correlations have been reported between total organic C (Baggs and Blum, 2004), 'available' C (Stanford *et al.*, 1975), water-soluble and 'mineralizable' C (Burford and Bremner, 1975; Paul and Beauchamp, 1989) and soluble C-to-N ratio of applied residues (Millar and Baggs, 2005). However, little is known about the effect of individual C compounds on the regulation of the enzymes producing or reducing N_2O, and we are still unsure whether the type of C compounds is an important driver of the denitrifier community. In a recent study, Henry *et al.* (2008) applied artificial root exudates comprised of different combinations of sugars, organic acids and amino acids, to soil microcosms, and showed that a higher proportion of sugars resulted in a lower N_2O/N_2 ratio. Such results warrant further attention before we can propose that manipulating plant-derived C flow has the potential to increase reduction of N_2O to N_2 (Philippot *et al.*, 2009b).

Among the other factors influencing denitrification, a special emphasis has been placed on soil pH. Denitrification rates are usually thought to decrease with pH, but significant denitrification rates can be still observed below pH 4.9 (Ellis *et al.*, 1998). This can reflect differences in the composition of the denitrifier community (Enwall *et al.*, 2005) and adaptation of denitrifiers to low soil pH (Parkin *et al.*, 1985). Interestingly, soil pH does not only affect denitrification rates but also the N_2O/N_2 ratio. Thus, several studies reported that the N_2O reductase is more sensitive to low pH than the other denitrification reductases, resulting in higher ratios of N_2O to N_2 as pH declines (Firestone *et al.*, 1980; Nägele and Conrad, 1990; Thomsen *et al.*, 1994; Šimek and Cooper, 2002; Dannenmann *et al.*, 2008; Cuhel *et al.*, 2010).

Finally it has also been suggested that the composition and the abundance of the denitrifying community itself could influence denitrification (reviewed in Philippot and Hallin, 2005). As an example, relationships between denitrifier community composition and N_2O production were suggested by Cavigelli and Roberston (2000), who found that the enzymes involved in N_2O-production or reduction had different sensitivities to oxygen or pH in denitrifiers from a conventionally tilled agricultural field and a never-tilled successional field. Further studies have both confirmed and contradicted such links between denitrifier community composition and abundance, and denitrification rates. Hallin *et al.* (2009) found that differences in denitrification rates between fertilization regimes were related to differences in abundance but not in composition of the denitrifier community. Other studies also reported a positive correlation between denitrification activity and the size of the denitrifier community (Patra *et al.*, 2005; Throbäck *et al.*, 2007; Enwall *et al.*, 2010). Links between N_2O emissions and denitrifier community are more complicated to investigate due to interwowen effects of N_2O emissions by nitrifiers and balance between N_2O production and reduction by denitrifiers. The addition of different types of artificial root exudates to soil microcosms resulted in differences in the N_2O/N_2 ratio, which were not related to differences in denitrifier community size or composition (Henry *et al.*, 2008). Comparison of agricultural soil, riparian soil, and creek sediment showed that denitrifier community composition and N_2O production were uncoupled across these agroecosystems (Rich and Myrold, 2004). Similarly, changes in denitrification and N_2O emissions over time and spatial location in a potato field were not related to changes in the abundances of the denitrifier community (Dandie *et al.*, 2008). However, comparison a limited number of ecosystems or conditions prevents drawing robust conclusions about the putative relationships between activity and structure or size of the denitrifier community.

To overcome this drawback, it is of importance to use a large gradient of environmental conditions. Using a spatially explicit approach, Philippot *et al.* (2009a) found that while the denitrification activity and N_2O emissions were uncoupled from the abundance of the denitrifier community in a grassland, a strong correlation was found between the distribution of the percentage of bacteria capable of reducing N_2O within the total bacterial community and potential N_2O emissions, both being also correlated to soil pH. Manipulation of the abundance of denitrifiers lacking the genetic ability to synthesize the nitrous oxide reductase in soils established a direct causal link between the denitrifier community composition and potential N_2O emissions by denitrification (Philippot *et al.*, Global change Biology, In press). This indicates that the extent of the reduction of N_2O to N_2 by the denitrifying community can have a genetic basis.

It is important however to emphasize that since denitrification is a facultative respiratory process, the factors primarily regulating denitrification activity are the presence of nitrate and carbon and the absence of oxygen but not the abundance or the composition of the denitrifying community. Nevertheless, there is a growing body of literature showing that when environmental conditions are favourable to denitrification, the abundance or the composition of the denitrifying community could also contribute to the regulation of the denitrification activity and N_2O emissions.

Fungal denitrification

The potential for fungi to produce N_2O has been demonstrated in woodland (Laverman *et al.*, 2000) and in grassland soils (Laughlin and Stevens, 2002). N_2O release from acid coniferous forest soils is often low and attributed to nitrification (Sitaula and Bakken, 1993), but there is evidence that fungal denitrification may actually be important in these systems (Kester *et al.*, 1996; Laverman *et al.*, 2000). Several fungal strains have been shown to possess cytochromes $P450_{nir}$ or $P450_{nor}$ that enable them to denitrify and release N_2O when anaerobically incubated with NO_2^- in culture (Shoun *et al.*, 1992; Takaya and Shoun, 2000). The Nir of the fungus *Cylindrocarpon tonkinense* has been shown to be a copper-containing Nir (Kubota *et al.*, 1999). In fungal denitrification the enzymes accept two electrons from NADH or NADPH without involving a membrane-bound electron transfer chain as in bacterial denitrification. In most cases the main product of fungal denitrification is N_2O, as many fungi lack a N_2O reductase (Shoun *et al.*, 1992), although it is possible that the $P450_{nor}$ gene has two functions and may also facilitate N_2 production (Takaya and Shoun, 2000). The ecological importance of N_2O emissions by fungi in terrestrial environments is still a matter of debate. A major bottleneck is related to the ability of soil microorganisms to degrade fungicides and antibiotics added to soil for distinguishing fungal from bacterial denitrification and N_2O emissions.

Denitrification by ammonia-oxidizing bacteria (nitrifier denitrification)

In nitrifier denitrification the oxidation of NH_3 to NO_2^- by ammonia oxidizing bacteria is followed by the reduction of NO_2^- to N_2O and possibly also to N_2 (Wrage *et al.*, 2001), although there is still no direct evidence that ammonia oxidisers can produce N_2 during this process. The entire process is undertaken by ammonia oxidizing bacteria (Kuai and Verstraete, 1998) and therefore contrasts with 'coupled nitrification-denitrification' where coexisting nitrifiers and denitrifiers can together transform NH_3 to NO_3^- and NO_3^- to N_2, respectively. The enzymes involved in nitrifier denitrification are essentially those for ammonia oxidation and denitrification, namely ammonia monooxygenase, hydroxylamine oxidoreductase, nitrite reductase, nitric oxide reductase and, possibly, nitrous oxide reductase (Jiang and Bakken, 1999). This process has been demonstrated in a number of studies on mixed and pure cultures of *Nitrosomonas*, with NH_3 converted under oxygen limitation to N_2O and N_2, with NO_2^- and NO as intermediates (Bock *et al.*, 1995; Kuai and Verstraete, 1998). Shaw *et al.* (2006) provided direct evidence for reduction of exogenously applied $^{15}N\text{-}NO_2^-$ to $^{15}N\text{-}N_2O$ in seven strains of ammonia oxidizing bacteria representative of clusters 0, 2 and 3 in the cultured *Nitrosospira* lineage, with up to 13.5% of measured N_2O derived from the applied $^{15}N\text{-}NO_2^-$ (Table 11.1). This suggests that

Table 11.1 Total emissions of $^{14+15}N$-N_2O, denitrifier ^{15}N-N_2O and ammonia oxidiser ^{15}N-N_2O over 41 days after addition of ^{15}N-labelled NH_4NO_3 (20 g N m^{-2}; 10 atom% excess ^{15}N) to a loamy sand soil maintained at pH 4.5 and pH 7.0 for over 40 years (long-term) and limed with $CaCO_3$ or acidified with H_2SO_4 (data from Baggs *et al.*, 2010).

	$^{14+15}N$-N_2O (mg N_2O-N m^{-2} 41d^{-1})	Denitrifier ^{15}N-N_2O (μg ^{15}N-N_2O m^{-2} 41d^{-1})	Ammonia oxidiser ^{15}N-N_2O (μg ^{15}N-N_2O m^{-2} 41d^{-1})
Long-term pH 4.5	11.7	39.5	8.7
Long-term pH 7.0	2.6	0.5	1.9
pH 4.5 limed to pH 7.0	11.5	5.3	33.7
pH 7.0 acidified to pH 4.7	0.9	0.2	0.5

the ability to denitrify may be a widespread trait amongst betaproteobacterial ammonia oxidizing bacteria, but its significance as a N_2O source in soil has still to be proven as it requires advances in techniques for its quantification (see below). Ammonia-oxidizing archaea have been shown to possess a nitrite reductase enzyme, and although they have been demonstrated to oxidize ammonia (Prosser and Nicol, 2008), any ability to denitrify is as yet unproven, but may well be akin to nitrifier denitrification.

Little is known about the environmental regulation of nitrifier denitrification and effects on N_2O production during this process. Several studies provide evidence for greater N_2O production through nitrifier denitrification as oxygen concentration decreases (Goreau *et al.*, 1980; Lipschultz *et al.*, 1981; Hynes and Knowles, 1984; Kester *et al.*, 1996; Jiang and Bakken, 1999; Dundee and Hopkins, 2001). Decreasing pH may also influence nitrifier denitrification (Jiang and Bakken, 1999; Wrage *et al.*, 2001). From our ongoing experiments we estimate that nitrifier denitrifier-N_2O accounts for a daily loss of fertilizer-N of up to 0.2% and have found that although N_2O emissions were raised following amendment of soil with amino acids and artificial root exudates, the proportional contribution of nitrifier denitrification did not vary with the different C amendments.

Co-denitrification

Comparatively little is known about the regulation or environmental significance of this process in which one N atom from NO or N_2O can combine with one atom from another source (a co-substrate) forming a hybrid product (Tanimoto *et al.*, 1992; Su *et al.*, 2004). N_2 can also be produced solely from the co-substrate (Tanimoto *et al.*, 1992) although it is not known how common this is. Co-denitrification has been shown to occur in both fungi (Tanimoto *et al.*, 1992) and bacteria (Garber and Hollocher, 1982) but the reasons why, or the conditions under which, an organism would utilize a co-substrate are unknown. Co-denitrification and conventional denitrification have been shown to occur simultaneously in some denitrifying bacteria such as *Pseudomonas stutzeri* and denitrifying fungi (Shoun *et al.*, 1992; Tanimoto *et al.*, 1992). Su *et al.* (2004) demonstrated the $P450_{nor}$ in *Fusarium oxysporum* to be multifunctional and able to catalyse denitrification or co-denitrification. There is some evidence for co-substrate preference between different fungal strains, which may indicate different mechanisms. For example, *Fusarium oxysporum* only produced N_2O during co-denitrification when azide or salicylhydroxamic acid was available (Tanimoto *et al.*, 1992; Su *et al.*, 2004) but other fungi, for example *Fusarium solani* and *Cylindrocarpon tonkinense* (now *Fusarium lichenicola*), have been shown to emit N_2 as a co-denitrification product when amino acids are available (Shoun *et al.*, 1992).

The contribution of co-denitrification to N_2O emission from soils is unknown, as there has been no targeted source determination for this process *in situ*, in part due to difficulties in distinguishing co-denitrification from denitrification. Studies have mostly been limited to pure cultures of fungal strains. The only existing evidence from soil is from Laughlin and Stevens (2002), who estimated 92% of measured N_2 production in a grassland soil (sandy loam, pH 6.3) amended with glucose (up to 15 mg per g of soil) to result from co-denitrification, with the remaining 8% attributed to conventional denitrification. However, the significance of this process over conventional denitrification, and its regulation under a range of soil and environmental conditions remains to be determined.

Nitrate ammonification or dissimilatory nitrate reduction to ammonium (DNRA)

During nitrate-dependent ammonification NO_3^- is reduced to NO_2^- and NH_4^+ thereby providing a 'short circuit' in the nitrogen cycle bypassing denitrification and N_2 fixation (Mohan *et al.*, 2004). The process of nitrate ammonification is in many cases coupled to a respiratory electron transport system through which energy conservation (and thereby ATP synthesis) can be achieved. However, a range of electron donors can be used by ammonifiers, including fermentation products such as lactate, formate and hydrogen, as well as NADH produced from intermediary metabolism and the nature of electron donor influences the level of energy conservation. This process is undertaken by both Gram-negative and Gram-positive bacteria, including obligate anaerobes (e.g. *Clostridium* spp.), facultative anaerobes (e.g. *Enterobacter* spp.) and aerobes (e.g. *Bacillus* spp.) (Fazzolari *et al.*, 1990), which are widely distributed in many different environments. Some of these bacteria are capable of producing N_2O as by-product during ammonification (Kelso *et al.*, 1997). Two biochemically distinct nitrate reductases, the membrane-bound nitrate reductase (Nar) and the periplasmic nitrate reductase (Nap), are found in ammonifying bacteria. These two enzymes can also catalyse the first step of the denitrification pathway. Nitrite is then reduced to ammonium by NrfA, which can be N_2O-genic (Jackson *et al.*, 1991). However, the biochemistry that determines the amount of N_2O formed is still unknown. The contribution of nitrate ammonification to N_2O emissions from soils is also pretty much unknown, mainly because of technical limitations in its determination.

It is possible that both nitrate ammonification and denitrification can occur simultaneously in anaerobic microsites in soils (Fazzolari *et al.*, 1990) and produce N_2O at the same time (Smith and Zimmerman, 1981; Stevens *et al.*, 1998) and may compete for NO_3^-, depending on environmental conditions as well as the soil microbial population. Compared to denitrification, very little is known about the regulation of nitrate ammonification. Intensively reduced and C-rich environments are traditionally thought to favour ammonification, as it is a more efficient electron acceptor where C availability is non-limiting (Buresh and Patrick, 1978; Tiedje *et al.*, 1982). Addition of glucose to anoxic soil has been shown to result in ammonification of 43% of nitrate applied (Caskey and Tiedje, 1979), and there is evidence that nitrate ammonification may be significant in agricultural soils, even those to which no C has been added (Chen *et al.*, 1995a,b).

Schmidt *et al.* (2010) measured nitrate ammonification rates of 0.2–5 mg N per kg dry soil per day in a range of temperate arable soils, but detection was only possible after the addition of a labile and reducing compound to the soil. They conclude that nitrate ammonification is unlikely to significantly contribute to the N budget in bulk temperate arable soil. However, root derived C exudation, O_2 and nitrate consumption by roots in the rhizosphere may create more favourable conditions for the nitrate ammonifier community.

Ammonia oxidation

Ammonia oxidizing microorganisms convert ammonia to nitrite in a two-step process with hydroxylamine as an intermediate. The membrane-bound ammonia mono-oxygenase enzyme (AMO) catalyses the oxidation of ammonia to hydroxylamine, and hydroxylamine is oxidized to nitrite by the periplasm associated enzyme, hydroxylamine oxidoreductase (Hooper *et al.*, 1997). The production of N_2O occurs at this stage. Two of the four electrons derived from the oxidation of hydroxylamine to nitrite are required during the oxidation of ammonia to hydroxylamine, the other two are available for energy production and the reduction of oxygen to water (Colliver and Stephenson, 2000). All known bacterial autotrophic ammonia oxidizers in soils belong to the genus *Nitrosomonas* and *Nitrosospira,* which form a monophyletic cluster within the β-*Proteobacteria*. Recent findings have revolutionized the diversity of ammonia-oxidizers by showing that archaea are also capable to oxidize ammonia to nitrite and that they could dominate over ammonia oxidizing bacteria in soils (Prosser and Nicol, 2008). Ammonia oxidizing archaea are often more abundant than ammonia oxidizing bacteria (Schleper, 2010). However their contribution to nitrification is still under debate, with activity yet to be related to group I.1b archaea in the natural environment, and their role in N_2O production remains unknown.

The rate of ammonia oxidation is influenced by ammonia availability. This is closely linked to the protonation and lower availability of NH_3 at low pH. Surprisingly, ammonia oxidisers seem to be able to survive under conditions of NH_3 starvation although the ability to compete for NH_3 and the ability to respond to NH_3 after a period of starvation has been shown to vary between different AOB (Frijlink *et al.*, 1992; Gerards *et al.*, 1998; Bollmann *et al.*, 2002). Furthermore, the theoretical sensitivity to low pH also varies between different AOB, with some possessing mechanisms to overcome this problem so that nitrification is not always restricted by low pH environments (De Boer and Kowalchuk, 2001). Much of the information on the physiology of AOB comes from studies on *Nitrosomonas* spp. due to the relative ease of culturing this microorganism in the laboratory. However, molecular investigations into AOB communities have shown that the more predominant AOB in the soil, especially soils treated with ammonia based fertilizers, are members of the genus *Nitrosospira* (Kowalchuk and Stephen, 2001; Avrahami *et al.*, 2002).

Nitrifier-N_2O production has been measured in cultures of nitrifiers under reduced O_2 potential (Goreau *et al.*, 1980), and there is recent evidence to suggest that ammonia oxidation can significantly contribute to net N_2O emission from soil (e.g. Abbasi and Adams, 2000; Bateman and Baggs, 2005; Avrahami and Bohannan, 2009; Wan *et al.*, 2009). Bateman and Baggs (2005) showed nitrification to be the predominant N_2O-producing process in a silt loam soil held at 35–60% WFPS, accounting for up to 81% of N_2O emitted at 60% water-filled pore space, indicating the significance of this process for global warming, despite its role having often been underplayed compared to that of denitrification.

Nitrification can also be carried out by a wide range of heterotrophic microbes (bacteria and fungi) using organic substrates such as urea as well as NH_3 (Papen *et al.*, 1989). Several different pathways of heterotrophic nitrification have been proposed, including an inorganic pathway similar to that of the autotrophic nitrifiers (Killham, 1986), and this capacity has been demonstrated in pure culture studies of a range of heterotrophic microbes. For example, many heterotrophs isolated from soil (including *Absidia cylindrospora*, *Pseudomonas putida* and *Paracoccus denitrificans*) have the ability to nitrify NH_3 in culture (Stroo *et al.*, 1986; Moir *et al.*, 1996a; Daum *et al.*, 1998). Brierley and Wood (2001) isolated heterotrophic bacteria and fungi from an acid forest soil capable of nitrifying both inorganic (ammonium acetate) and organic (β-alanine, peptone) N in pure culture and in inoculated soil solution. The potential for heterotrophic nitrification is further supported by the characterization of enzymes (ammonia and hydroxylamine reductases) capable of catalysing oxidation reactions typical of the autotrophic pathway in heterotrophs (Moir *et al.*, 1996a,b). The exact mechanisms of both organic and mineral heterotrophic nitrification pathways remain to be clarified, and the evidence seems to suggest that a combination of an appropriate N source and suitable soil environmental conditions combine to dictate whether the process occurs (Killham, 1986). The regulation and magnitude of N_2O production during these pathways are less well constrained, although the production of N_2O by heterotrophic nitrifiers has been demonstrated in culture (Papen *et al.*, 1989), and it has been shown under aerobic culture conditions that some heterotrophic nitrifiers, such as *Alcaligenes faecalis*, can produce much more N_2O per cell than the autotrophic nitrifier *Nitrosomonas europaea* (Papen *et al.*, 1989; Anderson *et al.*, 1993). The oxygen concentration at which most N_2O was produced was higher for the heterotroph *Alcaligenes faecalis* than the autotroph *Nitrosomonas europaea* (Anderson *et al.*, 1993). To our knowledge the only direct investigation of heterotrophic nitrification production in soil was undertaken by Bateman and Baggs (2005), using C_2H_2 to inhibit ammonia oxidation. However, due to high variability, there was no conclusive evidence for heterotrophic contribution in their arable soil.

Only a few studies have investigated the relationships between activity and diversity of nitrifiers with any focus on nitrification rates and nitrifier-N_2O production being most often neglected (Kowalchuk *et al.*, 2000; de Bie *et al.*, 2001). In one of the most comprehensive studies, which used molecular and cultivation-based approaches of ammonia oxidiser diversity together with measurements of changes in nitrate and ammonium concentrations, Webster *et al.* (2005) demonstrated a direct link between bacterial community structure and nitrification activity. More recently, Avrahami and Bohanann (2009) observed a significant relationship between diversity of ammonia oxidizing bacteria and N_2O emission rates and suggested that N_2O emissions may vary locally depending on local AOB community composition. However, other studies relating nitrifier-N_2O production to the nitrifier community have concluded that any change in activity, for example following addition of N, or with differing water content, was probably not the result of a major change in the ammonia-oxidizing community but of a physiological response (Mendum *et al.*, 1999; Avrahami *et al.*, 2002; Bateman, 2005). Bateman (2005) found changes in the AOB community structure over time after application of NH_4NO_3 to soils of varying water-filled pore space, but these changes occurred after the main period of flux activity (Fig. 11.2).

Chemodenitrification

Whilst the predominant sources of N_2O in terrestrial systems are generally microbial, there is evidence for the occurrence of chemodenitrification, which encompasses all non-biological processes that produce NO, N_2O or N_2, under very specific conditions. It is thought to occur when NO_2^- accumulates and reacts with organic compounds (thought to be phenolics) to produce NO and N_2O (Bremner, 1997), and may be a significant source of N_2O in acid soils <pH 5.5 where the NO_2^-–HNO_2 equilibrium shifts in favour of nitrous acid (HNO_2) which either self decomposes to form NO, HNO_3 and H_2O, or reacts with constituents of soil organic matter to form N_2, N_2O and CH_3ONO (Chalk and Smith, 1983; Van Cleemput and Samater, 1996; Venterea and Rolston, 2000). It is possible that at low pH chemodenitrification produces more N_2O than during microbial activity (Nelson, 1982; Yamulki *et al.*, 1997). Morkved *et al.* (2007) measured a 0.4% conversion of NO_2^- to N_2O in sterilized peat soil at pH 4.1–4.2, which they attributed to chemodenitrification.

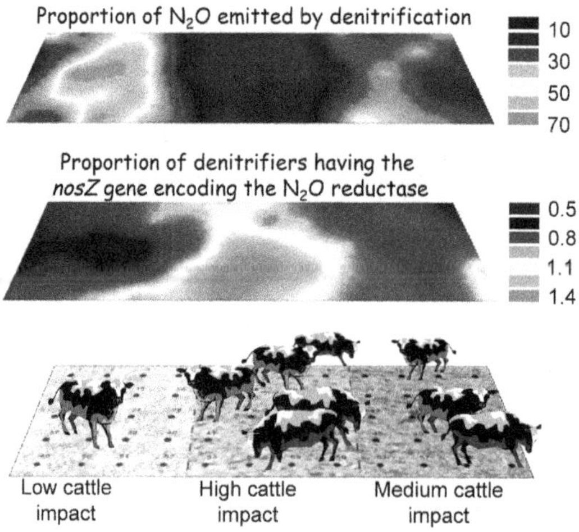

Figure 11.2 Spatial distribution of the proportion of greenhouse gas N_2O emitted by denitrification in relation to the proportion of denitrifiers genetically capable to reduce N_2O in a grassland field subjected to different cattle impacts.

Controls on N_2O production and reduction in soil

Nitrogen availability, oxygen partial pressure/soil water content, carbon availability and pH are considered the primary determinants on rates of N_2O production. N-fertilizer application results in short-term increased N_2O emissions (Bouwman, 1996; Mosier, 1994) that last between several days and up to a few weeks (VanCleemput *et al.*, 1994). This increase in N_2O emissions can be exacerbated by a raised denitrifier N_2O to- N_2 product ratio following nitrogen fertilizer application since nitrate is preferred over N_2O as an electron acceptor for denitrifiers at concentrations of >10 μg/g (Baggs *et al.*, 2003; Blackmer and Bremner, 1978). The magnitude of emission varies depending on the form of N applied, but emissions are generally greater following the application of organic rather than mineral sources (Clayton *et al.*, 1997). For example cattle slurries also provide organic C and additional moisture which can favour denitrification. Within the rhizosphere, ammonia oxidizing and nitrate reducing organisms may compete with plants for limited N resource, and when outcompeted, N_2O emissions may be lowered (Duxbury *et al., 1*982). On the other hand, higher C availability due to root exudation may also favour denitrification and N_2O emissions.

The aeration status of a soil, which depends upon soil water content, soil texture, and the partial pressure of oxygen in the gas phase, is a key factor affecting N_2O emission and the predominant microbial source of this N_2O. Soils in the range 60–80% water-filled pore space (WFPS) produce the most N_2O (Davidson, 1991; Bateman and Baggs, 2005), with denitrification the predominant N_2O producing process above 70–80% water-filled pore space (WFPS; Davidson, 1991), or at oxygen partial pressures below 0.5% (Parkin and Tiedje, 1984). At lower WFPS, ammonia oxidation was shown to be predominant process producing N_2O (Bateman and Baggs, 2005). Even aerobic soils may support anaerobic denitrification because aerobic–anaerobic interfaces within soil aggregates allow NO_3^- produced by nitrification to be available to denitrifiers, and there is also the possibility of aerobic denitrification occurring after a pre-anaerobic phase for enzyme synthesis (Morley and Baggs, 2010). Tillage has an important role to play in altering the aeration status of soil through modifying the soil structure, with typically higher N_2O emission from no-till soils compared to tilled soils (Baggs *et al.*, 2003, 2006; Linn and Doran, 1984). In terms of potential for denitrification, this is exacerbated by the often higher soil organic matter availability in the upper topsoil of no-till soils (Nieder *et al.*, 1989). Soil compaction, for example in tractor tramlines (Ruser *et al.*, 1998), or in areas of pasture subject to much animal trafficking, also creates conditions where the anaerobic volume is increased and more N_2O is produced. Fine textured soils become anoxic sooner than coarse soils as water content increases, therefore N_2O losses tend to be greater from fine textured clay soils.

The role of organic carbon in the regulation of N_2O production and reduction is poorly understood. The quality, or compound, as well as the quantity of C is now thought to be important in regulation N_2O-to-N_2 ratios (Henry *et al.*, 2008), and shows an interactive effect with O_2 availability (Morley and Baggs, 2010). The importance of root-derived organic compounds flow in the rhizosphere is dealt with in the next section where we consider the role of plants in mediating N_2O emissions from the plant–soil system.

Soil pH has been considered a 'master variable' of N transformations (Mørkved *et al.*, 2007), and N_2O emissions (Šimek and Hopkins, 1999). Long-term effects of pH on ammonia oxidation and denitrification are generally considered to be predominantly indirect rather than by direct toxicity, with ammonia oxidizer and denitrifier communities becoming adapted to environments of different pH (Nicol *et al.*, 2008; Šimek *et al.*, 2002). Nitrification rates are thought to be strongly controlled by soil pH, primarily through its effect on NH_3 availability, as the concentration of NH_3 substrate for the ammonia mono-oxygenase enzyme decreases exponentially compared to NH_4^+ as pH declines (Suzuki *et al.*, 1974). There are several reports that the denitrifier N_2O reductase is more sensitive to low pH than the other denitrification reductases (Thomsen *et al.*, 1994), which can result in higher ratios of N_2O-to-N_2 as pH declines (Dannenmann *et al.*, 2008; Nägele and Conrad, 1990; Šimek and Cooper, 2002; Stevens *et al., 1*998; Cuhel *et al.*, 2010). Baggs *et al.* (2010) report differences in net

N_2O emission and in the microbial source of this N_2O between short- and long-term pH changed soils (1), which they attribute to adaptation of the microbial community in the long-term soils, and impacts on the synthesis and activation of the N_2O reductase in denitrification.

The influence of plants on N_2O emission

Whilst direct emissions from plants are thought to account for up to 87% of the N_2O measured from plant-soil systems (Pihlatie *et al.*, 2005), validating this *in situ* has proved challenging, and whether this N_2O is produced within the plant, or whether the plant simply acts as a conduit for soil-produced N_2O, is unknown. In rice systems, aerenchyma are thought to be implicated in transfer of N_2O from soil (Rusch and Rennenberg, 1998; Yan *et al.*, 2000). It is also possible that N_2O is transported within the transpiration stream to leaves, from where it is emitted to the atmosphere (Chang *et al.*, 1998; Pihlatie *et al.*, 2005). If this is the case, where transpiration is high it should be possible to predict N_2O fluxes from the canopy based on measured soil N_2O and mineral N concentrations, and there is the possibility of developing mitigation practices at the canopy level in addition to the more traditional focus on soil management.

Plants may also produce N_2O, either in leaves during nitrite photoassimilation (Smart and Bloom, 2002; Hakata *et al.*, 2003) or by endophytic bacteria. If endophytic N_2O production occurs it is highly probable that this is denitrification as a number of bacteria that form associations with plants are denitrifiers, particularly in root nodules (Fernandez *et al.*, 2008). This also raises the possibility of N_2O produced within the plant, or being transported from the soil, being reduced to N_2 during transit. This is a relatively unexplored area of our understanding of the N_2O budget of plant-soil systems, but current perception is that most of the N_2O emission results from stimulation of the rhizosphere microbial community following inorganic N addition.

Plants also have an indirect effect on N_2O emissions from soils, through changing local conditions in the rhizosphere soil in numerous ways that influence microbial activity, abundance and community composition (Lynch, 1990; Sørensen, 1997). The main influence is thought to be through changes in rhizosphere biochemistry, pH, N and O_2 availability and provision of root derived C (Baggs and Blum, 2004; Bijay-Singh *et al.*, 1988; Paul and Beauchamp, 1989; Mounier *et al.*, 2004; Henry *et al.*, 2008). The rhizosphere has been shown to have a distinct microbial community in response to rhizodeposition and enhanced denitrification rates. Increased belowground organic compounds allocation by *Lolium perenne* swards under elevated partial pressure of carbon dioxide has been demonstrated to stimulate denitrifier N_2O production (Baggs and Blum, 2004; Baggs *et al.*, 2003), and Haller and Stolp (1985) provided evidence for rhizosphere stimulation of denitrification, with *Pseudomonas aeruginosa* producing 1.8 ml N_2O-N/day, which was equivalent to consumption of 72 mg glucose-C/day from root exudation. Such effects are now thought to vary between plant species (Patra *et al.*, 2006; Craine *et al.*, 2002; Bremer *et al.*, 2009) or with plant functional trait (Harrison and Bardgett, 2010), offering opportunities for managing belowground processes through changing above-ground diversity, although the potential for this to be used to mitigate N_2O emission has yet to be demonstrated. Whilst there is much published evidence for enhanced N_2O emissions in the presence of plants (Kilian and Werner ,1996; Klemedtsson *et al.*, 1987; Yang and Cai, 2006) the extent of rhizosphere denitrification stimulation reported is highly variable ranging from 2 to 22 times that of the bulk soil (Hojberg *et al.*, 1996; Kilian and Werner, 1996; Klemedtsson *et al.*, 1987; Bakken, 1988). Studies demonstrating a linkage between plants and ammonia oxidizer and denitrifier activity, and N_2O production are comparatively few. The impact of such plant traits on dynamic rhizosphere C and N availabilities will be integrated by the denitrifier community, however, the outcome with respect to denitrifier-N_2O production is not known. However, the techniques are now becoming available to address this.

The link between plants and the rhizosphere microbiology can now be more explicitly determined with the advent of approaches such as stable isotope probing (SIP), which provides an elegant means of distinguishing organisms contributing to the observed processes (Bull *et al.*, 2000; Morris *et al.*, 2002; Radajewski *et al.*, 2002; Whitby *et al.*, 2001). There is the potential for links to be made between plant-derived C flow and denitrifiers by identifying heterotrophic microorganisms capable of denitrification and utilizing the plant-C, through ^{13}C-CO_2 pulsing of vegetation (Johnson *et al.*, 2002; Rangel-Castro *et al.*, 2005), and recovery of the ^{13}C in nucleic acids or in phospholipid fatty acids (PLFA) (Waldrop and Firestone, 2004; Paterson *et al.*, 2006). Whilst issues about turnover and recovery of applied ^{13}C, and cross-feeding of this C (Manefield *et al.*, 2007) pose uncertainties for this approach, there are nevertheless some exciting opportunities for further characterization of active microorganisms through the use of stable isotopes to determine microbial sources. Coupling stable isotope and molecular approaches in this way will help us answer questions about spatially where N_2O is produced in the rhizosphere, and to revisit hypotheses from over four decades ago that denitrifier activity decreases with distance from roots (Smith and Tiedje, 1979).

Interactions between methane oxidation and N_2O emissions

In addition to the impact of plant low molecular weight C in driving denitrification in the rhizosphere, as described above, there are interactions between C and N cycles that can also impact on both nitrifier and denitrifier N_2O production. An example most pertinent to N_2O production is that of interactions between CH_4 oxidation and ammonia oxidizer N_2O production, and coupling between CH_4 oxidation and denitrification. Such interactions between different processes involved in GHG production mean that a better understanding may help guide mitigation practices for both these gases, which has to be preferable to just managing each GHG in isolation, but there is currently no soil management strategy that mitigates both CH_4 and N_2O.

Physiological, biochemical and ecological similarities between methane and ammonia oxidizing bacteria promote competition that affects both CH_4 and NH_3 oxidation rates, particularly after inorganic N application (Bodelier and Laanbroek, 2004). The mechanism involved in any inhibition of CH_4 oxidation is uncertain and may be related to the role of NH_3 in competing for methane monoxygenase enzymes (Holmes *et al.*, 1995), NO_2^- or hydroxylamine toxicity (King and Schnell, 1994), or osmotic effects (Hütsch *et al.*, 1996; Kravchenko *et al.*, 2002). Any N inhibition of CH_4 oxidation is thought to result in an increase in ammonia oxidation (Hütsch ,1998), meaning that there is the potential for N application to lead to a switch in function of methylotrophs away from their beneficial effect in lowering atmospheric loading of CH_4. Acton and Baggs (2010) report greater nitrifier-N_2O production (by ammonia oxidizing, and possibly also by methane oxidizing bacteria) occurring under conditions where CH_4 oxidation was lowered, or inhibited, after addition of 10, 20 and 30 g N/m^2 (NH_4NO_3) to a silt loam soil. However, although we have process evidence that hints towards a switch in function (Baggs and Blum, 2004; Khalil and Baggs, 2005; Acton and Baggs, 2010), the role of these bacteria in ammonia oxidation or N_2O production in soil has yet to be directly verified. Different enzyme affinities mean that it is unlikely that the converse situation, of ammonia oxidizers oxidizing CH_4 instead of NH_3, will significantly contribute to lowering both CH_4 and N_2O emission from soil.

The processes of CH_4 oxidation and denitrification may also be closely coupled, for example in methanotroph-dependent denitrification, which is usually attributed to provision of organic C during methane oxidation to denitrifiers (Islas-Lima *et al.*, 2004; Waki *et al.*, 2005; Zhang *et al.*, 2005). Again, there is still only indirect evidence that this occurs in soils, exemplified by strong positive relationships between denitrifier-N_2O production and methane oxidation rates (Khalil and Baggs, 2005). This provision of organic C may have a critical role to play in determining the product N species in denitrification particularly as C may, under certain conditions, be limiting for reduction of N_2O to N_2. There are reports of methanol, acetate (Costa *et al.*, 2000) and even methane itself (Islas-Lima *et al.*, 2004) being used as electron donors for denitrification but it is not known how this usage relates to denitrifier populations in soil, their production or reduction of N_2O, and whether the necessitated close coupling reflects the occurrence of anaerobic CH_4 oxidation (Raghoebarsing *et al.*, 2006) or aerobic denitrification (Lloyd *et al.*, 1987; Patureau *et al.*, 2000).

Opportunities for mitigation of N_2O

Reducing the application of inorganic N fertilizers, or increasing N-use efficiency, has traditionally been a key target option for lowering net emission of N_2O. Increasing fertilizer use efficiency can be achieved through plant density, split N applications, consideration of soil available N pools and specific crop N demand, combined organic and inorganic N applications (Frimpong and Baggs, 2010), slow release fertilizers (Rao, 1987), optimizing synchrony between N availability and crop N demand (Frimpong and Baggs, 2010; Gentile *et al.*, 2008), intercropping (Guo *et al.*, 2009), better adoption of animal manure inputs and improving animal diets (Johnson *et al.*, 2007), nitrification inhibitors (Di and Cameron, 2002) and plant breeding for N-use efficiency (Foulkes *et al.*, 2009).

The challenge is to develop technologies, management practices and incentives for farmers and land managers to adopt that address both the goals of society and the needs of the farmer. It will never be possible to completely reconcile competing objectives by manipulation of the N cycle, but there will always be new opportunities to improve the functioning of the system as our understanding and societal demands evolve. The economic effect of gaseous N losses in intensive agriculture on producers is often low and this represents a significant challenge to scientists in proposing management strategies to mitigate emissions. The concept of greenhouse gas intensity (GHGI) was introduced by Mosier *et al.* (2006), in which the global warming potential of a system is divided by grain yield. Thus, a negative GHGI indicates a net sink of GHG as CO_2 equivalents per kg grain yield. This sets the scene for adjustments to management practices that balance economic viability and lower negative effects on the environment, through selection of appropriate inputs, such as N fertilizer. Adopting this approach Van Groenigen *et al.* (2010) demonstrated that a zero fertilizer application resulted in higher N_2O emission (16.7 g N_2O/kg crop N) than where 233 kg N/ha fertilizer N was applied (6.1 g N_2O/kg crop N), due to high background emissions from soil, and reduced yield/total crop N where N is limits crop growth. However, above this rates of 233 kg N/ha there were diminishing yield returns, and associated non-linear increase in N_2O.

Given that denitrification is the predominant N_2O producing process in most soils, an alternative approach to mitigation is through better understanding of the regulation of the denitrifier N_2O reductase to inform development of management options to lower net N_2O emissions by enhancing its reduction to N_2, rather than to trying to eliminate denitrification (Richardson *et al.*, 2009). They consider the effect on the regulation of the N_2O reductase of management options such as application of nitrogenous or copper fertilizer to regulate Cu availability for this Cu-based enzyme, SOM management and liming of crops or grasslands with specific characterized carbon and nitrogen traits.

There are other 'new' options being proposed for mitigation of greenhouse gases that may also hold the potential for enhancing C sequestration, which may prove essential for a more holistic approach to global warming. One of these is application of biochar, and is a good example of where the ability to source partition N_2O becomes essential. Although the mechanism by which biochar can lower net emission of N_2O is still open to debate (Renner, 2007; Gaunt and Lehmann, 2008), its chemical composition and structure suggest that it may be due to uptake of NH_4^+ into the char structure, and a change in pH, rather than by enhancing the reduction of N_2O to N_2. If this is the case then the largest effect will be in lowering nitrifier-N_2O, with only indirect effects on nitrate reducing processes. Further studies are required before biochar can be recommended as a reliable management option for lowering greenhouse gas emissions.

In order to predict and reduce global warming it is necessary to devise research programs that consider CO_2 and CH_4 as well as N_2O and provide a global warming potential for systems and management practices investigated. Strategies proposed to reduce greenhouse gas emissions should also consider all three of these gases, with shared mitigation options where possible. This is a particular challenge as there can be interaction between C and N cycles, and between the processes of CH_4 oxidation, nitrification and denitrification that determine net N_2O and CH_4 emissions from soils. This is exemplified for WFPS in Fig. 11.3. Caution is required here as strategies most appropriate for mitigation of one greenhouse gas, may not necessarily be appropriate for a reduction in global warming potential. Additionally, successful mitigation strategies need to be adapted to the local socio-economic situation, and environmental conditions, agricultural productivity, and care taken to avoid the possibility of pollution swapping.

In order to design more targeted mitigation options we still require a better understanding of the regulation of all N_2O-genic processes, and particularly need to address uncertainties associated with the conditions conducive to N_2O production during nitrate ammonification, and conditions under which microbes may display flexibility in function pertaining to N_2O production, for example nitrifier denitrification, or NH_3 oxidation by methylotrophic bacteria. To this end we need to ensure that we unite the N and C cycles, to both understand the nature of interactions, to predict emissions and to formulate mitigation strategies for enhanced C storage as well as lowering GHG emissions. There is the potential for plants to self-regulate emissions, both directly and indirectly, which provides exciting possibilities for more integrated approaches to management of the plant-soil system. This needs to be founded on greater understanding of GHG production and reduction/oxidation at the enzyme, microbial, microsite, rhizosphere, field scales, with models of new techniques to bridge between and above measurable scales, and robust parameterization of the underpinning microbiology to ensure sustainability for the future, that any management option imposed to lower emissions has no adverse effect on the diversity and functioning of these microbial guilds.

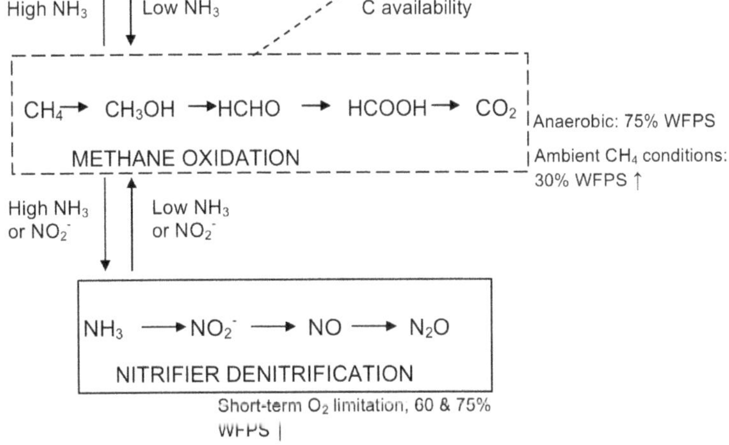

Figure 11.3 Conceptual diagram of the effect of soil water-filled pore space on the processes of nitrification, denitrification, nitrifier denitrification and interactions between them following addition of $^{13}C\text{-}CH_4$ at 10 µ/l. ↑ and ↓ signify an increase or decrease in process rate, respectively. Adapted from Khalil and Baggs (2005).

References

Abbasi, M.K., and Adams, W.A. (2000). Gaseous N emission during simultaneous nitrification-denitrification associated with mineral N fertilisation to a grassland soil under field conditions. Soil Biol. Biochem. *32,* 1251–1259.

Acton, S.D., and Baggs, E.M. (2010). Interactions between N application rate, CH_4 oxidation and N_2O production in soil. Biogeochemistry *103,* 15–26.

Ananyeva, N.D., Polyanskaya, L.M., Stolnikova, E.V., and Zvyagintzev, D.G. (2010). Fungal to bacterial biomass ratio in the forests soil profile. Biol. Bull. *37,* 254–262.

Anderson, I.C., Poth, M., Homstead, J., and Burdige, D. (1993). A comparison of, N.O., and N_2O production by the autotrophic nitrifier *Nitrosomonas europaea* and the heterotrophic nitrifier *Alcaligenes faecalis.* Appl. Environ. Microbiol. *59,* 3525–3533.

Avrahami, S., and Bohannan, B.J.M. (2009). N_2O emission rates in a California meadow soil are influenced by fertilizer level, soil moisture and the community structure of ammonia-oxidizing bacteria. Global Change Biol. *15,* 643–655.

Avrahami, S., Conrad, R., and Braker, G. (2002). Effect of soil ammonium concentration on N_2O release and on the community structure of ammonia oxidizers and denitrifiers. Appl. Environ. Microbiol. *68,* 5685–5692.

Azam, F., Muller, C., Weiske, A., Benckiser, G., and Ottow, J.C.G. (2002). Nitrification and denitrification as sources of atmospheric nitrous oxide – role of oxidizable carbon and applied nitrogen. Biol. Fertil. Soils 35, 54–61.

Baggs, E.M., and Blum, H. (2004). CH_4 oxidation and CH_4 and N_2O emissions from Lolium perenne swards under elevated atmospheric CO_2. Soil Biol. Biochem. *36,* 713–723.

Baggs, E.M., Rees, R.M., Smith K.A., and Vinten AJ.A. (2000). Nitrous oxide emissions from soils after incorporating plant material. Soil Use Manage. *16,* 82–87.

Baggs, E.M., Richter M, Cadish G, and Hartwig, U.A. (2003). Denitrification in grass swards is increased under elevated atmospheric CO_2. Soil Biol. Biochem. *35,* 729–732.

Baggs, E.M., Chebii J, and Ndufa, J.K. (2006). A short-term investigation of trace gas emissions following tillage and no-tillage of agroforestry residues in western Kenya. Soil Till Res. *90,* 69–76.

Baggs, E.M., Smales, C.L., and Bateman, E.J. (2010). Changing pH shifts the microbial source as well as the magnitude of N_2O emission from soil. Biol. Fertil. Soils *46,* 793–805.

Bakken, L.R. (1988). Denitrification under different cultivated plants: effects of soil moisture tension, nitrate concentration, and photosynthetic activity. Biol. Fertil. Soils 6, 271–278.

Bateman, E.J., and Baggs, E.M. (2005). Contributions of nitrification and denitrification to N_2O emissions from soils at different water-filled pore space. Biol. Fertil. Soils *41,* 379–388.

Blackmer, A.M., and Bremner, J.M. (1978). Inhibitory effect of nitrate on reduction of N_2O to N_2 by soil microorganisms. Soil Biol. Biochem. *10,* 187–191.

Bock, E., Schmidt, I., Stuven, R., and Zart, D. (1995). Nitrogen loss caused by denitrifying Nitrosomonas cells using ammonium or hydrogen as electron donors and nitrite as electron acceptor. Arch. Microbiol. *163,* 16–20.

Bodelier, P.L.E., and Laanbroek, H.J. (2004). Nitrogen as a regulatory factor of methane oxidation in soils and sediments. FEMS Microbiol. Ecol. *47,* 265–277.

Bollmann, A., Bar-Gilissen, M.J., and Laanbroek, H.J. (2002). Growth at low ammonium concentrations and starvation response as potential factors involved in niche differentiation among ammonia-oxidizing bacteria. Appl. Environ. Microbiol. *68,* 4751–4757.

Bouwman, A.F. (1996). Direct emission of nitrous oxide from agricultural soils. Nutr. Cycling Agroecosys. *46,* 53–70.

Bremer, C., Braker, G., Matthies, D., Beierkuhnlein, C., and Conrad, R. (2009). Plant presence and species combination, but not diversity, influence denitrifier activity and the composition of nirK-type denitrifier communities in grassland soil. FEMS Microbiol. Ecol. *70,* 377–87.

Bremner, J.M. (1997). Sources of nitrous oxide in soils. Nutr. Cycling Agroecosys. *49,* 7–16.

Brierley, E.D.R., and Wood, M. (2001). Heterotrophic nitrification in an acid forest soil: isolation and characterisation of a nitrifying bacterium. Soil Biol. Biochem. *33,* 1403–1409.

Bull, I.D., Parekh, N.R., Hall, G.H., Ineson, P., and Evershed, R.P. (2000). Detection and classification of atmospheric methane oxidizing bacteria in soil. Nature *405,* 175–178.

Buresh, R.J., and Patrick, W.H. (1978). Nitrate reduction to ammonium in anaerobic soil. Soil Sci. Soc. Am. J. *42,* 913–918.

Burford, J.R., and Bremner, J.M. (1975). Relationships between denitrification capacities of soils and total, water-soluble and readily decomposable soil organic matter. Soil Biol. Biochem. 7, 389–394

Caskey, W.H., and Tiedje, J.M. (1979). Nitrate stimulated mineralisation of ammonium in anaerobic soils. Oecologia *41,* 339–341.

Chalk, P.M., and Smith, C.J. (1983). Chemodenitrification. In Gaseous Loss of Nitrogen from Plant Soil Systems, Freney, J.R., and, Simpson, J.R., eds. (Martinus Nijhoff and W. Junk, The Hague), pp. 65–89.

Chang, C., Janzen, H.H., Cho, C.M., and Nakonechny, E.M. (1998). Nitrous oxide emission via plant transpiration. Soil Sci. Soc. Am. J. *62,* 35–38

Chen, D.L., Chalk, P.M., and Freney, J.R. (1995a). Distribution of reduced products of N-15-labelled nitrate in anaerobic soils. Soil Biol. Biochem. *27,* 1539–1545.

Chen, Y., Tessier, S., Mackenzie, A.F., and Laverdiere, M.R. (1995b). Nitrous oxide emission from an agricultural soil subjected to different freeze–thaw cycles. Agric. Ecosys. Environ. *55,* 123–128.

Clayton, H., McTaggart, I.P., Parker, J., Swan, L., Smith., K.A. (1997). Nitrous oxide emissions from fertilised grassland: A 2-year study of the effects of N fertiliser form and environmental conditions. Biol. Fertil. Soils *25,* 252–260.

Coffey, M.T. (1996). environmental challenges as related to animal agriculture – swine. In Nutrient Management of Food Animals to Enhance and Protect the Environment, Kornegay, E.T., ed. (Lewis Publishers, CRC, Boca Raton, FL), pp. 29–39.

Colliver, B.B., and Stephenson, T. (2000). Production of nitrogen oxide and dinitrogen oxide by autotrophic nitrifiers. Biotechnol. Advan.*18,* 219–232.

Conrad, R. (1996). Soil microorganisms as controllers of atmospheric trace gases (H_2, C.O., CH_4, OC.S., N_2O, and NO). Microbiol. Rev. *60,* 609–640.

Conrad, R., Seiler, W., and Bunse, G. (1983). Factors influencing the loss of fertiliser nitrogen into the atmosphere as nitrous oxide. J. Geophys. Res. *88,* 6709–6718.

Costa, C., Dijkema, C., Friedrich, M., Garcia-Encina, P., Fernandez-Polanco, F., Stams, A.J.M. (2000). Denitrification with methane as electron donor in oxygen-limited bioreactors. Appl. Microbiol. Biotechnol. *53,* 754–762.

Craine, J.M., Tilman, D., Wedin, D., Reich, P., Tjoelker, M., and Knops, J. (2002). Functional traits, productivity and effects on nitrogen cycling of 33 grassland species. Funct. Ecol. *16,* 563–574.

Crutzen, P.J. (1970). The influence of nitrogen oxides on the atmospheric ozone content. Q. J. R. Meteorol. Soc. *96,* 320–325.

Dannenmann, M., Butterbach-Bahl, K., Gasche, R., Willibald, G., and Papen, H. (2008). Dinitrogen emissions and the N_2:N_2O emission ratio of a Rendzic Leptosol as influenced by pH and forest thinning. Soil Biol. Biochem. *40,* 2317–2323.

Daum, M., Zimmer, W., Papen, H., Kloos, K., Nawrath, K., and Bothe, H. (1998). Physiological and molecular biological characterization of ammonia oxidation of the heterotrophic nitrifier *Pseudomonas putida*. Curr. Microbiol. *37,* 281–288.

Davidson, E.A. (1991). Fluxes of nitrous oxide and nitric oxide from terrestrial ecosystems. In Microbial Production and Consumption of Greenhouse Gases: Methane, Nitrogen Oxides, and Halomethanes, Rogers, J.E., and Whitman, W.B., eds. (American Society for Microbiology, Workshop, Georgia), 219–236.

De Boer, W., and Kowalchuk, G.A. (2001). Nitrification in acid soils: micro-organisms and mechanisms. Soil Biol. Biochem. *33,* 853–866.

Di, H.J., and Cameron, K.C. (2002). The use of a nitrification inhibitor, dicyandiamide (DCD), to decrease nitrate leaching and nitrous oxide emissions in a simulated grazed and irrigated grassland. Soil Use Manage. *18,* 395–403.

Dobbie, K.E., McTaggart, I.P., and Smith, K.A. (1999). Nitrous oxide emissions from intensive agricultural systems: Variations between crops and seasons, key driving variables, and mean emission factors. J. Geophys. Res. Atmospheres *104,* 26891–26899

Drury, C.F., McKenney, D.J., and Findlay, W.I. (1991). Relationships between denitrification, microbial biomass and indigenous soil properties. Soil Biol. Biochem. *23,* 751–755.

Dundee, L., and Hopkins, D.W. (2001). Different sensitivities to oxygen of nitrous oxide production by *Nitrosomonas europaea* and *Nitrosolobus multiformis*. Soil Biol. Biochem. *33,* 1563–1565.

Duxbury, J.M., and McConnaughey, P.K. (1986). Effect of fertiliser source on denitrification and nitrous oxide emisssions in a maize-field. Soil Sci. So. Am. J. *50,* 644–648.

Duxbury, J.M., Bouldin, D.R., Terry, R.E., Tate, R.L. (1982). Emissions of nitrous oxide from soils. Nature *298,* 462–464.

Eichner, M.J. (1990). Nitrous oxide emission from fertilized soils: Summary of available data. J. Environ. Qual. *19,* 272–280.

Ellis, S., Howe, M.T., Goulding, K.W.T., Mugglestone, M.A., and Dendooven, L. (1998). Carbon and nitrogen dynamics in a grassland soil with varying pH: Effect of pH on the denitrification potential and dynamics of the reduction enzymes. Soil Biol. Biochem. *30,* 359–367.

Enwall, K., Philippot, L., and Hallin, S. (2005). Activity and composition of the denitrifying bacterial community respond differently to long-term fertilization. Appl. Environ. Microbiol. *71,* 8335–8343.

Fazzolari, E., Mariotti, A., and Germon, J.C. (1990). Dissimilatory ammonia production vs denitrification invitro and in inoculated agricultural soil samples. Can. J. Microbiol. *36,* 786–793.

Fernandez, L.A., Perotti, E.B., Sagardoy, M.A., and Gomez, M.A. (2008). Denitrification activity of *Bradyrhizobium* sp isolated from Argentine soybean cultivated soils. World J. Microbiol. Biotechnol. *24,* 2577–2585.

Firestone, M.K., Firestone, R.B., and Tiedje, J.M. (1980). Nitrous oxide from soil denitrification – factors controlling its biological production. Science *208,* 749–751.

Foulkes, M.J., Hawkesford, M.J., Barraclough, P.B., Holdsworth, M.J., Kerr, S., Kightley, S., and Shewry, P.R. (2009). Identifying traits to improve the nitrogen economy of wheat: Recent advances and future prospects. Field Crops Res. *114,* 329–342.

Frijlink, M.J., Abee, T., Laanbroek, H.J., De Boer, W., and Konings, W.N. (1992). The bioenergetics of ammonia and hydroxylamine oxidation in *Nitrosomonas europaea* at acid and alkaline pH. Arch. Microbiol. *157,* 194–199.

Frimpong, K., and Baggs, E.M. (2010). Do combined applications of crop residues and inorganic fertiliser lower emission of N_2O from soil? Soil Use Manage. *26,* 412–424.

Garber, E.A., and Hollocher, T.C. (1982). Positional isotopic equivalence of nitrogen in nitrous oxide produced by the denitrifying bacterium *Pseudomonas stutzeri* indirect evidence for a nitroxyl pathway. J. Biol. Chem. *257,* 4705–4708.

Garcia-Ruiz, R., Baggs, E.M. (2007). N_2O emission from soil following combined application of fertiliser-N and ground weed residues. Plant Soil *299,* 263–274.

Gaunt, J.L., and Lehmann, J. (2008). Energy balance and emissions associated with biochar sequestration and pyrolysis bioenergy production. Environ. Sci. Technol. *42,* 4152–4158.

Gentile, R., Vanlauwe, B., Chivenge, P., and Six, J. (2008). Interactive effects from combining fertiliser and organic residue inputs on nitrogen transformations. Soil Biol. Biochem. *40,* 2375–2384.

Gerards, S., Duyts, H., and Laanbroek, H.J. (1998). Ammonium-induced inhibition of ammonium-starved *Nitrosomonas europaea* cells in soil and sand slurries. FEMS Microbiol. Ecol. *26,* 269–280.

Goreau, T.J., Kaplan, W.A., Wofsy, S.C., McElroy, M.B., Valois, F.W., and Watson, S.W. (1980). Production of NO_2^- and N_2O by nitrifying bacteria at reduced concentrations of oxygen. Appl. Environ. Microbiol. *40,* 526–532.

Gregory, L.G., Karakas-Sen, A., Richardson, D.J., and Spiro, S. (2000). Detection of genes for membrane-bound nitrate reductase in nitrate-respiring bacteria and in community DNA. FEMS Microbiol. Lett. *183,* 275–279.

Guckland A, Corre, M.D., Flessa H 2010. Variability of soil N cycling and N_2O emission in a mixed deciduous forest with different abundance of beech. Plant Soil *336,* 25–38.

Guo, Z.L., Cai, C.F., Li, Z.X., Wang, T.W., Zheng, M.J., 2009. Crop residue effect on crop performance, soil N_2O and CO_2 emissions in alley cropping systems in subtropical China. Agroforestry Syst. *76,* 67–80.

Hakata, M., Takahashi, M., Zumft, W., Sakamoto, A., and Morikawa, H. (2003). Conversion of the nitrate nitrogen and nitrogen dioxide to nitrous oxide in plants. Acta Biotechnol. *23,* 249–257.

Haller, T., and Stolp, H. (1985). Quantitative estimation of root exudation of maize plant. Plant Soil *86,* 207–216.

Hallin, S., and Lindgren, P.E. (1999). PC.R., detection of genes encoding nitrile reductase in denitrifying bacteria. Appl. Environ. Microbiol. *65,* 1652–1657.

Harper, S.H.T., and Lynch, J.M. (1981). The chemical components and decomposition of wheat straw leaves, internodes and nodes. J. Sci. Food Agriculture *32,* 1057–1062.

Harrison, K.A., and Bardgett, R.D. (2010). Influence of plant species and soil conditions on plant-soil feedback in mixed grassland communities. J. Ecol. *98,* 384–395.

Harrison, R., Ellis, S., Cross, R., and Hodgson, J.H. (2002). Emissions of nitrous oxide and nitric oxide associated with the decomposition of arable crop residues on a sandy loam soil in Eastern England. Agronomie *22,* 731–738.

Henry, S., Texier, S., Hallet, S., Bru, D., Dambreville, C., Chenehy, D., Bizouard, F., Germon, J.C., and Philip pot, L. (2008). Disentangling the rhizosphere effect on nitrate reducers and denitrifiers: insight into the role of root exudates. Environ. Microbiol. *10,* 3082–3092.

Hochstein, L., and Tomlinson, G.A. (1988). The enzymes associated with denitrification. Ann. Rev. Microbiol. *42,* 231–261.

Hojberg, O., Binnerup, S.J., and Sorensen, J. (1996). Potential rates of ammonium oxidation, nitrite oxidation, nitrate reduction and denitrification in the young barley rhizosphere. Soil Biol. Biochem. *28,* 47–54.

Holmes, A.J., Costello A, Lidstrom, M.E., and Murrell, J.C. (1995). Evidence that particulate methane monooxygenase and ammonia monooxygenase may be evolutionarily related. FEMS Microbiol. Lett. *132,* 203–208.

Holtan-Hartwig, L., Dorsch, P., and Bakken, L.R. (2000). Comparison of denitrifying communities in organic soils: kinetics of NO_3^- and N_2O reduction. Soil Biol. Biochem. *32,* 833–843.

Hooper, A.B., Vannelli, T., Bergmann, D.J., and Arciero, D.M. (1997). Enzymology of the oxidation of ammonia to nitrite by bacteria. Antonie Van Leeuwenhoek Int. J. Gen. Mol. Microbiol. *71,* 59–67.

Huang,Y., Zou, J., Zheng, X., Wang, Y., and Xu, X. (2004). Nitrous oxide emissions as influenced by amendments of plant residues with different C: N ratio. Soil Biol. Biochem. *36,* 973–981.

Hütsch, B.W. (1998). Methane oxidation in arable soil as inhibited by ammonium, nitrite and organic manure with respect to soil pH.Biol. Fertil. Soils *28,* 27–35.

Hütsch, B.W., Russell P, Mengel K (1996). CH_4 oxidation in two temperate arable soils as affected by nitrate and ammonium application. Biol. Fertil. Soils *23,* 86–92.

Hynes, R.K., and Knowles, R. (1984). Production of nitrous oxide by *Nitrosomonas europaea* – effects of acetylene, pH and oxygen. Can. J. Microbiol. *30,* 1397–1404.

IPCC. (2007). Climate Change 2007: Synthesis Report. Contribution of Working Groups I, I.I., and II.I., to the Fourth Assessment Report of the Intergovernmental Panel on Climate Change, Core Writing Team, Pachauri, R.K., and Reisinger, A., eds. (IPCC, Geneva, Switzerland).

Islas-Lima, S., Thalasso, F., and Gomez-Hernandez, J. (2004). Evidence of anoxic methane oxidation coupled to denitrification. Water Res. *38,* 13–16.

Jackson, M.A., Tiedje, J.M., and Averill, B.A. (1991). Evidence for an NO-rebound mechanism for production of N_2O from nitrite by the copper containing nitrite reductase from *Achromobacter cycloclastes.* FEBS Lett. *291,* 41–44.

Jiang, Q.Q., and Bakken, L.R. (1999). Nitrous oxide production and methane oxidation by different ammonia-oxidizing bacteria. Appl. Environ. Microbiol. *65,* 2679–2684.

Johnson, D., Leake, J.R., Ostle, N., Ineson, P., and Read, D.J. (2002). In situ (CO_2)-C-13 pulse-labelling of upland grassland demonstrates a rapid pathway of carbon flux from arbuscular mycorrhizal mycelia to the soil. New Phytol. *153,* 327–334.

Johnson, J.M.F., Franzluebbers, A.J., Weyers, S.L., and Reicosky, D.C. (2007). Agricultural opportunities to mitigate greenhouse gas emissions. Environ. Poll. *150,* 107–124.

Kelso, B.H.L., Smith, R.V., Laughlin, R.J., and Lennox, S.D. (1997). Dissimilatory nitrate reduction in anaerobic sediments leading to river nitrite accumulation. Appl. Environ. Microbiol. *63,* 4679–4685.

Kesik, M., Ambus, P., Baritz, R., Bruggemann, N.B., Butterbach-Bahl, K., Damm, M., Duyzer, J., Horvath, L., Kiese, R., Kitzler, B., *et al.* (2005). Inventories of N_2O and, N.O., emissions from European forest soils. Biogeosciences *2,* 353–375.

Kester, R.A., de Boer, W., and Laanbroek, H.J. (1996). Short exposure to acetylene to distinguish between nitrifier and denitrifier nitrous oxide production in soil and sediment samples. FEMS Microbiol. Ecol. *20,* 111–120.

Khalil, M.I., and Baggs, E.M. (2005). CH_4 oxidation and N_2O emissions at varied soil water-filled pore spaces and headspace CH_4 concentrations. Soil Biol. Biochem. *37,* 1785–1794.

Kilian, S., and Werner, D. (1996). Enhanced denitrification in plots of N_2-fixing faba beans compared to plots of a non-fixing legume and non-legumes. Biol. Fertil. Soils *21,* 77–83.

Killham, K. (1986). Heterotrophic nitrification. In Nitrification, Special Publications of the Society For General Microbiology, Vol. 20, Prosser, J.I., ed. (IRL Press Ltd, Oxford).

King, G.M., and Schnell, S. (1994). Ammonium and nitrite inhibition of methane oxidation by *Methylobacter albus* BG8 and *Methylosinus trichosporium* OB3b at low methane concentrations. Appl. Environ. Microbiol. *60,* 3508–3513.

de Klein, C.A.M., and Eckard, R.J. (2008). Targeted technologies for nitrous oxide abatement form animal agriculture. Aus. J. Exp. Agric. *48,* 14–20.

Klemedtsson, L., Svensson, B.H., Rosswall, T. (1987). Dinitrogen and nitrous oxide produced by denitrification and nitrification in soil with and without barley plants. Plant Soil *99,* 303–310.

Knowles, R. (1982). Denitrification. Microbiol. Rev. *46,* 43–70.

Kowalchuk, G.A., and Stephen, J.R. (2001). Ammonia-oxidizing bacteria: A model for molecular microbial ecology. Ann. Rev. Microbiol. *55,* 485–529.

Kravchenko, I., Boeckx, P., Galchenko, V., and Van Cleemput, O. (2002). Short- and medium-term effects of NH_4^+ on CH_4 and N_2O fluxes in arable soils with a different texture. Soil Biol. Biochem. *34,* 669–678.

Kuai, L.P., and Verstraete, W. (1998). Ammonium removal by the oxygen-limited autotrophic nitrification-denitrification system. Appl. Environ. Microbiol. *64,* 4500–4506.

Kubota, Y., Naoki Takaya, N., and Shoun, H. (1999). Membrane-associated, dissimilatory nitrite reductase of the denitrifying fungus *Cylindrocarpon tonkinense.* Arch. Microbiol. *171,* 210–213.

Laughlin, R.J., and Stevens, R.J. (2002). Evidence for fungal dominance of denitrification and codenitrification in a grassland soil. Soil Sci. Soc. Am. J. *66,* 1540–1548.

Laverman, A.M., Zoomer, H.R., Engelbrecht, D., Berg, M.P., van Straalen, N.M., van Verseveld, H.W., and Verhoef, H.A. (2000). Soil layer-specific variability in net nitrification and denitrification in an acid coniferous forest. Biol. Fertil. Soils *32,* 427–434.

Linn, D.M., and Doran, J.W. (1984). Effect of water-filled pore space on carbon dioxide and nitrous oxide production in tilled and nontilled soils. Soil Sci. Soc. Am. J. *48,* 1267–1272.

Lipschultz, F., Zafiriou, O.C., Wofsy, S.C., McElroy, M.B., Valois, F.W., and Watson, S.W. (1981). Production of, N.O., and N_2O by soil nitrifying bacteria. Nature *294,* 641–643.

Lloyd, D., Boddy, L., and Davies, K.J.P. (1987). Persistence of bacterial denitrification capacity under aerobic conditions – the rule rather than the exception. FEMS Microbiol. Ecol. *45,* 185–190.

Lynch, J.M. (1990). The Rhizosphere. (John Wiley, Chichester).

McKenney, D.J., Wang, S.W., Drury, C.F., and Findlay, W.I. (1995). Denitrification, immobilisation and mineralisation in nitrate limited and nonlimited residue-amended soil. Soil Sci. Soc. Am. J. *59,* 118–124.

McTaggart, I., Clayton, H., and Smith, K. (1994). Nitrous oxide flux from fertilised grassland: Strategies for reducing emissions. In Non CO_2 Greenhouse Gases, Van Ham, J., Jansson, L.J.H.M., Swart, R.J., eds. (Kluwer Academic Publishers, Dordrecht), pp. 421–426.

Manefield, M., Griffiths, R., McNamara, N.P., Sleep, D., Ostle, N., and Whiteley, A. (2007). Insights into the fate of a C-13 labelled phenol pulse for stable isotope probing (SIP). experiments. J. Microbiol. Methods *69,* 340–344.

Millar, N., and Baggs, E.M. (2004). Chemical composition, or quality, of agroforestry residues influences N_2O emissions after their addition to soil. Soil Biol. Biochem. *36,* 935–943.

Millar, N., and Baggs, E.M. (2005). Relationships between N_2O emissions and water-soluble C and N contents of agroforestry residues after their addition to soil. Soil Biol. Biochem. *37,* 605–608.

Millar, N., Ndufa, J.K., Cadisch, G., and Baggs, E.M. (2004). Nitrous oxide emissions following incorporation of improved-fallow residues in the humid tropics. Global Biogeochem. Cycles *18,* doi: 10.1029/2003GB002114.

Miller, L.A., Moorby, J.M., Davies, D.R., Humphreys, M.O., Scollan, N.D., MacRae, J.C., and Theodorou, M.K. (2001). Increased concentration of water-soluble carbohydrate in perennial ryegrass (*Lolium perenne,* L.): milk production from late-lactation dairy cows. Grass Forage Sci. *56,* 383–394.

Misselbrook, T.H., Powell, J.M., Broderick, G.A., and Grabber, J.H. (2005). Dietary manipulation in dairy cattle: Laboratory experiments to assess the influence on ammonia emissions. J. Dairy Sci. *88,* 1765–1777.

Mohan, S.B., Schmid, M., Jetten, M., and Cole, J. (2004). Detection and widespread distribution of the nrfA gene encoding nitrite reduction to ammonia, a short circuit in the biological nitrogen cycle that competes with denitrification. FEMS Microbiol. Ecol. *49,* 433–443.

Moir, J.W.B., Wehrfritz, J.M., Spiro, S., and Richardson, D.J. (1996a). The biochemical characterization of a novel non-haem-iron hydroxylamine oxidase from *Paracoccus denitrificans* GB17. Biochem. J. *319,* 823–827.

Moir, J.W.B., Crossman, L.C., Spiro, S., and Richardson, D.J. (1996b). The purification of ammonia monooxygenase from *Paracoccus denitrificans.* FEBS Lett. *387,* 71–74.

Morkved, P.T., Dorsch, P., and Bakken, L.R. (2007). The

N_2O product ratio of nitrification and its dependence on long-term changes in soil pH. Soil Biol. Biochem., *39,* 2048–2057.

Morley N, and Baggs, E.M. (2010). Carbon and oxygen controls on N_2O and N_2 production during nitrate reduction. Soil Biol. Biochem. *42,* 1864–1871.

Morley, N., Baggs, E.M., Dorsch, P., and Bakken, L. (2008). Production of, N.O., N_2O and N_2 by extracted soil bacteria, regulation by NO_2^- and O_2 concentrations. FEMS Microbiol. Ecol., *65,* 102–112.

Morris, S.A., Radajewski, S., Willison, T.W., and Murrell, J.C. (2002). Identification of the functionally active methanotroph population in a peat soil microcosm by stable-isotope probing. Appl. Environ. Microbiol. *68,* 1446–1453.

Mosier, A.R. (1994). Nitrous oxide emissions from agricultural soils. Fertil. Res. *37,* 191–200.

Mosier, A.R., and Hutchinson, G.L. (1981). Nitrous oxide emissions from cropped fields. J. Environ. Qual. *10,* 169–173.

Mosier, A.R., Hutchinson, G.L., Sabey, B.R., and Baxter, J. (1982). Nitrous oxide emissions from barley plots treated with ammonium nitrate or sewage sludge. J. Environ. Qual. *11,* 78–81.

Mosier, A.R., Parton, W.J., and Hutchinson, G.L. (1983). Modelling nitrous oxide evolution from cropped and native soils. Environ. Biogeochem. Ecol. Bull. *35,* 229–241.

Mosier, A.R., Halvorson, A.D., Reule, C.A., and Liu, X.J.J. (2006). Net global warming potential and greenhouse gas intensity in irrigated cropping systems in northeastern Colorado. J. Environ. Qual. *35,* 1584–1598.

Nägele, W., and Conrad, R. (1990). Influence of pH on the release of, N.O., and N_2O from fertilised and unfertilised soil. Biol. Fert. Soils *10,* 139–144.

Nelson, D.W. (1982). Gaseous losses of nitrogen other than through denitrification. Agronomy *22,* 327–346.

Nicol, G.W., Leininger, S., Schleper, C., and Prosser, J.I. (2008). The influence of soil pH on the diversity, abundance and transcriptional activity of ammonia oxidizing archaea and bacteria. Environ. Microbiol. *10,* 2966–2978.

Nieder, R., Schollmayer, G., and Richter, J. (1989). Denitrification in the rooting zone of cropped soils with regard to methodology and climate: a review. Biol Fertil. Soils *8,* 219–226.

Oenema, O., Wrage, N., Velthof, G.L., van Groenigen, J.W., Dolfing, J., and Kuikman, P.J. (2005). trends in global nitrous oxide emissions from animal production systems. Nutr. Cycling Agroecosystems *72,* 54–65.

Otte, S., Grobben, N.G., Robertson, L.A., Jetten, M.S.M., and Kuenen, J.G. (1996). Nitrous oxide production by *Alcaligenes faecalis* under transient and dynamic aerobic and anaerobic conditions. Appl. Environ. Microbiol. *62,* 2421–2426.

Papen, H., Vonberg, R., Hinkel, I., Thoene, B., and Rennenberg, H. (1989). Heterotrophic nitrification by *Alcaligenes faecalis* – NO_2^-, NO_3^-, N_2O and, N.O., production in exponentially growing cultures. Appl. Environ. Microbiol. *55,* 2068–2072.

Parkin, T.B., and Tiedje, J.M. (1984). Application of a soil core method to investigate the effect of oxygen concentration on denitrification. Soil Biol. Biochem. *16,* 331–334.

Parkin, T.B., Sexstone, A.J., and Tiedje, J.M. (1985). Adaptation of denitrifying populations to low soil pH. Appl. Environ. Microbiol. *49,* 1053–1056.

Paterson, E., Sim, A., Standing, D., Dorward, M., and McDonald, A.J.S. (2006). Root exudation from *Hordeum vulgare* in response to localized nitrate supply. J. Exp. Botany *57,* 2413–2420.

Patra, A.K., Clays-Josserand, A., Degrange, V., Grayston, S. J., Guillaumaud, N., Loiseau, P., Louault, F., Mahmood, S., Nazaret, S., Philippot, L., Poly, F., Prosser, J. I. and Le Roux, X. (2006). Effects of management regime and plant species on the enzyme activity and genetic structure of N-fixing, denitrifying and nitrifying bacterial communities in grassland soils. Environ. Microbiol. *8,* 1005–1016.

Patureau, D., Zumstein, E., Delgenes, J.P., and Moletta, R. (2000). Aerobic denitrifiers isolated from diverse natural and managed ecosystems. Microbial Ecol. *39,* 45–152.

Paul, J.W., and Beauchamp, E.G. (1989a). Denitrification and fermentation in plant-residue-amended soil. Biol. Fertil. Soils 7, 303–309.

Paul, J.W., and Beauchamp, E.G. (1989b). Effect of carbon constituents in manure on denitrification in soil. Can. J. Soil Sci. *69,* 49–61.

Philippot, L., Piutti, S., Martin-Laurent, F., Hallet, S., and Germon, J.C. (2002). Molecular analysis of the nitrate-reducing community from unplanted and maize-planted soils. Appl. Environ. Microbiol. *68,* 6121–6128.

Philippot, L., Hallin, S., Borjessen, G., and Baggs, E.M. (2009). Biochemical cycling in the rhizosphere having an impact on global change. Plant Soil DOI: 10.1007/s11104-008-9796-9.

Pihlatie, M., Ambus, P., Rinne, J., Pilegaard, K., and Vesala, T. (2005). Plant-mediated nitrous oxide emissions from beech (*Fagus sylvatica*). leaves. New Phytol. *168,* 93–98.

Piña-Ochoa, E., Høgslund, S., Geslin, E., Cedhagen, T., Revsbech, N.P., Nielsen, L.P., Schweizer, M., Jorissen, F., Rysgaard, S., Risgaard-Petersen, N. (2010). Widespread occurrence of nitrate storage and denitrification among Foraminifera and Gromiida. Proc. Natl. Acad. Sci. U. S. A. *107,* 1148–1153

Prendergast-Miller, Baggs, E.M., and Johnson, D.J. (2011). Nitrous oxide production by the ectomycorrhizal fungi *Paxillus involutus* and *Tylospora fibrillosa*. FEMS Microbiol. Lett. *316,* 31–35.

Prosser, J.I., and Nicol, G.W. (2008). Relative contributions of archaea and bacteria to aerobic ammonia oxidation in the environment. Environ. Microbiol. *10,* 2931–2941.

Radajewski, S., Webster, G., Reay, D.S., Morris, S.A., Ineson, P., Nedwell, D.B., Prosser, J.I., and Murrell, J.C. (2002). Identification of active methylotroph populations in an acidic forest soil by stable isotope probing. Microbiol. SGM *148,* 2331–2342.

Raghoebarsing, A.A., Pol, A., van de Pas-Schoonen,

K.T., Smolders, A.J.P., Ettwig, K.F., Rijpstra, W.I.C., Schouten, S., Damste, J.S.S., Op den Camp, H.J.M., Jetten, M.S.M., *et al.* (2006). A microbial consortium couples anaerobic methane oxidation to denitrification. Nature *440,* 918–921.

Rangel-Castro, J.I., Killham, K., Ostle, N., Nicol, G.W., Anderson, I.C., Scrimgeour, C.M., Ineson, P., Meharg, A., and Prosser, J.I. (2005). Stable isotope probing analysis of the influence of liming on root exudate utilization by soil microorganisms. Environ. Microbiol. *7,* 828–838.

Rao, D.L.N. (1987). Slow-release urea fertilisers- effect on floodwater chemistry, ammonia volatilistaion and rice growth in an alkali soil. Fertil. Res. *13,* 209–221.

Ravishankara, A.R., Daniel, J.S., and Portmann, R.W. (2009). Nitrous Oxide (N_2O): The dominant ozone-depleting substance emitted in the 21st Century. Science *326,* 123–125.

Reinertsen, S.A., Elliott, L.F., Cochran, V.L., and Campbell, G.S. (1984). Role of available carbon and nitrogen in determining the rate of wheat straw decomposition. Soil Biol. Biochem. *16,* 459–464.

Renner, R. (2007). Rethinking biochar. Environ. Sci. Technol. *41,* 5932–5933.

Richardson, D., Felgate, H., Watmough, N., Thomson, A., and Baggs, E.M. (2009). Mitigating release of the potent greenhouse gas N_2O from the nitrogen cycle: Could enzymic regulation hold the key? Trends Biotechnol. *27,* 388–397.

Risgaard-Petersen, N., Langezaal, A.M., Ingvardsen, S., Schmid, M.C., Jetten, M.S., Op den Camp, H.J., Derksen, J.W., Piña-Ochoa E, Eriksson, S.P., Nielsen, L.P., *et al.* (2006). Evidence for complete denitrification in a benthic foraminifer. Nature *443,* 93–6.

Robertson, G.P., and Tiedje, J.M. (1987). Nitrous oxide sources in aerobic soils – nitrification, denitrification and other biological processes. Soil Biol. Biochem. *19,* 187–193.

Rusch, H., and Rennenberg, H. (1998). Black Alder (*Alnus glutinosa* (L.). Gaertn.). trees mediate methane and nitrous oxide emission from the soil to the atmosphere. Plant Soil *201,* 1–7.

Ruser, R., Flessa, H., Schilling, R., Steindl, H., and Beese, F. (1998). Soil compaction and fertilization effects on nitrous oxide and methane fluxes in potato fields. Soil Sci. Soc. Am. J. *62,* 1587–1595.

Schils, R.L.M., Verhagen, A., Aarts, H.F.M., Kuikman, P.J., and Sebek, L.B.J.(2006). Effect of improved nitrogen management on greenhouse gas emissions from intensive dairy systems in the Netherlands. Global Change Biol. *12,* 382–391.

Schleper, C. (2010). Ammonia oxidation: different niches for bacteria and archaea? ISME J. *4,* 1092–1094.

Schmidt, C.S., Richardson, D.J., and Baggs, E.M. (2010). Constraining the conditions conducive to dissimilatory nitrate reduction to ammonium in soil. Soil Biol. Biochem. doi: 10.1016/j.soilbio.2011.02.015.

Schulte-Bisping, H., Brumme, R., and Priesack, E. (2003). Nitrous oxide emission inventory of German forest soils. J. Geophys. Res. Atmospheres *108,* doi:10.1029/2002JD002292.

Sehy, U., Ruser, R., and Munch, J. (2003). Nitrous oxide fluxes from maize fields: relationship to yield, site-specific fertilization, and soil conditions. Agri. Ecosys. Environ. *99,* 97–111.

Seiler, W., and Conrad, R. (1981). Field measurements of natural and fertiliser induced N_2O release rates from soils. J. Air Poll. Control Assoc. *31,* 767–772.

Shoun, H., Kim, D.H., Uchiyama, H., and Sugiyama, J. (1992). Denitrification by fungi. FEMS Microbiol. Lett.*73,* 277–281.

Šimek, M., and Hopkins, D.W. (1999). Regulation of potential denitrification by soil pH in long-term fertilized arable soil. Biol. Fertil. Soil *30,* 41–47.

Šimek, M., and Cooper, J.E. (2002). The influence of soil pH on denitrification: Progress towards the understanding of this interaction over the last 50 years. Euro. J. Soil Sci. *53,* 345–354.

Šimek, M., Jišová, L., and Hopkins, D.W. (2002). What is the so-called optimum pH for denitrification in soil? Soil Biol. Biochem. *34,* 1227–1234.

Sitaula, B.K., and Bakken, L.R. (1993). Nitrous oxide release from spruce forest soil – relationships with nitrification, methane uptake, temperature, moisture and fertilisation. Soil Biol. Biochem., *25,* 1415–1421.

Skiba, U., and Smith, K.A. (2000). The control of nitrous oxide emissions from agricultural and natural soils. Chemosphere – Global Change Sci. *2,* 379–386.

Skiba, U., McTaggart, I.P., Smith, K.A., Hargreaves, K.J., and Fowler, D. (1996). Estimates of nitrous oxide emissions from soil in the UK. Energy Conversion Manage. *37,* 1303–1308.

Smart, D.R., and Bloom, A.J. (2001). Wheat leaves emit nitrous oxide during nitrate assimilation. Proc. Natl. Acad. Sci.U.S.A. *98,* 7875–7878.

Smith, M.S., and Tiedje, J.M. (1979). Effect of roots on soil denitrification. Soil Sci. Soc. Am. J. *43,* 951–955.

Smith, M.S., and Zimmerman, K. (1981). Nitrous oxide production by non-denitrifying soil nitrate reducers. Soil Sci. Soc. Am. J. *45,* 865–871.

Solomon, S., Qin, D., Manning, M., Alley, R.B., Berntsen, T., Bindoff, N.L., Chen, Z., Chidthaisong, A., Gregory, J.M., Hegerl, G.C., *et al.* (2007). Technical Summary. In Climate Change 2007: The Physical Science Basis. Contribution of Working Group I to the Fourth Assessment Report of the Intergovernmental Panel on Climate Change, Solomon, S., Qin, D., Manning, M., Chen, Z., Marquis, M., Averyt, K.B., Tignor, M., and Miller, H.L. eds. (Cambridge University Press, Cambridge, UK, New York, USA).

Sørensen, J. (1997). The rhizosphere as a habitat for soil microorganisms. In: Modern Soil Microbiology. Elsas, J.D.V., Trevors, J.T., and Wellington, E.M.K., eds. (Marcel Dekker, New York).

Stanford, G., Vanderpol, R.A., and Dzienia, S. (1975). Denitrification rates in relation to total and extractable soil carbon. Soil Sci. Soc. Am. J. *39,* 284–289.

Steinfeld, H., Greber, P.T.W., Castel, V., Rosales, M., and de Haan, C. (2006). Livestock's Long Shadow – Environmental Issues and Options. (Food and Agriculture Organization of the United Nations, Rome), 390 pp.

Stevens, R.J., Laughlin, R.J., and Malone, J.P. (1998a).

Soil pH affects the processes reducing nitrate to nitrous oxide and di-nitrogen. Soil Biol. Biochem. *30,* 1119–1126.

Stevens, R.J., Laughlin, R.J., and Malone, J.P. (1998b). Measuring the mole fraction and source of nitrous oxide in the field. Soil Biol. Biochem. *30,* 541–543.

Stres, B., Mahne, I., Avgustin, G., and Tiedje, J.M. (2004). Nitrous oxide reductase (nosZ). gene fragments differ between native and cultivated Michigan soils. Appl. Environ. Microbiol. *70,* 301–309.

Stroo, H.F., Klein, T.M., and Alexander, M. (1986). Heterotrophic nitrification in an acid forest soil and by an acid-tolerant fungus. Appl. Environ. Microbiol. *52,* 1107–1111.

Su, F., Takaya, N., and Shoun, H. (2004). Nitrous oxide-forming codenitrification catalyzed by cytochrome P450nor. Biosci. Biotechnol. Biochem. *68,* 473–475.

Suzuki, I., Dular, U., and Kwok, S.C. (1974). Ammonia or ammonium ion as substrate for oxidation by *Nitrosomonas europaea* cells and extracts. J. Bacteriol. *120,* 556–558.

Takaya, N., and Shoun, H. (2000). Nitric oxide reduction, the last step in denitrification by *Fusarium oxysporum,* is obligatorily mediated by cytochrome P450nor. Mol. Gen. Genet. *263,* 342–348.

Takaya, N., Catalan-Sakairi, M.A.B., Sakaguchi, Y., Kato, I., Zhou, Z.M., and Shoun, H. (2003). Aerobic denitrifying bacteria that produce low levels of nitrous oxide. Appl. Environ. Microbiol. *69,* 3152–3157.

Tanimoto, T., Hatano, K., and Kim, D.H. (1992). Co-denitrification by the denitrifying system of the fungus *Fusarium oxysporum.* FEMS Microbiol. Lett.*93,* 177–180.

Thomsen, J.K., Geest, T., and Cox, R.P. (1994). Mass spectrometric studies of the effect of pH on the accumulation of intermediates in denitrification by *Paracoccus denitrificans.* Appl. Environ. Microbiol. *60,* 536–541.

Tiedje, J.M., Sexstone, A.J., Myrold, D.D., and Robinson, J.A. (1982). Denitrification – ecological niches, competition and survival. Antonie van Leeuwenhoek J. Microbiol. *48,* 569–583.

Van Cleemput, O., and Samater, A.H. (1996). Nitrite in soils: accumulation and role in the formation of gaseous N compounds. Fertil. Res. *45,* 81–89.

Van Cleemput, O., Vermoesen, A., De Groot, C.J., and Van Ryckeghem, K. (1994). Nitrous oxide emission out of grassland. Environ. Monitoring Assess. *31,* 145–152.

Van Groenigen, J.W., Velthof, G., Oenema, O., Van Groenigen, K.J., and Van Kessel, C. (2010). Toward an agronomic assessment of N_2O emissions: A case study for arable crops. Eur. J. Soil Sci. *61,* 903–913.

Venterea, R.T., and Rolston, D.E. (2000). Mechanisms and kinetics of nitric and nitrous oxide production during nitrification in agricultural soil. Global Change Biol. *6,* 303–316.

Waki, M., Suzuki, K., Osada, T., and Tanaka, Y. (2005). Nitrogen removal from purified swine wastewater using biogas by semi-partitioned reactor. Biores. Technol. *96,* 921–927.

Waldrop, M.P., and Firestone, M.K. (2004). Altered utilization patterns of young and old soil C by microorganisms caused by temperature shifts and N additions. Bioegeochem. *67,* 235–248.

Wan, Y.J., Ju, X.T., Ingwersen, J., Schwarz, U., Stange, C.F., Zhang, F.S., and Streck, T. (2009). Gross nitrogen transformations and related nitrous oxide emissions in an intensively used calcareous soil. Soil Sci. Soc. Am. J. *73,* 102–112.

Wang, W.C., Yung, Y.L., Lacis, A.A., Mo, T., and Hansen, J.E. (1976). Greenhouse effects due to man-made perturbations of trace gases. Science *194,* 685–690.

Weier, K.L., Doran, J.W., Power, J.F., and Walters, D.T. (1993). Denitrification and the dinitrogen nitrous oxide ratio as affected by soil water, available carbon and nitrate. Soil Sci. Soc. Am. J. *57,* 67–72.

Weymann, D., Well, R., Flessa, H., von der Heide, C., Deurer, M., Meyer, K., Konrad, C., and Walther, W. (2008). Groundwater N2O emission factors of nitrate-contaminated aquifers as derived from denitrification progress and N2O accumulation. Biogeosciences *5,* 1215–1226.

Whitby, C.B., Hall, G., Pickup, R., Saunders, J.R., Ineson, P., Parekh, N.R., and McCarthy, A. (2001). C-13 incorporation into DN.A., as a means of identifying the active components of ammonia-oxidizer populations. Lett. Appl. Microbiol.*32,* 398–401.

Wolf, I., and Brumme, R. (2002). Contribution of nitrification and denitrification sources for seasonal N_2O emissions in an acid German forest soil. Soil Biol. Biochem. *34,* 741–744.

Wrage, N., Velthof, G.L., van Beusichem, M.L., and Oenema, O. (2001). Role of nitrifier denitrification in the production of nitrous oxide. Soil Biol. Biochem. *33,* 1723–1732.

Yamulki, S., Harrison, R.M., Goulding, K.W.T., and Webster, C.P. (1997). N_2O, N.O., and NO_2 fluxes from a grassland: effect of soil pH. Soil Biol. Biochem. *29,* 1199–1208.

Yan, X., Du Shi, S.L., and Xing, G. (2000). Pathways of N_2O emission from rice paddy soil. Soil Biol. Biochem. *32,* 437–440.

Yang, L., Cai, Z. (2006). Effects of shading soybean plants on N_2O emission from soil. Plant Soil *283,* 265–274.

Zhang, D.J., Lu, P.L., Long, T.R., and Verstraete, W. (2005). The integration of methanogensis with simultaneous nitrification and denitrification in a membrane bioreactor. Process Biochem. *40,* 541–547.

Zumft, W.G. (1997). Cell biology and molecular basis of denitrification. Microbiol. Mol. Biol. Rev. *61,* 533.

Bacterial Nitrogen Cycling in the Human Body

12

James W. B. Moir

Abstract

Surfaces of the human body exposed to the environment are heavily colonized by bacteria. The bacteria that live in these environments are frequently exposed to anoxia and to nitric oxide which is generated by the host. Dealing with these two environmental factors often involves implementing nitrogen cycle processes to (i) maintain growth and survival by respiration in the absence of oxygen, and (ii) detoxify the free radical nitric oxide. In this chapter I explore the nitrogen cycling processes relevant to the human body environment. Whilst microbial colonization is part of the normal physiology of the human body, the body can also be exposed to pathogenic bacteria which also utilize nitrogen cycling processes. Specific sections deal with the processes and consequences of nitrogen cycling by key human pathogens *Pseudomonas aeruginosa*, *Mycobacterium tuberculosis* and the pathogenic *Neisseria* species *N. meningitidis* and *N. gonorrhoeae*.

Background

Traditionally, most study of the nitrogen cycle processes of microorganisms has been carried out in the context of global nitrogen cycle processes affecting the distribution of nitrogenous species in 'natural' environments (atmosphere, marine and aquatic environments, soil). More recently, increasing attention has been paid to the role of microorganisms in nitrogen transformations in animals, and their impact on the overall ecosystem of the body. Here I will put bacterial nitrogen cycling into the context of the microorganisms that inhabit the human body, under both normal physiological and pathophysiological conditions.

Nitrogen speciation in the mammalian body environment: some considerations

Nutrient availability

The human body is abundantly colonized by bacteria. In fact it has been calculated that there are around 100 trillion (10^{14}) bacterial cells in the human body, which is ten times more bacterial cells in a human than there are human cells (Ley *et al.*, 2006). This is due to the nutrient richness of the mammalian body and the stability of the physical and chemical environment. Nutrient richness is most obvious in the gastrointestinal tract where the intake of food leads to an environment rich in both carbon and nitrogen (chiefly from peptides and amino acids). Environments such as the skin are less rich in nutrients but inhabitants here are adapted to use products from the degradation of cells and mucus for nutrition. The predominant limiting elemental nutrient in the human body is likely to be iron in many tissues. Other major factors limiting growth are the limited availability of respiratory electron acceptor (see section 'Gases', below) and bacterial killing caused by the body's immune system.

Gases

The availability of two key diatomic gases has a significant impact on the microbial activity in human bodies: oxygen (O_2) and nitric oxide (NO).

Oxygen is required for human respiration and the body is perfused with oxygen via the blood to maintain aerobic conditions. However, in the GI tract and on mucosal surfaces colonized by bacterial biofilms oxygen tends to become depleted by bacterial aerobic respiratory activity to the point of anaerobiosis. This activity is reflected in the microbial composition of tissues: in the large intestine for example, where bacteria are present at around 10^{12} cells per gram of faecal matter (Savage, 1977), and greater than 90% of the species present are strict anaerobes belonging to the Firmicute and Bacteroides divisions (Eckburg *et al.*, 2005). Even the oral cavity which might be presumed to be well perfused with oxygen contains both aerobes and strict anaerobes (Schuster, 1999), reflecting the variation in available microenvironments in this heterogeneous location.

Nitric oxide is the smallest stable free radical gas in chemistry and is an important biological molecule in both the bacterial nitrogen cycle and human physiology. NO is generated by mammalian cells from arginine in a reaction catalysed by nitric oxide synthase (Moncada and Higgs, 1993). There are several different NO synthase enzymes that operate in different tissues and have different regulatory properties (Forstermann *et al.*, 1991). Neuronal NOS (nNOS) generates NO in neuronal tissue where it acts as a signalling molecule in neurotransmission. Endothelial NOS (eNOS) generates NO which acts as a signal to increase vasodilation. Macrophages generate NO via an inducible enzyme iNOS in response to infection as part of the innate immune response. Generally, nNOS and eNOS are associated with nanomolar concentrations of NO that affect signalling cascades, whereas iNOS activity is associated with micromolar concentrations of NO which are associated with toxicity towards bacteria (and inhibition of mammalian enzymes such as cytochrome oxidase). NO is a freely diffusible gas, and can therefore migrate from the site of production to other sites nearby. Despite a short half-life for NO *in vivo* of less than 1 second (Kelm, 1999), it is clear that iNOS activity has a significant impact on microbial viability in human tissue. iNOS-deficient knockout mice are very sensitive to infection by bacteria (Chakravortty and Hensel, 2003). In addition to host NO synthases, NO may be generated chemically from nitrite. A major source of nitrite in the body is the microbial reduction of nitrate. Indeed, the increase in plasma nitrite concentrations observed following ingestion of nitrate-rich food is abolished by treatment of human volunteers with antimicrobial mouthwash (Govoni *et al.*, 2008). In addition to microbial routes for nitrite reduction to nitric oxide (e.g. by bacteria of the *Neisseria* genus), under acidic conditions nitrite disproportionates into nitrate and NO ($3HNO_2 \rightarrow HNO_3 + 2\,NO + H_2O$).

Ammonium, nitrate and nitrite

Ammonia is a cyto(neuro)toxic compound (Norenberg *et al.*, 2005) and hence the availability of free ammonia/ammonium in human tissue is kept low (measurements of 7–40 μM have been made for exhaled breath; Norenberg *et al.*, 2005; Wakabayashi *et al.*, 1997). Ammonia is rapidly assimilated into amino acids and so its availability for the inorganic biochemistry of the microbial nitrogen cycle is extremely limited. This limited availability, alongside the large excess of carbon in the mammalian host are hard to reconcile with lithotrophic nitrogen microbial metabolic processes such as nitrification (ammonia → nitrite → nitrate) or anammox (ammonia + nitrite → dinitrogen) that typically occur in carbon-poor environments as a means of providing energy to support autotrophic carbon fixation. These processes are probably insignificant in the human body.

Nitrate is ingested from food and can be reduced to nitrite by the normal microbial inhabitants of the oral cavity (see 'Nitrogen cycling in the oral cavity and upper respiratory tract', below). Its abundance varies between foods, but it is present in particularly high concentrations in green vegetables (e.g. spinach, broccoli) (van Velzen *et al.*, 2008). The levels of nitrate (and nitrite) in food and drinking water are tightly regulated due to their toxic potential. The nitrogen oxyanions are thought to be capable of causing serious deleterious effects on human health, by causing methaemoglobinaemia (Fan and Steinberg, 1996) and gastric cancer (Felley-Bosco, 1998). These conditions are dependent on the nitrogen oxyanions being reduced to nitric oxide in the gut

(McKnight *et al.*, 1997), and potentially to dinitrogen trioxide (N_2O_3) and S-nitrosocompounds (Felley-Bosco, 1998). However, there is an emerging body of opinion that nitrate and nitrite should not be seen as enemies of human health, since they have a positive role to play in providing a source for antimicrobial nitric oxide, and regulation of blood flow under hypoxic conditions, for example (see, for example, Lundberg *et al.*, 2009; Lundberg and Weitzberg, 2010; McKnight *et al.*, 1999).

Tissues and microbial diversity

Normal colonization of the human body is confined to external surfaces of the body (including the gastrointestinal tract). Factors affecting colonization include the availability of nutrients, which can enable the body to support very high bacterial numbers such as in the large intestine. Factors negatively affecting colonization include chemical, physical and biological stresses imposed in the tissues of the human body that prevent or limit bacterial colonization or growth. These include dryness, saltiness and low pH on the skin, for example, and immune surveillance and the mucociliary escalator that prevent colonization of the lungs under normal conditions. Three very different microbially colonized environments of the human body will be considered further below.

Microbially colonized environments in the body vary substantially. The colon, for example, contains a large number of microorganisms in a nutrient (carbon)-rich environment, leading to deprivation of oxygen. Hence, microorganisms colonizing here typically are usually strict anaerobes (although some notable facultative anaerobes such as *E. coli* are well-known residents). The major groups of microorganisms found in the large intestine are Bacteroidetes and Firmicutes particularly members of the genera *Bacteroides, Clostridium, Eubacterium* and *Ruminococcus* (Hattori and Taylor, 2009; Qin *et al.*, 2010) which are strict anaerobes. On the other hand, an environment like the skin is less nutrient rich and tends to be aerobic. The oral cavity and upper respiratory tract contain heterogeneous microenvironments from which both aerobes and strict anaerobes can be readily isolated (see, for example, Brook, 1988).

Which nitrogen cycle processes/ enzymes would be predicted to occur?

The biological nitrogen cycle processes of microorganisms are summarized in Fig. 12.1, with the reactions that are likely to be dominant in the environment of the human body highlighted. The major nitrogen cycling processes are likely to be associated with heterotrophic microbial growth, particularly anaerobic respiratory processes of denitrification and ammonification; chemolithotrophic processes are unlikely. Human diets are typically sufficiently rich in fixed nitrogen to make nitrogen fixation seem unlikely. Nitrogen fixation has been identified as potentially significant in some other mammalian species (Hobson *et al.*, 1973; Vecherskii *et al.*, 2006, 2009), but is apparently insignificant even in human populations with very nitrogen-poor diets (Huang *et al.*, 1979). Heterotrophic nitrification might potentially be used as a means of redox balancing in a carbon rich environment, rich in highly reduced carbon substrates such as butyric acid, and could potentially be a source of nitrite and nitrate biosynthesis but there is no experimental evidence to support the occurrence of this pathway in humans *in vivo*. One citation referring to the possibility of this pathway (Gomez *et al.*, 1980) precedes the discovery of nitric oxide generation via NO synthase which is currently reckoned to be the major source of nitrate synthesis in the body.

Nitrogen cycling potential in the gut

The highest abundance of microorganisms in the human body is found in the gastrointestinal tract, and more specifically the large intestine (colon). Here the high load of bacteria and the carbon and nitrogen richness due to digestion of food leads to an anaerobic environment. The major divisions of bacteria present in this tissue are the firmicutes and the bacteroidetes [constituting around 87% of all microorganisms in several metagenomic studies in which a total of $>$45,000 bacterial 16S rDNA samples were isolated and identified (Hattori and Taylor, 2009)]. The anaerobic nutrient rich environment might favour the existence of

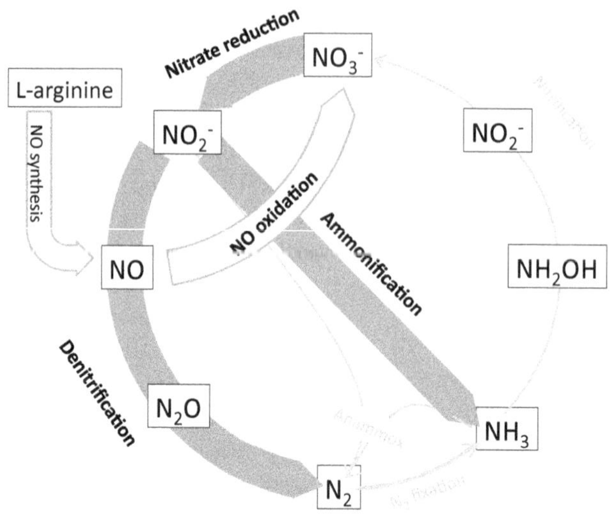

Figure 12.1 The nitrogen cycle. Interconversions between forms of inorganic nitrogen compounds of different redox states are shown. Reactions that are pertinent to the environments present in the human body are shown in thick arrows, thin arrows representing steps that are unlikely to be of significance in the body *in vivo* under normal conditions. The synthesis of nitric oxide from L-arginine is shown. Although this is mainly a reaction catalysed by the host rather than the microbes inhabiting the human body, there is evidence for NO synthase activity in bacteria (reviewed in Chapter 9).

denitrification and/or ammonification in this environment. As a way of analysing the potential for these processes in this tissue we examined the distribution of denitrification and ammonification genes in all sequences of firmicutes and bacteroidetes (307 complete genomes as of February 2010) using BLAST. *narG* was found commonly amongst the firmicutes, although if firmicutes of the order Bacillales were excluded [members of this order are not represented amongst the top 50 most abundant microbial genomes in human gut (Qin *et al.*, 2010)] *narG* homologues were only identifiable in 14 of the remaining 255 genomes (including *Desulfitobacterium, Mitsuokella, Selenomonas, Veillonella* and *Lactobacillus*). Genes for the periplasmic nitrate reductase *napA* subunit could not be identified. We analysed the distribution of nitrite reductases that generate ammonia as a product (periplasmic pentahaem nitrite reductase, *nrfA* and cytoplasmic sirohaem, *nirB*) and that generate NO as a product (cytochrome cd_1 nitrite reductase *nirS* and copper nitrite reductase *nirK*). 53 *nrfA* homologues were found amongst the 307 genomes widely distributed amongst bacteroidetes and firmicutes. *nirB* was found predominantly in the Bacillales; outside of this group only seven *nirB* homologues were found -within the genera *Algoriphagus, Pedobacter, Flavobacterium, Cytophaga, Chitinophaga, Dyadobacter* and *Spirosoma* (which are all bacteroidetes). Only one cytochrome cd_1 homologue was found: in *Rhodothermus marinus*. This is potentially of general interest as this genome also contains a homologue for the copper nitrite reductase, *nirK*, and previously no organism has been found that contains genes for both different types of NO generating nitrite reductase. However, *R. marinus* is not a gut resident. The copper nitrite reductase is found in 13 genomes including members of *Rhodothermus, Flavobacteriales, Chryseobacteirum, Capnocytophaga, Symbiobacterium, Paenibacillus* and *Geobacillus*. Nitric oxide reductases were identifiable in around 20 genomes belonging to genera *Veillonella, Flavobacterium, Capnocytophaga, Chitinophaga, Chryeobacterium, Geobacillus, Staphylococcus, Bacillus, Lactobacillus* and *Diethiobacter*. Homologues of the nitrous oxide reductase nosZ were found in 13 genomes from genera *Geobacillus, Desulfitobacterium, Salinibacter, Rhodothermus, Flavobacteriales, Dyadobacter, Capnocytophage, Gramella, Robinitalea* and *Psychroflexus*.

The major bacterial genera found in human intestines are *Bacteroides, Eubacterium, Clostridium* and *Ruminococcus* (Hattori and Taylor, 2009). Searches were performed with the 73 available genomes of members of these genera with the denitrification/ammonification enzymes as above. There appear to be no *nar* homologues amongst these organisms, and putative *nap* homologues are actually more closely related to the assimilatory nitrate reductase *nas* than *napA*. *nrfA* homologues were found in most the *Bacteroides* species, but not the clostridial genera (*Clostridium, Eubacterium* and *Ruminococcus*). The NADH-dependent *nirB* type nitrite reductase is found widely distributed amongst the clostridial genera. Denitrification (NO-generating) nitrite reductases, respiratory NO reductases and nitrous oxide reductase appear absent from the major anaerobic species occupying the gut. Thus, it appears that the major reductive nitrogen cycling reaction in the gut is nitrite reduction to ammonia. It is worth noting here that whilst the facultative anaerobes such as *E. coli* and other enterics have well characterized nitrite and nitrite reducing activities, these facultative organisms represent around 0.1% of the abundance of microorganisms in the gastrointestinal tract (Eckburg *et al.*, 2005, and references therein).

The nitrate reductase ('NarA') and NirB type nitrite reductases have been analysed in *Clostridium perfringens*. It was concluded that their functions are primarily in nitrate assimilation, not respiratory nitrate metabolism (Fujinaga *et al.*, 1999; Sekiguchi *et al.*, 1983). Unlike NirB, which is cytoplasmic and utilizes NADH as reductant, NrfA type nitrite reductases are periplasmic and linked to an electron transport chain. They are generally considered to have more of a role in energy conservation than in assimilation. NrfA has also been shown to have a role in nitric oxide detoxification in enteric bacteria (Mills *et al.*, 2008; Poock *et al.*, 2002). However, others have presented evidence to indicate that ammonification leads to an elevated production of NO (Vermeiren *et al.*, 2009). There may be distinctive differences between bacterial species with respect to their potential to generate NO from nitrate and nitrite via dissimilatory pathways (Sobko *et al.*, 2005). At present, the precise role of NrfA in dissimilation, NO generation or NO detoxification cannot be asserted with complete confidence. Presuming that at least part of its role of NrfA is in dissimilatory/respiratory metabolism it is interesting to note that microorganisms in the gut appear to favour ammonification over denitrification. Ammonification is (typically) less effective at conservation of energy compared to denitrification. Ammonification may be favoured in the carbon rich environment of the gut as a means of supporting growth by respiration and as a redox valve that prevents over-reduction of the cell in an environment rich in highly reduced carbon substrates (such as butyric acid).

Homologues of MoFe-type nitrogenases are found in *Clostridium beijerinckii, C. kluyveri, C. thermocellum, C. butyricium, C. cellulolyticum* and *Ruminococcus flavefaciens*. In fact the first nitrogen-fixing bacterium to be isolated was a *Clostridium* species (Winogradsky, 1895). However, the clostridia operate in a variety of environments and whether nitrogenase genes are abundantly present, expressed and active in the human gut remains an open question.

Nitrogen cycling in the oral cavity and upper respiratory tract

Nitrate in the oral cavity has two main sources, dietary nitrate and nitrate from the oxygen dependent oxidation of host-derived nitric oxide. Nitrite availability is governed by the reduction of salivary nitrate to nitrite via microbial action and oxidation of host derived NO to nitrite. The oral cavity and upper respiratory tract contain sites for bacterial colonization with variable availability of oxygen. The microorganismal content of this tissue is a mixture of aerobes, facultative anaerobes and strict anaerobes, in support of the notion that oxygen concentration is highly variable between microenvironments. Nitrite is readily measurable [hundreds of micromolar (Xia *et al.*, 2003)] in saliva from the mouth, and this is predominantly due to nitrate reductase activity from the microbiota here, as judged by its ablation following treatment with antimicrobial mouthwash (van Maanen *et al.*, 1996). The major microorganisms responsible for nitrite production from nitrate in the oral cavity are probably *Veillonella, Actinomyces, Rothia* and *Staphylococcus* species (Doel *et al.*, 2005). The nitrite thus synthesized may be sufficiently significant to have an impact on the physiological activity of the host. The nitrite thus generated is a major source for NO generation in the acidic conditions in the stomach (McKnight *et al.*, 1997), but may also be metabolized directly by oral microbes to ammonia (chiefly via *Veillonella* species; Yordy and Delwiche, 1979) or via nitric oxide to inert nitrous oxide/dinitrogen by *Neisseria* species, as discussed in the following section.

Neisseria species

The genus *Neisseria* contains around 20 species, most of which inhabit the mucosal surfaces of mammals, and have been isolated most particularly from the human oral cavity and nasopharynx. Within the genus there are two pathogenic species, *N. meningitidis* and *N. gonorrhoeae*. *N. meningitidis* is a part of the normal flora of a significant minority (estimated carriage rates of around 10% are based on culturing from nasopharyngeal swabs (Cartwright *et al.*, 1987; Claus *et al.*, 2005), but these may underestimate actual carriage by around four-fold; Sim *et al.*, 2000) of humans and its only known natural habitat is the human oro/nasopharynx. Whilst usually harmless, *N. meningitidis* is able occasionally to cross into the bloodstream where it can multiply, causing septicaemia. From the blood the bacteria can cross the blood–brain barrier into the cerebrospinal fluid, where multiplication occurs, leading to meningitis (van Deuren M. *et al.*, 2000). The causation of disease is a dead-end for the bacterium, as it cannot be transmitted to a new host from internal bodily fluids. The appropriate environment for consideration of the normal physiology of *N. meningitidis* is the epithelial mucosa in the upper respiratory tract. *N. gonorrhoeae* is the closest relative of *N. meningitidis* and is the causative agent of the sexually transmitted infection gonorrhoea. It inhabits the mucosal surfaces of the urogenital tract, but can occasionally be found in the upper respiratory tract. The relative lack of genetic diversity between strains within the *N. gonorrhoeae* species relative to the diversity of *N. meningitidis* strains is consistent with *N. gonorrhoeae* having evolved from an ancestral *N. meningitidis*.

The *Neisseria* appear to be reliant on respiration to support growth. Oxygen respiration is carried out via a cytochrome cbb_3 oxidase in the pathogenic *Neisseria* species, and the genes encoding the oxidase appear to be essential (Aspholm *et al.*, 2010). Cytochrome cbb_3 oxidases are typically associated with oxygen respiration under microaerophilic conditions. Cytochrome cbb_3 is required by nitrogen-fixing bacteria such as *Bradyrhizobium japonicum* as a means for protecting the oxygen sensitive nitrogenase (Zufferey *et al.*, 1996; Preisig *et al.*, 1996). In *Paracoccus, Rhodobacter, Pseudomonas* and *Rhizobium* species, cbb_3 oxidase expression is upregulated in response to oxygen limitation (Granados-Baeza *et al.*, 2007; Marchal *et al.*, 1998; Mouncey and Kaplan, 1998; Pitcher *et al.*, 2002; Ugidos *et al.*, 2008). Furthermore, it has been established biochemically that cbb_3 oxidase from *B. japonicum* has a K_m for oxygen in the low nanomolar range (Preisig *et al.*, 1996), compared to the much weaker affinity around 1 μM found for conventional haem–copper oxidases such as the aa_3 oxidase. That *Neisseria* species only contain the cbb_3 oxidase indicates that members of this genus are particularly adapted for life under oxygen limitation. An analysis of the available genome sequence data within the family Neisseriaceae indicates that only cytochrome cbb_3 oxidase can be found in the nine *Neisseria* species with significant genome sequence coverage (as of March 2010), whereas other Neisseriaceae outside the genus *Neisseria* (*Laribacter hongkongensis, Lutiella nitroferrum, Chromobacterium violaceum*) do contain other haem–copper oxidases and a cytochrome *bd* oxidase. This is consistent with the genesis of *Neisseria* being associated with a contraction of metabolic versatility, accompanied by its habitat becoming restricted to epithelial mucosa. The *Neisseria* possess denitrification enzymes that can act as alternative respiratory pathways. The diversity of denitrification machinery amongst the Neisseria was recently reviewed by Barth and colleagues (2009) (Barth *et al.*, 2009) who showed that most *Neisseria* species and strains contain a core truncated denitrification pathway with a copper-type nitrite reductase (aniA) and quinol oxidizing nitric oxide reductase (qNOR). Most species also possess a nitrous oxide reductase (although this is absent from the pathogenic species and from *N. mucosa*) and one species contains the genes for a nitrate reductase (*N. mucosa*). The activities and regulation of the nitrite and nitric oxide reductases in *N. meningitidis* and *N. gonorrhoeae* have been analysed extensively (Householder *et al.*, 1999, 2000; Isabella *et al.*, 2008; Cardinale and Clark, 2005; Overton *et al.*, 2006; Rock *et al.*, 2005, 2007; Heurlier *et al.*, 2008), showing that *aniA* and *norB* are required for nitrite reduction to the inert gas nitrous oxide under oxygen limiting conditions. Whilst *aniA* expression is controlled by oxygen via the oxygen sensitive transcriptional activator FNR, *norB* is controlled by nitric oxide accumulation via the nitric oxide sensitive repressor NsrR. The regulation of these genes is more complex than this, but the key point to make here is that *norB* is not expressed until NO accumulates, thus cell cultures expose themselves to potentially toxic NO during the transition to steady state denitrification. This may give rise to a selection pressure to lose *aniA*. Pseudogenization of *aniA* is seen in a significant minority of *N. meningitidis* strains, but not in other *Neisseria* species (Barth *et al.*, 2009; Ku *et al.*, 2009; Stefanelli *et al.*, 2008; Thomson *et al.*, 2008). The precise reasons why *N. meningitidis* loses *aniA*, but other species do not are unclear, but presumably lies in the balance between the selective advantage of possessing an active denitrification pathway versus the disadvantage caused by synthesizing nitric oxide. The loss of nitrous oxide reductase, and partial loss of nitrite reductase in *N. meningitidis* suggests that this organism is in the process of losing metabolic versatility as it becomes adapted to a more aerobic niche than its non-pathogenic ancestors. That *N. meningitidis* is on an evolutionary trajectory away from facultative anaerobe and towards aerobe is further supported by the finding that oxidative stress dependent genes such as *sodC* (encoding Cu/Zn superoxide dismutase) (Seib *et al.*, 2004) and *gpxA* (encoding glutathione peroxidase) (Moore and Sparling, 1995) are present in *N. meningitidis* but not other *Neisseria* species. This may reflect a closer physical relationship with the aerobic mucosal surface, required for its action as a pathogen, rather than oxygen-limited sites favoured by commensal bacteria which are further from both the oxygen perfusion provided by the blood and patrolling macrophages. Recent studies with *N. gonorrhoeae* have shown that *aniA* (and *norB*) are upregulated in biofilms (Falsetta *et al.*, 2009). Although *N. meningitidis* can also form biofilms, biofilm formation is poor in capsulated *N. meningitidis* strains (Lappann *et al.*, 2006; O'Dwyer *et al.*, 2009; Yi *et al.*, 2004). *N. meningitidis* is the only *Neisseria* species to produce a polysaccharide capsule, and this may also help to explain why *N. meningitidis* is losing *aniA*.

Notably, the *norB* gene encoding nitric oxide reductase appears to be absolutely conserved between *N. meningitidis* strains, indicating that this gene continues to be of paramount importance for *N. meningitidis* in its natural habitat. In addition to providing a route for respiration, this is a detoxification pathway (Anjum *et al.*, 2002). Using a variety of tissue culture models for analysis of meningococcal growth, we have found that *norB* protects *N. meningitidis* during co-culture with macrophages and in nasal explant organ culture (Stevanin *et al.*, 2005). Furthermore, NorB activity depletes nitric oxide from these cultures, which subverts physiological signalling in the host tissue, affecting such processes as cytokine release (Stevanin *et al.*, 2007) and apoptosis (Tunbridge *et al.*, 2006). NorB also depletes nitrosothiol content in tissue culture, which may also impact on host processes (Laver *et al.*, 2010). In addition to *norB*, there is a subordinate role in NO detoxification of the mono-haem cytochrome c′ (Anjum *et al.*, 2002; Stevanin *et al.*, 2005; Tunbridge *et al.*, 2006). Other genes in Neisseria such as thioredoxin reductase (Potter *et al.*, 2009) and DNA repair protein PriA (Tala *et al.*, 2008) are also required for resistance to nitric oxide in *Neisseria*.

Bacterial metabolism impacting on host (NO) signalling mechanisms is observed via a different mechanism in *Helicobacter pylori*. Here, the bacterium expresses an arginase that limits the availability of the substrate L-arginine for the NO synthase. This has been shown to limit the capacity of host cells to generate NO during culturing with *H. pylori* (Gobert *et al.*, 2002).

It is intriguing that in the oral cavity different species catalyse the reduction of nitrate to nitrite, and the reduction of nitrite to nitrous oxide. Typically in natural environments organisms carry out the entire process of reduction of nitrate through to dinitrogen gas. However, denitrification need not necessarily be regarded as a 'metabolic pathway' as such. The function of the enzymes of denitrification is primarily for coupling redox processes to biochemical energy generation and only secondarily the limitation of possible toxicity of intermediates (and this is confined mainly to the highly toxic nitric oxide). Thus there is no imperative for organisms to contain all parts of the denitrification system. In the oral cavity it appears that bacteria have adapted to specialize and exploit discreet parts of the denitrification process. This may reflect both a tendency towards specialization rather than versatility and the fact that the availability of nitrate (largely from diet) is not tightly coupled to the availability of nitrite and nitric oxide (largely from host NO synthesis) in this environment.

Nitrogen cycling in pathogenic bacteria

Whilst metabolic interconversions of inorganic nitrogen compounds are used by many human commensal bacteria as a strategy to maintain growth and survival, particularly in low oxygen/high nitric oxide environments prevalent in the body, these pathways are also relevant to the lifestyle of some important pathogens in humans. This has been noted already (above) in discussions about nitrogen metabolism in (pathogenic) *Neisseria*. Nitrate and nitrite respiration and nitric oxide detoxification are important in enteric bacteria including pathogenic *E. coli* strains and *Salmonella* species but will not be dealt with here, as they are covered extensively elsewhere. The genus *Brucella* contains a number of important, mainly veterinary, pathogens, and is a very close relative of the model denitrifier *Paracoccus denitrificans*. *Brucella* species have a complete denitrification apparatus that may be significant for virulence (Philippot, 2005). It is likely that nitrate respiration is required for survival of *Brucella* in macrophages (Kohler *et al.*, 2002), and that *Brucella* nitrite and nitric oxide respiration is important for achieving full virulence (Baek *et al.*, 2004; Haine *et al.*, 2006). Below I will illustrate the importance of bacterial nitrogen cycling in enabling pathogen lifestyle using as examples two major human pathogens, *Mycobacterium tuberculosis* and *Pseudomonas aeruginosa.*

Mycobacterium tuberculosis

M. tuberculosis causes more deaths than any other bacterial pathogen. *M. tuberculosis* is airborne and the disease is characterized by colonization of the lungs for many years or decades leading to gradual degradation of the tissue and eventually death. The bacterial growth *in vivo* (and indeed *in vitro*) is very slow and the microorganism can survive in a dormant/non-replicating persistent state for long periods. Although *M. tuberculosis* is an obligate aerobe, it is able to survive for long periods under hypoxic conditions in the laboratory (Lim *et al.*, 1999), and it is thought that hypoxia is relevant to the lifestyle of the bacteria *in vivo* within hypoxic/anaerobic granulomas. *M. tuberculosis* and other members of the same genus possess nitrate reductase activity (a nitrate reductase screen is used as a routine method for identifying drug susceptibility in *M. tuberculosis* strains e.g. Syre *et al.*, 2010), and it has been shown that this is relevant to the long term survival of the bacteria *in vitro* and *in vivo*. Mycobacterial genomes contain a cluster of genes (*narGHJI*) encoding a membrane-bound nitrate reductase. *M. bovis* BCG is used as a vaccine against tuberculosis, but this strain is also capable of causing disease, particularly in the immunocompromised. Weber and colleagues (Weber *et al.*, 2000) generated mutants of *M. bovis* BCG lacking *narG* and found that that these mutants could not reduce nitrate under anaerobic conditions and furthermore, on infecting this mutant into immunodeficient mice, the infection produced smaller granulomas containing fewer bacteria than those infected with wild-type, and the mice infected with the mutant showed no signs of clinical disease after 200 days whereas all mice infected with wild-type all showed disease symptoms by 50 days. Nitrate promotes survival of *M. tuberculosis* during a non-replicating persistent state if cultures are made anaerobic rapidly, or if respiration is inhibited by antibiotic respiratory inhibitors such as thioridazine (Sohaskey, 2008) or the competitive inhibitor of the oxidase, nitric oxide (Sohaskey, 2005).

Nitrate reductase *narGHJI* expression is constitutive under aerobic or anaerobic conditions, but nitrate reductase activity increases markedly on the transition to anaerobiosis (Sohaskey and Wayne, 2003). In addition to *narGHJI* there are several other genes identifiable in the genomes of Mycobacteria with potential roles in nitrate reduction (Amon *et al.*, 2009). Four transporters of the NarK type (designated NarK1, NarK2, NarK3 and NarU) are found in the *M. tuberculosis* genome. *narK2* is located in an operon upstream of a gene designated *narX*, which encodes a fused NarG (alpha subunit containing the molybdopterin active site) – NarJ (chaperone) – NarI (membrane-spanning subunit) nitrate reductase chimera. (Note: *M. tuberculosis* NarX is in no way related to the nitrate sensing protein NarX from *E. coli*). Whether NarX can generate an active nitrate reducing enzyme is unclear, but it has not been found to contribute to nitrate reduction in intact *M. tuberculosis* following mutagenesis studies (Sohaskey and Wayne, 2003). Whilst NarGHJI expression is constitutive, the *narK2X* operon is induced by anaerobiosis in *M. bovis* and *M. tuberculosis* (Hutter and Dick, 1999) (Muttucumaru *et al.*, 2004). NarX is not required for anaerobic induction of nitrate reductase activity, but a *narK2* knockout strain had wild-type aerobic nitrate reductase activity but there was no induction of activity under anaerobic conditions (Sohaskey and Wayne, 2003). It appears that NarK2 is specifically required for nitrate uptake under anaerobic conditions in *M. tuberculosis*. In addition to induction of expression of NarK2 enabling enhanced nitrate reductase activity under anaerobic conditions, the rate of nitrate reduction is regulated by oxygen at a metabolic level. Oxygen inhibits nitrate reduction, an effect that can be overcome by the inhibitor nitric oxide. This may occur via a redox sensing mechanism in the nitrate transporter NarK2, or via competition for electrons from the available pool in the respiratory chain (Sohaskey, 2005). Interestingly, *M. tuberculosis* reduces nitrate anaerobically at a greater rate than other Mycobacterial species, a fact that may contribute to its persistence *in vivo* and hence its virulence. *Mycobacterium smegmatis* and *M. avium* subsp. *paratuberculosis* possess *narGHJI* but lack *narK2* (Amon *et al.*, 2009). *M. bovis* possesses *narK2* but the gene is not induced under hypoxic conditions in this species, due to a single nucleotide mutation in the –10 promoter element (Chauhan *et al.*, 2010; Stermann *et al.*, 2003; Stermann *et al.*, 2004). Transfer of *narGHJI* and *narK2X* from *M. tuberculosis* into *M. bovis* failed to lead to anaerobically inducible nitrate reductase activity, indicating that *trans* regulatory factors are also different between the two species (Chauhan *et al.*, 2010; Sohaskey and Modesti, 2009). Nitrate reduction during dormancy is greatly decreased in *M. bovis* relative to *M. tuberculosis* (Honaker *et al.*, 2008).

Why does *M. tuberculosis* have multiple different NarK-like proteins? Most attention to date has focused on the function of NarK2, and little is known about the roles of the other putative nitrate/nitrite transporters. The organism may need to regulate nitrate uptake and nitrite extrusion differentially depending on growth conditions, and different NarK family proteins have different biochemical functions (as either nitrate proton symporters or nitrate nitrite antiporters). *M. tuberculosis* NarK1 and NarK2 proteins are more similar to assimilatory nitrate transporter NasA from *Bacillus subtilis* (presumed from physiological function to be a nitrate uptake protein rather than a nitrate nitrite antiporter), whereas NarK3 and NarU are more similar to NarK/NarU from *E. coli* and 'type 2' NarK from denitrifiers such as *Paracoccus denitrificans* which probably have nitrate/nitrite antiporter activity (Clegg *et al.*, 2002; Goddard *et al.*, 2008; Jia *et al.*, 2009; Moir and Wood, 2001; Wood *et al.*, 2002). For anaerobic nitrate uptake NarK2 is required, but may work in concert with NarK3 and/or NarU the latter two potentially acting as nitrate/nitrite antiporters. In addition to a role in dissimilatory nitrate reduction (Sohaskey, 2008) *narGHJI* is required for assimilatory nitrate uptake (Malm *et al.*, 2009). Under these conditions nitrite is reduced (cytoplasmically) to ammonia via NirBD, for subsequent utilization in biomass generation. This activity occurs aerobically under nitrogen limitation. We speculate that for assimilation, nitrate is transported via a different transporter whose expression is regulated by nitrogen availability rather than anoxia. The most likely candidate is the putative nitrate proton symporter NarK1.

In addition to nitrate respiration and assimilation, *M. tuberculosis* contains a well characterized nitric oxide protection system. *M. tuberculosis* possesses two truncated haemoglobins, HbN and HbO. HbN from *M. bovis* and *M. tuberculosis* has been shown to protect from nitric oxide toxicity and to confer resistance to other bacterial species when expressed heterologously in *E. coli* (Ouellet *et al.*, 2002; Pathania *et al.*, 2002). The truncated haemoglobins scavenge NO and bind oxygen, and detoxify NO by catalysing the formation of nitrate (Ouellet *et al.*, 2002; Pathania *et al.*, 2002). The causative agent of leprosy, *M. leprae*, has a reduced repertoire of genes compared to other mycobacteria and lacks HbN, but in this organism HbO does protect the organism against nitric oxide toxicity (Fabozzi *et al.*, 2006). Expression of HbN in *S. enterica* protects this organism from killing in macrophages, indicating that HbN may have an important role in enabling the persistence of *M. tuberculosis* during intracellular infection (Pawaria *et al.*, 2007). In addition to the direct detoxification of NO, nitrosothiols generated by reaction of mycothiol with NO may be detoxified by a dedicated S-nitrosomycothiol reductase (Vogt *et al.*, 2003).

Pseudomonas aeruginosa

Pseudomonas aeruginosa is a major opportunistic human pathogen, being able to cause chronic infections in individuals with cystic fibrosis (CF) and causing acute infections in, for example, burns victims. In CF, *P. aeruginosa* is able to form thick biofilms in the mucosal tissue of the lung, and here the capacity to continue respiration in the absence of oxygen may be advantageous.

P. aeruginosa is a metabolically versatile bacterium with a relatively large genome in excess of 6 megabases. It contains the genes necessary for complete denitrification of nitrate through to nitrogen gas. The reduction of nitrate to nitrogen is employed by *P. aeruginosa* under anaerobic/microaerobic conditions to enable respiration to continue in the absence of sufficient oxygen. Denitrification gene expression is controlled by anoxia via the FNR-like regulator Anr, nitrogen oxyanions via the nitrate/nitrite sensing two-component system NarXL and the NO sensing regulator Dnr (Arai *et al.*, 1997, 1999; Schreiber *et al.*, 2007). Furthermore, nitrate reductase activity itself requires oxygen limitation (Noriega *et al.*, 2007), which may be due to the requirement for anoxia for nitrate transport or that competition for electrons by oxygen respiration pathways inhibits nitrate reduction. The formation of anaerobic biofilms of *P. aeruginosa* is dependent on the presence of genes for nitric oxide reductase (Yoon *et al.*, 2002) and *nor* mutants survive poorly in culture with activated macrophages (Kakishima *et al.*, 2007). The sputum of the CF lung infected with a *P. aeruginosa* biofilm is thought to be anaerobic, and it has been found that this sputum contains sufficient nitrate to support growth by denitrification (Palmer *et al.*, 2007). The membrane-bound nitrate reductase appears to be essential for growth in CF sputum (Palmer *et al.*, 2007) although a periplasmic nitrate reductase may be able to support anaerobic growth of *P. aeruginosa* in the absence of the membrane nitrate reductase under some conditions (Van Alst *et al.*, 2009a).

Dissimilatory nitrate metabolism appears to be important for biofilm production and motility in *P. aeruginosa*, and may impact on *P. aeruginosa* virulence. Indeed, nitrate reductase mutants are avirulent in a *Caenorhabditis elegans* model of *P. aeruginosa* infection (Van Alst *et al.*, 2007). In models of infection utilizing infection of *P. aeruginosa* into human monocyte cells, it was found that nitrate and nitrite reductases are important for cytotoxicity. *nir* mutants had defects in expression of a type 3 secretion system, which is an important virulence factor for *P. aeruginosa* (Van Alst *et al.*, 2009b). This lesion appears to be due to deficiency in nitric oxide production via the denitrification pathway (Van Alst *et al.*, 2009b). Indeed, nitric oxide signalling appears to be important for the physiology of *P. aeruginosa* biofilm dispersal (Barraud *et al.*, 2006). Dispersal is controlled by low concentrations of nitric oxide (below toxic concentrations) and may regulate biofilm dispersal in a di-cyclic-GMP dependent manner (Barraud *et al.*, 2006; Barraud *et al.*, 2009).

Like other metabolically versatile bacteria, *P. aeruginosa* has multiple mechanisms for controlling nitric oxide toxicity. In mutant strains of *P. aeruginosa* deficient in nitric oxide reductase, expression of *nar* and *nir* genes is down-regulated on nitric oxide accumulation by inactivation of the FNR homologue ANR (Yoon *et al.*, 2007). Furthermore, two dioxygenases (homogentissate-1,2-dioxygenase and 4-hydroxyphenylpyruvate dioxygenase) are upregulated and bound NO thus protecting *nor* mutant biofilms (Yoon *et al.*, 2007). *P. aeruginosa* also contains a flavohaemoglobin that is induced under aerobic conditions by NO via regulator FhpR (Arai *et al.*, 2005). The flavohaemoglobin appears to be a route for NO detoxification under aerobic conditions, completely independent of the nitric oxide reductase system (Arai *et al.*, 2005). *P. aeruginosa* also contains a gene for a putative cytochrome c′, which has been shown to be important for NO detoxification in other bacteria (Cross *et al.*, 2000; Anjum *et al.*, 2002).

Concluding remarks

In conclusion, nitrate/nitrite reduction and nitric oxide detoxification seem to be the dominant nitrogen cycle reactions in the microbes that occupy the human body. These play important roles in allowing survival of both commensal and pathogenic bacteria. This is a rich area for further research, in particular to explore the apparent relationships between the presence and/or activity of nitrogen cycle reactions relating to the evolution and pathogenic potential of microbes such as *M. tuberculosis, P. aeruginosa, N. meningitidis* and the enteric bacteria (see Chapter 9). These nitrogen cycle processes should be seen in the context of other anaerobic metabolic processes which have been recognized as important in pathogenicity, with recent findings being in the relationship between anaerobic regulator FNR and type three secretion systems in *Shigella* (Marteyn *et al.*, 2010) and role of the respiratory tetrathionate reductase in inflammation during infection of the gut by *Salmonella* species (Winter *et al.*, 2010).

References

Amon, J., Titgemeyer, F., and Burkovski, A. (2009). A genomic view on nitrogen metabolism and nitrogen control in mycobacteria. J. Mol. Microbiol. Biotechnol. *17*, 20–29.

Anjum, M.F., Stevanin, T.M., Read, R.C., and Moir, J.W. (2002). Nitric oxide metabolism in *Neisseria meningitidis*. J. Bacteriol. *184*, 2987–2993.

Arai, H., Hayashi, M., Kuroi, A., Ishii, M., and Igarashi, Y. (2005). Transcriptional regulation of the flavohemoglobin gene for aerobic nitric oxide detoxification by the second nitric oxide-responsive regulator of *Pseudomonas aeruginosa*. J. Bacteriol. *187*, 3960–3968.

Arai, H., Kodama, T., and Igarashi, Y. (1997). Cascade regulation of the two CRP/FNR-related transcriptional regulators (ANR and DNR) and the denitrification enzymes in *Pseudomonas aeruginosa*. Mol. Microbiol. *25*, 1141–1148.

Arai, H., Kodama, T., and Igarashi, Y. (1999). Effect of nitrogen oxides on expression of the nir and nor genes for denitrification in *Pseudomonas aeruginosa*. FEMS Microbiol. Lett. *170*, 19–24.

Aspholm, M., Aas, F.E., Harrison, O.B., Quinn, D., Vik, A., Viburiene, R., Tonjum, T., Moir, J., Maiden, M.C., and Koomey, M. (2010). Structural alterations in a component of cytochrome c oxidase and molecular evolution of pathogenic *Neisseria* in humans. PLoS. Pathog. *6*.

Baek, S.H., Rajashekara, G., Splitter, G.A., and Shapleigh, J.P. (2004). Denitrification genes regulate Brucella virulence in mice. J. Bacteriol. *186*, 6025–6031.

Barraud, N., Hassett, D.J., Hwang, S.H., Rice, S.A., Kjelleberg, S., and Webb, J.S. (2006). Involvement of nitric oxide in biofilm dispersal *of Pseudomonas aeruginosa*. J. Bacteriol. *188*, 7344–7353.

Barraud, N., Schleheck, D., Klebensberger, J., Webb, J.S., Hassett, D.J., Rice, S.A., and Kjelleberg, S. (2009). Nitric oxide signaling in *Pseudomonas aeruginosa* biofilms mediates phosphodiesterase activity, decreased cyclic di-GMP levels, and enhanced dispersal. J. Bacteriol. *191*, 7333–7342.

Barth, K.R., Isabella, V.M., and Clark, V.L. (2009). Biochemical and genomic analysis of the denitrification pathway within the genus *Neisseria*. Microbiology *155*, 4093–4103.

Brook, I. (1988). Aerobic and anaerobic bacteriology of purulent nasopharyngitis in children. J. Clin. Microbiol. *26*, 592–594.

Cardinale, J.A., and Clark, V.L. (2005). Determinants of nitric oxide steady-state levels during anaerobic respiration by *Neisseria gonorrhoeae*. Mol. Microbiol. *58*, 177–188.

Cartwright, K.A., Stuart, J.M., Jones, D.M., and Noah, N.D. (1987). The Stonehouse survey: nasopharyngeal carriage of meningococci and *Neisseria lactamica*. Epidemiol. Infect. *99*, 591–601.

Chakravortty, D., and Hensel, M. (2003). Inducible nitric oxide synthase and control of intracellular bacterial pathogens. Microbes. Infect. *5*, 621–627.

Chauhan, S., Singh, A., and Tyagi, J.S. (2010). A single-nucleotide mutation in the -10 promoter region inactivates the narK2X promoter in Mycobacterium bovis and Mycobacterium bovis BCG and has an application in diagnosis. FEMS Microbiol. Lett. *303*, 190–196.

Claus, H., Maiden, M.C., Wilson, D.J., McCarthy, N.D., Jolley, K.A., Urwin, R., Hessler, F., Frosch, M., and Vogel, U. (2005). Genetic analysis of meningococci carried by children and young adults. J. Infect. Dis. *191*, 1263–1271.

Clegg, S., Yu, F., Griffiths, L., and Cole, J.A. (2002). The roles of the polytopic membrane proteins NarK, NarU and NirC in *Escherichia coli* K-12: two nitrate and three nitrite transporters. Mol. Microbiol. *44*, 143–155.

Cross, R., Aish, J., Paston, S.J., Poole, R.K., and Moir, J.W. (2000). Cytochrome c′ from *Rhodobacter capsulatus* confers increased resistance to nitric oxide. J. Bacteriol. *182*, 1442–1447.

Doel, J.J., Benjamin, N., Hector, M.P., Rogers, M., and Allaker, R.P. (2005). Evaluation of bacterial nitrate reduction in the human oral cavity. Eur. J. Oral Sci. *113*, 14–19.

Eckburg, P.B., Bik, E.M., Bernstein, C.N., Purdom, E., Dethlefsen, L., Sargent, M., Gill, S.R., Nelson, K.E., and Relman, D.A. (2005). Diversity of the human intestinal microbial flora. Science *308*, 1635–1638.

Fabozzi, G., Ascenzi, P., Renzi, S.D., and Visca, P. (2006). Truncated hemoglobin GlbO from Mycobacterium leprae alleviates nitric oxide toxicity. Microb. Pathog. *40*, 211–220.

Falsetta, M.L., Bair, T.B., Ku, S.C., Vanden Hoven, R.N., Steichen, C.T., McEwan, A.G., Jennings, M.P., and Apicella, M.A. (2009). Transcriptional profiling identifies the metabolic phenotype of gonococcal biofilms. Infect. Immun. *77*, 3522–3532.

Fan, A.M., and Steinberg, V.E. (1996). Health implications of nitrate and nitrite in drinking water: an update on methemoglobinemia occurrence and reproductive and developmental toxicity. Regul. Toxicol. Pharmacol. *23*, 35–43.

Felley-Bosco, E. (1998). Role of nitric oxide in genotoxicity: implication for carcinogenesis. Cancer Metastasis Rev. *17*, 25–37.

Forstermann, U., Schmidt, H.H., Pollock, J.S., Sheng, H., Mitchell, J.A., Warner, T.D., Nakane, M., and Murad, F. (1991). Isoforms of nitric oxide synthase. Characterization and purification from different cell types. Biochem. Pharmacol. *42*, 1849–1857.

Fujinaga, K., Taniguchi, Y., Sun, Y., Katayama, S., Minami, J., Matsushita, O., and Okabe, A. (1999). Analysis of genes involved in nitrate reduction in *Clostridium perfringens*. Microbiology *145*, 3377–3387.

Gobert, A.P., Cheng, Y., Wang, J.Y., Boucher, J.L., Iyer, R.K., Cederbaum, S.D., Casero, R.A., Jr., Newton, J.C., and Wilson, K.T. (2002). Helicobacter pylori induces macrophage apoptosis by activation of arginase II. J. Immunol. *168*, 4692–4700.

Goddard, A.D., Moir, J.W., Richardson, D.J., and Ferguson, S.J. (2008). Interdependence of two NarK domains in a fused nitrate/nitrite transporter. Mol. Microbiol. *70*, 667–681.

Gomez, R.F., Tannenbaum, S.R., Savoca, J., Ralt, D., and Rockowitz, N. (1980). Heterotrophic nitrification by intestinal microorganisms. Cancer *45*, 1066–1067.

Govoni, M., Jansson, E.A., Weitzberg, E., and Lundberg, J.O. (2008). The increase in plasma nitrite after a dietary nitrate load is markedly attenuated by an antibacterial mouthwash. Nitric Oxide.*19*, 333–337.

Granados-Baeza, M.J., Gomez-Hernandez, N., Mora, Y., Delgado, M.J., Romero, D., and Girard, L. (2007). Novel reiterated Fnr-type proteins control the production of the symbiotic terminal oxidase cbb3 in *Rhizobium* etli CFN42. Mol. Plant Microbe Interact. *20*, 1241–1249.

Haine, V., Dozot, M., Dornand, J., Letesson, J.J., and De, B., X. (2006). NnrA is required for full virulence and regulates several *Brucella melitensis* denitrification genes. J. Bacteriol. *188*, 1615–1619.

Hattori, M., and Taylor, T.D. (2009). The human intestinal microbiome: a new frontier of human biology. DNA Res. *16*, 1–12.

Heurlier, K., Thomson, M.J., Aziz, N., and Moir, J.W. (2008). The nitric oxide (NO)-sensing repressor NsrR of *Neisseria meningitidis* has a compact regulon of genes involved in NO synthesis and detoxification. J. Bacteriol. *190*, 2488–2495.

Hobson, P.N., Summers, R., Postgate, J.R., and Ware, D.A. (1973). Nitrogen fixation in the rumen of a living sheep. J. Gen. Microbiol. *77*, 225–226.

Honaker, R.W., Stewart, A., Schittone, S., Izzo, A., Klein, M.R., and Voskuil, M.I. (2008). Mycobacterium bovis BCG vaccine strains lack narK2 and narX induction and exhibit altered phenotypes during dormancy. Infect. Immun. *76*, 2587–2593.

Householder, T.C., Belli, W.A., Lissenden, S., Cole, J.A., and Clark, V.L. (1999). cis- and trans-acting elements involved in regulation of aniA, the gene encoding the major anaerobically induced outer membrane protein in *Neisseria gonorrhoeae*. J. Bacteriol. *181*, 541–551.

Householder, T.C., Fozo, E.M., Cardinale, J.A., and Clark, V.L. (2000). Gonococcal nitric oxide reductase is encoded by a single gene, norB, which is required for anaerobic growth and is induced by nitric oxide. Infect. Immun. *68*, 5241–5246.

Huang, P.C., Lee, N.Y., and Chen, S.H. (1979). Evidences suggestive of no intestinal nitrogen fixation for improving protein nutrition status in sweet potato eaters. Am. J. Clin. Nutr. *32*, 1741–1750.

Hutter, B., and Dick, T. (1999). Up-regulation of narX, encoding a putative 'fused nitrate reductase' in anaerobic dormant *Mycobacterium bovis* BCG. FEMS Microbiol. Lett. *178*, 63–69.

Isabella, V., Wright, L.F., Barth, K., Spence, J.M., Grogan, S., Genco, C.A., and Clark, V.L. (2008). cis- and trans-acting elements involved in regulation of norB (norZ), the gene encoding nitric oxide reductase in *Neisseria gonorrhoeae*. Microbiology *154*, 226–239.

Jia, W., Tovell, N., Clegg, S., Trimmer, M., and Cole, J. (2009). A single channel for nitrate uptake, nitrite export and nitrite uptake by *Escherichia coli* NarU and a role for NirC in nitrite export and uptake. Biochem. J. *417*, 297–304.

Kakishima, K., Shiratsuchi, A., Taoka, A., Nakanishi, Y., and Fukumori, Y. (2007). Participation of nitric oxide reductase in survival of *Pseudomonas aeruginosa* in LPS-activated macrophages. Biochem. Biophys. Res. Commun. *355*, 587–591.

Kelm, M. (1999). Nitric oxide metabolism and breakdown. Biochim. Biophys. Acta *1411*, 273–289.

Kohler, S., Foulongne, V., Ouahrani-Bettache, S., Bourg, G., Teyssier, J., Ramuz, M., and Liautard, J.P. (2002). The analysis of the intramacrophagic virulome of

Brucella suis deciphers the environment encountered by the pathogen inside the macrophage host cell. Proc. Natl. Acad. Sci. U.S.A. *99*, 15711–15716.

Ku, S.C., Schulz, B.L., Power, P.M., and Jennings, M.P. (2009). The pilin O-glycosylation pathway of pathogenic *Neisseria* is a general system that glycosylates AniA, an outer membrane nitrite reductase. Biochem. Biophys. Res. Commun. *378*, 84–89.

Lappann, M., Haagensen, J.A., Claus, H., Vogel, U., and Molin, S. (2006). Meningococcal biofilm formation: structure, development and phenotypes in a standardized continuous flow system. Mol. Microbiol. *62*, 1292–1309.

Laver, J.R., Stevanin, T.M., Messenger, S.L., Lunn, A.D., Lee, M.E., Moir, J.W., Poole, R.K., and Read, R.C. (2010). Bacterial nitric oxide detoxification prevents host cell S-nitrosothiol formation: a novel mechanism of bacterial pathogenesis. FASEB J. *24*, 286–295.

Ley, R.E., Peterson, D.A., and Gordon, J.I. (2006). Ecological and evolutionary forces shaping microbial diversity in the human intestine. Cell *124*, 837–848.

Lim, A., Eleuterio, M., Hutter, B., Murugasu-Oei, B., and Dick, T. (1999). Oxygen depletion-induced dormancy in Mycobacterium bovis BCG. J. Bacteriol. *181*, 2252–2256.

Lundberg, J.O., Gladwin, M.T., Ahluwalia, A., Benjamin, N., Bryan, N.S., Butler, A., Cabrales, P., Fago, A., Feelisch, M., Ford, P.C., *et al.* (2009). Nitrate and nitrite in biology, nutrition and therapeutics. Nat. Chem. Biol. *5*, 865–869.

Lundberg, J.O., and Weitzberg, E. (2010). The biological role of nitrate and nitrite: the times they are a-changin'. Nitric Oxide *22*, 61–63.

Malm, S., Tiffert, Y., Micklinghoff, J., Schultze, S., Joost, I., Weber, I., Horst, S., Ackermann, B., Schmidt, M., Wohlleben, W., Ehlers, S., Geffers, R., Reuther, J., and Bange, F.C. (2009). The roles of the nitrate reductase NarGHJI, the nitrite reductase NirBD and the response regulator GlnR in nitrate assimilation of *Mycobacterium tuberculosis*. Microbiology *155*, 1332–1339.

Marchal, K., Sun, J., Keijers, V., Haaker, H., and Vanderleyden, J. (1998). A cytochrome cbb3 (cytochrome c) terminal oxidase in Azospirillum brasilense Sp7 supports microaerobic growth. J. Bacteriol. *180*, 5689–5696.

Marteyn, B., West, N.P., Browning, D.F., Cole, J.A., Shaw, J.G., Palm, F., Mounier, J., Prevost, M.C., Sansonetti, P., and Tang, C.M. (2010). Modulation of Shigella virulence in response to available oxygen *in vivo*. Nature *465*, 355–358.

McKnight, G.M., Duncan, C.W., Leifert, C., and Golden, M.H. (1999). Dietary nitrate in man: friend or foe? Br. J. Nutr. *81*, 349–358.

McKnight, G.M., Smith, L.M., Drummond, R.S., Duncan, C.W., Golden, M., and Benjamin, N. (1997). Chemical synthesis of nitric oxide in the stomach from dietary nitrate in humans. Gut *40*, 211–214.

Mills, P.C., Rowley, G., Spiro, S., Hinton, J.C., and Richardson, D.J. (2008). A combination of cytochrome c nitrite reductase (NrfA) and flavorubredoxin (NorV) protects *Salmonella enterica* serovar Typhimurium against killing by NO in anoxic environments. Microbiology *154*, 1218–1228.

Moir, J.W., and Wood, N.J. (2001). Nitrate and nitrite transport in bacteria. Cell Mol. Life Sci. *58*, 215–224.

Moncada, S., and Higgs, A. (1993). The L-arginine-nitric oxide pathway. N. Engl. J. Med. *329*, 2002–2012.

Moore, T.D., and Sparling, P.F. (1995). Isolation and identification of a glutathione peroxidase homolog gene, gpxA, present in Neisseria meningitidis but absent in *Neisseria gonorrhoeae*. Infect. Immun. *63*, 1603–1607.

Mouncey, N.J., and Kaplan, S. (1998). Oxygen regulation of the ccoN gene encoding a component of the cbb3 oxidase in *Rhodobacter sphaeroides* 2.4.1T: involvement of the FnrL protein. J. Bacteriol. *180*, 2228–2231.

Muttucumaru, D.G., Roberts, G., Hinds, J., Stabler, R.A., and Parish, T. (2004). Gene expression profile of *Mycobacterium tuberculosis* in a non-replicating state. Tuberculosis. (Edinb.) *84*, 239–246.

Norenberg, M.D., Rao, K.V., and Jayakumar, A.R. (2005). Mechanisms of ammonia-induced astrocyte swelling. Metab Brain Dis. *20*, 303–318.

Noriega, C.E., Sharma, V., and Rowe, J.J. (2007). Artificial control of nitrate respiration through the lac promoter permits the assessment of oxygen-mediated posttranslational regulation of the nar operon in *Pseudomonas aeruginosa*. J. Bacteriol. *189*, 6501–6505.

O'Dwyer, C.A., Li, M.S., Langford, P.R., and Kroll, J.S. (2009). Meningococcal biofilm growth on an abiotic surface – a model for epithelial colonization? Microbiology *155*, 1940–1952.

Ouellet, H., Ouellet, Y., Richard, C., LaBarre, M., Wittenberg, B., Wittenberg, J., and Guertin, M. (2002). Truncated hemoglobin HbN protects Mycobacterium bovis from nitric oxide. Proc. Natl. Acad. Sci. U.S.A. *99*, 5902–5907.

Overton, T.W., Whitehead, R., Li, Y., Snyder, L.A., Saunders, N.J., Smith, H., and Cole, J.A. (2006). Coordinated regulation of the Neisseria gonorrhoeae-truncated denitrification pathway by the nitric oxide-sensitive repressor, NsrR, and nitrite-insensitive NarQ-NarP. J. Biol. Chem. *281*, 33115–33126.

Palmer, K.L., Brown, S.A., and Whiteley, M. (2007). Membrane-bound nitrate reductase is required for anaerobic growth in cystic fibrosis sputum. J. Bacteriol. *189*, 4449–4455.

Pathania, R., Navani, N.K., Gardner, A.M., Gardner, P.R., and Dikshit, K.L. (2002). Nitric oxide scavenging and detoxification by the Mycobacterium tuberculosis haemoglobin, HbN in *Escherichia coli*. Mol. Microbiol. *45*, 1303–1314.

Pawaria, S., Rajamohan, G., Gambhir, V., Lama, A., Varshney, G.C., and Dikshit, K.L. (2007). Intracellular growth and survival of Salmonella enterica serovar Typhimurium carrying truncated hemoglobins of *Mycobacterium tuberculosis*. Microb. Pathog. *42*, 119–128.

Philippot, L. (2005). Denitrification in pathogenic bacteria: for better or worst? Trends Microbiol. *13*, 191–192.

Pitcher, R.S., Brittain, T., and Watmough, N.J. (2002). Cytochrome cbb(3) oxidase and bacterial microaerobic metabolism. Biochem. Soc. Trans. *30*, 653–658.

Poock, S.R., Leach, E.R., Moir, J.W., Cole, J.A., and Richardson, D.J. (2002). Respiratory detoxification of nitric oxide by the cytochrome c nitrite reductase of Escherichia coli. J. Biol. Chem. *277*, 23664–23669.

Potter, A.J., Kidd, S.P., Edwards, J.L., Falsetta, M.L., Apicella, M.A., Jennings, M.P., and McEwan, A.G. (2009). Thioredoxin reductase is essential for protection of *Neisseria gonorrhoeae* against killing by nitric oxide and for bacterial growth during interaction with cervical epithelial cells. J. Infect. Dis. *199*, 227–235.

Preisig, O., Zufferey, R., Thony-Meyer, L., Appleby, C.A., and Hennecke, H. (1996). A high-affinity cbb3-type cytochrome oxidase terminates the symbiosis-specific respiratory chain of Bradyrhizobium japonicum. J. Bacteriol. *178*, 1532–1538.

Qin, J., Li, R., Raes, J., Arumugam, M., Burgdorf, K.S., Manichanh, C., Nielsen, T., Pons, N., Levenez, F., Yamada, T., *et al.* (2010b). A human gut microbial gene catalogue established by metagenomic sequencing. Nature *464*, 59–65.

Rock, J.D., Mahnane, M.R., Anjum, M.F., Shaw, J.G., Read, R.C., and Moir, J.W. (2005). The pathogen *Neisseria meningitidis* requires oxygen, but supplements growth by denitrification. Nitrite, nitric oxide and oxygen control respiratory flux at genetic and metabolic levels. Mol. Microbiol. *58*, 800–809.

Rock, J.D., Thomson, M.J., Read, R.C., and Moir, J.W. (2007). Regulation of denitrification genes in *Neisseria meningitidis* by nitric oxide and the repressor NsrR. J. Bacteriol. *189*, 1138–1144.

Savage, D.C. (1977). Microbial ecology of the gastrointestinal tract. Annu. Rev. Microbiol. *31*, 107–133.

Schreiber, K., Krieger, R., Benkert, B., Eschbach, M., Arai, H., Schobert, M., and Jahn, D. (2007). The anaerobic regulatory network required for *Pseudomonas aeruginosa* nitrate respiration. J. Bacteriol. *189*, 4310–4314.

Schuster, G.S. (1999). Oral flora and pathogenic organisms. Infect. Dis. Clin. North Am. *13*, 757–74, v.

Seib, K.L., Tseng, H.J., McEwan, A.G., Apicella, M.A., and Jennings, M.P. (2004). Defenses against oxidative stress in *Neisseria gonorrhoeae* and *Neisseria meningitidis*: distinctive systems for different lifestyles. J. Infect. Dis. *190*, 136–147.

Sekiguchi, S., Seki, S., and Ishimoto, M. (1983). Purification and some properties of nitrite reductase from *Clostridium perfringens*. J. Biochem. *94*, 1053–1059.

Sim, R.J., Harrison, M.M., Moxon, E.R., and Tang, C.M. (2000). Underestimation of meningococci in tonsillar tissue by nasopharyngeal swabbing. Lancet *356*, 1653–1654.

Sobko, T., Reinders, C.I., Jansson, E., Norin, E., Midtvedt, T., and Lundberg, J.O. (2005). Gastrointestinal bacteria generate nitric oxide from nitrate and nitrite. Nitric. Oxide. *13*, 272–278.

Sohaskey, C.D. (2005). Regulation of nitrate reductase activity in Mycobacterium tuberculosis by oxygen and nitric oxide. Microbiology *151*, 3803–3810.

Sohaskey, C.D. (2008). Nitrate enhances the survival of Mycobacterium tuberculosis during inhibition of respiration. J. Bacteriol. *190*, 2981–2986.

Sohaskey, C.D., and Modesti, L. (2009). Differences in nitrate reduction between Mycobacterium tuberculosis and Mycobacterium bovis are due to differential expression of both narGHJI and narK2. FEMS Microbiol. Lett. *290*, 129–134.

Sohaskey, C.D., and Wayne, L.G. (2003). Role of narK2X and narGHJI in hypoxic upregulation of nitrate reduction by *Mycobacterium tuberculosis*. J. Bacteriol. *185*, 7247–7256.

Stefanelli, P., Colotti, G., Neri, A., Salucci, M.L., Miccoli, R., Di, L.L., and Ippoliti, R. (2008). Molecular characterization of nitrite reductase gene (aniA) and gene product in Neisseria meningitidis isolates: is aniA essential for meningococcal survival? IUBMB Life *60*, 629–636.

Stermann, M., Bohrssen, A., Diephaus, C., Maass, S., and Bange, F.C. (2003). Polymorphic nucleotide within the promoter of nitrate reductase (NarGHJI) is specific for *Mycobacterium tuberculosis*. J. Clin. Microbiol. *41*, 3252–3259.

Stermann, M., Sedlacek, L., Maass, S., and Bange, F.C. (2004). A promoter mutation causes differential nitrate reductase activity of *Mycobacterium tuberculosis* and *Mycobacterium bovis*. J. Bacteriol. *186*, 2856–2861.

Stevanin, T.M., Laver, J.R., Poole, R.K., Moir, J.W., and Read, R.C. (2007). Metabolism of nitric oxide by *Neisseria meningitidis* modifies release of NO-regulated cytokines and chemokines by human macrophages. Microbes. Infect. *9*, 981–987.

Stevanin, T.M., Moir, J.W., and Read, R.C. (2005). Nitric oxide detoxification systems enhance survival of *Neisseria meningitidis* in human macrophages and in nasopharyngeal mucosa. Infect. Immun. *73*, 3322–3329.

Syre, H., Ovreas, K., and Grewal, H.M. (2010). Determination of the susceptibility of *Mycobacterium tuberculosis* to pyrazinamide in liquid and solid media assessed by a colorimetric nitrate reductase assay. J. Antimicrob. Chemother. *65*, 704–712.

Tala, A., De, S.M., Bucci, C., and Alifano, P. (2008). Reverse transcriptase-PCR differential display analysis of meningococcal transcripts during infection of human cells: up-regulation of priA and its role in intracellular replication. BMC. Microbiol. *8*, 131.

Thomson, M.J., Stevanin, T.M., and Moir, J.W. (2008). Measuring nitric oxide metabolism in the pathogen *Neisseria meningitidis*. Methods Enzymol. *437*, 539–560.

Tunbridge, A.J., Stevanin, T.M., Lee, M., Marriott, H.M., Moir, J.W., Read, R.C., and Dockrell, D.H. (2006). Inhibition of macrophage apoptosis by *Neisseria meningitidis* requires nitric oxide detoxification mechanisms. Infect. Immun. *74*, 729–733.

Ugidos, A., Morales, G., Rial, E., Williams, H.D., and Rojo, F. (2008). The coordinate regulation of multiple terminal oxidases by the *Pseudomonas putida* ANR global regulator. Environ. Microbiol. *10*, 1690–1702.

Van Alst, N.E., Picardo, K.F., Iglewski, B.H., and Haidaris, C.G. (2007). Nitrate sensing and metabolism modulate motility, biofilm formation, and virulence in *Pseudomonas aeruginosa*. Infect. Immun. *75*, 3780–3790.

Van Alst, N.E., Sherrill, L.A., Iglewski, B.H., and Haidaris, C.G. (2009a). Compensatory periplasmic nitrate reductase activity supports anaerobic growth of *Pseudomonas aeruginosa* PAO1 in the absence of membrane nitrate reductase. Can. J. Microbiol. *55*, 1133–1144.

Van Alst, N.E., Wellington, M., Clark, V.L., Haidaris, C.G., and Iglewski, B.H. (2009b). Nitrite reductase NirS is required for type III secretion system expression and virulence in the human monocyte cell line THP-1 by *Pseudomonas aeruginosa*. Infect. Immun. *77*, 4446–4454.

van Deuren M., Brandtzaeg, P., and van der Meer, J.W. (2000). Update on meningococcal disease with emphasis on pathogenesis and clinical management. Clin. Microbiol. Rev. *13*, 144–66, table.

van Maanen, J.M., van Geel, A.A., and Kleinjans, J.C. (1996). Modulation of nitrate-nitrite conversion in the oral cavity. Cancer Detect. Prev. *20*, 590–596.

van Velzen, A.G., Sips, A.J., Schothorst, R.C., Lambers, A.C., and Meulenbelt, J. (2008). The oral bioavailability of nitrate from nitrate-rich vegetables in humans. Toxicol. Lett. *181*, 177–181.

Vecherskii, M.V., Naumova, E.I., Kostina, N.V., and Umarov, M.M. (2006). Some specific features of nitrogen fixation in the digestive tract of the European beaver (*Castor fiber*). Dokl. Biol. Sci. *411*, 452–454.

Vecherskii, M.V., Naumova, E.I., Kostina, N.V., and Umarov, M.M. (2009). [Assimilation of biological nitrogen by European beaver]. Izv. Akad. Nauk Ser. Biol. 108–112.

Vermeiren, J., Van de, W.T., Verstraete, W., Boeckx, P., and Boon, N. (2009). Nitric oxide production by the human intestinal microbiota by dissimilatory nitrate reduction to ammonium. J. Biomed. Biotechnol. *2009*, 284718.

Vogt, R.N., Steenkamp, D.J., Zheng, R., and Blanchard, J.S. (2003). The metabolism of nitrosothiols in the Mycobacteria: identification and characterization of S-nitrosomycothiol reductase. Biochem. J. *374*, 657–666.

Wakabayashi, H., Kuwabara, Y., Murata, H., Kobashi, K., and Watanabe, A. (1997). Measurement of the expiratory ammonia concentration and its clinical significance. Metab Brain Dis. *12*, 161–169.

Weber, I., Fritz, C., Ruttkowski, S., Kreft, A., and Bange, F.C. (2000). Anaerobic nitrate reductase (narGHJI) activity of *Mycobacterium bovis* BCG *in vitro* and its contribution to virulence in immunodeficient mice. Mol. Microbiol. 35, 1017–1025.

Winogradsky, S. (1895) Recherches sur l'assimilation de l'azote libre de l'atmosphere par les microbes. Arch. Sci. Biol. 3, 297–352.

Winter, S.E., Thiennimitr, P., Winter, M.G., Butler, B.P., Huseby, D.L., Crawford, R.W., Russell, J.M., Bevins, C.L., Adams, L.G., Tsolis, R.M., Roth, J.R., and Baumler, A.J. (2010). Gut inflammation provides a respiratory electron acceptor for *Salmonella*. Nature *467*, 426–429.

Wood, N.J., Alizadeh, T., Richardson, D.J., Ferguson, S.J., and Moir, J.W. (2002). Two domains of a dual-function NarK protein are required for nitrate uptake, the first step of denitrification in *Paracoccus pantotrophus*. Mol. Microbiol. *44*, 157–170.

Xia, D., Deng, D., and Wang, S. (2003). Alterations of nitrate and nitrite content in saliva, serum, and urine in patients with salivary dysfunction. J. Oral Pathol. Med. *32*, 95–99.

Yi, K., Rasmussen, A.W., Gudlavalleti, S.K., Stephens, D.S., and Stojiljkovic, I. (2004). Biofilm formation by *Neisseria meningitidis*. Infect. Immun. *72*, 6132–6138.

Yoon, S.S., Hennigan, R.F., Hilliard, G.M., Ochsner, U.A., Parvatiyar, K., Kamani, M.C., Allen, H.L., DeKievit, T.R., Gardner, P.R., Schwab, U., Rowe, J.J., Iglewski, B.H., McDermott, T.R., Mason, R.P., Wozniak, D.J., Hancock, R.E., Parsek, M.R., Noah, T.L., Boucher, R.C., and Hassett, D.J. (2002). Pseudomonas aeruginosa anaerobic respiration in biofilms: relationships to cystic fibrosis pathogenesis. Dev. Cell 3, 593–603.

Yoon, S.S., Karabulut, A.C., Lipscomb, J.D., Hennigan, R.F., Lymar, S.V., Groce, S.L., Herr, A.B., Howell, M.L., Kiley, P.J., Schurr, M.J., Gaston, B., Choi, K.H., Schweizer, H.P., and Hassett, D.J. (2007). Two-pronged survival strategy for the major cystic fibrosis pathogen, *Pseudomonas aeruginosa*, lacking the capacity to degrade nitric oxide during anaerobic respiration. EMBO J. *26*, 3662–3672.

Yordy, D.M., and Delwiche, E.A. (1979). Nitrite reduction in *Veillonella alcalescens*. J. Bacteriol. *137*, 905–911.

Zufferey, R., Preisig, O., Hennecke, H., and Thony-Meyer, L. (1996). Assembly and function of the cytochrome cbb3 oxidase subunits in *Bradyrhizobium japonicum*. J. Biol. Chem. *271*, 9114–9119.

Index

T

www.ingramcontent.com/pod-product-compliance
Lightning Source LLC
LaVergne TN
LVHW062122200926
848484LV00007B/1739
9781904455868